INTERNATIONAL TABLE OF ATOMIC WEIGHTS (1987)

Based on relative atomic mass of ^{12}C = 12.

The following values apply to elements as they exist in materials of terrestrial origin and to certain artificial elements. Values in parentheses are the mass number of the isotope of longest half-life.

Name	Symbol	Atomic Number	Atomic Weight
Actinium[d,e]	Ac	89	(227)
Aluminum[a]	Al	13	26.981539
Americium[d,e]	Am	95	(243)
Antimony (Stibium)	Sb	51	121.75
Argon[b,g]	Ar	18	39.948
Arsenic[a]	As	33	74.92159
Astatine[d,e]	At	85	(210)
Barium	Ba	56	137.327
Berkelium[d,e]	Bk	97	(247)
Beryllium[a]	Be	4	9.012182
Bismuth[a]	Bi	83	208.98037
Boron[b,c,g]	B	5	10.811
Bromine	Br	35	79.904
Cadmium	Cd	48	112.411
Calcium[g]	Ca	20	40.078
Californium[d,e]	Cf	98	(251)
Carbon	C	6	12.011
Cerium[b,g]	Ce	58	140.115
Cesium[a]	Cs	55	132.90543
Chlorine	Cl	17	35.4527
Chromium	Cr	24	51.9961
Cobalt[a]	Co	27	58.93320
Copper	Cu	29	63.546
Curium[b]	Cm	96	(247)
Dysprosium[e,g]	Dy	66	162.50
Einsteinium[d,e]	Es	99	(252)
Erbium[g]	Er	68	167.26
Europium[g]	Eu	63	151.965
Fermium[d,e]	Fm	100	(257)
Fluorine[a]	F	9	18.9984032
Francium[d,e]	Fr	87	(223)
Gadolinium[g]	Gd	64	157.25
Gallium	Ga	31	69.723
Germanium	Ge	32	72.61
Gold[a]	Au	79	196.96654
Hafnium	Hf	72	178.49
Helium[g]	He	2	4.002602
Holmium[a,b]	Ho	67	164.93032
Hydrogen[b,c,g]	H	1	1.00794
Indium	In	49	114.82
Iodine[a]	I	53	126.90447
Iridium	Ir	77	192.22
Iron	Fe	26	55.847
Krypton[c,g]	Kr	36	83.80
Lanthanum[g]	La	57	138.9055
Lawrencium[d,e]	Lr	103	(260)
Lead[b,g]	Pb	82	207.2
Lithium[b,c,g]	Li	3	6.941
Lutetium[g]	Lu	71	174.967
Magnesium	Mg	12	24.3050
Manganese[a]	Mn	25	54.93805
Mendelevium[d,e]	Md	101	(258)
Mercury	Hg	80	200.59
Molybdenum	Mo	42	95.94
Neodymium[g]	Nd	60	144.24
Neon[c,g]	Ne	10	20.1797
Neptunium[d,e]	Np	93	(237)
Nickel	Ni	28	58.69
Niobium[a]	Nb	41	92.90638
Nitrogen[b,g]	N	7	14.00674
Nobelium[d,e]	No	102	(259)
Osmium[g]	Os	76	190.2
Oxygen[b,g]	O	8	15.9994
Palladium[g]	Pd	46	106.42
Phosphorus[a]	P	15	30.973762
Platinum	Pt	78	195.08
Plutonium[d,e]	Pu	94	(244)
Polonium[d,e]	Po	84	(209)
Potassium (Kalium)	K	19	39.0983
Praseodymium[a]	Pr	59	140.90765
Promethium[d,e]	Pm	61	(145)
Protactinium[f]	Pa	91	231.03588
Radium[d,e,g]	Ra	88	(226)
Radon[d,e]	Rn	86	(222)
Rhenium	Re	75	186.207
Rhodium[a]	Rh	45	102.90550
Rubidium[g]	Rb	37	85.4678
Ruthenium[g]	Ru	44	101.07
Samarium[g]	Sm	62	150.36
Scandium[a]	Sc	21	44.955910
Selenium	Se	34	78.96
Silicon[b]	Si	14	28.0855
Silver[g]	Ag	47	107.8682
Sodium (Natrium)[a]	Na	11	22.989768
Strontium[b,g]	Sr	38	87.62
Sulfur[b]	S	16	32.066
Tantalum	Ta	73	180.9479
Technetium[d,e]	Tc	43	(98)
Tellurium[g]	Te	52	127.60
Terbium[a]	Tb	65	158.92534
Thallium	Tl	81	204.3833
Thorium[b,f,g]	Th	90	232.0381
Thulium[a,b]	Tm	69	168.93421
Tin[g]	Sn	50	118.710
Titanium	Ti	22	47.88
Tungsten (Wolfram)	W	74	183.85
Unnilquadium[d]	Unq	104	(261)
Unnilpentium[d]	Unp	105	(262)
Unnilhexium[d]	Unh	106	(263)
Unnilseptium[d]	Uns	107	(262)
Uranium[c,f,g]	U	92	238.0289
Vanadium	V	23	50.9415
Xenon[a,c,g]	Xe	54	131.29
Ytterbium[g]	Yb	70	173.04
Yttrium[a]	Y	39	88.90585
Zinc	Zn	30	65.39
Zirconium[g]	Zr	40	91.224

[a] Elements with only one stable nuclide.
[b] Element for which known variation in isotopic abundance in terrestrial samples limits the precision of the atomic weight given.
[c] Element for which users are cautioned against the possibility of large variations in atomic weight due to inadvertent or undisclosed artificial separation in commercially available materials.
[d] Element has no stable nuclides.
[e] Radioactive element that lacks a characteristic terrestrial isotopic composition.
[f] An element, without stable nuclide(s), exhibiting a range of characteristic terrestrial compositions of long-lived radionuclide(s) such that a meaningful atomic weight can be given.
[g] In some geological specimens this element has an anomalous isotopic composition, corresponding to an atomic weight significantly different from that given.

General Chemistry

Fourth Edition

Kenneth W. Whitten
University of Georgia, Athens

Kenneth D. Gailey
Late of University of Georgia, Athens

Raymond E. Davis
University of Texas, Austin

Saunders College Publishing
Harcourt Brace Jovanovich College Publishers
Fort Worth Philadelphia San Diego New York
Orlando Austin San Antonio
Toronto Montreal
London Sydney Tokyo

To the memory of
Kenneth Durwood Gailey

To the Professor

AT&T Bell Laboratories

Some Comments about the Fourth Edition

In revising GENERAL CHEMISTRY and GENERAL CHEMISTRY WITH QUALITATIVE ANALYSIS, we have incorporated many helpful suggestions that we received from professors who used the earlier editions. Facts, explanations, and concepts are presented in a direct and concise fashion. The text is easier to read, while the scientific rigor has been improved significantly. The full-color presentation includes more than 500 photographs of substances, reactions, procedures, and applications. In the artwork and in the textual material, the pedagogic use of color has been expanded.

The fourth edition of the text is larger than the third edition, but the size of the *basic text* has changed very little. New features that make the text more useful and more attractive to students include the following:

1. A **plan,** i.e., a brief outline of the principles and logic used to solve each illustrative example, is given before the solution.
2. Even **more illustrative examples** (308) are worked out in detail.
3. Approximately **2400 end-of-chapter exercises** are included (300 more than the 3rd edition); of these more than 1400 are new.
4. **Enrichment boxes** provide more insight into selected topics for better prepared students. These can be omitted with no loss of continuity.
5. **Chemistry in Use** essays, several by guest authors, relate chemistry to topics of current or historical interest.
6. A list of **objectives** is given at the beginning of each chapter.
7. A more **open format** makes reading easier.

We have exerted great effort to make our text more **flexible.** Some examples follow.

1. We have clearly delineated the parts of Chapter 15, **Thermodynamics,** that can be moved forward for those who wish to cover **thermochemistry** (Sections 15-1 through 15-10) after **stoichiometry** (Chapters 2 and 3).
2. Chapter 4, **Some Types of Chemical Reactions,** is based on the periodic table. This material has been thoroughly reorganized and rewritten to introduce chemical reactions just after stoichiometry. Reactions are classified into the following classes: (a) precipitation, (b) acid–base, (c) displacement, and (d) oxidation–reduction reactions. There is no loss in continuity when **thermochemistry** is covered before Chapter 4. Chapter

4 can be moved to several positions later in text, e.g., after **structure and bonding,** for those who prefer this order.

3. Some professors prefer to discuss **gases** (Chapter 12) after **stoichiometry.** Chapter 12 can be moved into that position with no difficulty.
4. Chapters 5, (**The Structure of Atoms**), 6 (**Chemical Periodicity**) and 7 (**Chemical Bonding and Inorganic Nomenclature**) provide comprehensive coverage of these key topics.
5. As in earlier editions, **Molecular Structure and Covalent Bonding Theories** (Chapter 8) includes parallel comprehensive VSEPR and VB descriptions of simple molecules. This approach has been widely accepted. However, some professors prefer to present separate descriptions of covalent bonding. The chapter has been carefully *organized into numbered subdivisions* to accommodate these professors; detailed suggestions are included at the beginning of the chapter.
6. Chapter 9 (**Molecular Orbitals in Chemical Bonding**) is a "stand alone chapter" that may be omitted or moved with no loss in continuity.
7. Chapter 10 (**Reactions in Aqueous Solutions I: Acids, Bases, and Salts**) and Chapter 11 (**Reactions in Aqueous Solutions II: Calculations**) include (a) comprehensive discussions of acid–base and redox reactions in aqueous solutions, and (b) solution stoichiometry calculations for acid–base and redox reactions.

We have used color extensively to make the text easier to read and comprehend. A detailed description of our pedagogical use of color is given on page xiv. The result of all these changes is an improved clarity, accuracy, and simplicity of expression throughout the text.

We have kept in mind that chemistry is an experimental science and have emphasized the important role of theory in science. We have presented many of the classical experiments followed by interpretations and explanations of these milestones in the development of scientific thought.

We have defined each new term as accurately as possible and illustrated its meaning as early as practical. We begin each chapter at a very fundamental level and then progress through a series of carefully graded steps to a reasonable level of sophistication. *Numerous* illustrative examples are provided throughout the text and keyed to end-of-chapter exercises (EOC). The first examples in each section are quite simple, the last considerably more complex. The unit–factor method has been emphasized where appropriate.

We have used a blend of SI and the more traditional metric units, because many students are planning careers in areas in which SI units are not yet widely used. The health-care fields, the biological sciences, home economics, and agriculture are typical examples. We have used the joule rather than the calorie in nearly all energy calculations.

We have included throughout the text some interesting historical notes. Marginal notes have been used to point out historical facts, to provide additional bits of information, to further emphasize important points, to relate information to ideas developed earlier, and to note the "relevancy" of various discussions.

We welcome suggestions for improvements in future editions.

Organization

There are thirty-two chapters in GENERAL CHEMISTRY, and GENERAL CHEMISTRY WITH QUALITATIVE ANALYSIS includes eight additional chapters.

We present stoichiometry (**Chapters 2** and **3**) before atomic structure and bonding (**Chapters 5–9**) to establish a sound foundation for a laboratory program as early as possible. However, these chapters are as nearly self-contained as possible to provide flexibility for those who wish to cover structure and bonding before stoichiometry.

Because much of chemistry involves chemical reactions, we have introduced chemical reactions in a simplified, systematic way early in the text (**Chapter 4**). A logical, orderly introduction to formula unit, total ionic, and net ionic equations is included so that this information can be used throughout the remainder of the text. There are many references to this material in later chapters. Solubility rules are presented in this chapter, so that students can use them in writing chemical equations and in their laboratory work.

Because many students have difficulty in *systematizing* and *using* information, we have done our utmost to assist them. At many points throughout the text we summarize the results of recent discussions or illustrative examples in tabular form to help students see the "big picture." The basic ideas on chemical periodicity are introduced early (**Chapters 4** and **6**) and are used throughout the text. A detailed discussion of inorganic nomenclature is included at the end of **Chapter 7.** The simplified classification of acids and bases introduced in **Chapter 4** is expanded in **Chapter 10,** acids, bases, and salts, after the appropriate background on structure and bonding.

References are made to the classification of acids and bases and to the solubility rules throughout the text to emphasize the importance of systematizing and using previously covered information.

Chapter 11 covers solution stoichiometry for both acid–base and redox reactions. The qualitative aspects of redox reactions were presented in **Chapter 4.**

After our excursion through Gases and the Kinetic–Molecular Theory (**Chapter 12**), Liquids and Solids (**Chapter 13**), and Solutions (**Chapter 14**), we have covered sufficient material that students have appropriate background for a wide variety of laboratory experiments.

Comprehensive chapters are presented on Chemical Thermodynamics (**Chapter 15**) and Chemical Kinetics (**Chapter 16**). The distinction between the roles of standard and nonstandard Gibbs free energy change in predicting reaction spontaneity is clearly discussed. Chapter 16, Chemical Kinetics, has been thoroughly reorganized to provide early and consistent emphasis on the experimental basis of kinetics.

These chapters provide the necessary background for a strong introduction to Chemical Equilibrium in **Chapter 17.** This is followed by three chapters on Equilibria in Aqueous Solutions. A chapter on Electrochemistry (**Chapter 21**) completes the "common core" of the text except for Nuclear Chemistry, (**Chapter 30**), which is self-contained and may be studied at any point in the course.

A group of basically descriptive chapters follow. However, we have been careful to include appropriate applications of the principles that have been evolved in the first part of the text to explain descriptive chemistry. **Chapters 22,** The Metals and Metallurgy, **23,** The Representative Metals, and **28,** The Transition Metals, give broad coverage to the chemistry of the metals. **Chapter 29,** Coordination Compounds, is a sound introduction to that field.

Chapters 24–27 give a comprehensive introduction to the chemistry of the nonmetals. Again, care has been taken to explain descriptive chemistry in terms of the principles that have been developed earlier.

The section on organic chemistry has been reorganized and rewritten. **Chapter 31 (Organic Chemistry I: Compounds)** presents the classes of compounds, their structures, and nomenclature with major emphasis on the principal functional groups. **Chapter 32 (Organic Chemistry II: Molecular Geometry and Reactions)** is a highly structured, concise, well-illustrated discussion of the geometries of organic molecules, the three fundamental classes of organic reactions, and some reactions of key functional groups. This material provides a broad overview for students who will not take a course in organic chemistry. It also provides the introduction to important concepts for those who will study organic chemistry.

Eight additional chapters are included in GENERAL CHEMISTRY WITH QUALITATIVE ANALYSIS. In **Chapter 33,** the important properties of the metals of the five cation groups are tabulated, their properties are discussed, the sources of the elements are listed, their metallurgies are described, and a few uses of each metal are given.

Chapter 34 is a detailed introduction to the laboratory procedures used in semimicro qualitative analysis.

Chapters 35–39 cover the analysis of the five groups of cations. Each chapter includes a discussion of the important oxidation states of the metals, an introduction to the analytical procedures, and comprehensive discussions of the chemistry of each cation group. Detailed laboratory directions, set off in color, follow. Students are alerted to pitfalls in advance, and alternate confirmatory tests and "clean-up" procedures are described for troublesome cations. A set of exercises accompanies each chapter.

Chapter 40 contains a discussion of some of the more sophisticated ionic equilibria of qualitative analysis. The material is presented in a single chapter for the convenience of the instructor.

A Complete Ancillary Package

A number of ancillary materials have been prepared to assist the student in his or her study of GENERAL CHEMISTRY and to aid the instructor in presenting the course. Each supplement can be used with either version of this text.

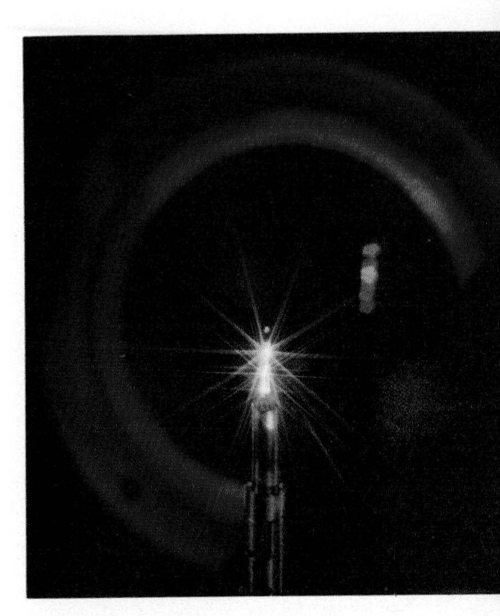
University of California, Lawrence Livermore Laboratory

1. LECTURE OUTLINE FOR GENERAL CHEMISTRY, 4th ed., Kenneth W. Whitten, Kenneth D. Gailey, and Richard M. Hedges (Texas A&M University). A comprehensive lecture outline that allows professors to use valuable classroom time more effectively. It provides great flexibility for the professor and makes available more time for special topics, increased drill, or whatever the professor chooses to do.
2. SOLUTIONS MANUAL FOR GENERAL CHEMISTRY, 4th ed., Yi-Noo Tang and Wendy Keeney-Kennicutt (both of Texas A&M Univer-

sity). It includes detailed answers and solutions for *all even-numbered* end-of-chapter exercises. In-depth answers for discussion questions and helpful comments that reinforce basic concepts are included, as well as references to specific examples and appropriate sections of chapters in the text.

3. STUDY GUIDE FOR GENERAL CHEMISTRY, 4th ed., Raymond E. Davis. It includes brief summaries of important ideas in each chapter, study goals with references to text sections and exercises, and simple preliminary tests (averaging more than 80 short questions per chapter, all with answers) that reinforce basic skills and vocabulary and encourage students to think about important ideas.
4. INSTRUCTOR'S MANUAL TO ACCOMPANY GENERAL CHEMISTRY, 4th ed., Raymond E. Davis. Also includes solutions to *odd-numbered* end-of-chapter exercises and may be made available to students, if the professor chooses.
5. EXPERIMENTAL GENERAL CHEMISTRY, 2nd Edition, Carl B. Bishop, Muriel B. Bishop (Clemson University), K. D. Gailey, and K. W. Whitten. A modern laboratory manual with excellent variety that includes descriptive, quantitative, and instrumental experiments. Designed for mainstream courses for science majors.
6. PROBLEM-SOLVING IN GENERAL CHEMISTRY, 4th ed., Leslie N. Kinsland (University of Southwestern Louisiana), K. W. Whitten, and K. D. Gailey. Covers the common core of general chemistry courses for science majors.
7. OVERHEAD TRANSPARENCIES. One hundred twenty-five four-color figures from the text.
8. TEST BANK AND COMPUTERIZED TEST BANK, Steven H. Albrecht (Ball State University). Over 1,100 questions, all completely new, and available in computerized versions for both IBM and Macintosh computers.
9. SHAKHASHIRI VIDEO TAPES, Bassam Shakhashiri (University of Wisconsin, Madison). Fifty 3–5 minute classroom experiments.
10. VIDEO DISK AND BARCODE MANUAL, containing all the Shakhashiri demonstrations and over 600 images drawn from the text and other sources. Barcode manual allows easy access.
11. PERIODIC TABLE VIDEODISC: REACTIONS OF THE ELEMENTS, JCE: Software. Alton J. Banks (Southwest Texas State University). A visual compilation of information about the chemical elements.

study this book. (Some of the following passages and pieces are displayed for example only, and may not make scientific sense without surrounding text and art.)

1. Important ideas, mathematical relationships, and summaries are displayed on pale blue screens, the width of the text.

> Different pure samples of a compound always contain the same elements in the same proportion by mass; this corresponds to atoms of these elements combined in fixed numerical ratios.

2. Answers to examples are shown on pale blue screens.

Solution

$$\underline{?}\ \frac{\text{lb}}{\text{ft}^3} = 13.59\ \frac{\text{g}}{\text{cm}^3} \times \frac{1\ \text{lb}}{453.6\ \text{g}} \times \left(\frac{2.54\ \text{cm}}{1\ \text{in}}\right)^3 \times \left(\frac{12\ \text{in}}{1\ \text{ft}}\right)^3 = \boxed{848.4\ \text{lb/ft}^3}$$

3. Intermediate steps (logic, guidance, and so on) are shown on gold screens.

Example 1-10

Express 1.0 gallon in milliliters.

Plan

We ask $\underline{?}$ mL = 1.0 gal and multiply by the appropriate factors.

gallons → quarts → liters → milliliters

Solution

$$\underline{?}\ \text{mL} = 1.0\ \text{gal} \times \frac{4\ \text{qt}}{1\ \text{gal}} \times \frac{1\ \text{L}}{1.06\ \text{qt}} \times \frac{1000\ \text{mL}}{1\ \text{L}} = \boxed{3.8 \times 10^3\ \text{mL}}$$

EOC 31, 44

---

4. Hybridization schemes and hybrid orbitals are emphasized in green.

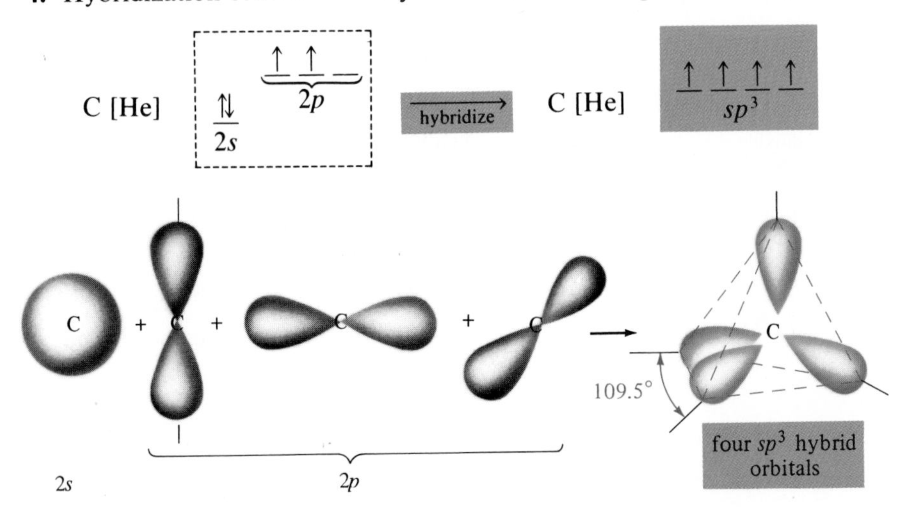

5. Acidic and basic properties are contrasted by using red and blue, respectively. Neutral solutions are indicated in pale purple.

Table 4-9
Bonding, Solubility, Electrolyte Characteristics, and Predominant Forms of Solutes in Contact with Water

	Acids		Bases			Salts	
	Strong acids	Weak acids	Strong soluble bases	Insoluble bases	Weak bases	Soluble salts	Insoluble salts
Examples	HCl HNO_3	CH_3COOH HF	NaOH $Ca(OH)_2$	$Mg(OH)_2$ $Al(OH)_3$	NH_3 CH_3NH_2	KCl, $NaNO_3$, NH_4Br	$BaSO_4$, AgCl, $Ca_3(PO_4)_2$
Pure compound ionic or covalent?	Covalent	Covalent	Ionic	Ionic	Covalent	Ionic	Ionic
Water soluble or insoluble?	Soluble*	Soluble*	Soluble	Insoluble	Soluble†	Soluble	Insoluble
~100% ionized or dissociated in dilute aqueous solution?	Yes	No	Yes	(footnote ‡)	No	Yes§	(footnote ‡)
Written in ionic equations as	Separate ions	Molecules	Separate ions	Complete formulas	Molecules	Separate ions	Complete formulas

* Most common inorganic acids and the low-molecular-weight organic acids (—COOH) are water soluble.
† The low-molecular-weight amines are water-soluble.
‡ The *very small concentrations* of "insoluble" metal hydroxides and insoluble salts in saturated aqueous solutions are nearly completely dissociated.
§ There are a few exceptions. A few soluble salts are molecular (and not ionic) compounds.

Table 18-1
Common Strong Acids and Strong Soluble Bases

Strong Acids	
HCl	HNO_3
HBr	$HClO_4$
HI	$HClO_3$
	H_2SO_4
Strong Soluble Bases	
LiOH	
NaOH	
KOH	$Ca(OH)_2$
RbOH	$Sr(OH)_2$
CsOH	$Ba(OH)_2$

$$NH_3(aq) + H_2O(\ell) \rightleftharpoons NH_4^+(aq) + OH^-(aq)$$

base₁ acid₂ acid₁ base₂

$$H:\ddot{N}(H)(H): + H:\ddot{O}(H): \rightleftharpoons [H:\ddot{N}(H)(H):H]^+ + :\ddot{O}(H):^-$$

6. Red and blue are also used in oxidation-reduction reactions and electrochemistry (Chapter 21).
   (a) Oxidation numbers are shown in red in red circles to avoid confusion with ionic charges.

	Nonmetal Oxide	+ Water	⟶	Ternary Acid	
carbon dioxide	(+4) $CO_2(g)$	$+ H_2O(\ell)$	⟶	(+4) $H_2CO_3(aq)$	carbonic acid
sulfur dioxide	(+4) $SO_2(g)$	$+ H_2O(\ell)$	⟶	(+4) $H_2SO_3(aq)$	sulfurous acid
sulfur trioxide	(+6) $SO_3(g)$	$+ H_2O(\ell)$	⟶	(+6) $H_2SO_4(aq)$	sulfuric acid

   (b) Oxidation is indicated by blue and reduction is indicated by red.

(+1) (+5) (−2) (+1) (−1) (0)

$$2KClO_3(s) \longrightarrow 2KCl(s) + 3O_2(g)$$

−6

+2

(c) In electrochemistry (Chapter 21) we learn that oxidation occurs at the *anode*; we use blue to indicate the anode and its half-reaction. Similarly, reduction occurs at the *cathode*; we use red to indicate the cathode and its half-reaction.

$$
\begin{array}{rll}
Cu \longrightarrow & Cu^{2+} + 2e^- & \text{(oxidation, anode)} \\
2(Ag^+ + e^- \longrightarrow & Ag) & \text{(reduction, cathode)} \\
\hline
Cu + 2Ag^+ \longrightarrow & Cu^{2+} + 2Ag & \text{(overall cell reaction)}
\end{array}
$$

7. In discussions of molecular orbitals (Chapter 9), bonding and antibonding molecular orbitals are shown in blue and red, respectively.

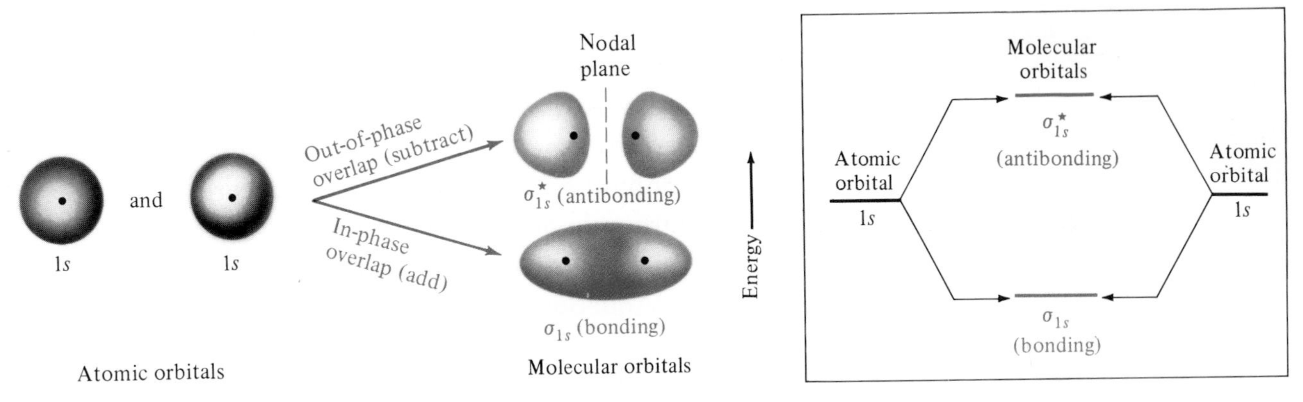

8. Color-coded periodic tables emphasize the classification of the elements as metals (blue), nonmetals (yellow), and metalloids (green). Please study the periodic table inside the front cover carefully so that you recognize this color scheme.

Table 4-2
The Periodic Table*

Metals / Nonmetals / Metalloids

	IA	IIA	IIIB	IVB	VB	VIB	VIIB	VIIIB			IB	IIB	IIIA	IVA	VA	VIA	VIIA	0
1	1 H																	2 He
2	3 Li	4 Be											5 B	6 C	7 N	8 O	9 F	10 Ne
3	11 Na	12 Mg											13 Al	14 Si	15 P	16 S	17 Cl	18 Ar
4	19 K	20 Ca	21 Sc	22 Ti	23 V	24 Cr	25 Mn	26 Fe	27 Co	28 Ni	29 Cu	30 Zn	31 Ga	32 Ge	33 As	34 Se	35 Br	36 Kr
5	37 Rb	38 Sr	39 Y	40 Zr	41 Nb	42 Mo	43 Tc	44 Ru	45 Rh	46 Pd	47 Ag	48 Cd	49 In	50 Sn	51 Sb	52 Te	53 I	54 Xe
6	55 Cs	56 Ba	57 La *	72 Hf	73 Ta	74 W	75 Re	76 Os	77 Ir	78 Pt	79 Au	80 Hg	81 Tl	82 Pb	83 Bi	84 Po	85 At	86 Rn
7	87 Fr	88 Ra	89 Ac †															

Contents Overview

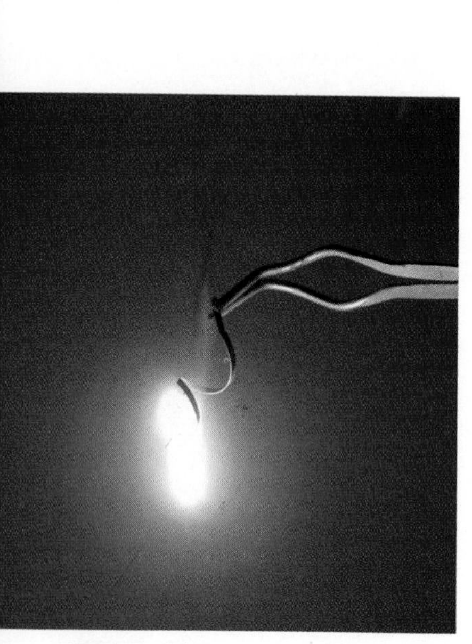

Charles D. Winters

Table of Contents

Charles D. Winters

Charles Steele

Robert W. Metz

Kip Peticolas, Fundamental Photographs, New York

Atlanta Gas Light Company

Charles Steele

Paul Silverman, Fundamental Photographs, New York

Courtesy American Cyanamid Company

Paul Silverman, Fundamental Photographs, New York

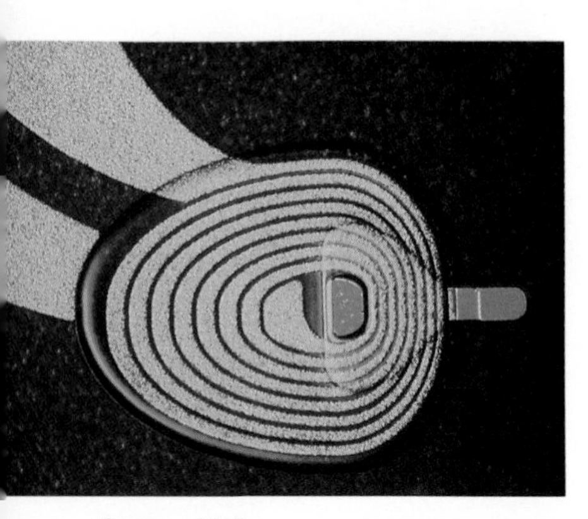

Courtesy IBM

U.S. Dept. Energy/Science Photo Library, Photo Researchers, Inc.

1 The Foundations of Chemistry

The earth is a huge chemical system, including innumerable reactions taking place constantly with some energy input from sunlight. The earth serves as the source of raw materials for *all* human activities, as well as the depository for the products of these activities. Maintaining life on the planet requires significantly improved understanding and wise use of these resources. Scientists can provide important information about the processes, but each of us must share in the responsibility for our environment.

Outline

Objectives

As you study this chapter, you should learn to

- ☐ Use the basic vocabulary of matter and energy
- ☐ Distinguish between chemical and physical properties and between chemical and physical changes
- ☐ Recognize various forms of matter: homogeneous and heterogeneous mixtures, substances, compounds, and elements
- ☐ Recognize some common methods of separation used in chemistry
- ☐ Apply the concept of significant figures
- ☐ Apply appropriate units to describe the results of measurement
- ☐ Use the unit factor method to carry out conversions among units
- ☐ Describe temperature measurements on various common scales, and convert between these scales
- ☐ Carry out calculations relating temperature change to heat absorbed or liberated

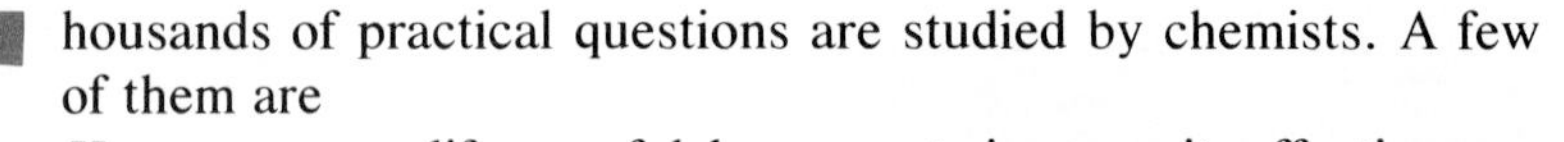

Thousands of practical questions are studied by chemists. A few of them are

How can we modify a useful drug so as to improve its effectiveness while minimizing harmful or unpleasant side effects?

How can we develop better materials to be used as synthetic bone in transplants?

Which substances could help to avoid rejection of foreign tissue in organ transplants?

What improvements in fertilizers or pesticides can improve agricultural yields? How can this be done with minimal environmental danger?

How can we get the maximum work from a fuel while producing the least harmful emissions possible?

Which really poses the greater environmental threat—the burning of fossil fuels and its contribution to the greenhouse effect and climatic change, or the use of nuclear power and the related radiation and disposal problems?

How can we develop suitable materials for the semiconductor and microelectronics industry? Can we develop a battery that is cheaper, lighter, and more powerful?

What changes in structural materials could help to make aircraft lighter and more economical, yet at the same time stronger and safer?

What relation is there between the substances we eat, drink, or breathe and the possibility of developing cancer? How can we develop substances that are effective in killing cancer cells preferentially over normal cells?

Can we economically produce fresh water from sea water for irrigation or consumption?

How can we slow down unfavorable reactions, such as corrosion of metals, while speeding up favorable ones, such as the growth of foodstuffs?

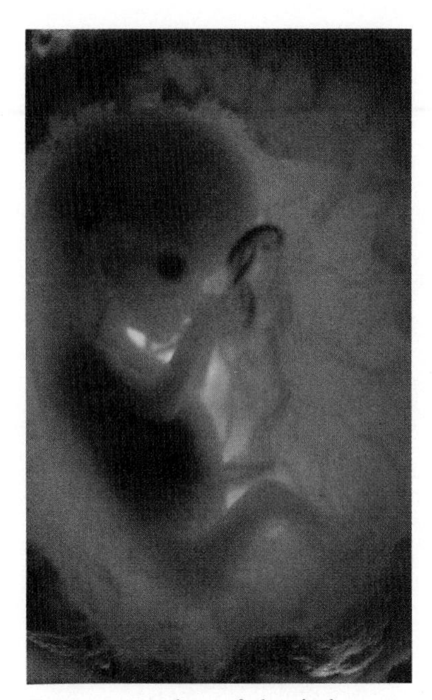

Enormous numbers of chemical reactions are necessary to produce a human embryo (here at 10 weeks, 6 cm long).

Chemistry touches almost every aspect of our lives, our culture, and our environment. Its scope encompasses the air we breathe, the food we eat, the fluids we drink, our clothing, dwellings, transportation and fuel supplies, and our fellow creatures.

Chemistry is the science that describes matter—its chemical and physical properties, the chemical and physical changes it undergoes, and the energy changes that accompany those processes. Matter includes everything that is tangible, from our bodies and the stuff of our everyday lives to the grandest objects in the universe. Some call chemistry the central science. It rests on the foundation of mathematics and physics and in turn underlies the life sciences—biology and medicine. To understand living systems fully, we must first understand the chemical reactions and chemical influences that operate within them. The chemicals of our bodies profoundly affect even the personal world of our thoughts and emotions.

We understand simple chemical systems well; they lie near chemistry's fuzzy boundary with physics. They can often be described exactly by mathematical equations. We fare less well with more complicated systems. Even where our understanding is fairly thorough, we must make approximations, and often our knowledge is far from complete. Each year researchers provide new insights into the nature of matter and its interactions. As chemists find answers to old questions, they learn to ask new ones. Our scientific knowledge has been described as an expanding sphere that, as it grows, encounters an ever-enlarging frontier.

In our search for understanding, we eventually must ask fundamental questions such as the following:

How do substances combine to form other substances? How are energy changes involved in chemical and physical changes?

How is matter constructed, in its intimate detail? How are atoms and the ways that they combine related to the properties of the matter that we can measure, such as color, hardness, chemical reactivity, and electrical conductivity?

What fundamental factors influence the stability of a substance? How can we force a desired (but energetically unfavorable) change to take place? What factors control the rate at which a chemical change takes place?

In your study of chemistry, you will learn about these and many other basic ideas that chemists have developed to help them describe and understand the behavior of matter. Along the way, we hope that you come to appreciate the development of this science, one of the grandest intellectual achievements of human endeavor. You will also learn how to apply these fundamental principles to solve real problems. One of your major goals in the study of chemistry should be to develop your ability to think critically and to solve problems (not just do numerical calculations!). In other words, you need to learn to manipulate not only numbers, but also quantitative ideas, words, and concepts.

In the first chapter, our main goals are (1) to begin to get an idea of what chemistry is about and the ways in which chemists view and describe the material world and (2) to acquire some skills that are useful and necessary in the understanding of chemistry and of science.

1-1 Matter and Energy

Matter is anything that has mass and occupies space. Mass is a measure of the quantity of matter in a sample of any material. The more massive an object is, the more force is required to put it in motion. All bodies consist of matter. Our senses of sight and touch usually tell us that an object occupies space. In the case of colorless, odorless, tasteless gases (such as air), our senses may fail us.

We might say that we can "touch" air when it blows in our faces, but we depend on other evidence to show that a still body of air fits our definition of matter.

Energy is defined as the capacity to do work or to transfer heat. We are familiar with many forms of energy, including mechanical energy, electrical energy, heat energy, and light energy. Light energy from the sun is used by plants as they grow; electrical energy allows us to light a room by flicking a switch; and heat energy cooks our food and warms our homes. Energy can be classified into two principal types: kinetic energy and potential energy.

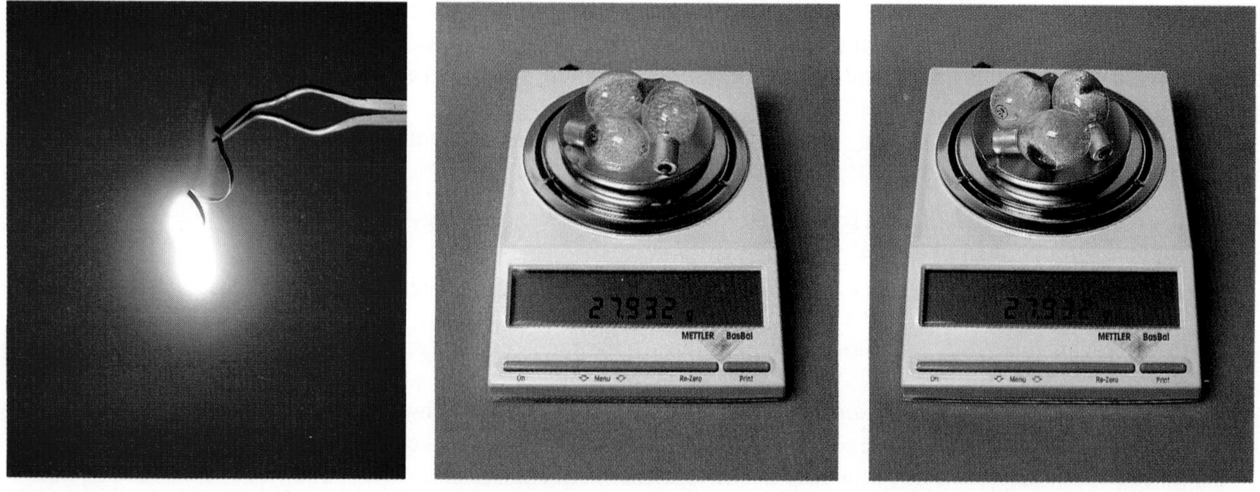

(a) (b) (c)

Figure 1-1
Magnesium burns in the oxygen of the air to form magnesium oxide, a white solid. This reaction occurs in photographic flashbulbs. There is no gain or loss of mass as the reaction occurs.

A body in motion, such as a rolling boulder, possesses energy because of its motion. Such energy is called **kinetic energy**. Kinetic energy represents the capacity for doing work directly. It is easily transferred between objects. **Potential energy** is the energy an object possesses because of its position or composition. Coal, for example, possesses chemical energy, a form of potential energy, because of its composition. Many electrical generating plants burn coal, producing heat and subsequently electrical energy. A boulder located atop a mountain possesses potential energy because of its height. It can roll down the mountainside and convert its potential energy into kinetic energy. We discuss energy because all chemical processes are accompanied by energy changes. As some processes occur, energy is released to the surroundings, usually as heat energy. We call such processes **exothermic**. Any combustion (burning) reaction is exothermic. However, some chemical reactions and physical changes are **endothermic**; i.e., they absorb energy from their surroundings. An example of a physical change that is endothermic is the melting of ice.

The term comes from the Greek word *kinein*, meaning "to move." The word "cinema" is derived from the same word.

Nuclear energy is an important kind of potential energy.

The Law of Conservation of Matter

When we burn a sample of metallic magnesium in the air, the magnesium combines with oxygen from the air (Figure 1-1) to form magnesium oxide, a white powder. This chemical reaction is accompanied by the release of large amounts of heat energy and light energy. When we weigh the product of the reaction, magnesium oxide, we find that it is heavier than the original piece of magnesium. The increase in mass of the solid is due to the combination of oxygen with magnesium to form magnesium oxide. Many experiments have shown that the mass of the magnesium oxide is exactly the sum of the masses of magnesium and oxygen that combined to form it. Similar statements can be made for all chemical reactions. These observations are summarized in the **Law of Conservation of Matter**:

> There is no observable change in the quantity of matter during a chemical reaction or during a physical change.

This statement is an example of a **scientific (natural) law**, a general statement based on the observed behavior of matter to which no exceptions are known. A nuclear reaction is *not* a chemical reaction.

The Law of Conservation of Energy

In exothermic chemical reactions, *chemical energy* usually is converted into *heat energy*. Some exothermic processes involve other kinds of energy changes. For example, some liberate light energy without heat, and others produce electrical energy without heat or light. In *endothermic* reactions, heat energy, light energy, or electrical energy is converted into chemical energy. Although chemical changes always involve energy changes, some energy transformations do not involve chemical changes at all. For example, heat energy may be converted into electrical energy or into mechanical energy without any simultaneous chemical changes. Many experiments have demonstrated that all of the energy involved in any chemical or physical change appears

Electricity is produced in hydroelectric plants by the conversion of mechanical energy (from flowing water) into electrical energy.

in some form after the change. These observations are summarized in the **Law of Conservation of Energy**:

> Energy cannot be created or destroyed in a chemical reaction or in a physical change. It can only be converted from one form to another.

The Law of Conservation of Matter and Energy

With the dawn of the nuclear age in the 1940s, scientists, and then the world, became aware that matter can be converted into energy. In nuclear reactions (Chapter 30) matter is transformed into energy. The relationship between matter and energy is given by Albert Einstein's now famous equation

Einstein formulated this equation in 1905 as a part of his theory of relativity. Its validity was demonstrated in 1939 with the first controlled nuclear reaction.

$$E = mc^2$$

This equation tells us that the amount of energy released when matter is transformed into energy is the product of the mass of matter transformed and the speed of light squared. At the present time, man has not (knowingly) observed the transformation of energy into matter on a large scale. It does, however, happen on an extremely small scale in "atom smashers," or particle accelerators, used to induce nuclear reactions. Now that the equivalence of matter and energy is recognized, the **Law of Conservation of Matter and Energy** can be stated in a single sentence:

> The combined amount of matter and energy in the universe is fixed.

1-2 States of Matter

Matter can be classified into three states (Figure 1-2), although everyone can think of examples that do not fit neatly into any of the three categories. In the **solid state**, substances are rigid and have definite shapes. Volumes of

(a)

(b)

(c)

Figure 1-2
(a) Iodine, a solid element. (b) Bromine, a liquid element. (c) Chlorine, a gaseous element.

solids do not vary much with changes in temperature and pressure. In many solids, called crystalline solids, the individual particles that make up the solid occupy definite positions in the crystal structure. The strengths of interaction between the individual particles determine how hard and how strong the crystals are. In the **liquid state**, the individual particles are confined to a given volume. A liquid flows and assumes the shape of its container up to the volume of the liquid. Liquids are very hard to compress. **Gases** are much less dense than liquids and solids. They occupy all parts of any vessel in which they are confined. Gases are capable of infinite expansion and are compressed easily. We conclude that they consist primarily of empty space; i.e., the individual particles are quite far apart.

1-3 Chemical and Physical Properties

To distinguish among samples of different kinds of matter, we determine and compare their **properties**. We recognize different kinds of matter by their properties, which are broadly classified into chemical properties and physical properties.

The properties of a person include height, weight, sex, skin and hair color, and the many subtle features that constitute that person's general appearance.

Chemical properties are properties exhibited by matter as it undergoes changes in composition. These properties of substances are related to the kinds of chemical changes that the substances undergo. For instance, we have already described the combination of metallic magnesium with gaseous oxygen to form magnesium oxide, a white powder. A chemical property of magnesium is that it can combine with oxygen, releasing energy in the process. A chemical property of oxygen is that it can combine with magnesium.

All substances also exhibit **physical properties** that can be observed in the *absence of any change in composition*. Color, density, hardness, melting point, boiling point, and electrical and thermal conductivities are physical properties. Some physical properties of a substance depend on the conditions, such as temperature and pressure, under which they are measured. For instance, water is a solid (ice) at low temperatures but is a liquid at higher temperatures. At still higher temperatures, it is a gas (steam). As water is converted from one state to another, its composition is constant. Its chemical properties change very little. On the other hand, the physical properties of ice, liquid water, and steam are different (Figure 1-3).

Properties of matter can be further classified according to whether or not they depend on the *amount* of substance present. The volume and the mass of a sample depend on, and are directly proportional to, the amount of matter in that sample. Such properties, which depend on the amount of material examined, are called **extensive properties**. By contrast, the color and the melting point of a substance are the same for a small sample and for a large one. Properties such as these, which are independent of the amount of material examined, are called **intensive properties**. All chemical properties are intensive properties.

Because no two substances have identical sets of chemical and physical properties under the same conditions, we are able to identify and distinguish among different substances. For instance, water is the only clear, colorless liquid that freezes at 0°C, boils at 100°C at one atmosphere of pressure, dissolves a wide variety of substances (including copper(II) sulfate), and

One atmosphere of pressure is the average atmospheric pressure at sea level.

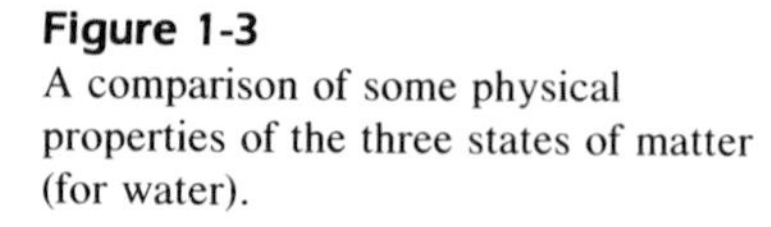

Figure 1-3
A comparison of some physical properties of the three states of matter (for water).

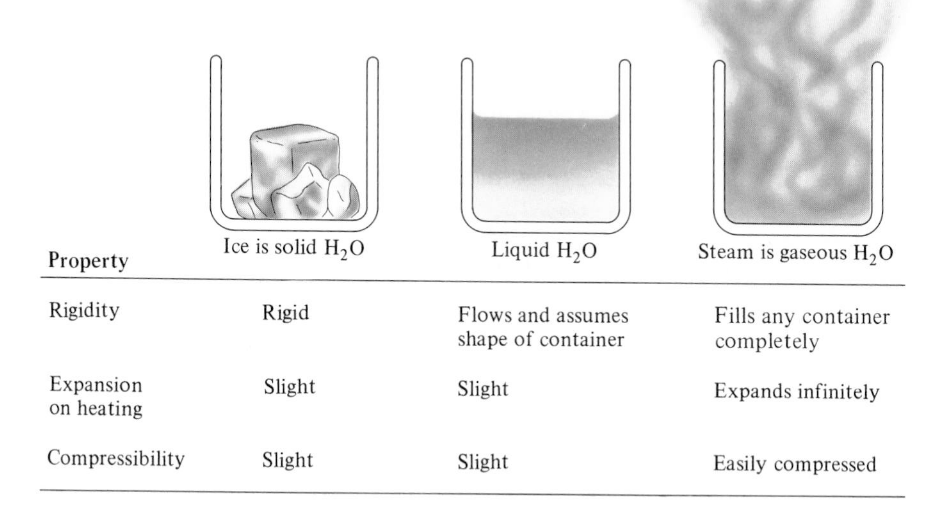

Property	Ice is solid H_2O	Liquid H_2O	Steam is gaseous H_2O
Rigidity	Rigid	Flows and assumes shape of container	Fills any container completely
Expansion on heating	Slight	Slight	Expands infinitely
Compressibility	Slight	Slight	Easily compressed

(a) (b) (c) (d)

Figure 1-4
Some physical and chemical properties of water. *Physical*: (a) It melts at 0°C; (b) it boils at 100°C (at normal atmospheric pressure); (c) it dissolves a wide range of substances, including copper(II) sulfate, a blue solid. *Chemical*: (d) It reacts with sodium to form hydrogen gas and a solution of sodium hydroxide. The solution is pink in the presence of the indicator phenolphthalein.

Table 1-1
Physical Properties of a Few Common Substances (at one atmosphere pressure)

Substance	Melting Pt. (°C)	Boiling Pt. (°C)	Solubility at 25°C (g/100 g) In water	In ethyl alcohol	Density (g/cm³)
acetic acid	16.6	118.1	infinite	infinite	1.05
benzene	5.5	80.1	0.07	infinite	0.879
bromine	−7.1	58.8	3.51	infinite	3.12
iron	1530	3000	insoluble	insoluble	7.86
methane	−182.5	−161.5	0.0022	0.033	6.67×10^{-4}
oxygen	−218.8	−183.0	0.0040	0.037	1.33×10^{-3}
sodium chloride	801	1473	36.5	0.065	2.16
water	0	100	—	infinite	1.00

reacts violently with sodium (Figure 1-4). Table 1-1 compares several physical properties of a few substances. A sample of any of these substances can be distinguished from the others by measurement of their properties.

1-4 Chemical and Physical Changes

We described the reaction of magnesium as it burns in the oxygen of the air (Figure 1-1). This reaction is a *chemical change*. In any **chemical change**, (1) one or more substances are used up (at least partially), (2) one or more new substances are formed, and (3) energy is absorbed or released. As substances undergo chemical changes they demonstrate their chemical properties. A **physical change**, on the other hand, occurs with *no change in chemical composition*. Physical properties are usually altered significantly as matter undergoes physical changes. In addition, a physical change *may* suggest that a chemical change has also taken place. For instance, a color change, a warming, or the formation of a solid when two solutions are mixed could indicate a chemical change.

Energy is always released or absorbed when chemical or physical changes occur. Energy is required to melt ice, and energy is required to boil water. Conversely, the condensation of steam to form liquid water always liberates

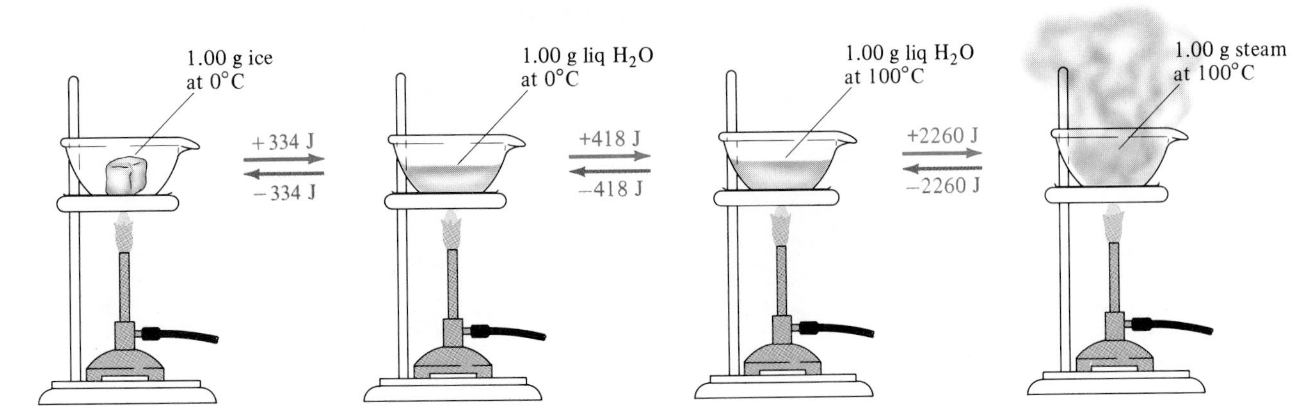

Figure 1-5
Changes in energy that accompany some physical changes for water. The energy unit joules (J) is defined in Section 1-13. The positive signs preceding joules above the arrows tell us that heat is *absorbed*. Negative signs below the arrows tell us that heat is *liberated*.

energy, as does the freezing of liquid water to form ice. The changes in energy that accompany these physical changes for water are shown in Figure 1-5. At a pressure of one atmosphere, ice always melts at the same temperature (0°C) and pure water always boils at the same temperature (100°C).

1-5 Mixtures, Substances, Compounds, and Elements

By "composition of a mixture" we mean both the identities of the substances present and their relative amounts in the mixture.

Mixtures are combinations of two or more pure substances in which each substance retains its own composition and properties. Almost every sample of matter that we ordinarily encounter is a mixture. The most easily recognized type of mixture is one in which different portions of the sample have recognizably different properties. Such a mixture, which is not uniform throughout, is called **heterogeneous**. Examples include mixtures of salt and charcoal (in which two components with different colors can be distinguished readily from one another by sight), foggy air (which includes a suspended mist of water droplets), and vegetable soup. Another kind of mixture has uniform properties throughout; such a mixture is described as a **homogeneous mixture** and is also called a **solution**. Examples include saltwater; some **alloys**, which are homogeneous mixtures of metals in the solid state; and air (free of particulate matter or mists). Air is a mixture of gases. It is mainly nitrogen, oxygen, argon, carbon dioxide, and water vapor. There are only trace amounts of other substances in the atmosphere.

An important characteristic of all mixtures is that they can have variable composition. (For instance, we can make an infinite number of different mixtures of salt and sugar by varying the relative amounts of the two components used.) Consequently, performing the same experiment over again on mixtures from different sources may give different results, whereas the same treatment of a pure sample will always give the same results. When the distinction between homogeneous mixtures and pure substances was realized and methods were developed (in the late 1700s) for separating mixtures and studying pure substances, consistent results could be obtained. This resulted in reproducible study of chemical properties, which formed the basis of real progress in the development of chemical theory.

Mixtures can be separated by physical means because each component retains its properties (see Figures 1-6 and 1-7). For example, a mixture of

(a)

(b)

Figure 1-6
(a) A mixture of iron and sulfur is a *heterogeneous* mixture. (b) Like any mixture, it can be separated by physical means, such as removing the iron with a magnet.

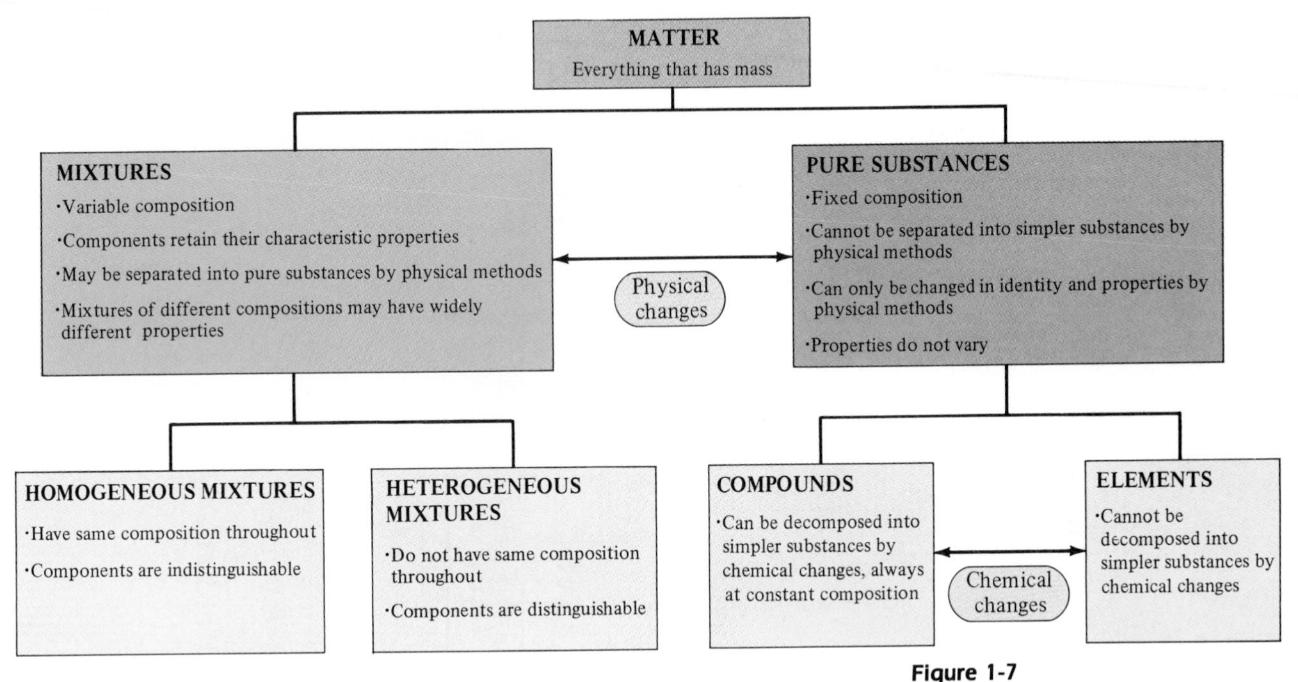

Figure 1-7
One scheme for classification of matter. Arrows indicate the general means by which matter can be separated.

salt and water can be separated by evaporating the water and leaving the solid salt behind. To separate a mixture of sand and salt, we could treat it with water to dissolve the salt, collect the sand by filtration, and then evaporate the water to obtain the solid salt. Very fine iron powder can be mixed with powdered sulfur to give what appears to the naked eye to be a homogeneous mixture of the two. However, separation of the components of this mixture is easy. The iron may be removed by a magnet, or the sulfur may be dissolved in carbon disulfide, which does not dissolve iron (Figure 1-6).

In *any* mixture, (1) the composition can be varied and (2) each component of the mixture retains its own properties.

Imagine that we have a sample of muddy river water (a heterogeneous mixture). We might first separate the suspended dirt from the liquid by filtration (Section 1-6). Then we could remove dissolved air by warming the water. Dissolved solids might be removed by cooling the sample until some of it freezes, pouring off the liquid, and then melting the ice. Other dissolved components might be separated by distillation (Section 1-6) or other methods. Eventually we would obtain a sample of pure water; it could not be further separated by any physical separation methods. No matter what the original source of the impure water—the ocean, the Mississippi River, a can of tomato juice, and so on—water samples obtained by purification all have identical composition and, under identical conditions, they all have identical properties. Any such sample is called a substance, or sometimes a pure substance.

The first ice that forms is quite pure. The dissolved solids tend to stay behind in the remaining liquid.

A **substance** cannot be further broken down by physical means.

If we use the definition given here of a *substance*, the phrase *pure substance* may appear to be redundant.

Figure 1-8
Electrolysis apparatus for small-scale decomposition of water by electrical energy. The volume of hydrogen produced (right) is twice that of oxygen (left). Some dilute sulfuric acid is added to increase the conductivity.

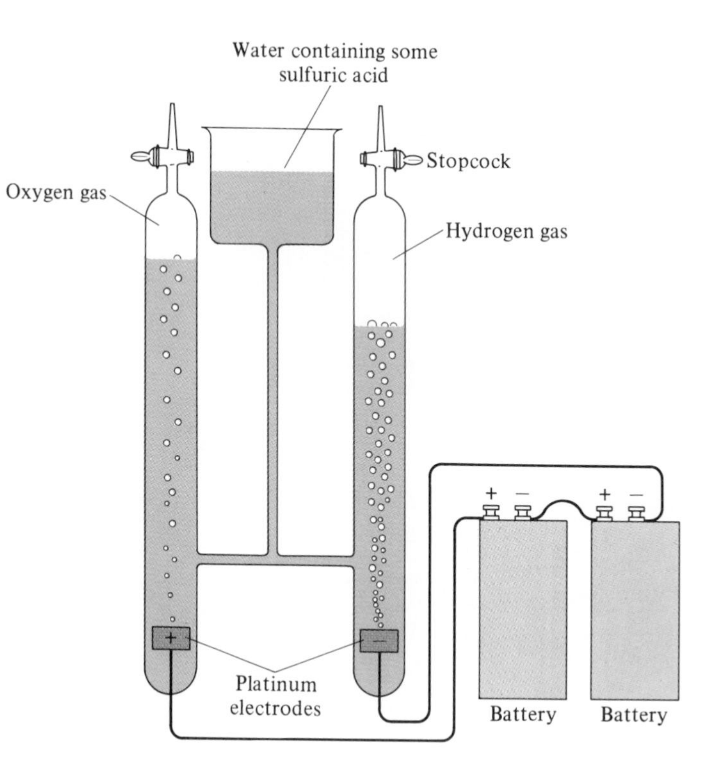

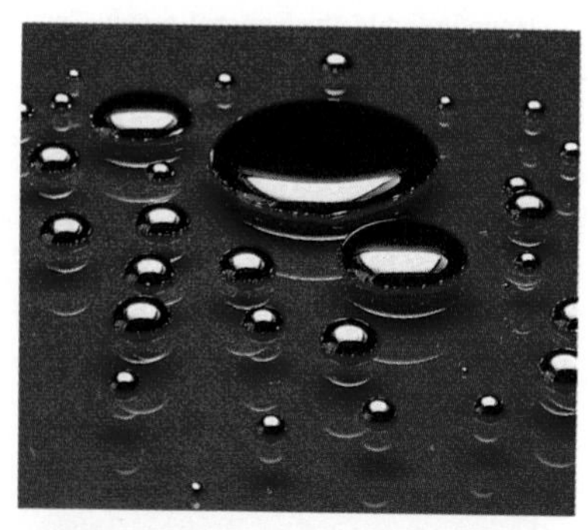
(a)

(b)

(a) Mercury is the only metal that is a liquid at room temperature. (b) The stable form of sulfur at room temperature is a solid.

Now suppose we decompose some water by passing electricity through it (Figure 1-8). (An *electrolysis* process is a chemical reaction.) We find that the water is converted into two simpler substances, hydrogen and oxygen; more significantly, hydrogen and oxygen are *always* present in the same ratio by mass, 11.1% to 88.9%. These observations allow us to identify water as a compound.

A **compound** is a substance that can be decomposed by chemical means into simpler substances, always in the same ratio by mass.

As we continue this process, starting with any substance, we eventually reach a stage at which the new substances formed cannot be further broken down by chemical means. The substances at the end of this chain are called elements.

An **element** is a substance that cannot be decomposed into simpler substances by chemical changes.

For instance, neither of the two gases obtained by the electrolysis of water hydrogen and oxygen—can be further decomposed, so they are elements.

As another illustration (see Figure 1-9), pure calcium carbonate (a white solid present in limestone and seashells) can be broken down by heating to give another white solid (call it A) and a gas (call it B) in the mass ratio 56.0:44.0. This observation tells us that calcium carbonate is a compound. The white solid A obtained from calcium carbonate can be further broken

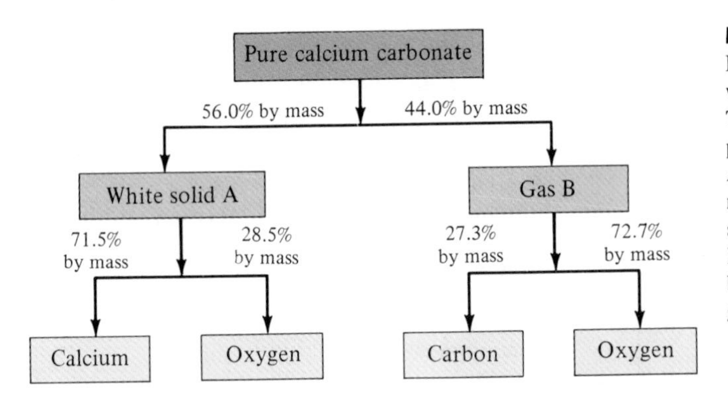

Figure 1-9
Diagram of the decomposition of calcium carbonate to give a white solid A (56.0% by mass) and a gas B (44.0% by mass). This decomposition into simpler substances at fixed ratio proves that calcium carbonate is a compound. The white solid A further decomposes to give the elements calcium (71.5% by mass) and oxygen (28.5% by mass). This proves that the white solid A is a compound; it is known as calcium oxide. The gas B also can be broken down to give the elements carbon (27.3% by mass) and oxygen (72.7% by mass). This establishes that gas B is a compound; it is known as carbon dioxide.

down into a solid and a gas in a definite ratio by mass, 71.5:28.5. But neither of these can be further decomposed, so they must be elements. The gas is identical to the oxygen obtained from the electrolysis of water; the solid is a metallic element called calcium. Similarly, the gas B originally obtained from calcium carbonate can be decomposed into two elements, carbon and oxygen in fixed mass ratio, 27.3:72.7. This sequence illustrates that a compound can be broken apart into simpler substances at fixed mass ratio, but those simpler substances may be either elements or simpler compounds.

How would you know that the white solid we called A is a compound?

Further, we may say that *a compound is a pure substance consisting of two or more different elements in a fixed ratio.* Water is 11.1% hydrogen and 88.9% oxygen by mass. Similarly, carbon dioxide is 27.3% carbon and 72.7% oxygen by mass, and calcium oxide (the white solid A above) is 71.5% calcium and 28.5% oxygen by mass. As we shall see presently, we could also combine the numbers in the previous paragraph to show that calcium carbonate is 40.1% calcium, 12.0% carbon, and 47.9% oxygen by mass. Observations such as these on innumerable pure compounds led to the statement of the **Law of Definite Proportions** (also known as the **Law of Constant Composition**):

> Different samples of any pure compound contain the same elements in the same proportions by mass.

The physical and chemical properties of a compound are different from the properties of its constituent elements. Sodium chloride is a white solid that we ordinarily use as table salt. This compound is produced by the combination of sodium (a soft, silvery white metal that reacts violently with water; Figure 1-4d) and chlorine (a pale green, corrosive, poisonous gas; Figure 1-2c). See Figure 1-10.

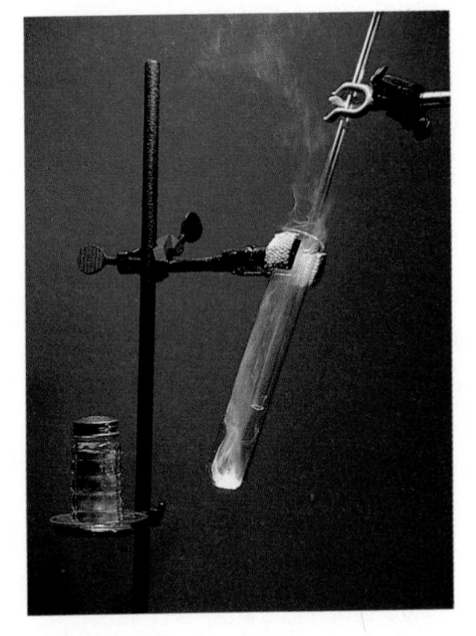

Figure 1-10
The reaction of sodium and chlorine to produce table salt, sodium chloride.

Recall that elements are substances that cannot be decomposed into simpler substances by chemical changes. Nitrogen, silver, aluminum, copper, gold, and sulfur are other examples of elements.

We use a set of **symbols** to represent the elements. These symbols can be written more quickly than names, and they occupy less space. The symbols for the first 103 elements consist of either a capital letter *or* a capital letter and a lowercase letter, such as C (carbon) or Ca (calcium). Symbols for elements beyond number 103 consist of three letters. A list of the known elements and their symbols is inside the front cover.

Table 1-2
Some Common Elements and Their Symbols

Symbol	Element	Symbol	Element	Symbol	Element
Ag	silver (*argentum*)	F	fluorine	Ni	nickel
Al	aluminum	Fe	iron (*ferrum*)	O	oxygen
Au	gold (*aurum*)	H	hydrogen	P	phosphorus
B	boron	He	helium	Pb	lead (*plumbum*)
Ba	barium	Hg	mercury	Pt	platinum
Bi	bismuth	I	iodine	Rb	rubidium
Br	bromine	K	potassium (*kalium*)	S	sulfur
C	carbon	Kr	krypton	Sb	antimony (*stibium*)
Ca	calcium	Li	lithium	Si	silicon
Cd	cadmium	Mg	magnesium	Sn	tin (*stannum*)
Cl	chlorine	Mn	manganese	Sr	strontium
Co	cobalt	N	nitrogen	U	uranium
Cr	chromium	Na	sodium (*natrium*)	W	tungsten (*Wolfram*)
Cu	copper (*cuprum*)	Ne	neon	Zn	zinc

A short list of symbols of common elements is given in Table 1-2. Learning this list will be helpful. Many symbols consist of the first one or two letters of the element's English name. Some are derived from the element's Latin name (indicated in parentheses in Table 1-2) and one, W for tungsten, is from the German *Wolfram*. Names and symbols for additional elements should be learned as they are encountered.

Most of the earth's crust is made up of a relatively small number of elements. Only 10 of the 88 naturally occurring elements make up more than 99% by mass of the earth's crust, oceans, and atmosphere (Table 1-3). Oxygen accounts for roughly half and silicon for approximately one fourth of the whole. Relatively few elements, approximately one fourth of the naturally occurring ones, occur in nature as free elements. The rest are always found chemically combined with other elements.

The other known elements have been made artificially in laboratories, as described in Chapter 30.

Figure 1-7 summarizes the classification of matter and the general means by which separations can be achieved.

Table 1-3
Abundance of Elements in the Earth's Crust, Oceans, and Atmosphere

Element	Symbol	% by Mass		Element	Symbol	% by Mass	
oxygen	O	49.5%	99.2%	chlorine	Cl	0.19%	0.7%
silicon	Si	25.7		phosphorus	P	0.12	
aluminum	Al	7.5		manganese	Mn	0.09	
iron	Fe	4.7		carbon	C	0.08	
calcium	Ca	3.4		sulfur	S	0.06	
sodium	Na	2.6		barium	Ba	0.04	
potassium	K	2.4		chromium	Cr	0.033	
magnesium	Mg	1.9		nitrogen	N	0.030	
hydrogen	H	0.87		fluorine	F	0.027	
titanium	Ti	0.58		zirconium	Zr	0.023	
				All others combined			~0.1%

Chemistry in Use. . .
The Resources of the Ocean

As is apparent to anyone who has swum in the ocean, sea water is not pure water but contains a large amount of dissolved solids. In fact, each cubic kilometer of seawater contains about 3.6×10^{10} kilograms of dissolved solids. Nearly 71% of the earth's surface is covered with water. The oceans cover an area of 361 million square kilometers at an average depth of 3729 meters, and hold approximately 1.35 billion cubic kilometers of water. This means that the oceans contain a total of more than 4.8×10^{21} kilograms of dissolved material (or more than 100,000,000,000,000,000,000 pounds). Rivers flowing into the oceans and submarine volcanoes constantly add to this storehouse of minerals. The formation of sediment and the biological demands of organisms constantly remove a similar amount.

Sea water is a very complicated solution of many substances. The main dissolved component of sea water is sodium chloride, common salt. Besides sodium and chlorine, the main elements in sea water are magnesium, sulfur, calcium, potassium, bromine, carbon, nitrogen, and strontium. Together these ten elements make up more than 99% of the dissolved materials in the oceans. In addition to sodium chloride, they combine to form such compounds as magnesium chloride, potassium sulfate, and calcium carbonate (lime). Animals absorb the latter from the sea and build it into bones and shells.

Many other substances exist in smaller amounts in sea water. In fact, most of the 92 naturally occurring elements have been measured or detected in sea water, and the remainder will probably be found as more sensitive analytical techniques become available. From an economic standpoint, there are staggering amounts of valuable metals in sea water, including approximately 1.3×10^{11} kilograms of copper, 4.2×10^{12} kilograms of uranium, 5.3×10^{9} kilograms of gold, 2.6×10^{9} kilograms of silver, and 6.6×10^{8} kilograms of lead. Other elements of economic importance include 2.6×10^{12} kilograms of aluminum, 1.3×10^{10} kilograms of tin, 2.6×10^{11} kilograms of manganese, and 4.0×10^{10} kilograms of mercury.

One would think that with such a large reservoir of dissolved solids, considerable "chemical mining" of the ocean would occur. At present (1991) only four elements are commercially extracted in large quantities. They are sodium and chlorine, which are produced from the sea by solar evaporation; magnesium; and bromine. In fact, most of the U.S. production of magnesium is derived from sea water, and the ocean is one of the principal sources of bromine. Most of the other elements are so thinly scattered through the ocean that the cost of their recovery would be much higher than their economic value. However, it is probable that as resources become more and more depleted from the continents, and as recovery techniques become more efficient, mining of sea water will become a much more desirable and feasible prospect.

One promising method of extracting elements from sea water uses marine organisms. Many marine animals concentrate certain elements in their bodies at levels many times higher than the levels in sea water. Vanadium, for example, is taken up by the mucus of certain tunicates and can be concentrated in these animals to more than 280,000 times its concentration in sea water. Other marine organisms can concentrate copper and zinc by a factor of about 1 million. If these animals could be cultivated in large quantities without endangering the ocean ecosystem, they could become a valuable source of trace metals.

In addition to dissolved materials, sea water holds a great store of suspended particulate matter that floats through the water. Some 15% of the manganese in sea water is present in particulate form, as are appreciable amounts of lead and iron. Similarly, most of the gold in sea water is thought to adhere to the surfaces of clay minerals in suspension. As in the case of dissolved solids, the economics of filtering these very fine particles from sea water is not favorable at present. However, because many of the particles suspended in sea water carry an electric charge, ion exchange techniques and modifications of electrostatic processes may someday provide important methods for the recovery of trace metals.

Beth A. Trust
Graduate student in chemistry
University of Texas Marine Sciences Institute

A large distillation tower used to separate complex mixtures such as petroleum.

1-6 Separation of Mixtures

Pure, or even nearly pure, specimens of elements and compounds seldom occur in nature. It is usually necessary to separate them from the mixtures in which they occur. When a compound is prepared in the laboratory, several steps are usually necessary to separate the pure compound from the reaction mixture (by-products, unreacted starting materials, and solvent). Thus we see that separation of mixtures is very important. We shall describe a few methods of separating pure substances from mixtures.

Filtration

Filtration is the process of separating solids that are suspended in liquids by pouring the mixture into a filter funnel. As the liquid passes through the filter, the solid particles are held on the filter (Figure 1-11). The amount of silver in a solution can be found by adding hydrochloric acid to form solid silver chloride, which is collected on a filter, dried, and weighed.

Distillation

A liquid that vaporizes readily is called a *volatile liquid*. When a liquid is heated to a sufficiently high temperature it boils; i.e., it is converted to the gaseous, or vapor, state. **Distillation** is one method by which a mixture containing volatile substances can be separated into its components. For example, if a salt solution is heated, the more volatile water boils off, leaving behind the solid salt. A simple laboratory distillation apparatus is shown in Figure 1-12a.

The container in which the mixture is heated is called the distilling flask. The condenser is a double-walled glass tube. Cold water passes through the outer chamber to condense the hot vapor (gas) to a liquid.

(a)

(b)

Figure 1-11
(a) Silver chloride precipitates when silver ions are added to a solution containing chloride ions. (b) The solution, which is mixed with solid silver chloride, is poured through filter paper in the shape of a cone. Solid silver chloride remains on the paper. (On exposure to light, silver chloride turns dark.)

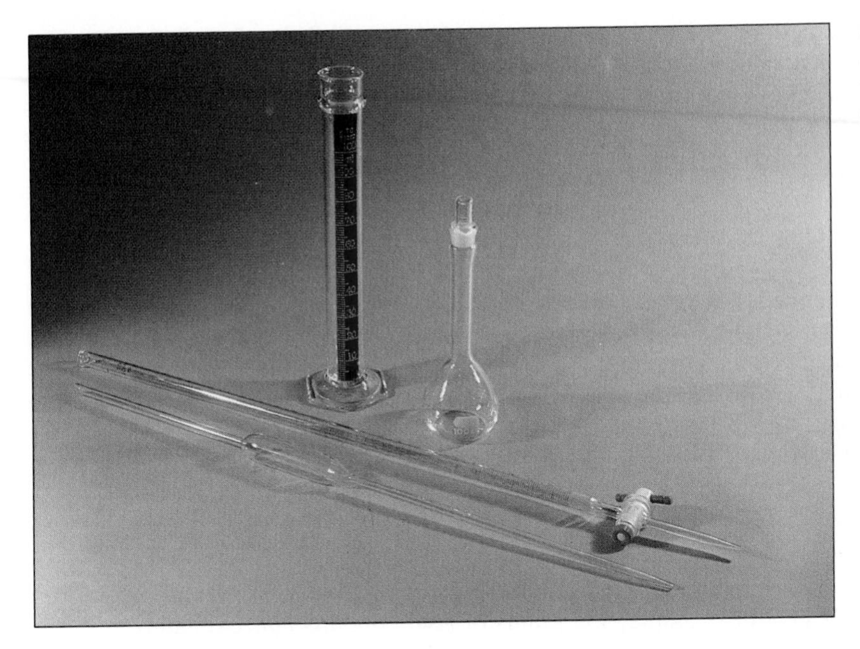

Figure 1-18
Some laboratory apparatus used to measure volumes of liquids: 100-mL graduated cylinder (left rear), 100-mL volumetric flask (right rear), 25-mL buret (center), and 25-mL volumetric pipet (front).

Volume

Volumes are measured in liters or milliliters in the metric system. One liter (1 L) is one cubic decimeter (1 dm^3), or 1000 cubic centimeters (1000 cm^3). One milliliter (1 mL) is 1 cm^3. In the SI the cubic meter is the basic volume unit, and the cubic decimeter replaces the metric unit, liter. Different kinds of glassware are used to measure the volume of liquids. The one we choose depends on the accuracy we desire. For example, the volume of a liquid dispensed can be measured more accurately with a buret than with a small graduated cylinder (Figure 1-18). Equivalences between common English units and metric units are summarized in Table 1-7.

Table 1-7
Conversion Factors Relating Length, Volume, and Mass (weight) Units

	Metric		English		Metric–English Equivalents	
Length	1 km	$= 10^3$ m	1 ft	= 12 in	2.54 cm	= 1 in
	1 cm	$= 10^{-2}$ m	1 yd	= 3 ft	39.37 in*	= 1 m
	1 mm	$= 10^{-3}$ m	1 mile	= 5280 ft	1.609 km*	= 1 mile
	1 nm	$= 10^{-9}$ m				
	1 Å	$= 10^{-10}$ m				
Volume	1 mL	$= 1\ cm^3 = 10^{-3}$ L	1 gal	= 4 qt = 8 pt	1 L	= 1.057 qt*
	1 m^3	$= 10^6\ cm^3 = 10^3$ L	1 qt	= 57.75 in^3*	28.32 L	= 1 ft^3
Mass	1 kg	$= 10^3$ g	1 lb	= 16 oz	453.6 g*	= 1 lb
	1 mg	$= 10^{-3}$ g			1 g	= 0.03527 oz*
	1 metric tonne	$= 10^3$ kg	1 short ton	= 2000 lb	1 metric tonne	= 1.102 short ton

* These conversion factors, unlike the others listed, are inexact. They are quoted to four significant figures, which is ordinarily more than sufficient.

Sometimes we must combine two or more units to describe a quantity. For instance, we might express the speed of a car as 60 mi/hr. Recall that the algebraic notation x^{-1} means $1/x$; applying this notation to units, we see that hr^{-1} means 1/hr, or "per hour." So the unit of speed could also be expressed as $mi \cdot hr^{-1}$.

1-9 Use of Numbers

In chemistry, we measure and calculate many things, so we must be sure we understand how to use numbers. In this section we discuss two aspects of the use of numbers: (1) the notation of very large and very small numbers and (2) an indication of how well we actually know the numbers we are using. You will carry out many calculations with calculators. Please refer to Appendix A for some instructions about the use of electronic calculators.

Scientific Notation

In exponential form these numbers are
6.02×10^{23} gold atoms

3.27×10^{-22} gram

We use **scientific notation** when we deal with very large and very small numbers. For example, 197 grams of gold contains approximately

602,000,000,000,000,000,000,000 gold atoms.

The mass of one gold atom is approximately

0.000 000 000 000 000 000 000 327 gram.

In using such large and small numbers, it is inconvenient to write down all the zeroes. In scientific (exponential) notation, we place one nonzero digit to the left of the decimal.

$$4{,}300{,}000. = 4.3 \times 10^6$$

6 places to the left, ∴ exponent of 10 is 6

$$0.000348 = 3.48 \times 10^{-4}$$

4 places to the right, ∴ exponent of 10 is −4

The reverse process converts numbers from exponential to decimal form. See Appendix A for more detail, if necessary.

Significant Figures

An *exact* number may be thought of as containing an *infinite* number of significant figures.

There are two kinds of numbers. **Exact numbers** may be *counted* or *defined*. They are known to be absolutely accurate. For example, the exact number of people in a closed room can be counted, and there is no doubt about the number of people. A dozen eggs is defined as exactly 12 eggs, no more, no fewer (Figure 1-19).

There is some uncertainty in all measurements.

Significant figures indicate the *uncertainty* in measurements.

Numbers obtained from measurements are not exact. Every measurement involves an estimate. For example, suppose you are asked to measure the length of this page to the nearest 0.1 mm. How do you do it? The smallest divisions (calibration lines) on a meter stick are 1 mm apart (Figure 1-17). An attempt to measure to 0.1 mm requires estimation. If three different people measure the length of the page to 0.1 mm, will they get the same answer? Probably not. We deal with this problem by using significant figures.

Significant figures are digits believed to be correct by the person who makes a measurement. We assume that the person is competent to use the measuring device. Suppose one measures a distance with a meter stick and reports the distance as 343.5 mm. What does this number mean? In this person's judgment, the distance is greater than 343.4 mm but less than 343.6 mm, and the best estimate is 343.5 mm. The number 343.5 mm contains four significant figures. The last digit, 5, is a *best estimate* and is therefore doubtful, but it is considered to be a significant figure. In reporting numbers obtained from measurements, *we report one estimated digit, and no more*. Because the person making the measurement is not certain that the 5 is correct, it would be meaningless to report the distance as 343.53 mm.

To see more clearly the part significant figures play in reporting the results of measurements, consider Figure 1-20a. Graduated cylinders are used to measure volumes of liquids when a high degree of accuracy is not necessary. The calibration lines on a 50-mL graduated cylinder represent 1-mL increments. Estimation of the volume of liquid in a 50-mL cylinder to within 0.2 mL ($\frac{1}{5}$ of one calibration increment) with reasonable certainty is possible. We might measure a volume of liquid in such a cylinder and report the volume as 38.6 mL, i.e., to three significant figures.

Burets are used to measure volumes of liquids when higher accuracy is required. The calibration lines on a 50-mL buret represents 0.1-mL increments, allowing us to make estimates to within 0.02 mL ($\frac{1}{5}$ of one calibration increment) with reasonable certainty (Figure 1-20b). Experienced individuals estimate volumes in 50-mL burets to 0.01 mL with considerable reproducibility. For example, using a 50-mL buret, we can measure out 38.57 mL (four significant figures) of liquid with reasonable accuracy.

Accuracy refers to how closely a measured value agrees with the correct value. **Precision** refers to how closely individual measurements agree with each other. Ideally, all measurements should be both accurate and precise. Measurements may be quite precise, yet quite inaccurate, because of some *systematic error*, which is an error repeated in each measurement. (A faulty balance, for example, might produce a systematic error.) Very accurate measurements are seldom imprecise.

(a)

(b)

Figure 1-19
(a) A dozen eggs is exactly 12 eggs. (b) A swarm of honey bees contains an *exact* number of live bees. Could you count them easily?

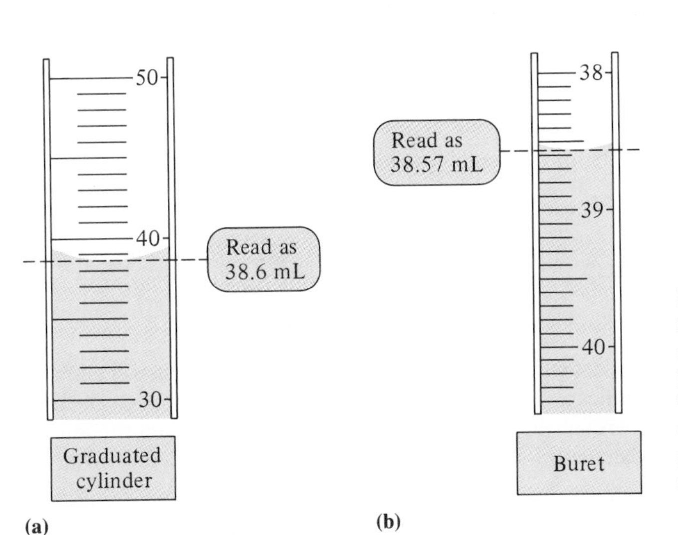

Figure 1-20
Measurement of the volume of water using two different pieces of volumetric glassware. For consistency, we always read the bottom of the meniscus (the curved surface of the water). (a) The level in a 50-mL graduated cylinder can be estimated to within 0.2 mL. The level here is 38.6 mL (three significant figures). (b) The level in a 50-mL buret can be read to within 0.02 mL. The level here is 38.57 mL (four significant figures).

Measurements are frequently repeated to improve accuracy and precision. Average values obtained from several measurements are usually more reliable than individual measurements. Significant figures indicate how accurately measurements have been made (assuming the person who made the measurements was competent).

Some simple rules govern the use of significant figures in calculations.

1. Zeroes used just to position the decimal point are not significant figures.

For example, the number 0.0234 g contains only three significant figures, because the two zeroes are used to place the decimal point. The number could also be reported as 2.34×10^{-2} g in scientific notation (Appendix A). When zeroes precede the decimal point, but come after other digits, we may have some difficulty in deciding whether the zeroes are significant figures or not. How many significant figures does the number 23,000 contain? We are given insufficient information to answer the question. If all three of the zeroes are used just to place the decimal point, the number should appear as 2.3×10^4 (two significant figures). If only two of the zeroes are being used to place the decimal point, the number is 2.30×10^4 (three significant figures). In the unlikely event that the number is actually known to be 23,000 $\pm$ 1, it should be written as 2.3000×10^4 (five significant figures).

When we wish to specify that all of the zeroes in such a number *are* significant, we may indicate this by placing a decimal point after the number. For instance, 130. grams can represent a mass known to *three* significant figures, that is, 130 $\pm$ 1 gram.

2. In multiplication and division, an answer contains no more significant figures than the least number of significant figures used in the operation.

Example 1-1

What is the area of a rectangle 1.23 cm wide and 12.34 cm long?

Plan

The area of a rectangle is its length times its width. We should first check to see that the width and length are expressed in the same units. (They are—but if they were not, one would first have to be converted to the units of the other.) Then we multiply the width by the length. We then follow Rule 2 for significant figures to find the correct number of significant figures. The units for the result are equal to the product of the units for the individual terms in the multiplication.

Solution

$$A = \ell \times w = (12.34 \text{ cm})(1.23 \text{ cm}) = 15.2 \text{ cm}^2$$

(calculator result = 15.1782)

Because three is the smallest number of significant figures used, the answer should contain only three significant figures. The number generated by an electronic calculator (15.1782) is wrong; the result cannot be more accurate than the information that led to it. Calculators have no judgment, so you must exercise yours.

The step-by-step calculation in the margin demonstrates why the area is reported as 15.2 cm^2 rather than 15.1782 cm^2. The length, 12.34 cm, contains four significant figures, whereas the width, 1.23 cm, contains only three. If we underline each uncertain figure, as well as each figure obtained from an uncertain figure, the step-by-step multiplication gives the result reported in Example 1-1. We see that there are only two certain figures (15) in the result. We report the first doubtful figure (.2), but no more. Division is just the reverse of multiplication, and the same rules apply.

$$
\begin{array}{r}
12.3\underline{4}\ \text{cm} \\
\times\ 1.2\underline{3}\ \text{cm} \\
\hline
\underline{37}\ \underline{02} \\
2\ 46\ \underline{8} \\
12\ 3\underline{4} \\
\hline
15.\underline{17}\ \underline{82}\ \text{cm}^2 = 15.2\ \text{cm}^2
\end{array}
$$

3. In addition and subtraction, the last digit retained in the sum or difference is determined by the position of the first doubtful digit.

Example 1-2

(a) Add 37.24 mL and 10.3 mL. (b) Subtract 21.2342 g from 27.87 g.

Plan

Again we first check to see that the quantities to be added or subtracted are expressed in the same units. We carry out the addition or subtraction. Then we follow Rule 3 for significant figures to express the answer to the correct number of significant figures.

Solution

Doubtful digits are underlined in this example.

(a)

$$
\begin{array}{r}
37.2\underline{4}\ \text{mL} \\
+10.\underline{3}\ \ \text{mL} \\
\hline
\end{array}
$$

$47.\underline{54}$ mL is reported as 47.5 mL (calculator gives 47.54)

(b)

$$
\begin{array}{r}
27.8\underline{7}\ \ \ \text{g} \\
-21.234\underline{2}\ \text{g} \\
\hline
\end{array}
$$

$6.6\underline{358}$ g is reported as 6.64 g (calculator gives 6.6358)

EOC 25, 26

With many examples we suggest selected exercises from the "end of chapter" (EOC). These exercises use the skills or concepts from that example. Now you should work Exercises 25 and 26 from the end of this chapter.

In the three simple arithmetic operations we have performed, the number combination generated by an electronic calculator is not the "answer" in a single case! However, the correct result of each calculation can be obtained by "rounding off." The rules of significant figures tell us where to round off.

In rounding off, certain conventions have been adopted. When the number to be dropped is less than 5, the preceding number is left unchanged (e.g., 7.34 rounds off to 7.3). When it is more than 5, the preceding number is increased by 1 (e.g., 7.37 rounds off to 7.4). When the number to be dropped is 5, the preceding number is not changed when it (the preceding number) is even (e.g., 7.45 rounds off to 7.4). When the preceding number is odd, it is increased by one (e.g., 7.35 rounds off to 7.4).

The even–odd rules are intended to reduce the accumulation of errors in chains of calculations.

1-10 The Unit Factor Method (Dimensional Analysis)

Many chemical and physical processes can be described by numerical relationships. In fact, many of the most useful ideas in science must be treated mathematically. Let us devote a little time to reviewing problem-solving skills.

First, multiplication by unity (by one) does not change the value of an expression. If we represent "one" in the right way, we can do many conversions by just "multiplying by one." This method of performing calculations is known as **dimensional analysis**, the **factor-label method**, or the **unit factor method**. Regardless of the name chosen, it is a very powerful mathematical tool that is almost foolproof.

Unit factors may be constructed from any two terms that describe the same or equivalent "amounts" of whatever we may consider. For example, 1 foot is equal to exactly 12 inches, by definition. We may write an equation to describe this equality:

$$1 \text{ ft} = 12 \text{ in}$$

Dividing both sides of the equation by 1 ft gives

$$\frac{1 \cancel{\text{ft}}}{1 \cancel{\text{ft}}} = \frac{12 \text{ in}}{1 \text{ ft}} \quad \text{or} \quad 1 = \frac{12 \text{ in}}{1 \text{ ft}}$$

The factor (fraction) 12 in/1 ft is a unit factor because the numerator and denominator describe the same distance. Dividing both sides of the original equation by 12 in gives 1 = 1 ft/12 in, a second unit factor that is the reciprocal of the first. *The reciprocal of any unit factor is also a unit factor.* Stated differently, division of an amount by the same amount always yields one!

In the English system we can write many unit factors, such as

$$\frac{1 \text{ yd}}{3 \text{ ft}}, \quad \frac{1 \text{ yd}}{36 \text{ in}}, \quad \frac{1 \text{ mile}}{5280 \text{ ft}}, \quad \frac{4 \text{ qt}}{1 \text{ gal}}, \quad \frac{2000 \text{ lb}}{1 \text{ ton}}$$

Unless otherwise indicated, a "ton" refers to a "short ton," 2000 lb. There are also the "long ton," which is 2240 lb, and the metric tonne, which is 1000 kg.

The reciprocal of each of these is also a unit factor. Items in retail stores are frequently priced with unit factors, such as 39¢/lb and \$3.98/gal. When all the quantities in a unit factor come from definitions, the unit is known to an unlimited (infinite) number of significant figures. For instance, if you bought eight 1-gallon jugs of something priced at \$3.98/gal, the total cost would be 8 × \$3.98, or \$31.84; the merchant would not round this to \$31.80, let alone to \$30.

In science, nearly all numbers have units. What does 12 mean? Usually we must supply appropriate units, such as 12 eggs or 12 people. In the unit factor method, the units guide us through calculations in a step-by-step process, because all units except those in the desired result cancel.

Example 1-3

Express 1.47 miles in inches.

Plan

First we write down the units of what we wish to know preceded by a question mark. Then we set it equal to whatever we are given:

$$\underline{?}\ \text{in} = 1.47\ \text{miles}$$

Then we choose unit factors to convert the given units (miles) to the desired units (inches):

$$\text{miles} \rightarrow \text{feet} \rightarrow \text{inches}$$

We relate (a) miles to feet and then (b) feet to inches.

Solution

$$\underline{?}\ \text{in} = 1.47\ \cancel{\text{miles}} \times \frac{5280\ \cancel{\text{ft}}}{1\ \cancel{\text{mile}}} \times \frac{12\ \text{in}}{1\ \cancel{\text{ft}}} = 9.31 \times 10^4\ \text{in} \quad \text{(calculator gives 93139.2)}$$

In the interest of clarity, cancellation of units will be omitted in the remainder of this book. You may find it useful to continue the practice.

Note that both miles and feet cancel, leaving only inches, the desired unit. Thus there is no ambiguity as to how the unit factors should be written. The answer contains three significant figures because there are three significant figures in 1.47 miles. The quantities 5280 ft, 1 mile, 12 in, and 1 ft all come from definitions, so each of them is known to an unlimited number of significant figures.

EOC 30

It is often helpful to consider whether an answer is reasonable. In Example 1-3, the distance involved is more than a mile; we expect this distance to be many inches, so a large answer is not surprising. Suppose we had mistakenly multiplied by the unit factor $\frac{1\ \text{mile}}{5280\ \text{ft}}$ (and not noticed that the units did not cancel properly); we would have gotten the answer 3.34×10^{-3} inches (0.00334 inches), which we would have immediately recognized as nonsense!

Conversions within the SI and metric systems are easy, because measurements of a particular kind are related to each other by powers of ten.

Example 1-4

The Ångstrom (Å) is a unit of length, 1×10^{-10} meter, that provides a convenient scale on which to express the radii of atoms. Radii of atoms are often expressed in nanometers. The radius of a phosphorus atom is 1.10 Å. What is the distance expressed in centimeters and nanometers?

Plan

We use the equalities 1 Å = 1×10^{-10} m, 1 cm = 1×10^{-2} m, and 1 nm = 1×10^{-9} m to construct the unit factors that convert 1.10 Å to the desired units.

Solution

$$\underline{?}\ \text{cm} = 1.10\ \text{Å} \times \frac{1 \times 10^{-10}\ \text{m}}{1\ \text{Å}} \times \frac{1\ \text{cm}}{1 \times 10^{-2}\ \text{m}} = 1.10 \times 10^{-8}\ \text{cm}$$

$$\underline{?}\ \text{nm} = 1.10\ \text{Å} \times \frac{1.0 \times 10^{-10}\ \text{m}}{1\ \text{Å}} \times \frac{1\ \text{nm}}{1 \times 10^{-9}\ \text{m}} = 0.110\ \text{nm}$$

All the unit factors used in this example contain only exact numbers.

EOC 34

Example 1-5
Assuming a phosphorus atom is spherical, calculate its volume in $Å^3$, cm^3, and nm^3. The volume of a sphere is $V = (\frac{4}{3})\pi r^3$. Refer to Example 1-4.

Plan

$1\ Å = 10^{-10}\ m = 10^{-8}\ cm$

We use the results of Example 1-4 to calculate the volume in each of the desired units.

Solution

$$\underline{?}\ Å^3 = (\tfrac{4}{3})\pi(1.10\ Å)^3 = 5.58\ Å^3$$

$$\underline{?}\ cm^3 = (\tfrac{4}{3})\pi(1.10 \times 10^{-8}\ cm)^3 = 5.58 \times 10^{-24}\ cm^3$$

$$\underline{?}\ nm^3 = (\tfrac{4}{3})\pi(1.10 \times 10^{-1}\ nm)^3 = 5.58 \times 10^{-3}\ nm^3$$

EOC 42

Example 1-6
A sample of gold has a mass of 0.234 mg. What is its mass in g? in cg?

Plan
We use the relationships 1 g = 1000 mg and 1 cg = 10 mg to write the required unit factors.

Solution

$$\underline{?}\ g = 0.234\ mg \times \frac{1\ g}{1000\ mg} = 2.34 \times 10^{-4}\ g$$

$$\underline{?}\ cg = 0.234\ mg \times \frac{1\ cg}{10\ mg} = 0.0234\ cg \quad \text{or} \quad 2.34 \times 10^{-2}\ cg$$

Again, we have used unit factors that contain only exact numbers.

EOC 43

Unity raised to *any* power is one. *Any* unit factor raised to a power is still a unit factor, as the next two examples show.

Example 1-7
How many square decimeters are there in 215 square centimeters?

Plan

We would multiply by the unit factor $\frac{1\ dm}{10\ cm}$ to convert cm to dm. Here we require the *square* of this unit factor.

Solution

$$\underline{?}\ dm^2 = 215\ cm^2 \times \left(\frac{1\ dm}{10\ cm}\right)^2 = 215\ cm^2 \times \left(\frac{1\ dm^2}{100\ cm^2}\right) = \boxed{2.15\ dm^2}$$

$$\left(\frac{1\ dm}{10\ cm}\right)^2 = \frac{1\ dm^2}{100\ cm^2} = 1$$

Example 1-7 shows that a unit factor *squared* is still a unit factor.

Example 1-8

How many cubic centimeters are there in 8.34×10^5 cubic decimeters?

Plan

We use the *cube* of the unit $\frac{10\ cm}{1\ dm}$ to carry out the desired conversion.

Solution

$$\underline{?}\ cm^3 = 8.34 \times 10^5\ dm^3 \times \left(\frac{10\ cm}{1\ dm}\right)^3 = 8.34 \times 10^5\ dm^3 \times \left(\frac{1000\ cm^3}{1\ dm^3}\right)$$

$$= \boxed{8.34 \times 10^8\ cm^3}$$

$$\left(\frac{10\ cm}{1\ dm}\right)^3 = \frac{1000\ cm^3}{1\ dm^3} = 1$$

EOC 33

Example 1-8 shows that a unit factor *cubed* is still a unit factor.

Example 1-9

A common unit of energy is the erg. Convert 3.74×10^{-2} erg to the SI units of energy, joules and kilojoules. One erg is exactly 1×10^{-7} joule.

Plan

The definition that relates ergs and joules is used to generate the needed unit factor. The second conversion uses a unit factor that is based on the definition of the prefix *kilo-*.

Solution

$$\underline{?}\ J = 3.74 \times 10^{-2}\ erg \times \frac{1 \times 10^{-7}\ J}{1\ erg} = \boxed{3.74 \times 10^{-9}\ J}$$

$$\underline{?}\ kJ = 3.74 \times 10^{-9}\ J \times \frac{1 \times 10^{-3}\ kJ}{J} = \boxed{3.74 \times 10^{-12}\ kJ}$$

EOC 38

Conversions between the English and SI (metric) systems are conveniently made by the unit factor method. Several conversion factors are listed in Table 1-7. It may be helpful to remember one each for

Six substances with different densities. The liquid layers are gasoline (top), water (middle), and mercury (bottom). A cork floats on gasoline. A piece of oak wood sinks in gasoline, but floats on water. Brass sinks in water, but floats on mercury.

Example 1-12

A 47.3-mL sample of ethyl alcohol (ethanol) has a mass of 37.32 g. What is its density?

Plan

We use the definition of density.

Solution

$$D = \frac{m}{V} = \frac{37.32\text{ g}}{47.3\text{ mL}} = 0.789\text{ g/mL}$$

EOC 50

Example 1-13

If 103 g of ethanol is needed for a chemical reaction, what volume of liquid would you use?

Plan

We determined the density of ethanol in Example 1-12. Here we are given the mass, m, of a sample of ethanol. So we know values for D and m in the relationship

$$D = \frac{m}{V}$$

We rearrange this relationship to solve for V, put in the known values, and carry out the calculation. Alternatively, we can use the unit factor method to solve the problem.

Solution

The density of ethanol is 0.789 g/mL (Table 1-8).

$$D = \frac{m}{V}, \quad \text{so} \quad V = \frac{m}{D} = \frac{103\text{ g}}{0.789\text{ g/mL}} = 130\text{ mL}$$

Alternatively,

$$\underline{?}\text{ mL} = 103\text{ g} \times \frac{1\text{ mL}}{0.789\text{ g}} = 130\text{ mL}$$

EOC 53

Example 1-14

Express the density of mercury in lb/ft³.

Plan

The density of mercury is 13.59 g/cm³ (Table 1-8). To convert this value to the desired unit, we can use unit factors constructed from the conversion factors in Table 1-7.

Solution

$$\underline{?}\ \frac{\text{lb}}{\text{ft}^3} = 13.59\ \frac{\text{g}}{\text{cm}^3} \times \frac{1\text{ lb}}{453.6\text{ g}} \times \left(\frac{2.54\text{ cm}}{1\text{ in}}\right)^3 \times \left(\frac{12\text{ in}}{1\text{ ft}}\right)^3 = 848.4\text{ lb/ft}^3$$

It would take a very strong person to lift a cubic foot of mercury!

The **specific gravity** (Sp. Gr.) of a substance is the ratio of its density to the density of water, both at the same temperature. Specific gravities are dimensionless numbers.

Density and specific gravity are both intensive properties; i.e., they do not depend upon the size of the sample.

$$\text{Sp. Gr.} = \frac{D_{\text{substance}}}{D_{\text{water}}}$$

The density of water is 1.000 g/mL at 3.98°C, the temperature at which the density of water is greatest. However, variations in the density of water with changes in temperature are small enough that we may use 1.00 g/mL up to 25°C without introducing significant errors into our calculations.

Example 1-15

The density of table salt is 2.16 g/mL at 20°C. What is its specific gravity?

Plan

We use the definition of specific gravity given above. The numerator and denominator have the same units, so the result is dimensionless.

Solution

$$\text{Sp. Gr.} = \frac{D_{\text{salt}}}{D_{\text{water}}} = \frac{2.16 \text{ g/mL}}{1.00 \text{ g/mL}} = 2.16$$

EOC 57

This example also demonstrates that the density and specific gravity of a substance are numerically equal near room temperature if density is expressed in g/mL (g/cm^3).

Labels on commercial solutions of acids and bases give specific gravities and the percentage by mass of the acid or base present in the solution. From this information, the amount of acid or base present in a given volume of the solution can be calculated.

At this point you need not be concerned if you do not know the chemical characteristics of an acid or a base. They are described in Chapters 4 and 10.

Example 1-16

Battery acid is 40.0% sulfuric acid, H_2SO_4, and 60.0% water by mass. Its specific gravity is 1.31. Calculate the mass of pure H_2SO_4 in 100.0 mL of battery acid.

Plan

The percentages given are on a mass basis, so we must first convert the 100.0 mL of acid solution to mass. To do this, we need a value for the density. We have demonstrated that density and specific gravity are numerically equal at 20°C because the density of water is 1.00 g/mL. We can use the density as a unit factor to convert the given volume of solution to mass of solution. Then we use the percentage by mass to convert the mass of solution to mass of acid.

Solution

From the given value for specific gravity, we may write

$$\text{density} = 1.31 \text{ g/mL}$$

The solution is 40.0% H_2SO_4 and 60.0% H_2O by mass. From this information we may construct the desired unit factor:

$$\frac{40.0 \text{ g } H_2SO_4}{100 \text{ g soln}} \leftarrow \text{because 100 g of solution contains 40.0 g of } H_2SO_4$$

We can now solve the problem:

$$\underline{?}\ H_2SO_4 = 100.0 \text{ mL soln} \times \frac{1.31 \text{ g soln}}{1 \text{ mL soln}} \times \frac{40.0 \text{ g } H_2SO_4}{100 \text{ g soln}} = 52.4 \text{ g } H_2SO_4$$

EOC 59

1-12 Heat and Temperature

In Section 1-1 you learned that heat is one form of energy. You also learned that the many different forms of energy can be interconverted and that in chemical processes, chemical energy is converted to heat energy or vice versa. The amount of heat a process uses (*endothermic*) or gives off (*exothermic*) can tell us a great deal about that process (Chapters 13 and 15). For this reason it is important for us to be able to measure intensity of heat.

Temperature measures the intensity of heat, the "hotness" or "coldness" of a body. A piece of metal at 100°C feels hot to the touch, while an ice cube at 0°C feels cold. Why? Because the temperature of the metal is higher, and that of the ice cube lower, than body temperature. *Heat always flows spontaneously from a hotter body to a colder body*—never in the reverse direction.

Temperatures are commonly measured with mercury-in-glass thermometers. A mercury thermometer consists of a reservoir of mercury at the base of a glass tube, open to a very thin (capillary) column extending upward. Mercury expands more than most other liquids as its temperature rises. As it expands, its movement up into the evacuated column can be seen.

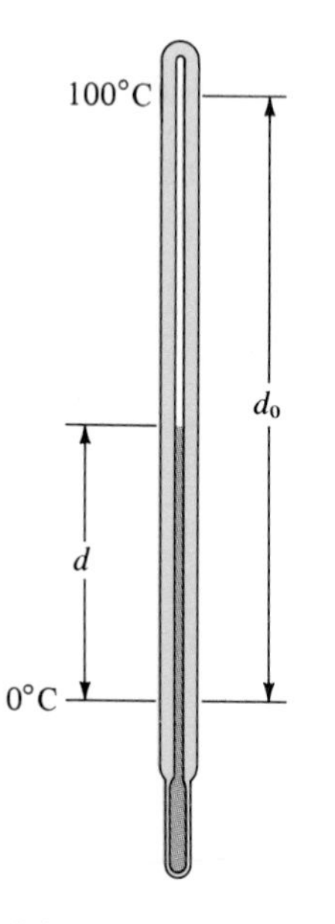

Figure 1-21
At 45°C, as read on a mercury-in-glass thermometer, d equals $0.45d_0$ where d_0 is the distance from the mercury level at 0°C to the level at 100°C.

Anders Celsius, a Swedish astronomer, developed the Celsius temperature scale, formerly called the centigrade temperature scale. When we place a Celsius thermometer in a beaker of crushed ice and water, the mercury level stands at exactly 0°C, the lower reference point. In a beaker of water boiling at one atmosphere pressure, the mercury level stands at 100°C, the higher reference point. There are 100 equal steps between these two mercury levels. They correspond to an interval of 100 degrees between the melting point of ice and the boiling point of water at one atmosphere. Figure 1-21 shows how temperature marks between the reference points are established.

In the United States, temperatures are frequently measured on the temperature scale devised by Gabriel Fahrenheit, a German instrument maker. On this scale the freezing and boiling points of water are defined as 32°F and 212°F, respectively. In scientific work, temperatures are often expressed on the **Kelvin** (absolute) temperature scale. As we shall see in Section 12-5, the zero point of the Kelvin temperature scale is *derived* from the observed behavior of all matter.

Relationships among the three temperature scales are illustrated in Figure 1-22. Between the freezing point of water and the boiling point of water, there are 100 steps (degrees C or kelvins, respectively) on the Celsius and Kelvin scales. Thus the "degree" is the same size on the Celsius and Kelvin

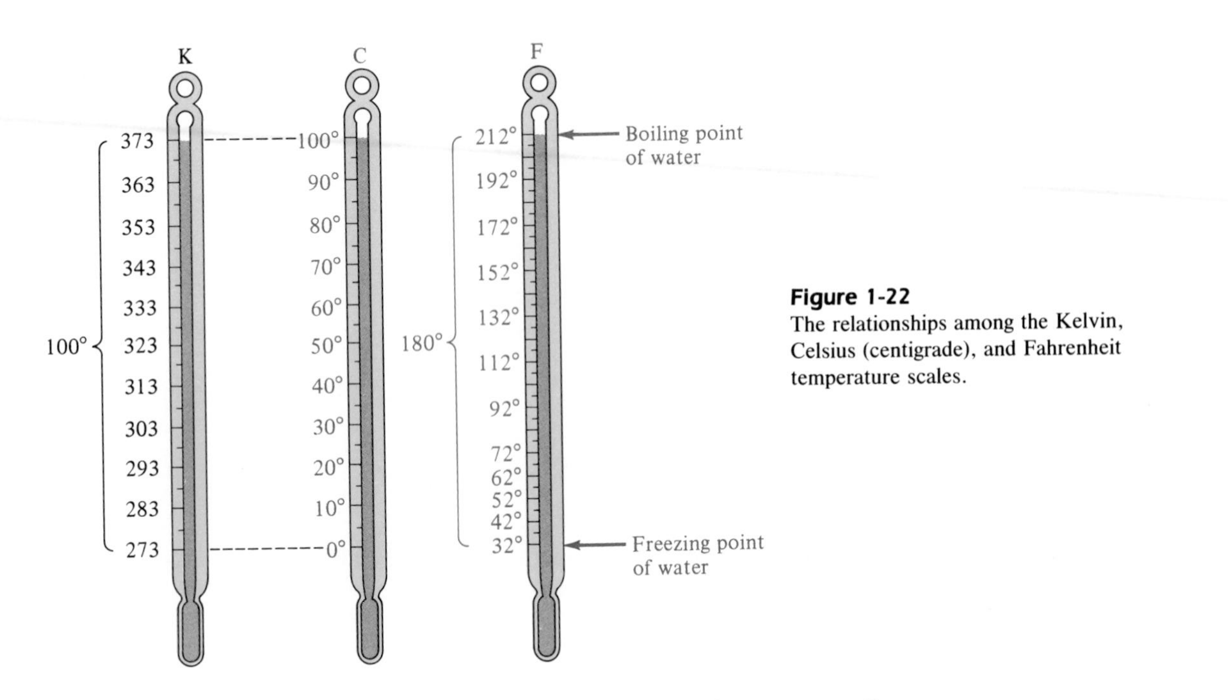

Figure 1-22
The relationships among the Kelvin, Celsius (centigrade), and Fahrenheit temperature scales.

scales. But every Kelvin temperature is 273.15 units above the corresponding Celsius temperature. The relationship between these two scales is as follows:

$$\underline{?}\ \text{K} = {}^\circ\text{C} + 273.15^\circ \quad \text{or} \quad \underline{?}{}^\circ\text{C} = \text{K} - 273.15^\circ$$

We shall usually round 273.15 to 273.

In the SI system, "degrees Kelvin" are abbreviated simply as K rather than °K and are called **kelvins**.

Please recognize that any temperature *change* has the same numerical value whether expressed on the Celsius scale or on the Kelvin scale. For example, a change from 25°C to 59°C represents a *change* of 34 Celsius degrees. Converting these to the Kelvin scale, the same change is expressed as (273 + 25) = 298 K to (59 + 273) = 332 K, or a *change* of 34 kelvins.

Comparing the Fahrenheit and Celsius scales, we find that the intervals between the same reference points are 180 Fahrenheit degrees and 100 Celsius degrees, respectively. Thus a Fahrenheit degree must be smaller than a Celsius degree. It takes 180 Fahrenheit degrees to cover the same temperature *interval* as 100 Celsius degrees. From this information, we can construct the unit factors for temperature *changes*:

$$\frac{180^\circ\text{F}}{100^\circ\text{C}} \quad \text{or} \quad \frac{1.8^\circ\text{F}}{1.0^\circ\text{C}} \quad \text{and} \quad \frac{100^\circ\text{C}}{180^\circ\text{F}} \quad \text{or} \quad \frac{1.0^\circ\text{C}}{1.8^\circ\text{F}}$$

But the starting points of the two scales are different, so we *cannot convert* a temperature on one scale to a temperature on the other just by multiplying by the unit factor. In converting from °F to °C, we must add 32 Fahrenheit degrees to reach the zero point on the Celsius scale (Figure 1-22).

$$\underline{?}{}^\circ\text{F} = \left(x^\circ\text{C} \times \frac{1.8^\circ\text{F}}{1.0^\circ\text{C}}\right) + 32^\circ\text{F} \quad \text{and} \quad \underline{?}{}^\circ\text{C} = \frac{1.0^\circ\text{C}}{1.8^\circ\text{F}}(x^\circ\text{F} - 32^\circ\text{F})$$

These are often remembered in abbreviated form:

$$^\circ\text{F} = 1.8^\circ\text{C} + 32^\circ$$

$$^\circ\text{C} = \frac{(^\circ\text{F} - 32^\circ)}{1.8}$$

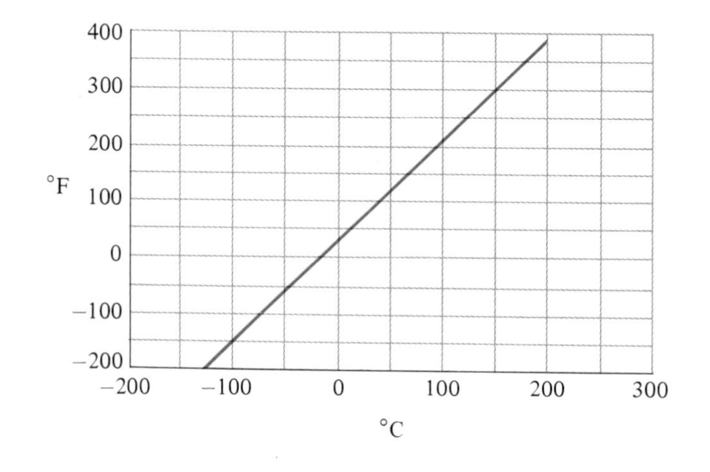

Figure 1-23
A graphical representation of the relationship between the Fahrenheit and Celsius temperature scales.

The relationship between temperatures on the Fahrenheit and Celsius scales is shown graphically in Figure 1-23.

Example 1-17
When the temperature reaches "100.°F in the shade," it's hot. What is this temperature on the Celsius scale?

Plan

We use the relationship $\underline{?}°C = \frac{1.0°C}{1.8°F}(x°F - 32°F)$ to carry out the desired conversion.

Solution

A temperature of 100°F is 38°C.

$$\underline{?}°C = \frac{1.0°C}{1.8°F}(100.°F - 32°F) = \frac{1.0°C}{1.8°F}(68°F) = 38°C$$

Example 1-18
When the absolute temperature is 400 K, what is the Fahrenheit temperature?

Plan

We first use the relationship $\underline{?}°C = K - 273.15°$ to convert from kelvins to degrees Celsius; then we carry out the further conversion from degrees Celsius to degrees Fahrenheit.

Solution

$$\underline{?}°C = (400\ K - 273\ K)\frac{1.0°C}{1.0\ K} = 127°C$$

$$\underline{?}°F = \left(127°C \times \frac{1.8°F}{1.0°C}\right) + 32°F = 261°F$$

EOC 62

1-13 Heat Transfer and the Measurement of Heat

Chemical reactions and physical changes occur with either the simultaneous evolution of heat (**exothermic processes**) or the absorption of heat (**endothermic processes**). The amount of heat transferred in a process is usually expressed in calories or in the SI unit joules. The **calorie** was originally defined as the amount of heat necessary to raise the temperature of one gram of water at one atmosphere from 14.5°C to 15.5°C. One calorie is *now defined* as exactly 4.184 joules. The amount of heat necessary to raise the temperature of one gram of liquid water varies slightly with temperature and pressure, so it was necessary to specify a particular temperature increment and constant pressure in describing the calorie. For our purposes the variations are sufficiently small that we are justified in ignoring them. The so-called "large calorie," used to indicate the energy content of foods, is really one kilocalorie, or 1000 calories. We shall do most calculations in joules.

In terms of electrical energy, one joule is equal to one watt/second. Thus one joule is enough energy to operate a 10-watt light bulb for 1/10 second.

The SI unit of energy and work is the **joule** (J), which is defined as $1\ kg \cdot m^2/s^2$. The kinetic energy (KE) of a body of mass m moving at speed v is given by $\frac{1}{2}mv^2$. A 2-kg object moving at one meter per second has KE $= \frac{1}{2}(2\ kg)(1\ m/s)^2 = 1\ kg \cdot m^2/s^2 = 1$ joule. You may find it more convenient to think in terms of the amount of heat required to raise the temperature of one gram of water from 14.5°C to 15.5°C, which is 4.184 joules.

In English units this corresponds to a 4.4-pound object moving at 197 feet per minute, or 2.2 miles per hour.

The **specific heat** of a substance is the amount of heat required to raise the temperature of one gram of the substance one degree C (also one kelvin) with no change in phase. Changes in phase (physical state) absorb or liberate relatively large amounts of energy (Figure 1-5). The specific heat of each substance, a physical property, is different for the solid, liquid, and gaseous phases of the substance. For example, the specific heat of ice is 2.09 J/g·°C near 0°C; for liquid water it is 4.18 J/g·°C; and for steam it is 2.03 J/g·°C near 100°C. The specific heat for water is quite high. A table of specific heats is provided in Appendix E.

The specific heat of a substance varies *slightly* with temperature and pressure. These variations can be ignored for calculations in this text.

$$\text{specific heat} = \frac{\text{(amount of heat in J)}}{\text{(mass of substance in g) (temperature change in °C)}}$$

The **heat capacity** of a body is the amount of heat required to raise its temperature 1°C. The heat capacity of a body is its mass in grams times its specific heat.

Example 1-19

How much heat, in joules, is required to raise the temperature of 205 g of water from 21.2°C to 91.4°C?

In this example, we calculate the amount of heat needed to prepare a cup of hot tea.

Plan

The specific heat of a substance is the amount of heat required to raise the temperature of 1 g of substance 1°C:

$$\text{specific heat} = \frac{\text{(amount of heat in J)}}{\text{(mass of substance in g) (temperature change in °C)}}$$

We can rearrange the equation so that

$$\text{amount of heat} = \text{(mass of substance) (specific heat) (temperature change)}$$

Alternatively, we can use the unit factor approach.

Solution

$$\text{amount of heat} = (205\ \text{g})\ (4.18\ \text{J/g}\cdot{}^\circ\text{C})\ (70.2^\circ\text{C}) = \boxed{6.02 \times 10^4\ \text{J}}$$

By the unit factor approach,

$$\underline{?}\ \text{J} = (205\ \text{g})\ (4.18\ \text{J/g}\cdot{}^\circ\text{C})\ (70.2^\circ\text{C}) = \boxed{6.02 \times 10^4\ \text{J}} \quad \text{or} \quad \boxed{60.2\ \text{kJ}}$$

All units except joules cancel. To cool 205 g of water from 91.4°C to 21.2°C, it would be necessary to remove exactly the same amount of heat, 60.2 kJ.

EOC 65

Example 1-20

How much heat, in calories, kilocalories, joules, and kilojoules, is required to raise the temperature of 205 g of iron from 294.2 K to 364.4 K? The specific heat of iron is 0.106 cal/g · °C, or 0.444 J/g · °C.

This is the same temperature change for the same mass of iron that we used for water in Example 1-19, except that here the temperatures are given in kelvins. You may wish to check the result of this example by converting the starting and ending temperatures to °C and reworking the problem on the Celsius scale.

Plan

First we recall (Section 1-12) that a temperature *change* expressed in kelvins has the *same* numerical value expressed in degrees Celsius. Remembering that specific heat is in terms of temperature change, we can write the specific heat of iron as 0.106 cal/g · K or 0.444 J/g · K. Then we can solve this problem with the temperature change expressed in kelvins, and avoid the work of converting temperatures to °C.

Solution

$$\text{temperature change} = 364.4\ \text{K} - 294.2\ \text{K} = 70.2\ \text{K}$$

$$\underline{?}\ \text{cal} = (205\ \text{g})\ (0.106\ \text{cal/g}\cdot\text{K})\ (70.2\ \text{K}) = \boxed{1.52 \times 10^3\ \text{cal}} \quad \text{or} \quad \boxed{1.52\ \text{kcal}}$$

$$\underline{?}\ \text{J} = (205\ \text{g})\ (0.444\ \text{J/g}\cdot\text{K})\ (70.2\ \text{K}) = \boxed{6.39 \times 10^3\ \text{J}} \quad \text{or} \quad \boxed{6.39\ \text{kJ}}$$

EOC 66, 67

The specific heat of iron is much smaller than the specific heat of water.

$$\frac{\text{specific heat of iron}}{\text{specific heat of water}} = \frac{0.444\ \text{J/g}\cdot\text{K}}{4.18\ \text{J/g}\cdot\text{K}} = 0.106$$

As a result, the amount of heat required to raise the temperature of 205 g of iron by 70.2 K (70.2°C) is less than that required to do the same for 205 g of water, by the same ratio.

$$\frac{\text{amount of heat for iron}}{\text{amount of heat for water}} = \frac{6.39\ \text{kJ}}{60.2\ \text{kJ}} = 0.106$$

Key Terms

Accuracy How closely a measured value agrees with the correct value.

Calorie The amount of heat required to raise the temperature of one gram of water from 14.5°C to 15.5°C. 1 calorie = 4.184 joules.

Chemical change A change in which one or more new substances are formed.

Chemical property See *Properties*.

Compound A substance composed of two or more elements in fixed proportions. Compounds can be decomposed into their constituent elements.

Density Mass per unit volume, $D = m/V$.

Element A substance that cannot be decomposed into simpler substances by chemical means.

Endothermic Describes processes that absorb heat energy.

Energy The capacity to do work or transfer heat.

Exothermic Describes processes that release heat energy.

Extensive property A property that depends upon the amount of material in a sample.

Heat A form of energy that flows between two samples of matter because of their difference in temperature.

Heat capacity The amount of heat required to raise the temperature of a body (of whatever mass) one degree Celsius.

Heterogeneous mixture A mixture that does not have uniform composition and properties throughout.

Homogeneous mixture A mixture that has uniform composition and properties throughout.

Intensive property A property that is independent of the amount of material in a sample.

Joule A unit of energy in the SI system. One joule is $1\ kg \cdot m^2/s^2$, which is also 0.2390 calorie.

Kinetic energy Energy that matter possesses by virtue of its motion.

Law of Conservation of Energy Energy cannot be created or destroyed in a chemical reaction or in a physical change; it may be changed from one form to another.

Law of Conservation of Matter There is no detectable change in the quantity of matter during a chemical reaction or during a physical change.

Law of Conservation of Matter and Energy The combined amount of matter and energy available in the universe is fixed.

Law of Constant Composition See *Law of Definite Proportions*.

Law of Definite Proportions Different samples of any pure compound contain the same elements in the same proportions by mass; also known as the *Law of Constant Composition*.

Mass A measure of the amount of matter in an object. Mass is usually measured in grams or kilograms.

Matter Anything that has mass and occupies space.

Mixture A sample of matter composed of variable amounts of two or more substances, each of which retains its identity and properties.

Physical change A change in which a substance changes from one physical state to another, but no substances with different compositions are formed.

Physical property See *Properties*.

Potential energy Energy that matter possesses by virtue of its position, condition, or composition.

Precision How closely repeated measurements of the same quantity agree with each other.

Properties Characteristics that describe samples of matter. Chemical properties are exhibited as matter undergoes chemical changes. Physical properties are exhibited by matter with no changes in chemical composition.

Significant figures Digits that indicate the precision of measurements—digits of a measured number that have uncertainty only in the last digit.

Specific gravity The ratio of the density of a substance to the density of water at the same temperature.

Specific heat The amount of heat required to raise the temperature of one gram of a substance one degree Celsius.

Substance Any kind of matter all specimens of which have the same chemical composition and physical properties.

Symbol A letter or group of letters that represents (identifies) an element.

Temperature A measure of the intensity of heat, i.e., the hotness or coldness of a sample or object.

Unit factor A factor in which the numerator and denominator are expressed in different units but represent the same or equivalent amounts. Multiplying by a unit factor is the same as multiplying by one.

Weight A measure of the gravitational attraction of the earth for a body.

Exercises

Basic Ideas

1. Define the following terms and illustrate each with a specific example: (a) matter; (b) energy; (c) heat energy; (d) exothermic process.
2. Define the following terms and illustrate each with a specific example: (a) mass; (b) potential energy; (c) kinetic energy; (d) endothermic process.
3. State the following laws and illustrate each.
   (a) the Law of Conservation of Matter
   (b) the Law of Conservation of Energy
   (c) the Law of Conservation of Matter and Energy
4. List the three states of matter and some characteristics of each. How are they alike? different?

5. Distinguish between the following pairs of terms and give two specific examples of each.
(a) chemical properties and physical properties
(b) intensive properties and extensive properties
(c) chemical changes and physical changes
6. Are the following chemical properties or physical properties? How can you tell? (a) the color of a liquid; (b) the odor of a gas; (c) the boiling point of water; (d) the ability to react with nitric acid; (e) the softness of paraffin.
7. Are the following chemical properties or physical properties? How can you tell? (a) the volume of a metal sphere; (b) the density of a gas; (c) ease of corrosion; (d) flammability; (e) electrical conductivity.
8. Are the following chemical changes or physical changes? How can you tell? (a) freezing water to make ice; (b) emission of light by an electric light bulb; (c) emission of light by a candle; (d) dissolving a penny in nitric acid.
9. Are the following chemical changes or physical changes? How can you tell? (a) corrosion of steel; (b) burning of gasoline; (c) boiling of water; (d) condensation of moisture on a cold window.
10. Which of the following processes are exothermic? endothermic? How can you tell? (a) combustion; (b) freezing water; (c) melting ice; (d) boiling water; (e) condensing steam.
11. Which of the following properties of a sample of matter are extensive? Which are intensive? (a) density; (b) melting point; (c) volume; (d) mass; (e) ability to conduct electricity; (f) temperature.
12. Define the following terms clearly and concisely. Give two illustrations of each. (a) substance; (b) mixture; (c) element; (d) compound.
13. Does each of the following describe a mixture or a pure substance? Justify your classification.
(a) A piece of "dry ice," commonly used as a coolant, sublimes (is transformed directly from solid to gas) at −78.5°C under normal atmospheric pressure. As the dry ice continues to sublime, it maintains a steady −78.5°C temperature.
(b) An alcoholic liquid labeled "100 proof neutral spirit" is divided into two equal portions, each of which is placed in an open dish. When a lighted match is applied to the first dish, the liquid burns with a pale blue flame. The liquid in the second dish is allowed to stand undisturbed until much of it has evaporated. This liquid residue cannot be ignited with a match.
(c) The label on a bottle of "hydrogen peroxide" includes the statement, "Contains 3% hydrogen peroxide. Inert ingredients 97%."
14. Classify each of the following as an element, a compound, or a mixture. Justify your classification. (a) coffee; (b) silver; (c) calcium carbonate; (d) a piece of concrete from a sidewalk; (e) shoe polish.
15. Classify each of the following as an element, a compound, or a mixture. Justify your classification. (a) a soft drink; (b) water; (c) air; (d) chicken noodle soup; (e) table salt; (f) popcorn.
16. What is a heterogeneous mixture? Which of the following are heterogeneous mixtures? Explain your answers. (a) salt and sulfur; (b) milk; (c) clean air; (d) gasoline; (e) a chocolate chip cookie.
17. What is a homogeneous mixture? Which of the following are homogeneous mixtures? Explain your answers. (a) sugar dissolved in water; (b) coffee; (c) french onion soup; (d) mud; (e) a clear liquid with no internal boundaries, consisting of corn oil and olive oil.
18. How could you separate the components of each of the following mixtures? Explain. (a) vinegar and oil; (b) sugar dissolved in water; (c) a suspension of sand in water.
19. How could you separate the components of each of the following mixtures? Explain. (a) charcoal and sugar; (b) baking soda dissolved in water; (c) ether (boiling point 36°C) and acetone (boiling point 56°C).
20. Describe paper chromatography. Describe column chromatography.

Scientific Notation and Significant Figures

21. Express the following numbers in scientific notation. (a) 6500.; (b) 0.0041; (c) 860 (assume that this number is measured to ±10); (d) 860 (assume that this number is measured to ±1); (e) 186,000; (f) 0.0516.
22. Express the following exponentials as ordinary numbers. (a) 5.26×10^4; (b) 4.10×10^{-6}; (c) 1.00×10^2; (d) 8.206×10^{-2}; (e) 9.346×10^3; (f) 9.346×10^{-3}.
23. Which of the following are likely to be exact numbers? Why? (a) 227 inches; (b) 7 computers; (c) $20,335.47; (d) 5 pounds of sugar; (e) 14.7 gallons of diesel fuel; (f) 25,446 ants.
24. To which of the quantities appearing in the following statements would the concept of significant figures apply? Where it would apply, indicate the number of significant figures. (a) The density of platinum at 20°C is 21.45 g/cm³. (b) Wilbur Shaw won the Indianapolis 500-mile race in 1940 with an average speed of 114.277 mi/h. (c) A mile is defined as 5280 feet. (d) The International Committee for Weights and Measures "accepts that the curie be . . . retained as a unit of radioactivity, with the value $3.7 \times 10^{10}\ s^{-1}$." (This resolution was passed in 1964.)

In Exercises 25–26, perform the indicated operations, and round off your answers to the proper number of significant figures. Apply appropriate units to your answers. Assume that all numbers were obtained from measurements.

25. (a) 2.68 ft + 11.4 ft; (b) 142 mi/2.2 h; (c) Three men work for 1.25 hours each. How many man-hours of work did they do? (d) $(1.54 \times 10^2\ cm)(2.336 \times 10^3\ cm)$; (e) What is the area of a square 4.62 m on edge?

26. (a) 0.3135 ft × 0.0669 ft; (b) 423.1 in + 0.256 in − 116 in; (c) The volume of a sphere is given by $\frac{4}{3}\pi r^3$, where r is the radius of the sphere. Note that the numbers 4 and 3 are exact numbers. Calculate the volume of a sphere with radius 2.27 in. Take the value of π as 3.141593. (d) 6.057×10^3 m − 9.35 m; (e) (8.54×10^5 mi)/(22 days).
27. How many significant figures are there in each of the following measured quantities? (a) 5.0045 g; (b) 0.087000 L; (c) 5.5×10^{22} atoms; (d) 2.25×10^3 kg; (e) 0.001 mL.
28. Which word prefix indicates the SI multiplier in each of the following numbers? (a) 1×10^3; (b) 1×10^{-3}; (c) 1×10^6; (d) 1×10^{-1}; (e) 0.01; (f) 0.1; (g) 0.001; (h) 1×10^{-6}.
29. Indicate the multiple or fraction of 10 by which a quantity is multiplied when it is preceded by each of the following prefixes. (a) M; (b) m; (c) c; (d) d; (e) k; (f) μ.

Conversions and Dimensional Analysis

30. Carry out each of the following conversions. (a) 10.3 m to km; (b) 10.3 km to m; (c) 247 kg to g; (d) 4.32 L to mL; (e) 85.9 dL to L; (f) 4567 L to cm³.
31. Express 3.72 yards in millimeters, centimeters, meters, and kilometers.
32. (a) Express 65 miles per hour in kilometers per hour. (b) If you are traveling at 65 miles per hour, how many feet are you traveling per second?
33. Express: (a) 2.00 gallons in cm³; (b) 1.00 pint in milliliters; (c) 3.75 cubic yards in cubic inches; (d) 3.75 cubic feet in milliliters.
34. For each of the following pairs, determine which quantity is larger. (a) 24.0 mg or 24.0 cg; (b) 500 cm or 0.5 m; (c) 0.8 nm or 8 Å; (d) 5 L or 3.2 m³.
35. A sample is marked as containing 27.3% calcium carbonate by mass. (a) How many grams of calcium carbonate are contained in 64.33 grams of the sample? (b) How many grams of the sample would contain 11.4 grams of calcium carbonate?
36. An iron ore is found to contain 8.77% hematite (a compound that contains iron). (a) How many tons of this ore would contain 5.50 tons of hematite? (b) How many kilograms of this ore would contain 5.50 kilograms of hematite?

*37. A foundry releases 5.0 tons of gas into the atmosphere each day. The gas contains 2.4% sulfur dioxide by mass. What mass of sulfur dioxide is released in one week?

*38. A certain chemical process requires 75 gallons of pure water each day. The available water contains 11 parts per million by mass of salt (i.e., for every 1,000,000 parts of available water, 11 parts of salt). What mass of salt must be removed each day? A gallon of water weighs 3.67 kg.

39. The radius of a hydrogen atom is about 0.37 Å, and the average radius of the earth's orbit around the sun is about 1.5×10^8 km. Find the ratio of the average radius of the earth's orbit to the radius of the hydrogen atom.
40. If the price of gasoline is $1.059 per gallon, what is its price in cents per liter?
41. Suppose your automobile gas tank holds 18 gallons and the price of gasoline is $0.293 per liter. How much would it cost to fill your gas tank?
42. A particular medicine dropper delivers 25 drops of water to make 1.0 mL. (a) What is the volume of one drop in cubic centimeters? In microliters? (b) If the drop were spherical, what would be its diameter in millimeters? (Volume of a sphere = $\frac{4}{3}\pi r^3$.)
43. Express the following masses or weights in grams and in kilograms. (a) 4.2 ounces; (b) 3.15 short tons; (c) 2.7×10^6 milligrams; (d) 7.33×10^4 centigrams.
44. Express: (a) 215 kilograms in pounds; (b) 215 pounds in kilograms; (c) 215 ounces in centigrams.

*45. At a given point in its orbit, the earth is 92.98 million miles from the sun (center to center). The radius of the sun is 432,000 miles and the radius of the earth is 3960 miles. How long does it take for light from the surface of the sun to reach the earth's surface? The speed of light is 3.00×10^8 m/s.

*46. Cesium atoms are the largest naturally occurring atoms. The radius of a cesium atom is 2.62 Å. How many cesium atoms would have to be laid side by side to give a row of cesium atoms 1.00 inch long? Assume that the atoms are spherical.

*47. The radius of a bromine atom is 1.04 Å. If 1.00 million bromine atoms were laid side by side to make a row, how long would the row be in inches? Assume that the atoms are spherical.

Density and Specific Gravity

48. Mercury poured into a glass of water sinks to the bottom, while gasoline poured into the same glass floats on the surface of the water. A piece of paraffin dropped into the mixture comes to rest between the water and the gasoline, while a piece of iron comes to rest between the water and the mercury. List these five substances in order of increasing density (least dense first).
49. Which is more dense at 0°C, ice or water? Which has the higher specific gravity? How do you know?
50. A 20.4-cubic-centimeter piece of a metal has a mass of 124.3 grams. What is its density?
51. What is the mass of a rectangular piece of copper 21.3 cm × 11.4 cm × 7.9 cm? The density of copper is 8.92 g/cm³.
52. A small crystal of sucrose (table sugar) had a mass of 2.236 mg. The dimensions of the box-like crystal were 1.11 × 1.09 × 1.12 mm. What is the density of sucrose expressed in g/cm³?
53. Vinegar has a density of 1.0056 g/cm³. What is the mass of one liter of vinegar?

*54. The radius of a neutron is approximately 1.5×10^{-15} m and its mass is 1.675×10^{-24} g. Find the density of a neutron, in g/cm^3.

*55. A container has a mass of 68.31 g when empty and 93.34 g when filled with water. The density of water is 1.0000 g/cm^3. (a) Calculate the volume of the container. (b) When filled with an unknown liquid, the container had a mass of 88.42 g. Calculate the density of the unknown liquid.

*56. The mass of an empty container is 66.734 g. The mass of the container filled with water is 91.786 g. (a) Calculate the volume of the container, using a density of 1.0000 g/cm^3 for water. (b) A piece of metal was added to the empty container and the combined mass was 87.807 g. Calculate the mass of the metal. (c) The container with the metal was filled with water and the mass of the entire system was 105.408 g. What mass of water was added? (d) What volume of water was added? (e) What is the volume of the piece of metal? (f) Calculate the density of the metal.

57. What is the specific gravity of a liquid if 225 mL of the liquid has the same mass as 396 mL of water?

58. The specific gravity of silver is 10.5. (a) What is the volume, in cm^3, of an ingot of silver with mass 0.555 kg? (b) If this sample of silver is a cube, how long is each edge in cm? (c) How long is the edge of this cube in inches?

*59. The acid in an automobile battery is sulfuric acid. The density of a particular sulfuric acid solution is 1.36 g/mL. This solution is 46.3% H_2SO_4 by mass, the remainder being water.
(a) 245 grams of the solution contains ______ grams of H_2SO_4 and ______ grams of water.
(b) 245 mL of the solution contains ______ grams of H_2SO_4 and ______ grams of water.

Temperature Scales

60. Which represents a larger temperature interval: (a) a Celsius degree or a Fahrenheit degree? (b) a kelvin or a Fahrenheit degree?

61. Express: (a) 373°C in K; (b) 10.15 K in °C; (c) −32.0°C in °F; (d) 100.0°F in K.

62. Express: (a) 0°F in °C; (b) 98.6°F in K; (c) 298 K in °F; (d) 18.5°C in °F.

*63. Use the graph in Figure 1-23 to make each of the following temperature conversions: (a) 50°C to °F, (b) −50°C to °F, and (c) 100°F to °C. Then check each of the conversions by calculation.

*64. On the Réamur scale, which is no longer used, water freezes at 0°R and boils at 80°R. (a) Derive an equation that relates this to the Celsius scale. (b) Derive an equation that relates this to the Fahrenheit scale. (c) Mercury is a liquid metal at room temperature. It boils at 356.6°C (673.9°F). What is the boiling point of mercury on the Réamur scale?

Heat and Heat Transfer

65. Calculate the amount of heat required to raise the temperature of 35.0 grams of water from 10.0°C to 35.0°C.

66. The specific heat of aluminum is 0.895 J/g · °C. Calculate the amount of heat required to raise the temperature of 15.8 grams of aluminum from 27.0°C to 41.0°C.

67. How much heat must be removed from 75.0 grams of water at 90.0°C to cool it to 23.0°C?

*68. In some solar-heated homes, heat from the sun is stored in rocks during the day, and then released during the cooler night. (a) Calculate the amount of heat required to raise the temperature of 85.0 kg of rocks from 25.0°C to 45.0°C. Assume that the rocks are limestone, which is essentially pure calcium carbonate. The specific heat of calcium carbonate is 0.818 J/g · °C. (b) Suppose that when the rocks in part (a) cool to 30.0°C, all the heat released goes to warm the 10,000 cubic feet (2.83×10^5 liters) of air in the house, originally at 10.0°C. To what final temperature would the air be heated? The specific heat of air is 1.004 J/g · °C, and its density is 1.20×10^{-3} g/mL.

*69. A small immersion heater is used to heat water for a cup of coffee. We wish to use it to heat 225 mL of water (about a teacupful) from 25°C to 90°C in 2.00 minutes. What must be the heat rating of the heater, in kJ/min, to accomplish this? Neglect the heat that goes to heat the teacup itself. The density of water is 0.997 g/mL.

Mixed Exercises

70. A 20.0-gram sample of Ca initially at 26.1°C absorbs 905 J. What is the final temperature of the sample? The specific heat of calcium is 0.628 J/g · °C.

71. If you ran a mile in 4.00 minutes, what would be your average speed in (a) km/h, (b) cm/s, and (c) mi/h?

72. A student found the following list of properties of iodine in an encyclopedia: (a) grayish black granules, (b) metallic luster, (c) characteristic odor, (d) forms a purple vapor, (e) density = 4.93 g/cm^3, (f) melting point = 113.5°C, (g) soluble in alcohol, (h) insoluble in water, (i) noncombustible, (j) forms ions in aqueous solutions, (k) poisonous. Which of these properties are chemical properties?

73. Which of the properties listed in Exercise 72 are intensive properties?

*74. (a) Draw a graph (similar to that in Figure 1-23) that relates temperatures on the Kelvin and Fahrenheit scales. (Use the Kelvin scale as the horizontal axis.) (b) Use this graph to convert 300 K to the Fahrenheit scale. (c) Use the graph to convert 300°F to the Kelvin scale. (d) Calculate the conversions in parts (b) and (c).

*75. At what temperature will a Fahrenheit thermometer give (a) the same reading as a Celsius thermometer? (b) a reading that is twice that on the Celsius thermometer? (c) a reading that is numerically the same but opposite in sign from that on the Celsius thermometer?

76. Use data from Table 1-8. (a) What would be the mass of a rectangular block of aluminum 1.70 in × 6.25 in × 12.00 in? (b) Calculate the volume in cm^3 of 1.00 lb of mercury.

77. At Angel Falls in Venezuela, the water falls 3212 feet. How many meters is this?

***78.** The lethal dose of potassium cyanide (KCN), taken orally is 1.6 milligrams per kilogram of body weight. Calculate the lethal dose of potassium cyanide taken orally by a 145-pound person.

79. The mass of a glass object is 395 grams. When the object is immersed in water at 25°C, it displaces 134 mL. What is the density of the glass?

80. The distance light travels through space in one year is called one light-year. Using the speed of light in vacuum listed in Appendix D, and assuming that one year is 365 days, determine the distance of a light-year in kilometers and in miles.

Chemical Formulas and Composition Stoichiometry

Outline

Objectives

As you study this chapter, you should learn to

- ☐ Understand an early concept of atoms
- ☐ Use chemical formulas to solve various kinds of chemical problems
- ☐ Relate names to formulas and charges of simple ions
- ☐ Combine simple ions to write formulas and names of ionic compounds
- ☐ Recognize and use formula weights and mole relationships
- ☐ Interconvert masses, moles, and formulas in problems
- ☐ Determine percent compositions in compounds
- ☐ Determine formulas from composition
- ☐ Perform calculations about purity of substances

Pure quartz crystals (silicon dioxide, SiO_2) are clear and colorless. Many gemstones consist of quartz that contains traces of certain metal ions. Amethyst, shown here, is quartz that contains small amounts of Fe^{3+} ions. Its color can range from pale lilac to royal purple depending on the amounts of Fe^{3+} ions present. Quartz is widely distributed over the earth's surface. White sand is relatively pure quartz that has been weathered into small pieces.

The language that chemists use to describe the forms of matter and the possible changes in its composition appears throughout the scientific world. Chemical symbols, formulas, and equations are used in such diverse areas as agriculture, home economics, engineering, geology, physics, biology, medicine, and dentistry. In this chapter we shall describe the simplest atomic theory. We shall use it as we represent chemical formulas of elements and compounds. Later this theory will be expanded when we discuss chemical changes.

It is important to learn this fundamental material well so that you can use it correctly and effectively.

The word "stoichiometry" is derived from the Greek *stoicheion,* which means "first principle or element," and *metron,* which means "measure." **Stoichiometry** describes the quantitative relationships among elements in compounds (composition stoichiometry) and among substances as they undergo chemical changes (reaction stoichiometry). In this chapter we shall be concerned with chemical formulas and composition stoichiometry. In Chapter 3 we shall discuss chemical equations and reaction stoichiometry.

2-1 Atoms and Molecules

Around 400 BC, the Greek philosopher Democritus suggested that all matter is composed of tiny, discrete, indivisible particles that he called atoms. His ideas, based entirely on philosophical speculation rather than experimental

The term "atom" comes from the Greek language and means "not divided" or "indivisible."

Figure 2-1
Relative sizes of atoms of the noble gases.

He Ne Ar Kr Xe Rn

evidence, were rejected for 2000 years. By the late 1700s, scientists began to realize that the concept of atoms provided an explanation for many experimental observations about the nature of matter.

By the early 1800s, the Law of Conservation of Matter (Section 1-1) and the Law of Definite Proportions (Section 1-5) were both accepted as general descriptions of how matter behaves. John Dalton, an English schoolteacher, tried to explain why matter behaves in such simple and systematic ways as those expressed above. In 1808, he published the first "modern" ideas about the existence and nature of atoms. He summarized and expanded the nebulous concepts of early philosophers and scientists; more importantly, his ideas were based on *reproducible experimental results* of measurements by many scientists. Taken together, these ideas form the core of **Dalton's Atomic Theory**, one of the highlights of scientific thought. In condensed form, Dalton's ideas may be stated as follows:

The radius of a calcium atom is only 1.97×10^{-8} cm, and its mass is 6.66×10^{-23} g.

Statement 3 is true for *chemical* reactions. However, it is not true for *nuclear* reactions (Chapter 30).

1. An element is composed of extremely small indivisible particles called atoms.
2. All atoms of a given element have identical properties, which differ from those of other elements.
3. Atoms cannot be created, destroyed, or transformed into atoms of another element.
4. Compounds are formed when atoms of different elements combine with each other in small whole-number ratios.
5. The relative numbers and kinds of atoms are constant in a given compound.

Dalton believed that atoms were solid indivisible spheres, an idea we now reject. But he showed remarkable insight into the nature of matter and its interactions. Some of his ideas could not be verified (or refuted) experimentally at the time. They were based on the limited experimental observations of his day. Even with their shortcomings, Dalton's ideas provided a framework that could be modified and expanded by later scientists. Thus John Dalton is the father of modern atomic theory.

The smallest particle of an element that maintains its chemical identity through all chemical and physical changes is called an **atom** (Figure 2-1). In nearly all **molecules**, two or more atoms are bonded together in very small, discrete units (particles) that are electrically neutral. A **molecule** is the small-

For Group 0 elements, the noble gases, a molecule contains only one atom and so an atom and a molecule are the same (Figure 2-1).

H_2 (hydrogen)

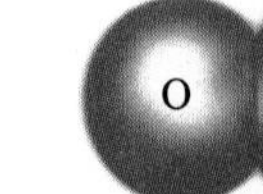

O_2 (oxygen)

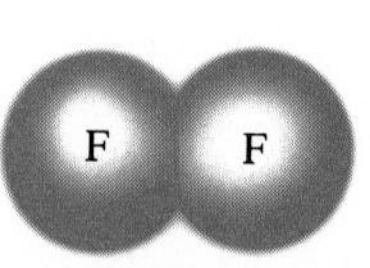

F_2 (fluorine)

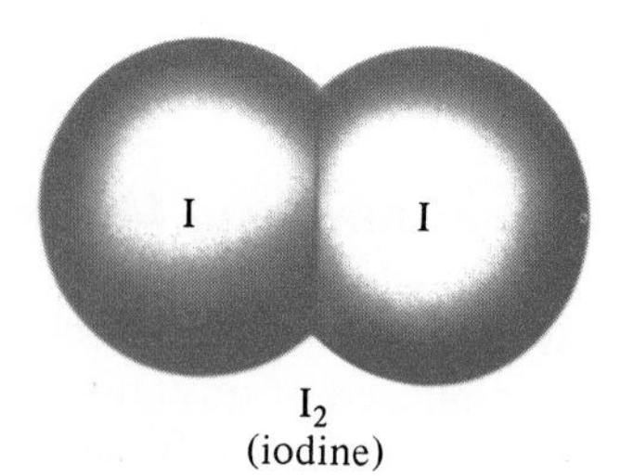

I_2 (iodine)

Figure 2-2
Models of diatomic molecules of some elements, approximately to scale.

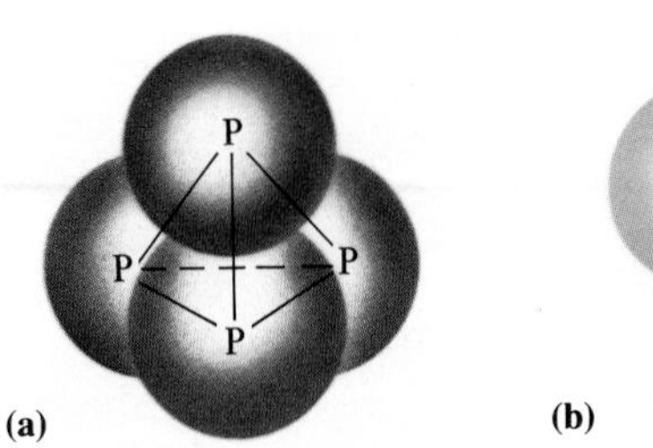

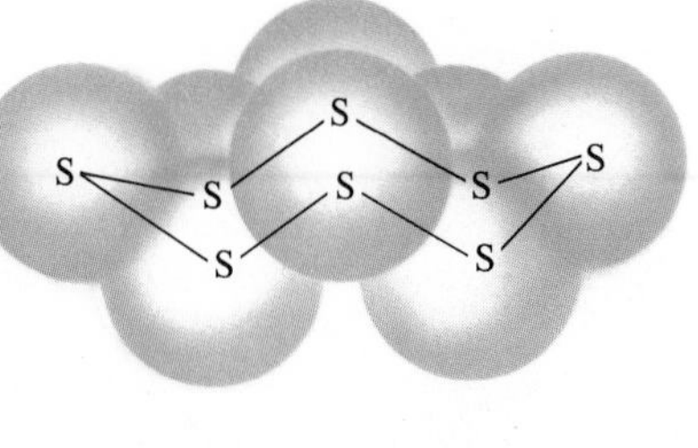

Figure 2-3
(a) A model of the P_4 molecule of white phosphorus. (b) A model of the S_8 ring found in rhombic sulfur.

est particle of an element or compound that can have a stable independent existence.

Individual oxygen atoms are not stable at room temperature and atmospheric pressure. Hence, single atoms of oxygen mixed under these conditions quickly combine to form pairs. The oxygen with which we are all familiar is made up of two atoms of oxygen; it is a *diatomic* molecule, O_2. Hydrogen, nitrogen, fluorine, chlorine, bromine, and iodine are other examples of diatomic molecules (Figure 2-2).

You should remember the common elements that occur as diatomic molecules: H_2, N_2, O_2, F_2, Cl_2, Br_2, I_2.

Some other elements exist as more complex molecules. Phosphorus molecules consist of four atoms, while sulfur exists as eight-atom molecules at ordinary temperatures and pressures. Molecules that contain two or more atoms are called *polyatomic* molecules. See Figure 2-3.

In modern terminology, O_2 is named dioxygen, H_2 is dihydrogen, P_4 is tetraphosphorus, and so on. Even though such terminology is officially preferred, it has not yet gained wide acceptance. Most chemists still refer to O_2 as oxygen, H_2 as hydrogen, P_4 as phosphorus, and so on.

Molecules of compounds are composed of more than one kind of atom. A water molecule consists of two atoms of hydrogen and one atom of oxygen. A molecule of methane consists of one carbon atom and four hydrogen atoms. The shapes of these and a few other molecules are shown in Figure 2-4.

The compound water, H_2O, is made up of water molecules, which are made up of hydrogen and oxygen atoms. Methane is the principal component of natural gas.

Atoms are the components of molecules, and molecules are the components of elements and most compounds. We are able to see samples of compounds and elements that consist of large numbers of atoms and mol-

H_2O
(water)

CO_2
(carbon dioxide)

CH_4
(methane)

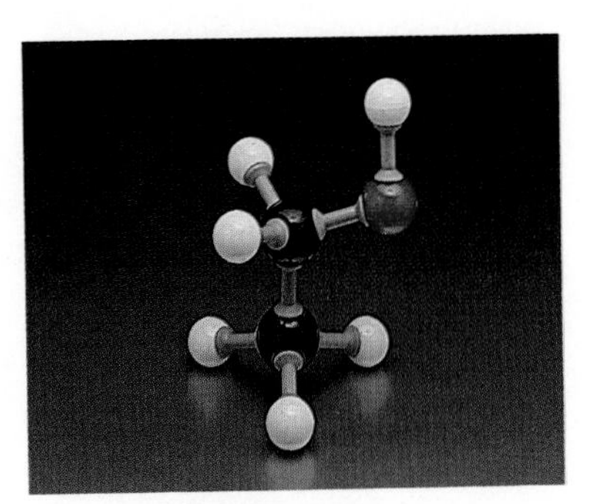

C_2H_5OH
(ethyl alcohol)

Figure 2-4
Formulas and models for some molecules.

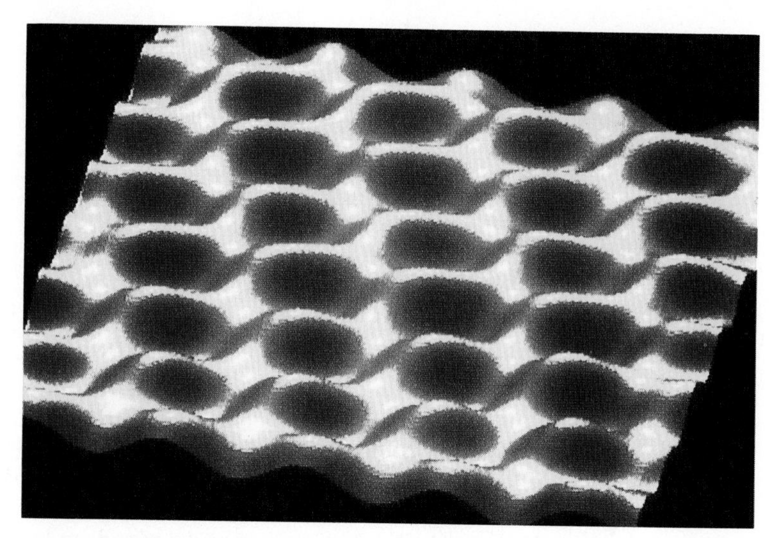

Figure 2-5
A computer reconstruction of the surface of a sample of highly ordered graphite, as observed with a scanning tunnelling electron microscope (STM), reveals the regular pattern of individual carbon atoms. Many important reactions occur on the surfaces of solids. Observations of the atomic arrangements on surfaces help chemists understand such reactions. New information available with the STM will give many details about chemical bonding in solids.

ecules. With the scanning tunnelling microscope it is now possible to "see" atoms (Figure 2-5). It would take 217 million silicon atoms to make a row 1 inch long.

2-2 Chemical Formulas

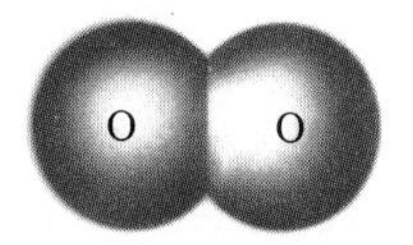

An O_2 molecule.

The **chemical formula** for a substance shows its chemical composition. This represents the elements present as well as the ratio in which the atoms of the elements occur. The formula for a single atom is the same as the symbol for the element. Thus, Na can represent a single sodium atom. It is unusual to find such isolated atoms in nature, with the exception of the noble gases (He, Ne, Ar, Kr, Xe, and Rn). A subscript following the symbol of an element indicates the number of atoms in a molecule. For instance, F_2 indicates a molecule containing two fluorine atoms, and P_4 a molecule containing four phosphorus atoms.

Some elements exist in more than one form. Familiar examples include (1) oxygen, found as O_2 molecules, and ozone, found as O_3 molecules, and (2) two different crystalline forms of carbon, diamond, and graphite (Figure 13-31). Different forms of the same element in the same physical state are called **allotropic modifications** or **allotropes**.

Compounds contain two or more elements in chemical combination in fixed proportions. Hence, each molecule of hydrogen chloride, HCl, contains one atom of hydrogen and one atom of chlorine; each molecule of carbon tetrachloride, CCl_4, contains one carbon atom and four chlorine atoms. An aspirin molecule, $C_9H_8O_4$, contains nine carbon atoms, eight hydrogen atoms, and four oxygen atoms.

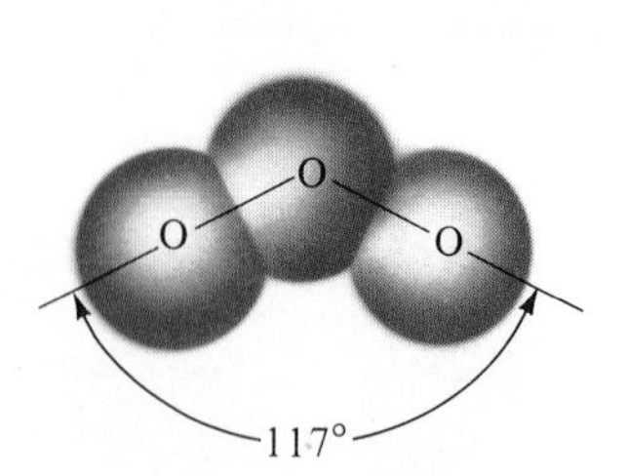

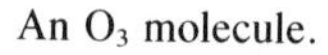

An O_3 molecule.

Some groups of atoms behave chemically as single entities. For instance, one nitrogen atom and two oxygen atoms may combine to form a *nitro* group that is a part of a molecule. In formulas of compounds containing two or more of the same group, the group formula is enclosed in parentheses. Thus, 2,4,6-trinitrotoluene (often abbreviated TNT) contains three *nitro* groups,

Table 2-1
Formulas and Names of Some Common Molecular Compounds

Formula	Name	Formula	Name
HCl	hydrogen chloride (or hydrochloric acid if dissolved in water)	SO_2	sulfur dioxide
		SO_3	sulfur trioxide
		CO	carbon monoxide
H_2SO_4	sulfuric acid	CO_2	carbon dioxide
HNO_3	nitric acid	CH_4	methane
CH_3COOH	acetic acid	C_2H_6	ethane
NH_3	ammonia	C_3H_8	propane

and its formula is $C_7H_5(NO_2)_3$ (see margin). When you count up the number of atoms in this molecule from its formula, you must multiply the numbers of nitrogen and oxygen atoms in the NO_2 group by 3. There are *seven* carbon atoms, *five* hydrogen atoms, *three* nitrogen atoms, and *six* oxygen atoms in a molecule of TNT.

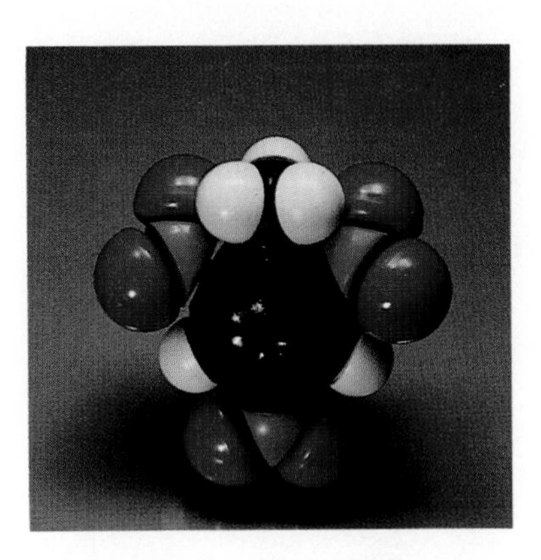

A space-filling model of a TNT molecule, $C_7H_5(NO_2)_3$.

Compounds were first recognized as distinct substances because of their different physical properties and because they could be separated from one another. Once the concept of atoms and molecules was established, the reason for these differences in properties could be understood: Two compounds differ from one another because their molecules are different. Conversely, if two molecules contain the same number of the same kinds of atoms, arranged the same way, then both are molecules of the same compound. Thus the atomic theory explains the **Law of Definite Proportions** (Section 1-5).

This law, also known as the **Law of Constant Composition**, can now be extended to include its interpretation in terms of atoms. It is so important for performing the calculations in this chapter that we restate it here:

> Different pure samples of a compound always contain the same elements in the same proportion by mass; this corresponds to atoms of these elements combined in fixed numerical ratios.

Throughout your study of chemistry you will have many occasions to refer to compounds by name. Table 2-1 includes a few examples for molecular compounds.

So we see that for a substance composed of molecules, the *chemical formula* gives the number of atoms of each type in the molecule. But this formula does not express the order in which the atoms in the molecules are bonded together. The **structural formula** shows how the atoms are connected. The lines connecting atomic symbols represent chemical bonds between atoms. The bonds are actually forces that tend to hold atoms at certain distances and angles from one another. For instance, the structural formula of propane shows that the three C atoms are linked in a chain, with three H atoms bonded to each of the end C atoms and two H atoms bonded to the center C. **Ball-and-stick** molecular models and **space-filling** molecular models help us to see the shapes and relative sizes of molecules. These four representations are shown in Figure 2-6.

Chemical Formula	Structural Formula	Ball-and-Stick Model	Space-Filling Model
H_2O, water	H—O—H		
H_2O_2, hydrogen peroxide	H—O—O—H		
CCl_4, carbon tetrachloride	Cl | Cl—C—Cl | Cl		
C_3H_8, propane	H H H |	| H—C—C—C—H |	| H H H		
C_2H_5OH, ethanol	H H | | H—C—C—O—H | | H H		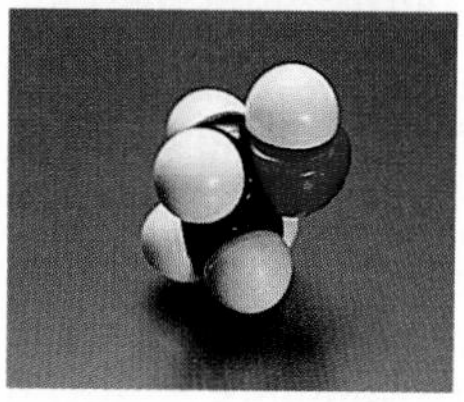

Figure 2-6
Formulas and models for some molecules. Structural formulas show the order in which atoms are connected, but do not represent true molecular shapes. Ball-and-stick models use balls of different colors to represent atoms and sticks to represent bonds; they show the three-dimensional shapes of molecules. Space-filling models show the (approximate) relative sizes of atoms and the shapes of molecules.

2-3 Ions and Ionic Compounds

So far we have discussed only compounds that exist as discrete molecules. Some compounds, such as sodium chloride, NaCl, consist of ions. An **ion** is an atom or group of atoms that carries an electrical charge. Ions that possess a *positive* charge, such as the sodium ion, Na^+, are called **cations**. Those carrying a *negative* charge, such as the chloride ion, Cl^-, are called

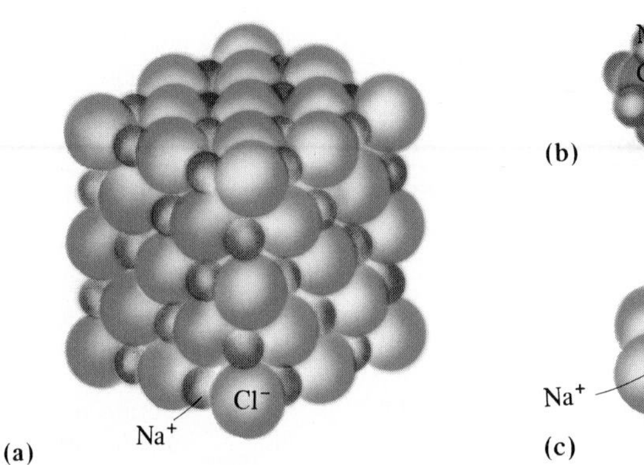

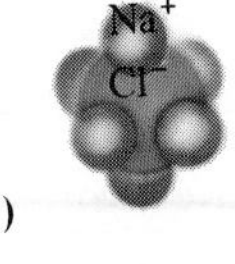

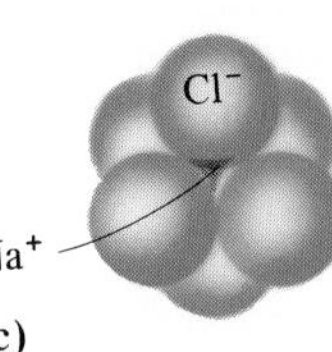

Figure 2-7
The arrangement of ions in NaCl. (a) A crystal of sodium chloride consists of an extended array that contains equal numbers of sodium ions (small spheres) and chloride ions (large spheres). Within the crystal, each chloride ion is surrounded by six sodium ions (b), and each sodium ion is surrounded by six chloride ions (c).

anions. The charge on an ion *must* be included as a superscript on the right side of the chemical symbol(s) when we write the formula for the individual ion.

As we shall see in Chapter 5, an atom consists of a very small, very dense, positively charged *nucleus* surrounded by a diffuse distribution of negatively charged particles called *electrons*. The number of positive charges in the nucleus defines the identity of the element to which the atom corresponds. Electrically neutral atoms contain the same number of electrons outside the nucleus as positive charges (protons) within the nucleus. Ions are formed when neutral atoms lose or gain electrons. An Na^+ ion is formed when a sodium atom loses one electron, and a Cl^- ion is formed when a chlorine atom gains one electron.

The compound NaCl consists of an extended array of Na^+ and Cl^- ions (Figure 2-7). Within the crystal (though not on the surface) each Na^+ ion is surrounded at equal distances by six Cl^- ions, and each Cl^- ion is similarly surrounded by six Na^+ ions. *Any* compound, whether ionic or molecular, is electrically neutral; i.e., it has no net charge. In NaCl this means that the Na^+ and Cl^- ions are present in a 1:1 ratio, and this is indicated by the formula NaCl.

Because there are no "molecules" of ionic substances, we should not refer to "a molecule of NaCl." Instead, we refer to a **formula unit (FU)** of NaCl, consisting of one Na^+ ion and one Cl^- ion. Similarly, we speak of the formula unit of all ionic compounds. It is also acceptable to refer to a molecule of a molecular compound as a formula unit. One formula unit of C_3H_8, which is the same as one molecule, contains three C atoms and eight H atoms.

The general term "formula unit" applies to molecular or ionic compounds, whereas the more specific term "molecule" applies only to elements and compounds that exist as discrete molecules.

For the present, we shall tell you which substances are ionic and which are covalent when it is important to know. Later you will learn to make the distinction yourself.

Electron pairs are shared by atoms in *covalent* compounds.

Polyatomic ions are groups of atoms that bear an electrical charge. Examples include the ammonium ion, NH_4^+, the sulfate ion, SO_4^{2-}, and the nitrate ion, NO_3^-. Table 2-2 shows the formulas, ionic charges, and names of some common ions. You should learn the formulas and names of these frequently encountered ions. They can be used to write the formulas and names of many ionic compounds. We write the formula of an ionic compound by adjusting the relative numbers of positive and negative ions so their total charges cancel (i.e., add to zero). The name is formed by giving the names of the ions, with the positive ion named first (by convention).

Table 2-2
Formulas, Ionic Charges, and Names of Some Common Ions

Common Cations (positive ions)			Common Anions (negative ions)		
Formula	Charge	Name	Formula	Charge	Name
Na^+	1+	sodium ion	F^-	1−	fluoride ion
K^+	1+	potassium ion	Cl^-	1−	chloride ion
NH_4^+	1+	ammonium ion	Br^-	1−	bromide ion
Ag^+	1+	silver ion	OH^-	1−	hydroxide ion
Mg^{2+}	2+	magnesium ion	CH_3COO^-	1−	acetate ion
Ca^{2+}	2+	calcium ion	NO_3^-	1−	nitrate ion
Zn^{2+}	2+	zinc ion			
Cu^+	1+	copper(I) or cuprous ion	O^{2-}	2−	oxide ion
			S^{2-}	2−	sulfide ion
Cu^{2+}	2+	copper(II) or cupric ion	SO_3^{2-}	2−	sulfite ion
			SO_4^{2-}	2−	sulfate ion
Fe^{2+}	2+	iron(II) or ferrous ion	CO_3^{2-}	2−	carbonate ion
Fe^{3+}	3+	iron(III) or ferric ion			
Al^{3+}	3+	aluminum ion	PO_4^{3-}	3−	phosphate ion

As we shall see, some metals can form more than one kind of ion with positive charge. For such metals we specify which ion we mean with a Roman numeral—e.g., iron(II) or iron(III). Because zinc forms no stable ions other than Zn^{2+}, we do not need to use Roman numerals in its name.

Example 2-1

Write the formulas for the following ionic compounds: (a) sodium fluoride, (b) calcium fluoride, (c) iron(II) sulfate, (d) zinc phosphate.

Plan

In each case, we identify the chemical formulas of the ions from Table 2-2. These ions must be present in a ratio that gives the compound *no net charge*. The formulas and names of ionic compounds are written by giving the positively charged ion first.

Solution

(a) The formula for the sodium ion is Na^+ and the formula for fluoride ion is F^- (Table 2-2). Because the charges on these two ions are equal in magnitude, the ions must be present in equal numbers, or in a 1:1 ratio. Thus the formula for sodium fluoride is NaF.

(b) The formula for the calcium ion is Ca^{2+} and the formula for fluoride ion is F^-. Now each positive ion (Ca^{2+}) provides twice as much charge as each negative ion (F^-). So there must be twice as many F^- ions as Ca^{2+} ions to equalize the charge. This means that the ratio of calcium to fluoride ions is 1:2. So the formula for calcium fluoride is CaF_2.

(c) The iron(II) ion is Fe^{2+} and the sulfate ion is SO_4^{2-}. As in (a), the equal magnitudes of positive and negative charges tell us that the ions must be present in equal numbers, or in a 1:1 ratio. The formula for iron(II) sulfate is $FeSO_4$.

(d) The zinc ion is Zn^{2+} and the phosphate ion is PO_4^{3-}. Now it will take *three* Zn^{2+} ions to account for as much charge (6+ total) as would be present in *two* PO_4^{3-} ions (6− total). So the formula for zinc phosphate is $Zn_3(PO_4)_2$.

EOC 12,14

Example 2-2

Name the following ionic compounds: (a) $(NH_4)_2S$, (b) $Cu(NO_3)_2$, (c) $ZnCl_2$, (d) $Fe_2(CO_3)_3$.

Plan

In naming ionic compounds, it is helpful to inspect the formula for atoms or groups of atoms that we recognize as representing familiar ions.

Solution

(a) The presence of the polyatomic grouping NH_4 in the formula suggests to us the presence of the ammonium ion, NH_4^+. There are two of these, each accounting for 1+ in charge. To balance this, the single S must account for 2− in charge, or S^{2-}, which we recognize as the sulfide ion. Thus the name of the compound is

ammonium sulfide.

(b) The NO_3 grouping in the formula tells us that the nitrate ion, NO_3^-, is present. Two of these nitrate ions account for $2 \times 1- = 2-$ in negative charge. To balance this, copper must account for 2+ charge and be the copper(II) ion. The name of the compound is copper(II) nitrate or, alternatively, cupric nitrate.

(c) The positive ion present is zinc ion, Zn^{2+}, and the negative ion is chloride, Cl^-. The name of the compound is zinc chloride.

(d) Each CO_3 grouping in the formula must represent the carbonate ion, CO_3^{2-}. The presence of *three* such ions accounts for a total of 6− in negative charge, so there must be a total of 6+ present in positive charge to balance this. It takes *two* iron ions to provide this 6+, so each ion must have a charge of 3+ and be Fe^{3+}, the iron(III) ion or ferric ion. The name of the compound is

iron(III) carbonate or ferric carbonate.

EOC 11,13

A more extensive discussion on naming compounds appears in Sections 7-11 and 7-12.

2-4 Atomic Weights

As the chemists of the eighteenth and nineteenth centuries painstakingly sought information about the compositions of compounds and tried to systematize their knowledge, it became apparent that each element has a characteristic mass relative to every other element. Although these early scientists did not have the experimental means to measure the mass of each kind of atom, they succeeded in defining a *relative* scale of atomic masses.

An early observation was that carbon and hydrogen have relative atomic masses, also traditionally called **atomic weights, AW**, of approximately 12 and 1, respectively. Thousands of experiments on the compositions of compounds have resulted in the establishment of a scale of relative atomic weights based on the **atomic mass unit (amu)**, which is defined as *exactly* $\frac{1}{12}$ *of the mass of an atom of a particular kind of carbon atom, called carbon-12.*

The term "atomic weight" is widely accepted because of its traditional use, although it is properly a mass rather than a weight. "Atomic mass" is often used.

On this scale, the atomic weight of hydrogen (H) is 1.00794 amu, that of sodium (Na) is 22.989768 amu, and that of magnesium (Mg) is 24.3050 amu. This tells us that Na atoms have nearly 23 times the mass of H atoms, while Mg atoms are about 24 times heavier than H atoms.

2-5 The Mole

Even the smallest bit of matter that can be handled reliably contains an enormous number of atoms. So we must deal with large numbers of atoms in any real situation, and some unit for conveniently describing a large number of atoms is desirable. The idea of using a unit to describe a particular number (amount) of objects has been around for a long time. You are no doubt already familiar with the dozen (12 items) and the gross (144 items).

"Mole" is derived from the Latin word *moles,* which means "a mass." "Molecule" is the diminutive form of this word and means "a small mass."

The SI unit for amount is the **mole**, abbreviated mol. It is *defined* as the amount of substance that contains as many entities (atoms, molecules, or other particles) as there are atoms in 0.012 kg of pure carbon-12 atoms. Many experiments have refined the number, and the currently accepted value is

$$1 \text{ mole} = 6.022045 \times 10^{23} \text{ particles}$$

This number, often rounded off to 6.022×10^{23}, is called **Avogadro's number** in honor of Amedeo Avogadro (1776–1856), whose contributions to chemistry are discussed in Section 12-8.

According to its definition, the mole unit refers to a fixed number of entities, whose identities must be specified. Just as we speak of a dozen eggs or a dozen automobiles, we refer to a mole of atoms or a mole of molecules (or a mole of ions, electrons, or other particles). We could even think about a mole of eggs, although the size of the required carton staggers the imagination! Helium exists as discrete He atoms, so one mole of helium consists of 6.022×10^{23} He *atoms*. Hydrogen commonly exists as diatomic (two-atom) molecules, so one mole of hydrogen contains 6.022×10^{23} H_2 *molecules* and $2(6.022 \times 10^{23})$ H atoms.

Every kind of atom, molecule, and ion has a definite characteristic mass. It follows that one mole of a given pure substance also has a definite mass, regardless of the source of the sample. This idea is of central importance in many calculations throughout the study of chemistry and the related sciences.

Because the mole is defined as the number of atoms in 0.012 kg (or 12 grams) of carbon-12, and the atomic mass unit is defined as $\frac{1}{12}$ of the mass of a carbon-12 atom, the following convenient relationship is true:

Atomic weights of the elements are listed inside the front cover.

> The mass of one mole of atoms of a pure element in grams is numerically equal to the atomic weight of that element in amu. This is also called the **molar mass** of the element; its units are grams/mole.

For instance, if you obtain a pure sample of the metallic element titanium (Ti), whose atomic weight is 47.88 amu, and measure out 47.88 grams of it, you will have one mole, or 6.022×10^{23} atoms, of titanium.

The symbol for an element can (1) identify the element, (2) represent one atom of the element, or (3) represent one mole of atoms of the element. The last interpretation will be extremely useful in calculations in the next chapter.

12 eggs
or
1 dozen eggs
or
24 ounces of eggs

6.022×10^{23} Fe atoms
or
1 mole of Fe atoms
or
55.847 grams of iron

Figure 2-8
Three different ways of representing amounts.

A quantity of a substance may be expressed in a variety of ways. For example, consider a dozen eggs and 55.847 grams of iron filings, or one mole of iron (Figure 2-8). We can express the amount of eggs or iron filings present in any of several different units. We can then construct unit factors to relate an amount of the substance expressed in one kind of unit to the same amount expressed in another unit.

The atomic weight of iron (Fe) is 55.847 amu.

Unit Factors for Eggs	Unit Factors for Iron
$\frac{12 \text{ eggs}}{1 \text{ doz eggs}}$	$\frac{6.022 \times 10^{23} \text{ Fe atoms}}{1 \text{ mol Fe atoms}}$
$\frac{12 \text{ eggs}}{24 \text{ ounces of eggs}}$	$\frac{6.022 \times 10^{23} \text{ Fe atoms}}{55.847 \text{ g Fe}}$
and so on	and so on

As Table 2-3 suggests, the concept of a mole as applied to atoms is especially useful. It provides a convenient basis for comparing equal numbers of atoms of different elements.

Table 2-3
Mass of One Mole of Atoms of Some Common Elements

Element	A Sample with a Mass of	Contains
carbon	12.011 g C	6.022×10^{23} C atoms or 1 mole of C atoms
titanium	47.88 g Ti	6.022×10^{23} Ti atoms or 1 mole of Ti atoms
gold	196.96654 g Au	6.022×10^{23} Au atoms or 1 mole of Au atoms
hydrogen	1.00794 g H_2	6.022×10^{23} H atoms or 1 mole of H atoms (3.011×10^{23} H_2 molecules or 0.5 mole of H_2 molecules)
sulfur	32.066 g S_8	6.022×10^{23} S atoms or 1 mole of S atoms (0.7528×10^{23} S_8 molecules or 0.1250 mole of S_8 molecules)

Chemistry in Use. . .
Names of the Elements

If you were to discover a new element, how would you name it? Throughout history, scientists have answered this question in different ways. Most have chosen to honor a person or place or to describe the new substance. Even elements known long ago, whose discoverers are unknown, have names with etymological significance. Looked at from a historical point of view, these names tell us much about the nature of chemistry and scientific discovery. They also tell us about the nature of scientists—their values, heroes, and practices. Many elements were unearthed by teams rather than individuals. Chemists of different nationalities or schools of thought who had worked cooperatively were sometimes reduced to bickering enemies when the time came to choose a name for their discovery!

Until the Middle Ages only nine elements were known: gold, silver, tin, mercury, copper, lead, iron, sulfur, and carbon. The metals' chemical symbols are taken from descriptive Latin names: *aurum* (''yellow''), *argentum* (''shining''), *stannum* (''dripping'' or ''easily melted''), *hydrargyrum* (''silvery water''), *cuprum* (Cyprus, where many copper mines were located), *plumbum* (exact meaning unknown—possibly ''heavy''), and *ferrum* (also unknown). Some of these were derived from even earlier Sanskrit words. The English names are derived from old Anglo-Saxon terms. Mercury is named after the planet, one reminder that the ancients associated metals with gods and celestial bodies. In turn, both the planet, which moves rapidly across the sky, and the element, which is the only metal that is liquid at room temperature and thus flows rapidly, are named for the fleet god of messengers in Roman mythology. In English, mercury is nicknamed ''quicksilver.''

Prior to the reforms of Antoine Lavoisier, chemistry was a largely nonquantitative, unsystematic science in which experimenters had little contact with each other. There were few rules for documenting and sharing information. Thus, elements discovered prior to Lavoisier's contributions in the late 18th century have names whose sources are hard to identify. They include the following. ''Zinc'' may have originated from the Persian *seng* (''stone'') or the German *Zinke* (''spike''). ''Antimony'' is thought to have come from the Arabic *al ithmid*, the name for the compound Sb_2S_3, which was used to darken women's eyebrows (its Latin name, *stibium*, means mark). Arsenic is another element with an ambiguous etymology. The Greek word *arsenikos*, meaning male, is one possible source, as alchemists believed that metals were either male or female. The Persian *zarnik* (''golden'') is another.

In 1787 Lavoisier published his *Methode de Nomenclature Chimique*, which proposed, among other changes, that all new elements be named descriptively. For the next 125 years, most elements were given names that corresponded to their properties. Greek roots were one popular source, as evidenced by hydrogen (*hydros-gen*, ''water-producing''), oxygen (*oksys-gen*, ''acid-producing''), nitrogen (*nitron-gen*, ''soda-producing''), bromine (*bromos*, ''stink''), and argon (*a-ergon*, ''no reaction''). The discoverers of argon, Ramsay and Rayleigh, originally proposed the name *aeron* (from *aer* or air) but critics thought is was too close to the biblical name Aaron! Latin roots such as *radius* (''ray'') were also used (radium and radon are both naturally radioactive elements that emit ''rays''). Color was often the determining property, especially after the invention of the spectroscope in 1859, because different elements (or the light that they emit) have prominent characteristic colors. Cesium, indium, iodine, rubidium, and thallium were all named in this manner. Their respective Greek and Latin roots denote blue-gray, indigo, violet, red, and green (*thallus* means ''tree sprout''). Because of the great variety of colors of its compounds. iridium takes its name from the Latin *iris*, meaning rainbow. Alternatively, an element name might suggest a mineral or the ore that contained it. One example is wolfram or tungsten (W), which was isolated from wolframite. Two other ''inconsistent'' elemental symbols, K and Na, arose from occurrence as well. *Kalium* was first obtained from the saltwort plant, *Salsola kali*, and *natrium* from niter. Their English names, potassium and sodium, are derived from the ores potash and soda.

Other elements, contrary to Lavoisier's suggestion, were named after planets, mythological figures, places, or superstitions, ''Celestial elements'' include helium (sun), tellurium (earth), selenium (moon—the element was discovered in close proximity to tellurium), cerium (the asteroid Ceres, which was discovered only two years before the element), and uranium (the planet Uranus, discovered a few years earlier). The first two transuranium elements (those *beyond* uranium) to be produced were named neptunium and plutonium for the next two planets, Neptune and Pluto. The names promethium (Prometheus, who stole fire from heaven), vanadium (Scandinavian goddess, Vanadis), titanium (Titans, the first sons of the earth), tantalum (Tantalos, father of Niobe), and thorium (Thor, Scandinavian god of war) all arise from Greek or Norse mythology. Cobalt was named for Kobold, German evil spirit, when its presence interfered with the mining of copper (as did nickel, from *Kupfernickel*, or false copper).

"Geographical elements," shown on the map, sometimes honored the discoverer's native country or workplace. The Latin names for Russia (*ruthenium*), France (*gallium*), Paris (*lutetium*), and Germany (*germanium*) were among those used. Marie Sklodowska Curie named one of the elements that she discovered polonium, after her native Poland. Often the locale of discovery lends its name to the element; the record holder is certainly the Swedish village Ytterby, the site of ores from which the four elements terbium, erbium, ytterbium, and yttrium were isolated. Elements honoring important scientists include curium, einsteinium, nobelium, fermium, and lawrencium.

Sometimes the name of an element contains a history of its discovery. In 1839 Mosander gave a new element he had extracted as a minor component in a cerium compound the name lanthanum (Greek, "to lie hidden"). Two years later he thought that he had found another new element from the same source, and named it didymium (Greek, "twin") because it was "an inseparable twin brother of lanthanum." But in 1885, von Welsbach separated didymium into two new elements, which he named neodymium (Greek, "new twin") and praseodymium (Greek, "green twin").

Most of the 109 elements now known were given titles peacefully, but a few were not. Niobium, isolated in 1803 by Ekeberg from an ore that also contained tantalum, and named after the Greek goddess Niobe (daughter of Tantalus), was later found to be identical to an 1802 discovery of Hatchett, columbium. (Interestingly, Hatchett first found the element in an ore sample that had been sent to England more than a century earlier by John Winthrop, the first governor of Connecticut.) While "niobium" became the accepted designation in Europe, the Americans, not surprisingly, chose "columbium." It was not until 1949—when the International Union of Pure and Applied Chemistry (IUPAC) ended more than a century of controversy by ruling in favor of mythology—that element 41 received a unique name. Current arguments over the proper names of elements 104 and 105 have prompted the IUPAC to begin hearing claims of priority to numbers 104 through 110. Some Russian and Scandinavian texts refer to 104 as kurchatovium; some American and English texts, as rutherfordium. Similarly, 105 is known as both hahnium and nielsbohrium. In 1978, the IUPAC recommended that, at least for now, the elements beyond element 103 be known by systematic names based on numerical roots; element 104 is unnilquadium (*un* for 1, *nil* for 0, *quad* for 4, plus the *-ium* ending), followed by unnilpentium, unnilhexium, and so on.

The number and variety of element names indicate something of the long and colorful history of chemistry. Examining them shows us how scientists' values and beliefs have changed over the years. As there will probably not be many more new elements synthesized, we might hope that the IUPAC will allow nonsystematic titles to be given to the new, artificially created elements. It would be a shame if discoverers were not allowed to participate in naming their discoveries. Even when the last possible transuranium element has been made, synthetic chemists will still be in need of monikers for their products. Judging from the abundance of names such as "buckminsterfullerene" (C_{60}) for novel substances, it is doubtful that scientists' creativity and sense of humor in such matters will be easily stifled.

Lisa L. Saunders
Chemistry major
University of Texas at Austin

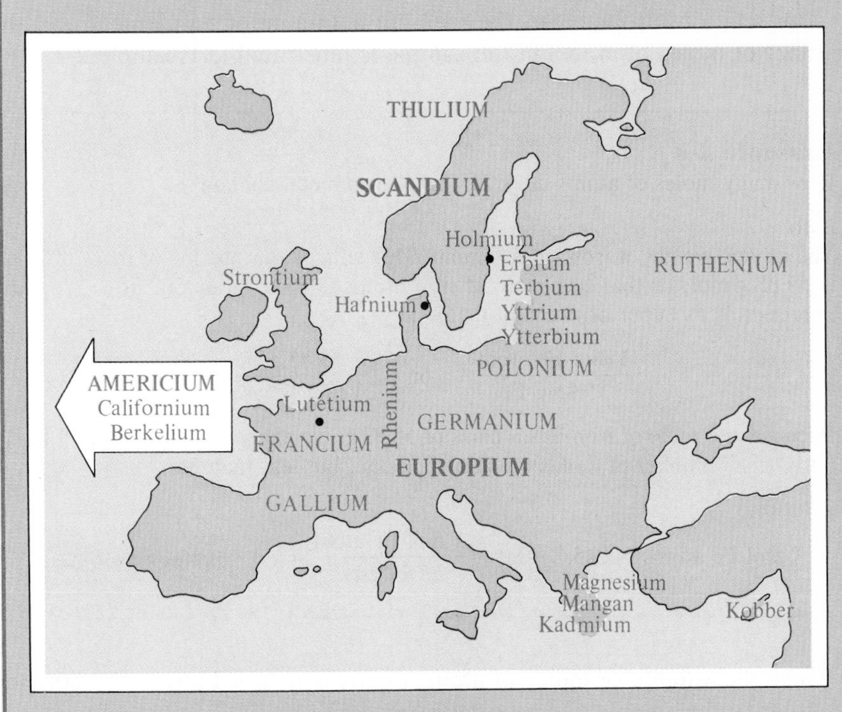

Many chemical elements were named after places.

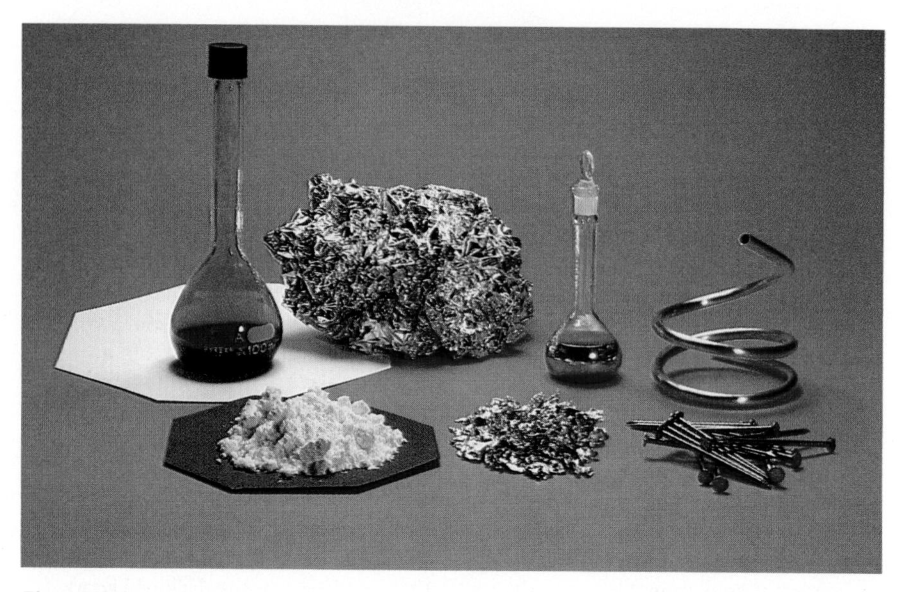

Figure 2-9
One mole of atoms of some common elements. Back row (left to right): bromine, aluminum, mercury, copper. Front row (left to right): sulfur, zinc, iron.

Figure 2-9 shows what one mole of atoms of each of some common elements looks like. Each of the examples in Figure 2-9 represents 6.022×10^{23} *atoms* of the element.

The relationship between the mass of a sample of an element and the number of moles of atoms in the sample is illustrated in Example 2-3.

Example 2-3
How many moles of atoms does 245.2 g of iron metal contain?

Plan
The atomic weight of iron is 55.85 amu. This tells us that the molar mass of iron is 55.85 g/mol, or that one mole of iron atoms is 55.85 grams of iron. We can express this as either of two unit factors:

$$\frac{1 \text{ mol Fe atom}}{55.85 \text{ g Fe}} \quad \text{or} \quad \frac{55.85 \text{ g Fe}}{1 \text{ mol Fe atom}}$$

Because one mole of iron has a mass of 55.85 g, we expect that 245.2 g will be a fairly small number of moles (greater than one, but less than ten).

Solution

$$\underline{?} \text{ mol Fe atoms} = 245.2 \text{ g Fe} \times \frac{1 \text{ mol Fe atom}}{55.85 \text{ g Fe}} = \boxed{4.390 \text{ mol Fe atoms}}$$

Once the number of moles of atoms of an element is known, the number of atoms in the sample can be calculated, as Example 2-4 illustrates.

Example 2-4
How many atoms are contained in 4.390 moles of iron atoms?

Plan

One mole of atoms of an element contains Avogadro's number of atoms, or 6.022×10^{23} atoms. This lets us generate the two unit factors

$$\frac{6.022 \times 10^{23} \text{ atoms}}{1 \text{ mol atoms}} \quad \text{or} \quad \frac{1 \text{ mol atoms}}{6.022 \times 10^{23} \text{ atoms}}$$

We expect that the number of atoms in more than four moles of atoms is a very large number.

Solution

$$\underline{?} \text{ Fe atoms} = 4.390 \text{ mol Fe atoms} \times \frac{6.022 \times 10^{23} \text{ Fe atoms}}{1 \text{ mol Fe atoms}}$$

$$= 2.644 \times 10^{24} \text{ Fe atoms}$$

EOC 26

If we know the atomic weight of an element on the carbon-12 scale, we can use the mole concept and Avogadro's number to calculate the *average* mass of one atom of that element in grams (or any other mass unit we choose).

Example 2-5

Calculate the mass of one iron atom in grams.

Plan

We expect that the mass of a single atom in grams would be a *very* small number. We know that one mole of Fe atoms has a mass of 55.85 g and contains 6.022×10^{23} Fe atoms. We use this information to generate unit factors to carry out the desired conversion.

Solution

$$\frac{\underline{?} \text{ g Fe}}{\text{Fe atom}} = \frac{55.85 \text{ g Fe}}{1 \text{ mol Fe atoms}} \times \frac{1 \text{ mol Fe atoms}}{6.022 \times 10^{23} \text{ Fe atoms}}$$

$$= 9.274 \times 10^{-23} \text{ g Fe/Fe atom}$$

Thus, we see that the mass of one Fe atom is only 9.274×10^{-23} g.

To gain some appreciation of how little this is, write 9.274×10^{-23} gram as a decimal fraction and try to name the fraction.

This example demonstrates how small atoms are and why it is necessary to use large numbers of atoms in practical work. Let's calculate the number of atoms in an iron micrometeorite that weighs only one billionth of a gram. The radius of a spherical micrometeorite of this mass is 3.0×10^{-4} cm.

Micrometeorites (mostly Ni and Fe) of this size constantly shower down on the earth from space.

Example 2-6

Calculate the number of atoms in 1.0 billionth of a gram of iron metal to two significant figures.

Plan

We use unit factors that we can construct from the atomic weight of iron and Avogadro's number.

g of Fe ⟶ mol of Fe atoms ⟶ number of Fe atoms

Solution

Even though the number 1.0×10^{-9} has two significant figures, we carry the other numbers to more significant figures. Then we round at the end to the appropriate number of significant figures.

$$\underline{?}\text{ Fe atoms} = 1.0 \times 10^{-9}\text{ g} \times \frac{1\text{ mol Fe atoms}}{55.85\text{ g Fe}} \times \frac{6.022 \times 10^{23}\text{ Fe atoms}}{1\text{ mol Fe atoms}}$$

$$= 1.1 \times 10^{13}\text{ Fe atoms}$$

One billionth of a gram of iron contains about 11,000,000,000,000 Fe atoms.

2-6 Formula Weights, Molecular Weights, and Moles

The **formula weight (FW)** of a substance is the sum of the atomic weights (AW) of the elements in the formula, each taken the number of times the element occurs. Hence a formula weight gives the mass of one formula unit in amu.

Formula weights, like the atomic weights on which they are based, are relative masses. The formula weight for sodium hydroxide, NaOH, (rounded to the nearest 0.01 amu) is found as follows.

No. of Atoms of Stated Kind		× Mass of One Atom	= Mass Due to Elements
1 × Na =	1	× 23.00 amu	= 23.00 amu of Na
1 × H =	1	× 1.01 amu	= 1.01 amu of H
1 × O =	1	× 16.00 amu	= 16.00 amu of O
		Formula weight of NaOH	= 40.01 amu

The term **molecular weight (MW)** is used interchangeably with "formula weight" when reference is made to molecular (nonionic) substances, i.e., substances that exist as discrete molecules.

Example 2-7

Calculate the formula weight (molecular weight) of 2,4,6-trinitrotoluene (TNT), $C_7H_5(NO_2)_3$, using the precisely known values for atomic weights given in the International Table of Atomic Weights (1987) inside the front cover of the text.

Plan

We add the atomic weights of the elements in the formula, each multiplied by the number of times the element occurs. Because the least precisely known atomic weight (12.011 amu for C) is known to three significant figures past the decimal point, the result is known to only that number of significant figures.

Solution

No. of Atoms of Stated Kind		× Mass of One Atom	= Mass Due to Element
7 × C =	7	× 12.011 amu	= 84.077 amu of C
5 × H =	5	× 1.00794 amu	= 5.09370 amu of H
3 × N =	3	× 14.00674 amu	= 42.02022 amu of N
6 × O =	6	× 15.9994 amu	= 95.9964 amu of O

Formula weight of 2,4,6-trinitrotoluene (TNT) = 227.187 amu

EOC 22

The term "formula weight" is correctly used for either ionic or molecular substances. When we refer specifically to molecular (nonionic) substances, i.e., substances that exist as discrete molecules, we often substitute the term **molecular weight (MW)**. We could say that the *molecular weight* of 2,4,6-trinitrotoluene is 227.187 amu.

The amount of substance that contains the mass in grams numerically equal to its formula weight in amu contains 6.022×10^{23} formula units, or *one mole* of the substance. This is sometimes called the **molar mass** of the substance. Molar mass is *numerically equal* to the formula weight of the substance (the atomic weight for atoms of elements), and has the units grams/mole.

One mole of sodium hydroxide is 40.01 g of NaOH, and one mole of TNT is 227.187 g of $C_7H_5(NO_2)_3$. One mole of any molecular substance contains 6.022×10^{23} molecules of the substance, as Table 2-4 illustrates.

Table 2-4
One Mole of Some Common Molecular Substances

Substance	A Sample with a Mass of	Contains
hydrogen	2.016 g H_2	6.022×10^{23} H_2 molecules or 1 mol of H_2 molecules (contains $2 \times 6.022 \times 10^{23}$ H atoms or 2 mol of H atoms)
oxygen	32.00 g O_2	6.022×10^{23} O_2 molecules or 1 mol of O_2 molecules (contains $2 \times 6.022 \times 10^{23}$ O atoms or 2 mol of O atoms)
methane	16.04 g CH_4	6.022×10^{23} CH_4 molecules or 1 mol of CH_4 molecules (contains 6.022×10^{23} C atoms and $4 \times 6.022 \times 10^{23}$ H atoms)
2,4,6-trinitro-toluene (TNT)	227.19 g $C_7H_5(NO_2)_3$	6.022×10^{23} $C_7H_5(NO_2)_3$ molecules or 1 mol of $C_7H_5(NO_2)_3$ molecules

Figure 2-10
One mole of some compounds. The clear liquid is water, H_2O (1 mol = 18.0 g = 18.0 mL). The white solid (left) is *anhydrous* oxalic acid, $(COOH)_2$ (1 mol = 90.0 g). The second white solid is *hydrated* oxalic acid, $(COOH)_2 \cdot 2H_2O$ (1 mol = 126.0 g). The blue solid is hydrated copper(II) sulfate, $CuSO_4 \cdot 5H_2O$ (1 mol = 249.68 g). The red solid is mercury(II) oxide (1 mol = 216.59 g).

Heating blue $CuSO_4 \cdot 5H_2O$ forms anhydrous $CuSO_4$, which is white. Some blue $CuSO_4 \cdot 5H_2O$ is visible in the cooler center portion of the crucible.

The physical appearance of one mole of each of some compounds is illustrated in Figure 2-10. Two different forms of oxalic acid are shown. The formula unit (molecule) of oxalic acid is $(COOH)_2$ (FW = 90.04 amu; molar mass = 90.04 g/mol). However, when oxalic acid is obtained by crystallization from a water solution, two molecules of water are present for each molecule of oxalic acid, even though it appears dry. The formula of this **hydrate** is $(COOH)_2 \cdot 2H_2O$ (FW = 126.06 amu; molar mass = 126.06 g/mol). The dot shows that the crystals contain two H_2O molecules per $(COOH)_2$ molecule. The water can be driven out of the crystals by heating to leave **anhydrous** oxalic acid, $(COOH)_2$. Anhydrous means "without water." Copper(II) sulfate, an *ionic* compound, shows similar behavior. Anhydrous copper(II) sulfate ($CuSO_4$; FW = 159.60 amu; molar mass = 159.60 g/mol) is almost white. Hydrated copper(II) sulfate ($CuSO_4 \cdot 5H_2O$; FW = 249.68 amu; molar mass = 249.68 g/mol) is deep blue.

Because there are no simple NaCl molecules at ordinary temperatures, it is inappropriate to refer to the "molecular weight" of NaCl or any ionic compound. One mole of an ionic compound contains 6.022×10^{23} *formula units* (FU) of the substance. Recall that one formula unit of sodium chloride consists of one sodium ion, Na^+, and one chloride ion, Cl^-. One mole, or 58.44 grams, of NaCl contains 6.022×10^{23} Na^+ ions and 6.022×10^{23} Cl^- ions. See Table 2-5.

Table 2-5
One Mole of Some Ionic Compounds

Compound	A Sample with a Mass of 1 Mole	Contains
sodium chloride	58.44 g NaCl	6.022×10^{23} Na^+ ions or 1 mole of Na^+ ions 6.022×10^{23} Cl^- ions or 1 mole of Cl^- ions
calcium chloride	111.0 g $CaCl_2$	6.022×10^{23} Ca^{2+} ions or 1 mole of Ca^{2+} ions $2(6.022 \times 10^{23})$ Cl^- ions or 2 moles of Cl^- ions
aluminum sulfate	342.1 g $Al_2(SO_4)_3$	$2(6.022 \times 10^{23})$ Al^{3+} ions or 2 moles of Al^{3+} ions $3(6.022 \times 10^{23})$ SO_4^{2-} ions or 3 moles of SO_4^{2-} ions

The mole concept, together with Avogadro's number, provides important connections among the extensive properties mass of substance, number of moles of substance, and number of molecules or ions. These are summarized as follows.

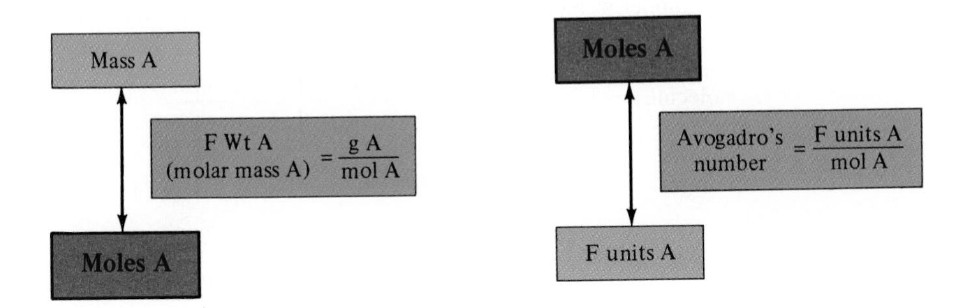

The following examples show the relations between numbers of molecules, atoms, or formula units and their masses.

Example 2-8
What is the mass in grams of 10.0 billion SO_2 molecules?

Plan
One mole of SO_2 contains 6.02×10^{23} SO_2 molecules and has a mass of 64.1 grams.

When fewer than four significant figures are used in calculations, Avogadro's number is rounded off to 6.02×10^{23}.

Solution

$$\underline{?}\ \text{g } SO_2 = 10.0 \times 10^9\ SO_2 \text{ molecules} \times \frac{64.1\ \text{g } SO_2}{6.02 \times 10^{23}\ SO_2 \text{ molecules}}$$

$$= 1.06 \times 10^{-12}\ \text{g } SO_2$$

Ten billion SO_2 molecules have a mass of only 0.00000000000106 gram. The most commonly used analytical balances are capable of weighing to ± 0.0001 gram.

EOC 30

Example 2-9
How many (a) moles of O_2, (b) O_2 molecules, and (c) O atoms are contained in 40.0 grams of oxygen gas (dioxygen) at 25°C?

Plan
We construct the needed unit factors from the following equalities: (a) the mass of one mole of O_2 is 32.0 g (molar mass O_2 = 32.0 g/mol); (b) one mole of O_2 contains 6.02×10^{23} O_2 molecules; (c) one O_2 molecule contains two O atoms.

Solution
One mole of O_2 contains 6.02×10^{23} O_2 molecules, and its mass is 32.0 g.

(a) $$\underline{?}\ \text{mol } O_2 = 40.0\ \text{g } O_2 \times \frac{1\ \text{mol } O_2}{32.0\ \text{g } O_2} = 1.25\ \text{mol } O_2$$

(b)

$$\underline{?}\ O_2 \text{ molecules} = 40.0 \text{ g } O_2 \times \frac{6.02 \times 10^{23}\ O_2 \text{ molecules}}{32.0 \text{ g } O_2}$$

$$= 7.52 \times 10^{23} \text{ molecules}$$

Or, we can use the number of moles of O_2 calculated in (a) to find the number of O_2 molecules.

$$\underline{?}\ O_2 \text{ molecules} = 1.25 \text{ mol } O_2 \times \frac{6.02 \times 10^{23}\ O_2 \text{ molecules}}{1 \text{ mol } O_2}$$

$$= 7.52 \times 10^{23}\ O_2 \text{ molecules}$$

(c)

$$\underline{?}\ \text{O atoms} = 40.0 \text{ g } O_2 \times \frac{6.02 \times 10^{23}\ O_2 \text{ molecules}}{32.0 \text{ g } O_2} \times \frac{2 \text{ O atoms}}{1\ O_2 \text{ molecule}}$$

$$= 1.50 \times 10^{24} \text{ O atoms}$$

EOC 32

Example 2-10

Calculate the number of hydrogen atoms in 39.6 grams of ammonium sulfate, $(NH_4)_2SO_4$.

Plan

One mole of $(NH_4)_2SO_4$ is 6.02×10^{23} formula units (FU) and has a mass of 132 g.

In Example 2-10, we relate (a) g to mol, (b) mol to FU, and (c) FU to H atoms.

g of $(NH_4)_2SO_4$ ⟶ mol of $(NH_4)_2SO_4$ ⟶ FU of $(NH_4)_2SO_4$ ⟶ H atoms

Solution

$$\underline{?}\ \text{H atoms} = 39.6 \text{ g } (NH_4)_2SO_4 \times \frac{1 \text{ mol } (NH_4)_2SO_4}{132 \text{ g } (NH_4)_2SO_4}$$

$$\times \frac{6.02 \times 10^{23} \text{ FU } (NH_4)_2SO_4}{1 \text{ mol } (NH_4)_2SO_4} \times \frac{8 \text{ H atoms}}{1 \text{ FU } (NH_4)_2SO_4}$$

$$= 1.44 \times 10^{24} \text{ H atoms}$$

EOC 31

Table 2-6
Comparison of Moles and Millimoles

Grams can be converted to milligrams by shifting the decimal three places to the right.

Compound	1 Mole	1 Millimole
NaOH	40.0 g	40.0 mg or 0.0400 g
H_3PO_4	98.1 g	98.1 mg or 0.0981 g
SO_2	64.1 g	64.1 mg or 0.0641 g
C_3H_8	44.1 g	44.1 mg or 0.0441 g

The term "millimole" (mmol) is useful in laboratory work. As the prefix indicates, one **mmol** is 1/1000 of a mole. Small masses are frequently expressed in milligrams (mg) rather than grams. The relation between millimoles and milligrams is the same as that between moles and grams (Table 2-6).

Example 2-11
Calculate the number of millimoles of sulfuric acid in 0.147 gram of H_2SO_4.

Plan
1 mol H_2SO_4 = 98.1 g H_2SO_4; 1 mmol H_2SO_4 = 98.1 mg H_2SO_4, or 0.0981 g H_2SO_4. We can use these equalities to solve this problem by either of two methods. Method 1: Express formula weight in g/mmol, then convert g H_2SO_4 to mmol H_2SO_4. Method 2: Convert g H_2SO_4 to mg H_2SO_4, then use unit factor mg/mmol to convert to mmol H_2SO_4.

Solution
Method 1:

$$\underline{?}\text{ mmol } H_2SO_4 = 0.147 \text{ g } H_2SO_4 \times \frac{1 \text{ mmol } H_2SO_4}{0.0981 \text{ g } H_2SO_4} = 1.50 \text{ mmol } H_2SO_4$$

Method 2: Using 0.147 g H_2SO_4 = 147 mg H_2SO_4, we have

$$\underline{?}\text{ mmol } H_2SO_4 = 147 \text{ mg } H_2SO_4 \times \frac{1 \text{ mmol } H_2SO_4}{98.1 \text{ mg } H_2SO_4} = 1.50 \text{ mmol } H_2SO_4$$

EOC 33

2-7 Percent Composition and Formulas of Compounds

If the formula of a compound is known, its chemical composition can be expressed as the mass percent of each element in the compound. For example, one carbon dioxide molecule, CO_2, contains one C atom and two O atoms. Percentage is the part divided by the whole times 100 percent (or simply parts per 100), so we can represent the percent composition of carbon dioxide as follows:

$$\%\text{ C} = \frac{\text{mass of C}}{\text{mass of } CO_2} \times 100\% = \frac{\text{AW of C}}{\text{MW of } CO_2} \times 100\%$$

$$= \frac{12.0 \text{ amu}}{44.0 \text{ amu}} \times 100\% = 27.3\%$$

$$\%\text{ O} = \frac{\text{mass of O}}{\text{mass of } CO_2} \times 100\% = \frac{2 \times \text{AW of O}}{\text{MW of } CO_2} \times 100\%$$

$$= \frac{2(16.0 \text{ amu})}{44.0 \text{ amu}} \times 100\% = 72.7\% \text{ O}$$

One *mole* of CO_2 (44.0 g) contains one *mole* of C atoms (12.0 g) and two *moles* of O atoms (32.0 g). Therefore, we could have used these masses in the preceding calculation. These numbers are the same as the ones used—

only the units are different. In Example 2-12 we shall base our calculation on one *mole* rather than one *molecule*.

Example 2-12

Calculate the percent composition of HNO_3 by mass.

Plan

We first calculate the mass of one mole as in Example 2-7. Then we express the mass of each element as a percent of the total.

Solution

The molar mass of HNO_3 is calculated first.

No. of Mol of Atoms		× Mass of One Mol of Atoms	= Mass Due to Element
1 × H =	1	× 1.0 g	= 1.0 g of H
1 × N =	1	× 14.0 g	= 14.0 g of N
3 × O =	3	× 16.0 g	= 48.0 g of O
		Mass of 1 mol of HNO_3	= 63.0 g

Now, its percent composition is

When chemists use the % notation, they mean percent by mass unless they specify otherwise.

$$\% \text{ H} = \frac{\text{mass of H}}{\text{mass of HNO}_3} \times 100\% = \frac{1.0 \text{ g}}{63.0 \text{ g}} \times 100\% = 1.6\% \text{ H}$$

$$\% \text{ N} = \frac{\text{mass of N}}{\text{mass of HNO}_3} \times 100\% = \frac{14.0 \text{ g}}{63.0 \text{ g}} \times 100\% = 22.2\% \text{ N}$$

$$\% \text{ O} = \frac{\text{mass of O}}{\text{mass of HNO}_3} \times 100\% = \frac{48.0 \text{ g}}{63.0 \text{ g}} \times 100\% = 76.2\% \text{ O}$$

Total = 100.0%

Percentages must add to 100%. However, round-off errors may not cancel, and totals such as 99.9% or 100.1% may be obtained in calculations.

EOC 39

Nitric acid is 1.6% H, 22.2% N, and 76.2% O by mass. All samples of pure HNO_3 have this composition, according to the Law of Definite Proportions.

2-8 Derivation of Formulas from Elemental Composition

Each year thousands of new compounds are made in laboratories or discovered in nature. One of the first steps in characterizing a new compound is the determination of its percent composition. A *qualitative* analysis is performed to determine *which* elements are present in the compound. Then a *quantitative* analysis is performed to determine the *amount* of each element.

Once the percent composition of a compound (or its elemental composition by mass) is known, the simplest formula can be determined. The **simplest** or **empirical formula** for a compound is the smallest whole-number ratio of atoms present. For molecular compounds the **molecular formula** indicates the *actual* numbers of atoms present in a molecule of the compound. It may be the same as the simplest formula or else some whole-number multiple of it. For example, the simplest and molecular formulas for water are both H_2O. However, for hydrogen peroxide, they are HO and H_2O_2, respectively.

Example 2-13

Compounds containing sulfur and oxygen are serious air pollutants; they represent the major cause of acid rain. Analysis of a sample of a pure compound reveals that it contains 50.1% sulfur and 49.9% oxygen by mass. What is the simplest formula of the compound?

Plan

The ratio of moles of atoms in any sample of a compound is the same as the ratio of atoms in that compound, because one mole of atoms of any element is 6.022 $\times$ 10^{23} atoms. This calculation is carried out in two steps. Step 1: Let's consider 100.0 grams of compound, which contains 50.1 grams of S and 49.9 grams of O. We calculate the number of moles of atoms of each. Step 2: We then obtain a whole-number ratio between these numbers that gives the ratio of atoms in the sample, and hence in the simplest formula for the compound.

Solution

Step 1: $$\underline{?} \text{ mol S atoms} = 50.1 \text{ g S} \times \frac{1 \text{ mol S atoms}}{32.1 \text{ g S}} = 1.56 \text{ mol S atoms}$$

$$\underline{?} \text{ mol O atoms} = 49.9 \text{ g O} \times \frac{1 \text{ mol O atoms}}{16.0 \text{ g O}} = 3.12 \text{ mol O atoms}$$

Step 2: Now we know that 100.0 grams of compound contains 1.56 moles of S atoms and 3.12 moles of O atoms. We obtain a whole-number ratio between these numbers that gives the ratio of atoms in the simplest formula.

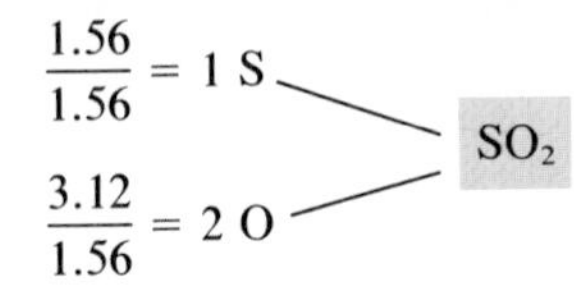

$$\frac{1.56}{1.56} = 1 \text{ S} \qquad \frac{3.12}{1.56} = 2 \text{ O} \qquad \rightarrow SO_2$$

A simple and useful way to obtain whole-number ratios among several numbers follows. (1) Divide each number by the smallest, and then, (2) if necessary, multiply all of the resulting numbers by the smallest whole number that will eliminate fractions.

EOC 41, 42

The solution for Example 2-13 can be set up in tabular form:

Element	Relative Mass of Element	Relative Number of Atoms (divide mass by AW)	Divide by Smaller Number	Smallest Whole-Number Ratio of Atoms
S	50.1	$\frac{50.1}{32.1} = 1.56$	$\frac{1.56}{1.56} = 1.00$	SO_2
O	49.9	$\frac{49.9}{16.0} = 3.12$	$\frac{3.12}{1.56} = 2.00$	

This tabular format provides a convenient way to solve simplest-formula problems, as the next example illustrates.

The "Relative Mass" column is proportional to the mass of each element in grams. With this interpretation, the next column could be headed "Relative Number of *Moles* of Atoms." Then the last column would represent the smallest whole-number ratios of *moles* of atoms. But because a mole is always the same number of items (atoms), that ratio is the same as the smallest whole-number ratio of atoms.

Example 2-14

A 20.882-gram sample of an ionic compound is found to contain 6.072 grams of Na, 8.474 grams of S, and 6.336 grams of O. What is its simplest formula?

Plan

We reason as in Example 2-13, calculating the number of moles of each element and the ratio among them. Here we use the tabular format that was introduced above.

Solution

Element	Relative Mass of Element	Relative Number of Atoms (divide mass by AW)	Divide by Smallest Number	Convert Fractions to Whole Numbers (multiply by integer)	Smallest Whole-Number Ratio of Atoms
Na	6.072	$\frac{6.072}{23.0} = 0.264$	$\frac{0.264}{0.264} = 1.00$	$1.00 \times 2 = 2$	
S	8.474	$\frac{8.474}{32.1} = 0.264$	$\frac{0.264}{0.264} = 1.00$	$1.00 \times 2 = 2$	$Na_2S_2O_3$
O	6.336	$\frac{6.336}{16.0} = 0.396$	$\frac{0.396}{0.264} = 1.50$	$1.50 \times 2 = 3$	

In this procedure we often obtain numbers such as 0.99 and 1.52. Because there is always some error in results obtained by analysis of samples (as well as round-off errors), we would interpret 0.99 as 1.0 and 1.52 as 1.5.

The ratio of atoms in the simplest formula *must be a whole-number ratio* (by definition). To convert the ratio 1:1:1.5 to a whole-number ratio, each number in the ratio was multipled by 2, which gave the simplest formula $Na_2S_2O_3$.

EOC 43

2-9 Determination of Molecular Formulas

Percent composition data yield only simplest formulas. To determine the molecular formula for a molecular compound, *both* its simplest formula and its molecular weight must be known. Methods for experimental determination of molecular weights are introduced in Chapter 12.

Millions of compounds are composed of carbon, hydrogen, and oxygen. Analyses for C and H can be performed in a C-H combustion system (Figure 2-11). An accurately known mass of a compound is burned in a furnace in a stream of oxygen. The carbon and hydrogen in the sample are converted to carbon dioxide and water vapor, respectively. The resulting increases in

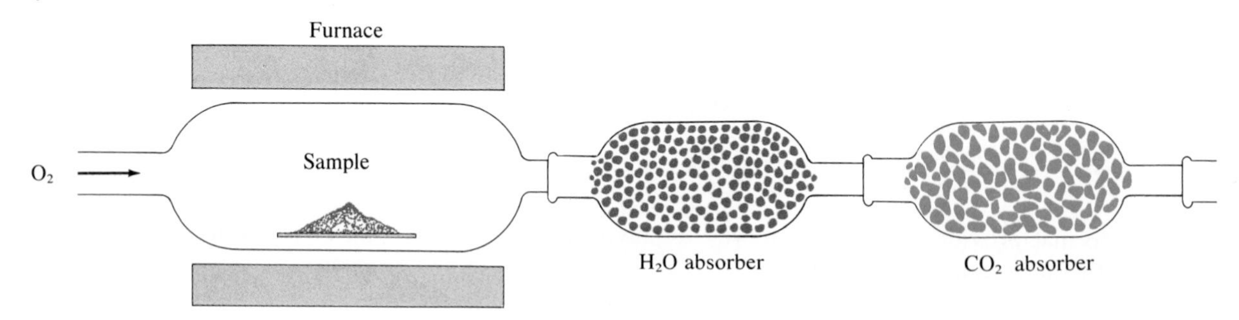

Figure 2-11
A combustion train used for carbon-hydrogen analysis. The absorbent for water is magnesium perchlorate, $Mg(ClO_4)_2$. Carbon dioxide is absorbed by finely divided sodium hydroxide supported on glass wool. Only a few milligrams of sample are needed for an analysis.

masses of the CO_2 and H_2O absorbers can then be related to the masses and percentages of carbon and hydrogen in the original sample.

Example 2-15

Hydrocarbons are organic compounds composed entirely of hydrogen and carbon. A 0.1647-gram sample of a pure hydrocarbon was burned in a C-H combustion train to produce 0.5694 gram of CO_2 and 0.0826 gram of H_2O. Determine the masses of C and H in the sample and the percentages of these elements in this hydrocarbon.

Plan

Step 1: We use the observed mass of CO_2, 0.5694 grams, to determine the mass of carbon in the original sample. There is one mole of carbon atoms, 12.01 grams, in each mole of CO_2, 44.01 grams; we use this information to construct the unit factor

$$\frac{12.01 \text{ g C}}{44.01 \text{ g CO}_2}$$

Step 2: Likewise, we can use the observed mass of H_2O, 0.0826 grams, to calculate the amount of hydrogen in the original sample. We use the fact that there are two moles of hydrogen atoms, 2.016 grams, in each mole of H_2O, 18.02 grams, to construct the unit factor

$$\frac{2.016 \text{ g H}}{18.02 \text{ g H}_2\text{O}}$$

We could calculate the mass of H by subtracting mass of C from mass of sample. However, it is good experimental practice, if possible, to base both on experimental measurements as we have done here. This would help to check for errors in the analysis or calculation.

Step 3: Then we calculate the percentages by mass of each element in turn, using the relationship

$$\% \text{ element} = \frac{\text{g element}}{\text{g sample}} \times 100\%$$

Solution

Step 1: $$\underline{?} \text{ g C} = 0.5694 \text{ g CO}_2 \times \frac{12.01 \text{ g C}}{44.01 \text{ g CO}_2} = 0.1554 \text{ g C}$$

Step 2: $$\underline{?} \text{ g H} = 0.0826 \text{ g H}_2\text{O} \times \frac{2.016 \text{ g H}}{18.02 \text{ g H}_2\text{O}} = 0.00924 \text{ g H}$$

Step 3: $$\% \text{ C} = \frac{0.1554 \text{ g C}}{0.1647 \text{ g sample}} \times 100\% = 94.4\% \text{ C}$$

$$\% \text{ H} = \frac{0.00924 \text{ g H}}{0.1647 \text{ g sample}} \times 100\% = 5.61\% \text{ H}$$

Total = 100.0%

EOC 43

When the compound to be analyzed contains oxygen, the calculation of the amount or percentage of oxygen in the sample is somewhat different. Part of the oxygen that goes to form CO_2 and H_2O comes from the sample and part comes from the oxygen stream supplied. Therefore we cannot directly determine the amount of oxygen already in the sample. The approach

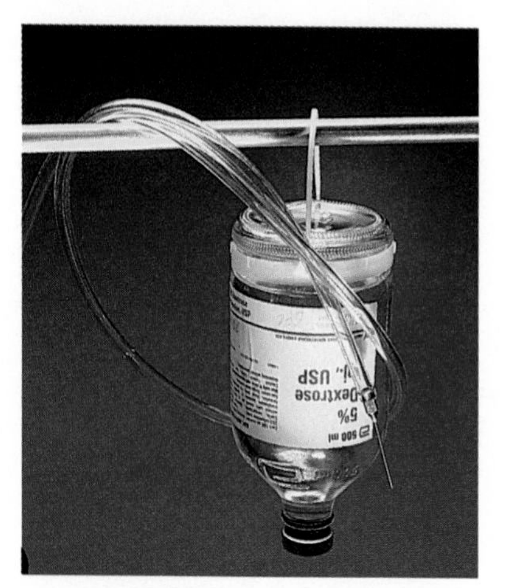

Glucose, a simple sugar, is the main component of intravenous feeding liquids. Its common name is dextrose. It is also one of the products of carbohydrate metabolism.

is to analyze as we did in Example 2-15 for all elements *except* oxygen. Then we subtract the sum of their masses from the mass of the original sample. The next example illustrates such a calculation.

Example 2-16

A 0.1014-gram sample of purified glucose was burned in a C-H combustion train to produce 0.1486 gram of CO_2 and 0.0609 gram of H_2O. An elemental analysis showed that glucose contains only carbon, hydrogen, and oxygen. Determine the masses of C, H, and O in the sample and the percentages of these elements in glucose.

Plan

Steps 1 and 2: We first calculate the masses of carbon and hydrogen as we did in Example 2-15. Step 3: The rest of the sample must be oxygen because glucose has been shown to contain only C, H, and O. So we subtract the masses of C and H from the total mass of sample. Step 4: Then we calculate the percentage by mass for each element.

Solution

Step 1: $$\underline{?}\text{ g C} = 0.1486\text{ g }CO_2 \times \frac{12.01\text{ g C}}{44.01\text{ g }CO_2} = 0.0406\text{ g C}$$

Step 2: $$\underline{?}\text{ g H} = 0.0609\text{ g }H_2O \times \frac{2.016\text{ g H}}{18.02\text{ g }H_2O} = 0.00681\text{ g H}$$

Step 3: $$\underline{?}\text{ g O} = 0.1014\text{ g sample} - [0.0406\text{ g C} + 0.00681\text{ g H}]$$

$$= 0.0540\text{ g O}$$

We say that the mass of O in the sample is calculated by *difference*.

Step 4: Now we can calculate the percentages by mass for each element:

$$\%\text{ C} = \frac{0.0406\text{ g C}}{0.1014\text{ g}} \times 100\% = 40.0\%\text{ C}$$

$$\%\text{ H} = \frac{0.00681\text{ g H}}{0.1014\text{ g}} \times 100\% = 6.72\%\text{ H}$$

$$\%\text{ O} = \frac{0.0540\text{ g O}}{0.1014\text{ g}} \times 100\% = 53.3\%\text{ O}$$

Total = 100.0%

EOC 44

For many compounds the molecular formula is a multiple of the simplest formula. Consider butane, C_4H_{10}. The simplest formula for butane is C_2H_5, but the molecular formula contains twice as many atoms; i.e., $(C_2H_5)_2 = C_4H_{10}$. Benzene, C_6H_6, is another example. The simplest formula for benzene is CH, but the molecular formula contains six times as many atoms; i.e., $(CH)_6 = C_6H_6$.

The molecular formula for a compound is *either* the same as, *or* a whole-number multiple of, the simplest formula. So we can write

$$\text{molecular weight} = n \times \text{simplest formula weight}$$

$$n = \frac{\text{molecular weight}}{\text{simplest formula weight}}$$

where n is the number of simplest formula units in a molecule of the compound. The subscripts in the molecular formula are obtained by multiplying the subscripts in the simplest formula by n.

Example 2-17

In Example 2-16 we found the elemental composition of glucose. Other experiments show that its molecular weight is approximately 180 amu. Determine the simplest formula and the molecular formula of glucose.

Plan

Step 1: We first use the masses of C, H, and O found in Example 2-16 to determine the simplest formula. Step 2: We can use the simplest formula to calculate the simplest formula weight. Because the molecular weight of glucose is known (approximately 180 amu), we can determine the molecular formula by dividing the molecular weight by the simplest formula weight.

As an alternative, we could have used the percentages by mass from Example 2-16. Using the earliest available numbers helps to minimize the effects of rounding-off errors.

$$n = \frac{\text{molecular weight}}{\text{simplest formula weight}}$$

The molecular weight is n times the simplest formula weight, so the molecular formula of glucose is n times the simplest formula.

Solution

Step 1:

Element	Mass of Element	Moles of Element (divide mass by AW)	Divide by Smallest	Smallest Whole-Number Ratio of Atoms
C	0.0406 g	$\frac{0.0406}{12.01} = 0.00338$ mol	$\frac{0.00338}{0.00338} = 1.00$	
H	0.00681 g	$\frac{0.00681}{1.008} = 0.00676$ mol	$\frac{0.00676}{0.00338} = 2.00$	CH_2O
O	0.0540 g	$\frac{0.0540}{16.00} = 0.00338$ mol	$\frac{0.00338}{0.00338} = 1.00$	

Step 2: The simplest formula is CH_2O, which has a formula weight of 30.02 amu. Because the molecular weight of glucose is approximately 180 amu, we can determine the molecular formula by dividing the molecular weight by the simplest formula weight.

$$n = \frac{180 \text{ amu}}{30.02 \text{ amu}} = 6.00$$

Traces of impurities and round-off errors may give numbers such as 5.96, 6.02, and so on, which are very close to integers.

The molecular weight is six times the simplest formula weight, $(CH_2O)_6 = C_6H_{12}O_6$, so the molecular formula of glucose is $C_6H_{12}O_6$.

EOC 50

As we shall see when we discuss the composition of compounds in some detail, two (and sometimes more) elements may form more than one compound. The **Law of Multiple Proportions** summarizes many experiments on such compounds. It is usually stated: When two elements, A and B, form more than one compound, the ratio of the masses of element B that combine with a given mass of element A in each of the compounds can be expressed by small whole numbers. Water, H_2O, and hydrogen peroxide, H_2O_2, provide an example. The ratio of masses of oxygen that combine with a given mass of hydrogen is 1:2 in H_2O and H_2O_2. Many similar examples, such as CO and CO_2 (1:2 ratio) and SO_2 and SO_3 (2:3 ratio), are known. The Law of Multiple Proportions had been recognized from studies of elemental composition before the time of Dalton. It provided additional support for his atomic theory.

Example 2-18

What is the ratio of the masses of oxygen that are combined with 1.00 gram of nitrogen in the compounds NO and N_2O_3?

Plan

First we calculate the mass of O that combines with one gram of N in each compound. Then we determine the ratio of the values of $\frac{\text{g O}}{\text{g N}}$ for the two compounds.

Solution

In NO: $$\frac{?\text{ g O}}{\text{g N}} = \frac{16.0\text{ g O}}{14.0\text{ g N}} = 1.14\text{ g O/g N}$$

In N_2O_3: $$\frac{?\text{ g O}}{\text{g N}} = \frac{48.0\text{ g O}}{28.0\text{ g N}} = 1.71\text{ g O/g N}$$

The ratio is $$\frac{\frac{\text{g O}}{\text{g N}}\text{ (in } N_2O_3)}{\frac{\text{g O}}{\text{g N}}\text{ (in NO)}} = \frac{1.71\text{ g O/g N}}{1.14\text{ g O/g N}} = \frac{1.5}{1.0} = \frac{3}{2}$$

We see that the ratio is 3 mass units of O (in N_2O_3) to 2 mass units of O (in NO).

EOC 55, 56

2-10 Some Other Interpretations of Chemical Formulas

Once we master the mole concept and the meaning of chemical formulas, we can use them in many other ways. The examples in this section illustrate a few additional kinds of information we can get from a chemical formula and the mole concept.

Example 2-19

What mass of chromium is contained in 35.8 grams of $(NH_4)_2Cr_2O_7$?

Plan

Let us first solve the problem in several steps. Step 1: The formula tells us that each mole of $(NH_4)_2Cr_2O_7$ contains two moles of Cr atoms, so we first find the number of moles of $(NH_4)_2Cr_2O_7$, using the unit factor

$$\frac{1 \text{ mol } (NH_4)_2Cr_2O_7}{252.0 \text{ g } (NH_4)_2Cr_2O_7}$$

Step 2: Then we convert the number of moles of $(NH_4)_2Cr_2O_7$ into the number of moles of Cr atoms it contains, using the unit factor

$$\frac{2 \text{ mol Cr atoms}}{1 \text{ mol } (NH_4)_2Cr_2O_7}$$

Step 3: We then use the atomic weight of Cr to convert the number of moles of chromium atoms to mass of chromium.

$$\text{Mass } (NH_4)_2Cr_2O_7 \longrightarrow \text{mol } (NH_4)_2Cr_2O_7 \longrightarrow \text{mol Cr} \longrightarrow \text{Mass Cr}$$

Solution

Step 1: $$\underline{?} \text{ mol } (NH_4)_2Cr_2O_7 = 35.8 \text{ g } (NH_4)_2Cr_2O_7 \times \frac{1 \text{ mol } (NH_4)_2Cr_2O_7}{252.0 \text{ g } (NH_4)_2Cr_2O_7}$$

$$= 0.142 \text{ mol } (NH_4)_2Cr_2O_7$$

Step 2: $$\underline{?} \text{ mol Cr atoms} = 0.142 \text{ mol } (NH_4)_2Cr_2O_7 \times \frac{2 \text{ mol Cr atoms}}{1 \text{ mol } (NH_4)_2Cr_2O_7}$$

$$= 0.284 \text{ mol Cr atoms}$$

Step 3: $$\underline{?} \text{ g Cr} = 0.284 \text{ mol Cr atoms} \times \frac{52.0 \text{ g Cr}}{1 \text{ mol Cr atoms}} = 14.8 \text{ g Cr}$$

If you understand the reasoning in these conversions, you should be able to solve this problem in a single setup:

$$\underline{?} \text{ g Cr} = 35.8 \text{ g } (NH_4)_2Cr_2O_7 \times \frac{1 \text{ mol } (NH_4)_2Cr_2O_7}{252.0 \text{ g } (NH_4)_2Cr_2O_7}$$

$$\times \frac{2 \text{ mol Cr atoms}}{1 \text{ mol } (NH_4)_2Cr_2O_7} \times \frac{52.0 \text{ g Cr}}{1 \text{ mol Cr}} = 14.8 \text{ g Cr}$$

EOC 60

Example 2-20

What mass of potassium chlorate, $KClO_3$, would contain 40.0 grams of oxygen?

Plan

The formula $KClO_3$ tells us that each mole of $KClO_3$ contains three moles of oxygen atoms. Each mole of oxygen atoms weighs 16.0 grams. So we can set up the solution to convert:

$$\text{Mass } O_2 \longrightarrow \text{mol } O_2 \longrightarrow \text{mol } KClO_3 \longrightarrow \text{Mass } KClO_3$$

Solution

$$\underline{?} \text{ g } KClO_3 = 40.0 \text{ g O} \times \frac{1 \text{ mol O atoms}}{16.0 \text{ g O atoms}} \times \frac{1 \text{ mol } KClO_3}{3 \text{ mol O}} \times \frac{122.6 \text{ g } KClO_3}{1 \text{ mol } KClO_3}$$

$$= 102 \text{ g } KClO_3$$

EOC 62

Example 2-21

(a) What mass of sulfur dioxide, SO_2, would contain the same mass of oxygen as is contained in 33.7 g of arsenic pentoxide, As_2O_5? (b) What mass of calcium chloride, $CaCl_2$, would contain the same number of chloride ions as are contained in 48.6 g of sodium chloride, NaCl?

Plan

(a) We could find explicitly the number of grams of O in 33.7 g of As_2O_5, and then find the mass of SO_2 that contains that same number of grams of O. But this method includes some unnecessary calculation. We need only convert to *moles* of O (because this is the same mass of O regardless of its environment) and then to SO_2.

$$\text{Mass } As_2O_5 \longrightarrow \text{mol } As_2O_5 \longrightarrow \text{mol O atoms} \longrightarrow \text{mol } SO_2 \longrightarrow \text{Mass } SO_2$$

(b) Because one mole always consists of the same number (Avogadro's number) of items, we can reason in terms of *moles* of Cl^- ions:

$$\text{Mass NaCl} \longrightarrow \text{mol NaCl} \longrightarrow \text{mol } Cl^- \text{ ions} \longrightarrow \text{mol } CaCl_2 \longrightarrow \text{Mass } CaCl_2$$

Solution

(a) $$\underline{?} \text{ g } SO_2 = 33.7 \text{ g } As_2O_5 \times \frac{1 \text{ mol } As_2O_5}{229.8 \text{ g } As_2O_5} \times \frac{5 \text{ mol O atoms}}{1 \text{ mol } As_2O_5}$$

$$\times \frac{1 \text{ mol } SO_2}{2 \text{ mol O atoms}} \times \frac{64.1 \text{ g } SO_2}{1 \text{ mol } SO_2} = 23.5 \text{ g } SO_2$$

(b) $$\underline{?} \text{ g } CaCl_2 = 48.6 \text{ g NaCl} \times \frac{1 \text{ mol NaCl}}{58.4 \text{ g NaCl}} \times \frac{1 \text{ mol } Cl^-}{1 \text{ mol NaCl}}$$

$$\times \frac{1 \text{ mol } CaCl_2}{2 \text{ mol } Cl^-} \times \frac{111.0 \text{ g } CaCl_2}{1 \text{ mol } CaCl_2} = 46.2 \text{ g } CaCl_2$$

EOC 64

We have already mentioned the existence of hydrates (for example, $(COOH)_2 \cdot 2H_2O$ and $CuSO_4 \cdot 5H_2O$ in Section 2-6). In such hydrates, two components, water and another compound are present in a definite integer ratio by moles. The following example illustrates how we might find and use the formula of a hydrate.

Example 2-22

A reaction requires pure anhydrous calcium sulfate, $CaSO_4$. Only an unidentified hydrate of calcium sulfate, $CaSO_4 \cdot xH_2O$, is available. (a) We heat 67.5 g of the unknown hydrate until all the water has been driven off. The resulting mass of pure $CaSO_4$ is 53.4 g. What is the formula of the hydrate, and what is its formula weight? (b) Suppose we wish to obtain enough of this hydrate to supply 95.5 grams of $CaSO_4$. How many grams should we weigh out?

Plan

(a) To find the formula of the hydrate, we must figure out the value of x in the formula $CaSO_4 \cdot xH_2O$. The mass of water removed from the sample is equal to the difference in the two masses given. The value of x is the number of moles of H_2O per mole of $CaSO_4$ in the hydrate.
(b) We use the formula weights of $CaSO_4$, 136.2 g/mol, and of $CaSO \cdot xH_2O$, $(136.2 + x18.0)$ g/mol, to write the conversion factor required for the calculation.

Solution

(a) $\underline{?}$ g water driven off = 67.5 g $CaSO_4 \cdot xH_2O$ − 53.4 g $CaSO_4$ = 14.1 g H_2O

$$x = \underline{?}\ \frac{\text{mol } H_2O}{\text{mol } CaSO_4} = \frac{14.1 \text{ g } H_2O}{53.4 \text{ g } CaSO_4} \times \frac{1.0 \text{ mol } H_2O}{18.0 \text{ g } H_2O} \times \frac{136.2 \text{ g } CaSO_4}{1 \text{ mol } CaSO_4}$$

$$= \frac{2.00 \text{ mol } H_2O}{\text{mol } CaSO_4}$$

Thus the formula of the hydrate is $CaSO_4 \cdot 2H_2O$. Its formula weight is

$$\text{FW} = 1 \times (\text{formula weight } CaSO_4) + 2 \times (\text{formula weight } H_2O)$$

$$= 136.2 \text{ g/mol} + 2(18.0 \text{ g/mol}) = 172.2 \text{ g/mol.}$$

(b) The formula weights of $CaSO_4$ (136.2 g/mol) and of $CaSO_4 \cdot 2H_2O$ (172.2 g/mol) allow us to write the unit factor

$$\frac{172.2 \text{ g } CaSO_4 \cdot 2H_2O}{136.2 \text{ g } CaSO_4}$$

We use this factor to perform the required conversion:

$$\underline{?} \text{ g } CaSO_4 \cdot 2H_2O = 95.5 \text{ g } CaSO_4 \text{ desired} \times \frac{172.2 \text{ g } CaSO_4 \cdot 2H_2O}{136.2 \text{ g } CaSO_4}$$

$$= 121 \text{ g } CaSO_4 \cdot 2H_2O$$

EOC 68

2-11 Purity of Samples

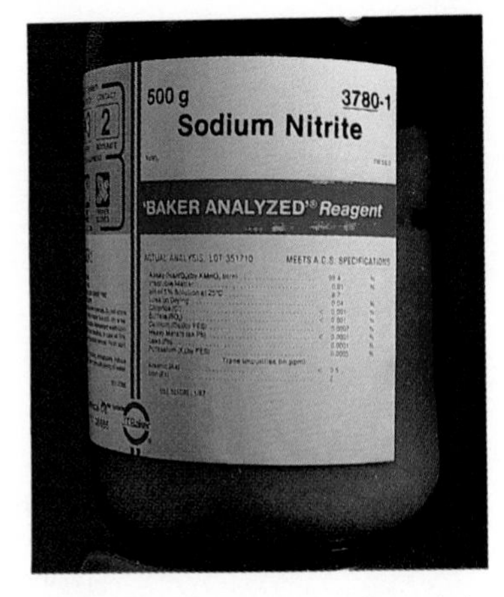

A label from a bottle of sodium nitrite.

Most substances obtained from laboratory reagent shelves are not 100% pure. When impure samples are used for precise work, account must be taken of impurities. The figure shows the label of a jar of reagent-grade sodium nitrite, $NaNO_2$, which is 99.4% pure by mass. From this information we know that total impurities represent 0.6% of the total mass of any sample. We can write several unit factors:

$$\frac{99.4 \text{ g } NaNO_2}{100 \text{ g sample}}, \quad \frac{0.6 \text{ g impurities}}{100 \text{ g sample}}, \quad \text{and} \quad \frac{0.6 \text{ g impurities}}{99.4 \text{ g } NaNO_2}$$

The inverse of each of these gives us a total of six unit factors.

Example 2-23

Calculate the masses of $NaNO_2$ and impurities in 45.2 g of 99.4% pure $NaNO_2$.

Plan

The percentage of $NaNO_2$ in the sample gives the unit factor $\frac{99.4 \text{ g } NaNO_2}{100 \text{ g sample}}$. The remainder of the sample is 100% − 99.4% = 0.6% impurities; this gives the unit factor $\frac{0.6 \text{ g impurities}}{100 \text{ g sample}}$.

Solution

$$\underline{?} \text{ g } NaNO_2 = 45.2 \text{ g sample} \times \frac{99.4 \text{ g } NaNO_2}{100 \text{ g sample}} = 44.9 \text{ g } NaNO_2$$

$$\underline{?} \text{ g impurities} = 45.2 \text{ sample} \times \frac{0.6 \text{ g impurities}}{100 \text{ g sample}} = 0.3 \text{ g impurities}$$

EOC 69, 72

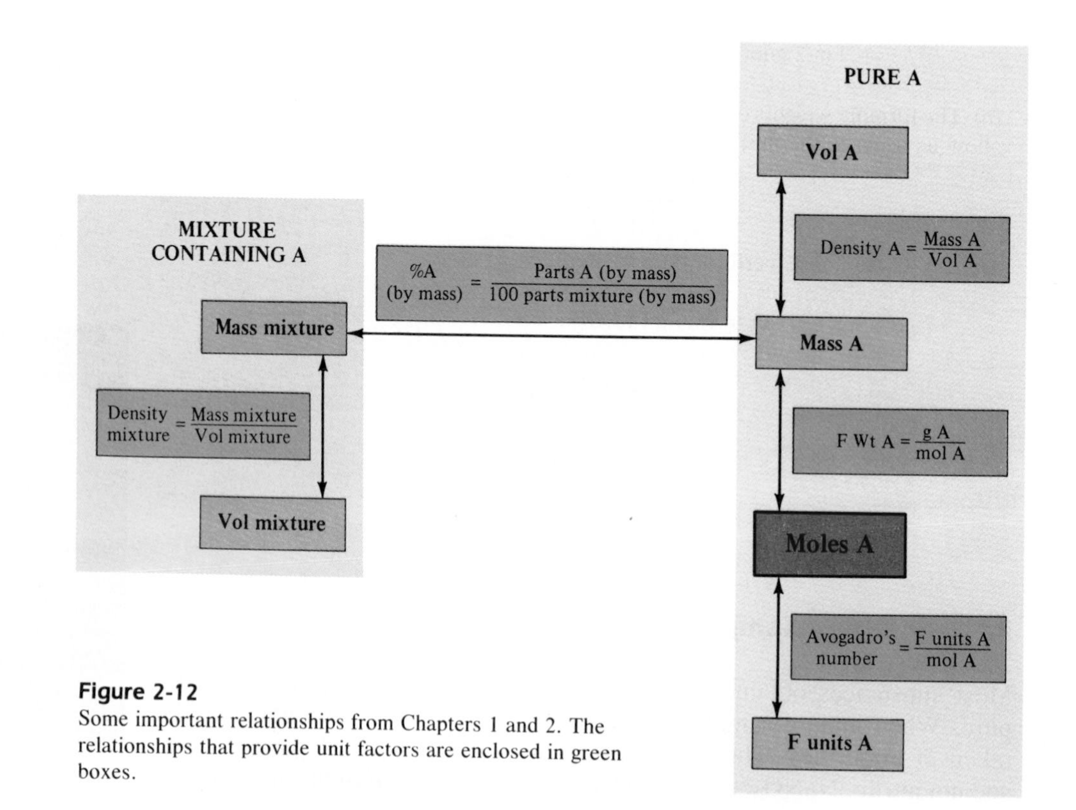

Figure 2-12
Some important relationships from Chapters 1 and 2. The relationships that provide unit factors are enclosed in green boxes.

Observe the beauty of the unit-factor approach to problem solving! Such questions as "Do we multiply by 0.994 or divide by 0.994?" never arise. The units always point toward the correct answer because we use unit factors constructed so that units *always* cancel until we arrive at the desired unit.

Many important relationships have been introduced in this chapter. Some of the most important transformations you have seen in Chapters 1 and 2 are summarized in Figure 2-12.

Key Terms

Allotropic modifications (allotropes) Different forms of the same element in the same physical state.

Atom The smallest particle of an element that maintains its chemical identity through all chemical and physical changes.

Atomic mass unit (amu) One twelfth of the mass of an atom of the carbon-12 isotope; a unit used for stating atomic and formula weights; also called dalton.

Atomic weight Weighted average of the masses of the constituent isotopes of an element; the relative masses of atoms of different elements.

Avogadro's number 6.022×10^{23} of the specified items. See *Mole*.

Composition stoichiometry Describes the quantitative (mass) relationships among elements in compounds.

Empirical formula See *Simplest formula*.

Formula Combination of symbols that indicates the chemical composition of a substance.

Formula unit The smallest repeating unit of a substance—for nonionic substances, the molecule.

Formula weight The mass, in atomic weight units, of one formula unit of a substance. Numerically equal to the mass, in grams, of one mole of the substance (see *Molar mass*). This number is obtained by adding the atomic weights of the atoms specified in the formula.

Hydrate A crystalline sample that contains water, H_2O, and another compound in a fixed mole ratio. Examples include $CuSO_4 \cdot 5H_2O$ and $(COOH)_2 \cdot 2H_2O$.

Ion An atom or group of atoms that carries an electrical charge. A positive ion is a *cation;* a negative ion is an *anion*.

Ionic compound A compound that is composed of ions. An example is sodium chloride, NaCl.

Law of Constant Composition See *Law of Definite Proportions*.

Law of Definite Proportions Different samples of a pure compound always contain the same elements in the same proportions by mass; this corresponds to atoms of these elements in fixed numerical ratios. Also called Law of Constant Composition.

Law of Multiple Proportions When two elements, A and B, form more than one compound, the ratio of the masses of element B that combine with a given mass of element A in each of the compounds can be expressed by small whole numbers.

Molar mass The mass of substance in one mole of the substance; numerically equal to the formula weight of the substance. See *Formula weight*; see *Molecular weight*.

Mole 6.022×10^{23} (Avogadro's number of) formula units (or molecules, for a nonionic substance) of the substance under discussion. The mass of one mole, in grams, is numerically equal to the formula (molecular) weight of the substance.

Molecular formula A formula that indicates the actual number of atoms present in a molecule of a molecular substance. Compare with *Simplest formula*.

Molecule The smallest particle of an element or compound that can have a stable independent existence.

Molecular weight The mass, in atomic mass units, of one molecule of a nonionic (molecular) substance. Numerically equal to the mass, in grams, of one mole of such a substance. This number is obtained by adding the atomic weights of the atoms specified in the formula.

Percent composition The mass percentage of each element in a compound.

Percent purity The percentage of a specified compound or element in an impure sample.

Simplest formula The smallest whole-number ratio of atoms present in a compound; also called empirical formula. Compare with *Molecular formula*.

Stoichiometry Description of the quantitative relationships among elements in compounds (composition stoichiometry) and among substances as they undergo chemical changes (reaction stoichiometry).

Exercises

Basic Ideas

1. (a) What is the origin of the word "stoichiometry"? (b) Distinguish between composition stoichiometry and reaction stoichiometry.
2. List the basic ideas of Dalton's atomic theory.
3. Give examples of molecules that contain (a) two atoms, (b) three atoms, (c) four atoms, and (d) eight atoms.
4. Give two examples of diatomic molecules and two examples of polyatomic molecules.
5. Write formulas for the following compounds: (a) sulfuric acid; (b) methane; (c) sulfur dioxide; (d) acetic acid.
6. Name the following compounds: (a) HNO_3; (b) C_2H_6; (c) NH_3; (d) CO_2.

Ions and Ionic Compounds

7. Define and illustrate the following: (a) ion; (b) cation; (c) anion; (d) polyatomic ion.
8. (a) There are no *molecules* in ionic compounds. Why not? (b) What is the difference between a formula unit of an ionic compound and a polyatomic molecule?
9. Name each of the following ions. Classify each as a monatomic or polyatomic ion. Classify each as a cation or an anion. (a) Na^+; (b) OH^-; (c) NO_3^-; (d) S^{2-}; (e) Fe^{2+}.
10. Write the chemical symbol for each of the following ions. Classify each as a monatomic or polyatomic ion. Classify each as a cation or an anion. (a) potassium ion; (b) sulfate ion; (c) copper(II) ion; (d) ammonium ion; (e) carbonate ion.
11. Name each of the following compounds: (a) AgCl; (b) $FeCl_2$; (c) K_2SO_4; (d) $Ca(OH)_2$; (e) $Fe_2(SO_4)_3$.
12. Write the chemical formula for each of the following ionic compounds: (a) calcium acetate; (b) ammonium carbonate; (c) zinc phosphate; (d) sodium hydroxide; (e) aluminum nitrate.
13. Write the chemical formula for the ionic compound formed between each of the following pairs of ions. Name each compound. (a) Ca^{2+} and Cl^-; (b) Al^{3+} and SO_4^{2-}; (c) K^+ and PO_4^{3-}; (d) Mg^{2+} and NO_3^-; (e) Fe^{3+} and CO_3^{2-}.
14. Write the chemical formula for the ionic compound formed between each of the following pairs of ions. Name each compound. (a) Cu^{2+} and SO_4^{2-}; (b) Mg^{2+} and OH^-; (c) NH_4^+ and CO_3^{2-}; (d) Zn^{2+} and Cl^-; (e) Fe^{3+} and CH_3COO^-.
15. Convert each of the following into a correct formula represented with correct notation. (a) $CaOH_2$; (b) $Mg(CO_3)$; (c) $Zn(CO_3)_2$; (d) $(NH_4)^2SO_4$; (e) $Mg_2(SO_4)_2$.

Atomic Weights

16. What is the mass ratio (4 significant figures) of one atom of P to 1 atom of F?
17. 4.05 g of magnesium combines exactly with 6.33 g of fluorine, forming magnesium fluoride, MgF_2. Find the relative masses of the atoms of magnesium and fluorine. Check your answer using a table of atomic weights. If the formula were not known, could you still do this calculation?

The Mole Concept

18. The mass of one round-head wood screw is 3.71 g. Find the mass of one dozen, one gross, and one mole of these wood screws. Compare the latter value with the mass of the earth, 5.98×10^{24} kg.
19. Sulfur molecules exist under various conditions as S_8, S_6, S_4, S_2, and S. (a) Is the mass of one mole of each of these molecules the same? (b) Is the number of molecules in one mole of each of these molecules the same? (c) Is the mass of sulfur in one mole of each of these molecules the same? (d) Is the number of atoms of sulfur in one mole of each of these molecules the same?
20. Complete the following table. You may refer to a table of atomic weights.

	Element	Atomic Weight	Mass of One Mole of Atoms
(a)	B	______	______
(b)	______	74.922 amu	______
(c)	Al	______	______
(d)	______	______	51.9961 g

21. Complete the following table. You may refer to a table of atomic weights.

	Element	Formula	Mass of One Mole of Molecules
(a)	Cl	Cl_2	______
(b)	______	H_2	______
(c)	______	P_4	______
(d)	______	______	4.0026 g
(e)	S	______	256.528 g
(f)	O	______	______

22. Determine the formula weight of each of the following substances: (a) chlorine, Cl_2; (b) water, H_2O; (c) saccharin, $C_7H_5NSO_3$; (d) sodium dichromate, $Na_2Cr_2O_7$.
23. Determine the formula weight of each of the following substances: (a) calcium sulfate, $CaSO_4$; (b) benzene, C_6H_6; (c) the sulfa drug sulfanilamide, $C_6H_4SO_2(NH_2)_2$; (d) uranyl phosphate, $(UO_2)_3(PO_4)_2$.
24. How many moles of substance are contained in each of the following samples? (a) 16.8 g of NH_3; (b) 3.25 kg of ammonium bromide; (c) 5.6 g of PCl_5; (d) 126.5 g of Fe.
25. How many moles of substance are contained in each of the following samples? (a) 24.5 g of formaldehyde, H_2CO; (b) 10.03 g of calcium carbonate; (c) 33.5 g of acetic acid; (d) 19.4 g of ethyl alcohol, C_2H_5OH.

26. The atomic weight of chlorine is 35.453 amu. Calculate the number of moles of chlorine atoms in (a) 1.00 g of chlorine, (b) 35.453 atomic mass units of chlorine, and (c) 5.66×10^{20} chlorine atoms.
27. Calculate the number of moles equivalent to each of the following: (a) 9.5×10^{21} atoms of Cs; (b) 4.7×10^{27} molecules of carbon dioxide; (c) 1.63×10^{23} formula units of $BaCl_2$; (d) 1.2×10^{22} atoms of Cu.
28. Calculate the number of moles equivalent to each of the following: (a) 5.5×10^{16} atoms of Fe; (b) 3.92×10^{18} molecules of CH_4; (c) 4.61×10^{25} molecules of O_2; (d) 4.61×10^{25} formula units of iron(III) nitrate.
29. What is the mass, in grams, of each of the samples of Exercise 27?
30. What is the mass, in grams, of each of the samples of Exercise 28?
31. How many atoms of C, H, and O are in each of the following? (a) 0.744 mol of glucose, $C_6H_{12}O_6$; (b) 2.50×10^{19} glucose ($C_6H_{12}O_6$) molecules; (c) 0.300 g of glucose.
32. How many molecules are in 28.0 g of each of the following substances? (a) CO; (b) N_2; (c) P_4; (d) P_2. (e) Do parts (c) and (d) contain the same number of atoms of phosphorus?
33. What mass, in grams, should be weighed for an experiment that requires 135 mmol $(NH_4)_2HPO_4$?
34. What mass, in grams, corresponds to each of the following? (a) 0.503 mol phenol, C_6H_5OH; (b) 1.01 mol of quartz, SiO_2; (c) 422 mmol of saccharin, $C_7H_5NSO_3$; (d) 125 μmol of saltpeter, KNO_3.
*35. Which of the following samples contains the smallest mass of silver? Which contains the largest? 0.0100 mol Ag; 0.0100 mol Ag_2O; 200 mg Ag; 3.01×10^{21} Ag atoms; 3.01×10^{21} Ag_2 molecules.
36. Complete the following table.

Moles of Compound	Moles of Cations	Moles of Anions
1 mol NaCl	______	______
2 mol Na_2SO_4	______	______
0.1 mol calcium nitrate	______	______
______	0.75 mol NH_4^+	0.25 mol PO_4^{3-}

*37. What volume of glycerine, $C_3H_8O_3$, density 1.26 g/mL, should be taken to obtain 2.50 moles of glycerine?

Percent Composition and Simplest Formulas

38. A 3.56-gram sample of iron powder was heated in gaseous chlorine, and 10.39 g of an iron chloride was formed. What is the percent composition of this compound?
39. Calculate the percent composition of each of the following compounds: (a) acetone, CH_3COCH_3; (b) carborundum, SiC; (c) aspirin, $CH_3COOC_6H_4COOH$.
*40. Copper is obtained from ores containing the following minerals: azurite, $Cu_3(CO_3)_2(OH)_2$; chalcocite, Cu_2S; chalcopyrite, $CuFeS_2$; covelite, CuS; cuprite, Cu_2O; and malachite, $Cu_2CO_3(OH)_2$. Which mineral has the highest copper content on a percent-by-mass basis?

41. Determine the simplest formula for each of the following compounds.
(a) copper(II) tartrate: 30.03% Cu; 22.70% C; 1.91% H; 45.37% O.
(b) nitrosyl fluoroborate: 11.99% N; 13.70% O; 9.25% B; 65.06% F.
42. The hormone epinephrine is released in the human body during stress and increases the body's metabolic rate. Like many biochemical compounds, epinephrine is composed of carbon, hydrogen, oxygen, and nitrogen. The percent composition of this hormone is 56.8% C, 6.56% H, 28.4% O, and 8.28% N. What is the simplest formula of epinephrine?
43. (a) A compound is found to contain 5.60 g N, 14.2 g Cl, and 0.800 g H. What is the simplest formula of this compound? (b) Another compound containing the same elements is found to be 26.2% N, 66.4% Cl, and 7.5% H. What is the simplest formula of this compound?
44. Combustion of 0.5707 mg of a hydrocarbon produces 1.759 mg of CO_2. What is the simplest formula of the hydrocarbon.
45. A 2.00-gram sample of a compound gave 4.86 g of CO_2 and 2.03 g of H_2O upon combustion in oxygen. The compound is known to contain only C, H, and O. What is its simplest formula?
46. A 1.000-gram sample of an alcohol was burned in oxygen to produce 1.913 g of CO_2 and 1.174 g of H_2O. The alcohol contained only C, H, and O. What is the simplest formula of the alcohol?
*47. Complicated chemical reactions occur at hot springs on the ocean floor. One compound obtained from such a hot spring consists of Mg, Si, H, and O. From a 0.334-gram sample, the Mg is recovered as 0.115 g of MgO; H is recovered as 25.7 mg of H_2O; and Si is recovered as 0.172 g of SiO_2. What is the simplest formula of this compound?

*48. A 2.31-gram sample of an oxide of iron, heated in a stream of H_2, produces 0.720 g of H_2O. What is the simplest formula of the oxide?

49. A 0.2360-gram sample of a white compound was analyzed and found to contain 0.0944 g of Ca, 0.0283 g of C, and 0.1133 g of O. (a) What is the percent composition by mass of this compound? (b) What is its simplest formula?

Determination of Molecular Formulas

50. Skatole is found in coal tar and in human feces. It contains three elements: C, H, and N. It is 82.40% C and 6.92% H by mass. Its simplest formula is its molecular formula. What are (a) the formula and (b) the molecular weight of skatole?

51. Testosterone, the male sex hormone, contains only C, H, and O. It is 9.79% H and 11.09% O by mass. Each molecule contains two O atoms. What are (a) the molecular weight and (b) the molecular formula for testosterone?

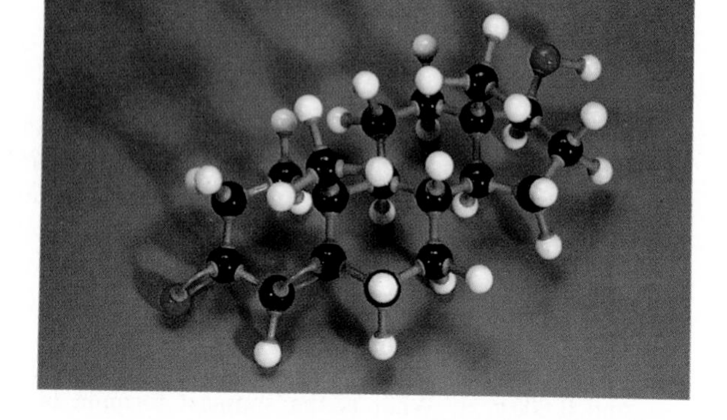

*52. More than 1 billion pounds of adipic acid (MW 146.1 g/mol) is manufactured in the United States each year. Most of it is used to make synthetic fabrics. Adipic acid contains only C, H, and O. Combustion of a 1.6380-gram sample of adipic acid gives 2.960 g of CO_2 and 1.010 g of H_2O. (a) What is the simplest formula for adipic acid? (b) What is its molecular formula?

53. Three allotropes of phosphorus are observed, with molecular weights of 62.0, 31.0, and 124.0. Write the molecular formula for each allotrope.

*54. The β-blocker drug, timolol, is expected to reduce the need for heart bypass surgery. Its composition by mass is 47.2% C, 6.55% H, 13.0% N, 25.9% O, and 7.43% S. The mass of 0.0100 mol of timolol is 4.32 g. (a) What is the simplest formula of timolol? (b) What is the molecular formula of timolol?

The Law of Multiple Proportions

55. Show that the compounds water, H_2O, and hydrogen peroxide, H_2O_2, obey the Law of Multiple Proportions.

56. Nitric oxide, NO, is produced in internal combustion engines. When NO comes in contact with air, it is quickly converted into nitrogen dioxide, NO_2, a very poisonous, corrosive gas. What mass of O is combined with 1.00 g of N in (a) NO and (b) NO_2? Show that NO and NO_2 obey the Law of Multiple Proportions.

57. Phosphorus forms two chlorides. A 30.00-gram sample of one chloride decomposes to give 4.35 g of P and 25.65 g of Cl. A 30.00-gram sample of the other chloride decomposes to give 6.61 g of P and 23.39 g of Cl. Show that these compounds obey the Law of Multiple Proportions.

58. What mass of oxygen is combined with 1.00 g of sulfur in (a) sulfur dioxide, SO_2, and in (b) sulfur trioxide, SO_3?

Interpretation of Chemical Formulas

59. One prominent ore of copper is chalcopyrite, $CuFeS_2$. How many tons of copper are contained in 255 tons of pure $CuFeS_2$?

60. Mercury occurs as a sulfide ore called *cinnabar*, HgS. How many grams of mercury are contained in 175.0 g of pure HgS?

61. (a) How many grams of copper are contained in 155 g of $CuSO_4$? (b) How many grams of copper are contained in 155 g of $CuSO_4 \cdot 5H_2O$?

62. What mass of $KMnO_4$ would contain 27.5 g of manganese?

63. What mass of azurite, $Cu_3(CO_3)_2(OH)_2$, would contain 435 g of copper?

64. Two ores that contain copper are chalcopyrite, $CuFeS_2$, and chalcocite, Cu_2S. What mass of chalcocite would contain the same mass of copper as is contained in 375 tons of chalcopyrite?

65. Tungsten is a very dense metal (19.3 g/cm³) with extremely high melting and boiling points (3370°C and 5900°C). When a small amount of it is included in steel, the resulting alloy is far harder and stronger than ordinary steel. Two important ores of tungsten are $FeWO_4$ and $CaWO_4$. How many grams of $CaWO_4$ would contain the same mass of tungsten that is contained in 874 g of $FeWO_4$?

66. What mass of NaCl would contain the same total number of ions as 245 g of $CaCl_2$?

67. Suppose we have equal masses of ammonium sulfate, $(NH_4)_2SO_4$, and aluminum sulfate, $Al_2(SO_4)_3$. (a) What is the ratio of the numbers of sulfate ions contained in these two samples? (b) What is the ratio of the masses of sulfur contained in these two samples?

*68. When $CuSO_4 \cdot 5H_2O$ is heated to 110°C, it loses only four moles of H_2O per mole of $CuSO_4$, to form $CuSO_4 \cdot H_2O$. When it is heated to temperatures above 150°C, the other mole of H_2O is lost. (a) How many grams of $CuSO_4 \cdot H_2O$ could be obtained by heating 665 g of $CuSO_4 \cdot 5H_2O$ to 110°C? (b) How many grams of anhydrous $CuSO_4$ could be obtained by heating 665 g of $CuSO_4 \cdot 5H_2O$ to 180°C?

Percent Purity

69. (a) How many grams of sodium chloride, NaCl, are contained in 33.0 g of saline solution that is 5.0% NaCl by mass? (b) Vinegar is 5.0% acetic acid, CH_3COOH, by mass. How many grams of acetic acid are contained in 33.0 g of vinegar? (c) How many pounds of acetic acid are contained in 33.0 pounds of vinegar?

70. (a) What is the percent by mass of oxalic acid, $(COOH)_2$, in a sample of pure oxalic acid dihydrate, $(COOH)_2 \cdot 2H_2O$? (b) What is the percent by mass of $(COOH)_2$ in a sample that is 72.4% $(COOH)_2 \cdot 2H_2O$ by mass?

71. What mass of each of the following elements is contained in 64.4 g of 88.2% pure $Ca(NO_3)_2$? Assume that the impurities do not contain the element mentioned. (a) Ca; (b) N

72. What weight of magnesium carbonate is contained in 775 pounds of an ore that is 24.3% magnesium carbonate by weight? (b) What weight of impurities is contained in the sample? (c) What weight of magnesium is contained in the sample? (Assume that no magnesium is present in the impurities.)

Mixed Examples

73. How many moles of bromine atoms are contained in each of the following? (a) 79.9×10^{23} Br atoms; (b) 79.9×10^{23} Br_2 molecules; (c) 79.9 g of bromine; (d) 79.9 mol of Br_2.

74. What is the *maximum* number of moles of CO_2 that could be obtained from the carbon in each of the following? (a) 1.00 mol of $Fe_2(CO_3)_3$; (b) 2.00 mol of $CaCO_3$; (c) 1.00 mol of $Ni(CN)_4$.

75. (a) How many formula units are contained in 238.1 g of K_2MoO_4? (b) How many potassium ions? (c) How many MoO_4^{2-} ions? (d) How many atoms of all kinds?

76. (a) How many moles of ozone molecules are contained in 64.0 g of ozone, O_3? (b) How many moles of oxygen atoms are contained in 64.0 g of ozone? (c) What mass of O_2 would contain the same number of oxygen atoms as 64.0 g of ozone? (d) What mass of oxygen gas, O_2, would contain the same number of molecules as 64.0 g of ozone?

77. Cocaine has the following percent composition by mass: 67.30% C, 6.930% H, 21.15% O, and 4.62% N. What is the simplest formula of cocaine?

78. What mass corresponds to each of the following? (a) 5.3 mol of C; (b) 0.12173 mol of N_2O_5; (c) 1.3 mmol of $Al_2(SO_4)_3$; (d) 1.0×10^{-10} mol of HCl.

79. How many moles of chloroform, $CHCl_3$, are contained in 338 mL of chloroform? The density of chloroform is 1.84 g/mL.

80. Find the number of moles of Ag needed to form each of the following: (a) 0.263 mol Ag_2S; (b) 0.263 mol Ag_2O; (c) 0.263 g Ag_2S; (d) 2.63×10^{20} formula units of Ag_2S.

***81.** (a) A sample contains 50.0% NaCl and 50.0% KCl by mass. What is the percent Cl, by mass, in this sample? (b) A second sample of NaCl and KCl contains 50.0% Cl by mass. What is the mass percent of NaCl in this sample?

***82.** Analysis of a 20.0-mg sample of an organic compound for H, N, and C yields 1.99 mg H_2O, 1.25 mg NH_3, and 6.47 mg CO_2. The Cl and Br are recovered as a mixture of AgCl and AgBr with a mass of 48.8 mg. Finally, when all the AgBr is converted to AgCl (by adding chloride from an outside source), the mass of AgCl formed from the AgBr plus the mass of AgCl originally present in the mixture are 42.3 mg. Calculate the empirical formula of the compound.

83. A metal, M, forms an oxide having the empirical formula M_2O_3. This oxide contains 68.4% of the metal by mass. (a) Calculate the atomic weight of the metal. (b) Identify the metal.

84. We have a 85.0-gram sample of "red lead," Pb_3O_4. (a) How many moles of lead are in the sample? (b) How many grams of lead are in the sample? (c) How many lead atoms are in the sample?

***85.** A 23.4-gram sample of a nickel sulfide is converted to 17.20 g of NiO. (a) What is the mass composition of the sulfide? (b) What is the empirical formula of the sulfide?

86. Three samples of magnesium oxide were analyzed to determine the mass ratios O/Mg, giving the following results: $\frac{1.60 \text{ g O}}{2.43 \text{ g Mg}}, \frac{0.658 \text{ g O}}{1.00 \text{ g Mg}}, \frac{2.29 \text{ g O}}{3.48 \text{ g Mg}}$. Which law of chemical combination is illustrated by these data?

***87.** The molecular weight of hemoglobin is about 65,000 g/mol. Hemoglobin contains 0.35% Fe by mass. How many iron atoms are in a hemoglobin molecule?

***88.** (a) We can drive off the water from copper sulfate pentahydrate, $CuSO_4 \cdot 5H_2O$, by heating. How many grams of anhydrous (meaning "without water") $CuSO_4$ could be obtained by heating 173 g of $CuSO_4 \cdot 5H_2O$? (b) An experiment calls for 2.50 mol of $CuSO_4$. How many grams of $CuSO_4 \cdot 5H_2O$ should we use to supply the required amount of $CuSO_4$?

***89.** During volcanic action, S_8 is converted to S, which is then converted to H_2S. In turn, the H_2S reacts with Fe, forming FeS_2. In water containing O_2, the FeS_2 reacts to form "mine acid," H_2SO_4. Find the maximum mass, in grams, of H_2SO_4 that can be formed from 0.366 mol of S_8.

90. One method of analyzing for the amount of Cr_2O_3 in a sample involves converting the chromium to $BaCrO_4$, and then weighing the amount of $BaCrO_4$ formed. Suppose that this process could be carried out with no loss of chromium. How many grams of Cr_2O_3 are present in the original sample for every gram of $BaCrO_4$ that could be isolated and weighed?

91. A 475-mg sample was found to contain 32.6% $PbSO_4$ (and no other lead). (a) How many milligrams of lead were contained in the sample? (b) What is the percent lead in the sample?

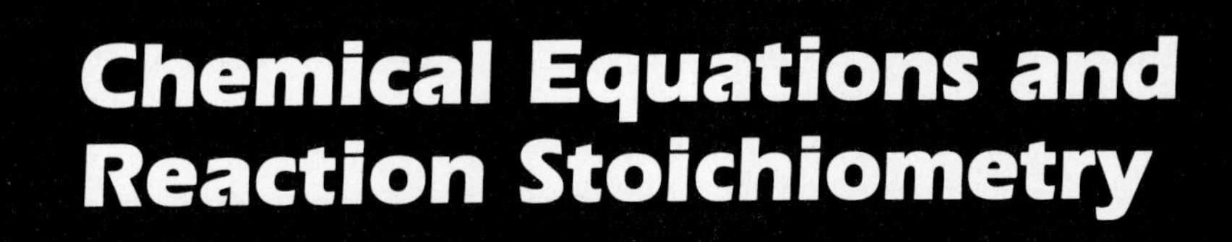

3 Chemical Equations and Reaction Stoichiometry

Sulfur burns in oxygen with a bright blue flame.

$$S_8 + 8O_2 \longrightarrow 8SO_2$$

This reaction is the first step in the commercial production of sulfuric acid, H_2SO_4, the most widely used industrial chemical.

Outline

Objectives

As you study this chapter, you should learn to

- ☐ Write a balanced chemical equation to describe a chemical reaction
- ☐ Interpret a balanced chemical equation to calculate the *moles* of reactants and products involved in the reaction
- ☐ Interpret a balanced chemical equation to calculate the *masses* of reactants and products involved in the reaction
- ☐ Determine which is the limiting reactant
- ☐ Use the limiting reactant concept in calculations with chemical equations
- ☐ Compare the amount of substance actually formed in a reaction (actual yield) with the predicted amount (theoretical yield), and determine the percent yield
- ☐ Understand sequential reactions
- ☐ Use the terminology of solutions—solute, solvent, concentration
- ☐ Calculate concentrations of solutions when they are diluted
- ☐ Understand the concept of, and calculations for, titrations

Methane, CH_4, is the main component of natural gas.

In the last chapter we studied composition stoichiometry, the quantitative relationships among elements in compounds. In this chapter we shall study reaction stoichiometry, the quantitative relationships among substances as they participate in chemical reactions. We ask several important questions. *How* can we describe the reaction of one substance with another? *How much* of one substance reacts with a given amount of another substance? *Which reactant* determines the amounts of products formed in a chemical reaction? *How* can we describe reactions in aqueous solutions?

Whether we are concerned with describing a reaction used in a chemical analysis, one used industrially in the production of a plastic, or one that occurs during metabolism in the body, we must describe it accurately. Chemical equations represent a very precise, yet a very versatile, language that describes chemical changes. We shall begin by studying chemical equations.

3-1 Chemical Equations

Chemical reactions always involve changing one or more substances into one or more different substances. That is, they involve regrouping atoms or ions to form other substances.

Chemical equations are used to describe chemical reactions, and they show (1) *the substances that react,* called **reactants**, (2) *the substances formed,* called **products**, and (3) *the relative amounts of the substances involved.* As a typical example, let's consider the combustion (burning) of natural gas, a reaction used to heat buildings and cook foods. Natural gas is a mixture of several substances, but the principal component is methane, CH_4. The equation that describes the reaction of methane with excess oxygen is

$$\underbrace{CH_4 + 2O_2}_{\text{reactants}} \longrightarrow \underbrace{CO_2 + 2H_2O}_{\text{products}}$$

Sometimes it is not possible to represent a chemical change with a single chemical equation. For example, under certain conditions, both CO_2 and CO are found as products, and a second chemical equation must be used. In the present case (excess oxygen), only one equation is required.

What does this equation tell us? In the simplest terms, it tells us that methane reacts with oxygen to produce carbon dioxide, CO_2, and water. More specifically, it says that for every CH_4 molecule that reacts, two molecules of O_2 also react, and that one CO_2 molecule and two H_2O molecules are formed. That is,

$$\underset{\text{1 molecule}}{CH_4} + \underset{\text{2 molecules}}{2O_2} \xrightarrow{\Delta} \underset{\text{1 molecule}}{CO_2} + \underset{\text{2 molecules}}{2H_2O}$$

The arrow may be read "yields."

Special conditions required for some reactions are indicated by notation over the arrow. The capital Greek letter delta (Δ) means that heat is necessary to start this reaction. Figure 3-1 shows the rearrangement of atoms described by this equation.

As we pointed out in Section 1-1, *there is no detectable change in the quantity of matter during an ordinary chemical reaction.* This guiding principle, the **Law of Conservation of Matter**, provides the basis for "balancing" chemical equations and for calculations based on those equations. Because matter is neither created nor destroyed during a chemical reaction, a balanced chemical equation must always include the same number of each kind of atom on both sides of the equation. Chemists usually write equations with the smallest possible whole-number coefficients.

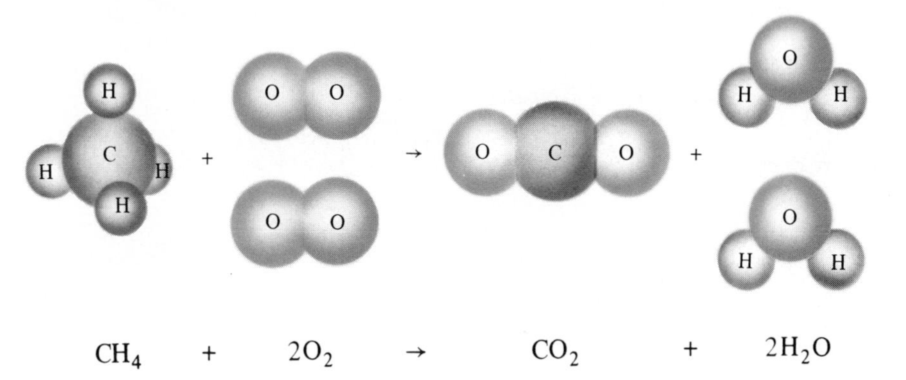

Figure 3-1
Representation of the reaction of methane with oxygen to form carbon dioxide and water. Some chemical bonds are broken and some new ones are formed.

Chemistry in Use. . .
Alchemy

During the Dark Ages, a pseudoscience known as *alchemy* flourished in Europe and the Middle East. Its practitioners, the predecessors of modern chemists, sought to turn base metals into gold and silver. They believed they could accomplish this by means of the Philosopher's Stone, which would transmute one element into another. Although their efforts proved fruitless, in the process of trying to make this magical substance the alchemists became very skillful at refining and alloying metals. They are credited with the discovery of arsenic, antimony, bismuth, zinc, and phosphorus, and they developed useful purification techniques such as distillation, sublimation, and crystallization. We also owe to alchemy the concept of scientific laboratories and the experimental approach to solving problems.

Practiced during a period when religion and mysticism affected all areas of life, the philosophy of alchemy dealt dually with matter and spirit. The desire to change ordinary metals into gold also represented man's wish to become better. Thus, alchemical processes and materials were frequently described in terms of astrology, mythology, the Church, or nature. Metals were identified with planets and gods; other substances were depicted as animals or people. Treatises were thick with allegory and had titles such as *The Crowne of Nature* and *The Triumphal Chariot of Antimony*. Further obscurity resulted from alchemists' desire to keep their "secrets" to themselves. Only *adepts,* or initiated experimenters, were able to decipher the complex codes and references. Because they often made up their own personal symbols and wrote in several languages, including Greek, Latin, Arabic, and Hebrew, it is unlikely that even an adept could have understood a recipe for the Philosopher's Stone written by a colleague!

One way of transmitting alchemical information was through pictures. In the figure, a method for refining gold is described. Antimony (wolf) is added to the impure gold (dead king) to "devour" any impurities present (foreground). When the two are thrown into a fire (background), the gold becomes pure (resurrected king).

One of the earliest alchemists was the Arab Jābir ibn Hayyān (also known as Geber), who originated a theory that all metals are composed of mercury and sulfur. This theory lasted for many centuries in Europe, where it was taken up by Christian scholars after being translated into Latin. (One of those scholars, a monk named Basil Valentine, authored the pompously named *The Triumphal Chariot of Antimony*.)

Other alchemists made great advances in science. Agricola, a German physician, fathered mineralogy with *De Re Metallica,* a clear and scholarly work classifying metals and describing their uses. His contemporary, Paracelsus, began the application of chemistry to medicine, called iatrochemistry, by noting the curative properties of several elements. As alchemists applied their chemical expertise in new directions, desires to synthesize gold slowly faded away. New findings were recorded clearly, and references to religion and the occult gradually vanished. A "scientific approach" prevailed, and in the 17th century chemistry came into being as an experimental science.

It can be said that in recent times the Philosopher's Stone *has* been discovered—through atomic reactions that split and fuse nuclei. As they explore ways to make new elements and transmute others, today's scientists could be considered "modern alchemists!"

Lisa L. Saunders
Chemistry major
University of Texas at Austin

The Death and Resurrection of the King (Atalanta Fugiens, 1617). Alchemists often represented chemicals with animals and mythological symbols.

All substances must be represented by formulas that describe them *as they exist* before we attempt to balance an equation. For instance, we must write H_2 to represent diatomic hydrogen molecules—not H, which represents hydrogen atoms. Once formulas are correct, the subscripts in the formulas may not be changed. Different subscripts in formulas specify different compounds, so the equation would no longer describe the same reaction if formulas were changed.

Let's generate the balanced equation for the reaction of aluminum metal with hydrochloric acid (hydrogen chloride dissolved in water) to produce aluminum chloride and hydrogen. The unbalanced "equation" is

$$Al + HCl \longrightarrow AlCl_3 + H_2$$

Atoms of	*In reactants*	*In products*
Al	1	1
H	1	2
Cl	1	3

Equation is not balanced.

As it now stands, the "equation" does not satisfy the Law of Conservation of Matter because there are two H atoms in the H_2 molecule and three Cl atoms in one formula unit of $AlCl_3$ (right side), but only one H atom and one Cl atom in the HCl molecule (left side).

Let us first balance chlorine by putting a coefficient of 3 in front of HCl.

$$Al + 3HCl \longrightarrow AlCl_3 + H_2$$

Atoms of	*In reactants*	*In products*
Al	1	1
H	3	2
Cl	3	3

Equation is not balanced.

Now there are 3H on the left and 2H on the right. The least common multiple of 3 and 2 is 6; to balance H, we multiply the 3HCl by 2 and the H_2 by 3.

$$Al + 6HCl \longrightarrow AlCl_3 + 3H_2$$

Atoms of	*In reactants*	*In products*
Al	1	1
H	6	6
Cl	6	3

Equation is still not balanced.

Now Cl is again unbalanced (6Cl on the left, 3 on the right), but we can fix this by putting a coefficient of 2 in front of $AlCl_3$ on the right.

$$Al + 6HCl \longrightarrow 2AlCl_3 + 3H_2$$

Atoms of	*In reactants*	*In products*
Al	1	2
H	6	6
Cl	6	6

Equation is still not balanced.

Now all elements except Al are balanced (1 on the left, 2 on the right); we complete the balancing by putting a coefficient of 2 in front of Al on the left.

$$\underset{\text{aluminum}}{2Al} + \underset{\text{hydrochloric acid}}{6HCl} \longrightarrow \underset{\text{aluminum chloride}}{2AlCl_3} + \underset{\text{hydrogen}}{3H_2}$$

Atoms of	*In reactants*	*In products*
Al	2	2
H	6	6
Cl	6	6

Now the equation is balanced.

When we think that we have finished the balancing, we should *always* do a complete check for each element, as shown in red in the margin.

Dimethyl ether, C_2H_6O, burns in an excess of oxygen to give carbon dioxide and water. Let's balance the equation for this reaction. In unbalanced form,

$$C_2H_6O + O_2 \longrightarrow CO_2 + H_2O$$

Atoms of	*In reactants*	*In products*
C	2	1
H	6	2
O	3	3

Equation is not balanced.

Carbon appears in only one compound on each side, and the same is true for hydrogen. We begin by balancing these elements:

$$C_2H_6O + O_2 \longrightarrow 2CO_2 + 3H_2O$$

Atoms of	*In reactants*	*In products*
C	2	2
H	6	6
O	3	7

Equation is still not balanced.

Now we have an odd number of atoms of O on each side. The single O in C_2H_6O balances one of the atoms of O on the right. We balance the other six by placing a coefficient of 3 before O_2 on the left.

Atoms of	*In reactants*	*In products*
C	2	2
H	6	6
O	7	7

Now the equation is balanced.

$$C_2H_6O + 3O_2 \longrightarrow 2CO_2 + 3H_2O$$

Balancing chemical equations "by inspection" is a *trial-and-error* approach. It requires a great deal of practice, but it is *very important!* Remember that we use the smallest whole-number coefficients.

3-2 Calculations Based on Chemical Equations

As we indicated earlier, chemical equations represent a very precise and versatile language. We are now ready to use them to calculate the relative *amounts* of substances involved in chemical reactions. Let us again consider the combustion of methane in excess oxygen. The balanced chemical equation for that reaction is

$$CH_4 + 2O_2 \xrightarrow{\Delta} CO_2 + 2H_2O$$

On a quantitative basis, at the molecular level, the equation says

A balanced chemical equation may be interpreted on a *molecular* basis.

CH_4	+	$2O_2$	$\longrightarrow$	CO_2	+	$2H_2O$
1 molecule of methane		2 molecules of oxygen		1 molecule of carbon dioxide		2 molecules of water

Example 3-1

How many O_2 molecules are required to react with 47 CH_4 molecules according to the above equation?

Plan

The *balanced* equation tells us that *one* CH_4 molecule reacts with *two* O_2 molecules. We can construct two unit factors from this fact:

$$\frac{1\ CH_4\ \text{molecule}}{2\ O_2\ \text{molecules}} \quad \text{and} \quad \frac{2\ O_2\ \text{molecules}}{1\ CH_4\ \text{molecule}}$$

These are unit factors for *this* reaction because the numerator and denominator are *chemically equivalent*. In other words, the numerator and the denominator represent the same amount of reaction. We convert CH_4 molecules to O_2 molecules.

Solution

$$\underline{?}\ O_2\ \text{molecules} = 47\ CH_4\ \text{molecules} \times \frac{2\ O_2\ \text{molecules}}{1\ CH_4\ \text{molecule}} = 94\ O_2\ \text{molecules}$$

EOC 8, 10

A chemical equation also indicates the relative amounts of each reactant and product in a given chemical reaction. We showed earlier that formulas can represent moles of substances. Suppose Avogadro's number of CH_4 molecules, rather than just one CH_4 molecule, undergo this reaction. Then the equation can be written as follows:

CH_4	+	$2O_2$	$\xrightarrow{\Delta}$	CO_2	+	$2H_2O$
6.02×10^{23} molecules		$2(6.02 \times 10^{23}$ molecules)		6.02×10^{23} molecules		$2(6.02 \times 10^{23}$ molecules)
1 mol		2 mol		1 mol		2 mol

This tells us that *one* mole of methane reacts with *two* moles of oxygen to produce *one* mole of carbon dioxide and *two* moles of water.

A balanced chemical equation may be interpreted in terms of *moles* of reactants and products.

Example 3-2

How many moles of water could be produced by the reaction of 3.5 moles of methane with excess oxygen?

Plan

The equation for the combustion of methane

$$\underset{1\ \text{mol}}{CH_4} + \underset{2\ \text{mol}}{2O_2} \longrightarrow \underset{1\ \text{mol}}{CO_2} + \underset{2\ \text{mol}}{2H_2O}$$

shows that one mole of methane reacts with two moles of oxygen to produce two moles of water. From this information we construct two *unit factors*:

$$\frac{1\ \text{mol}\ CH_4}{2\ \text{mol}\ H_2O} \quad \text{and} \quad \frac{2\ \text{mol}\ H_2O}{1\ \text{mol}\ CH_4}$$

We use the second factor in this calculation.

Solution

$$\underline{?}\ \text{mol}\ H_2O = 3.5\ \text{mol}\ CH_4 \times \frac{2\ \text{mol}\ H_2O}{1\ \text{mol}\ CH_4} = 7.0\ \text{mol}\ H_2O$$

EOC 18, 20

We know the mass of one mole of each of these substances, so we can also write

$$\begin{array}{cccc} CH_4 & + \ 2O_2 & \longrightarrow \ CO_2 & + \ 2H_2O \\ 1\ \text{mol} & 2\ \text{mol} & 1\ \text{mol} & 2\ \text{mol} \\ 16\ \text{g} & 2(32\ \text{g}) & 44\ \text{g} & 2(18\ \text{g}) \\ 16\ \text{g} & 64\ \text{g} & 44\ \text{g} & 36\ \text{g} \\ \multicolumn{2}{c}{\underbrace{\qquad\qquad}_{80\ \text{g reactants}}} & \multicolumn{2}{c}{\underbrace{\qquad\qquad}_{80\ \text{g products}}} \end{array}$$

The equation now tells us that 16 grams of CH_4 reacts with 64 grams of O_2 to form 44 grams of CO_2 and 36 grams of H_2O. The Law of Conservation of Matter is satisfied. Chemical equations describe **reaction ratios**, i.e., the *mole ratios* of reactants and products as well as the *relative masses* of reactants and products.

A balanced equation may be interpreted on a *mass* basis. We have rounded molecular weights to the nearest whole number of grams here.

Example 3-3

What mass of oxygen is required to react completely with 1.2 moles of CH_4?

Plan

The balanced equation

$$\begin{array}{cccc} CH_4 & + \ 2O_2 & \longrightarrow \ CO_2 & + \ 2H_2O \\ 1\ \text{mol} & 2\ \text{mol} & 1\ \text{mol} & 2\ \text{mol} \\ 16\ \text{g} & 2(32\ \text{g}) & 44\ \text{g} & 2(18\ \text{g}) \end{array}$$

gives the relationships among moles and grams of reactants and products.

$$\text{mol}\ CH_4 \longrightarrow \text{mol}\ O_2 \longrightarrow \text{g}\ O_2$$

Solution

$$\underline{?}\text{ g } O_2 = 1.2 \text{ mol } CH_4 \times \frac{2 \text{ mol } O_2}{1 \text{ mol } CH_4} \times \frac{32 \text{ g } O_2}{1 \text{ mol } O_2} = \boxed{77 \text{ g } O_2}$$

EOC 22, 24

Example 3-4

What mass of oxygen is required to react completely with 24 grams of CH_4?

Plan

Recall the balanced equation

$$\begin{array}{cccc} CH_4 + & 2O_2 \longrightarrow & CO_2 + & 2H_2O \\ 1 \text{ mol} & 2 \text{ mol} & 1 \text{ mol} & 2 \text{ mol} \\ 16 \text{ g} & 64 \text{ g} & 44 \text{ g} & 36 \text{ g} \end{array}$$

This shows that 16 grams of CH_4 reacts with 64 grams of O_2. These two quantities are chemically equivalent, so we can construct two *unit factors:*

$$\frac{16 \text{ g } CH_4}{64 \text{ g } O_2} \quad \text{and} \quad \frac{64 \text{ g } O_2}{16 \text{ g } CH_4}$$

Solution

$$\underline{?}\text{ g } O_2 = 24 \text{ g } CH_4 \times \frac{64 \text{ g } O_2}{16 \text{ g } CH_4} = \boxed{96 \text{ g } O_2}$$

EOC 26, 27

Another approach to the problem we have just solved is known as the **mole method**. In this method, the number of moles of reactant or product is calculated and then converted to grams (or other desired unit). Example 3-4 asked, "What mass of oxygen is required to react with 24 grams of CH_4?" The balanced equation and the calculation by the mole method follow:

$$\begin{array}{cccc} CH_4 + & 2O_2 \longrightarrow & CO_2 + & 2H_2O \\ 1 \text{ mol} & 2 \text{ mol} & 1 \text{ mol} & 2 \text{ mol} \end{array}$$

We convert
1. g CH_4 → mol CH_4
2. mol CH_4 → mol O_2
3. mol O_2 → g O_2

$$\underline{?}\text{ mol } CH_4 = 24 \text{ g } CH_4 \times \frac{1 \text{ mol } CH_4}{16 \text{ g } CH_4} = 1.5 \text{ mol } CH_4$$

$$\underline{?}\text{ mol } O_2 = 1.5 \text{ mol } CH_4 \times \frac{2 \text{ mol } O_2}{1 \text{ mol } CH_4} = 3.0 \text{ mol } O_2$$

$$\underline{?}\text{ g } O_2 = 3.0 \text{ mol } O_2 \times \frac{32 \text{ g } O_2}{1 \text{ mol } O_2} = \boxed{96 \text{ g } O_2}$$

All these steps could be combined into one setup in which we convert:

$$\text{g of } CH_4 \longrightarrow \text{mol of } CH_4 \longrightarrow \text{mol of } O_2 \longrightarrow \text{g of } O_2$$

$$\underline{?}\text{ g } O_2 = 24 \text{ g } CH_4 \times \frac{1 \text{ mol } CH_4}{16 \text{ g } CH_4} \times \frac{2 \text{ mol } O_2}{1 \text{ mol } CH_4} \times \frac{32 \text{ g } O_2}{1 \text{ mol } O_2} = \boxed{96 \text{ g } O_2}$$

All valid methods of calculations are based on balanced chemical equations.

The same answer, 96 grams of O_2, is obtained by both of these methods. The question may be reversed, as in Example 3-5.

Example 3-5

What mass of CH_4, in grams, is required to react with 96 grams of O_2?

Plan

We recall that one mole of CH_4 reacts with two moles of O_2.

Solution

$$\underline{?}\text{ g CH}_4 = 96\text{ g O}_2 \times \frac{1\text{ mol O}_2}{32\text{ g O}_2} \times \frac{1\text{ mol CH}_4}{2\text{ mol O}_2} \times \frac{16\text{ g CH}_4}{1\text{ mol CH}_4} = \boxed{24\text{ g CH}_4}$$

These unit factors are the reciprocals of those used in Example 3-4.

or, more simply,

$$\underline{?}\text{ g CH}_4 = 96\text{ g O}_2 \times \frac{16\text{ g CH}_4}{64\text{ g O}_2} = \boxed{24\text{ g CH}_4}$$

EOC 28

This is the amount of CH_4 in Example 3-4 that reacted with 96 grams of O_2.

Example 3-6

Most combustion reactions occur in excess O_2, i.e., more than enough O_2 to burn the substance completely. Calculate the mass of CO_2, in grams, that can be produced by burning 6.0 moles of CH_4 in excess O_2.

It is important to recognize that the reaction must stop when the 6.0 mol of CH_4 has been used up. Some O_2 will remain unreacted.

Plan

Recall the balanced equation

$$\underset{\substack{1\text{ mol}\\16\text{ g}}}{CH_4} + \underset{\substack{2\text{ mol}\\2(32\text{ g})}}{2O_2} \longrightarrow \underset{\substack{1\text{ mol}\\44\text{ g}}}{CO_2} + \underset{\substack{2\text{ mol}\\2(18\text{ g})}}{2H_2O}$$

which tells us that one mole of CH_4 produces one mole (44 g) of CO_2.

Solution

$$\underline{?}\text{ g CO}_2 = 6.0\text{ mol CH}_4 \times \frac{1\text{ mol CO}_2}{1\text{ mol CH}_4} \times \frac{44\text{ g CO}_2}{1\text{ mol CO}_2} = \boxed{2.6 \times 10^2\text{ g CO}_2}$$

From the mole interpretation of the chemical equation for the combustion of methane, we can see many chemically equivalent pairs of terms. Each pair gives a unit factor relating substances. Some of these factors are

$$\frac{1\text{ mol CH}_4}{2\text{ mol O}_2} \qquad \frac{1\text{ mol CH}_4}{64\text{ g O}_2} \qquad \frac{1\text{ mol CH}_4}{2(6.02 \times 10^{23})\text{ O}_2\text{ molecules}}$$

$$\frac{16\text{ g CH}_4}{2\text{ mol O}_2} \qquad \frac{16\text{ g CH}_4}{64\text{ g O}_2} \qquad \frac{16\text{ g CH}_4}{2(6.02 \times 10^{23})\text{ O}_2\text{ molecules}}$$

$$\frac{6.02 \times 10^{23}\text{ CH}_4\text{ molecules}}{2\text{ mol O}_2} \qquad \frac{6.02 \times 10^{23}\text{ CH}_4\text{ molecules}}{64\text{ g O}_2} \qquad \frac{6.02 \times 10^{23}\text{ CH}_4\text{ molecules}}{2(6.02 \times 10^{23})\text{ O}_2\text{ molecules}}$$

We have written down nine unit factors relating CH_4 and O_2 for this particular reaction. The nine factors obtained by inverting each of these give

Please don't try to memorize unit factors for chemical reactions; rather, learn the general *method* for constructing them from balanced chemical equations.

a total of 18 unit factors relating CH_4 and O_2 for *this* reaction in three kinds of units. In fact, we can write down many factors relating *any* two substances involved in any chemical reaction! Try writing down some that involve reactants and products.

Reaction stoichiometry usually involves interpreting a balanced chemical equation to relate a *given* bit of information to the *desired* bit of information.

Example 3-7

What mass of CH_4 produces 3.01×10^{23} H_2O molecules when burned in excess O_2?

Plan

The balanced equation tells us that one mole of CH_4 produces two moles of H_2O.

$$\underset{1\ \text{mol}}{CH_4} + \underset{2\ \text{mol}}{2O_2} \longrightarrow \underset{1\ \text{mol}}{CO_2} + \underset{2\ \text{mol}}{2H_2O}$$

$$\boxed{H_2O\ \text{molecules}} \longrightarrow \boxed{\text{mol of}\ H_2O} \longrightarrow \boxed{\text{mol of}\ CH_4} \longrightarrow \boxed{\text{g of}\ CH_4}$$

Solution

$$\underline{?}\ \text{g}\ CH_4 = 3.01 \times 10^{23}\ H_2O\ \text{molecules} \times \frac{1\ \text{mol}\ H_2O}{6.02 \times 10^{23}\ H_2O\ \text{molecules}}$$

$$\times \frac{1\ \text{mol}\ CH_4}{2\ \text{mol}\ H_2O} \times \frac{16\ \text{g}\ CH_4}{1\ \text{mol}\ CH_4} = \boxed{4.0\ \text{g}\ CH_4}$$

The possibilities for this kind of problem-solving go on and on. Before you continue, you should work Exercises 16–31 at the end of the chapter.

3-3 The Limiting Reactant Concept

In the problems we have worked thus far, the presence of an excess of one reactant was stated or implied. The calculations were based on the substance that was used up first, called the **limiting reactant**. Before we study the concept of the limiting reactant in stoichiometry, let's develop the basic idea by considering a simple but analogous nonchemical example.

Suppose you have four slices of ham and six slices of bread and you wish to make as many ham sandwiches as possible using only one slice of ham and two slices of bread per sandwich. Obviously, you can make only three sandwiches, at which point you run out of bread. (In a chemical reaction this would correspond to one of the reactants being used up—so the reaction would stop.) Therefore the bread is the "limiting reactant" and the extra slice of ham is the "excess reactant." The amount of product, ham sandwiches, is determined by the amount of the limiting reactant, bread in this case.

Example 3-8

What mass of CO_2 could be formed by the reaction of 8.0 g of CH_4 with 48 g of O_2?

Plan

Recall the balanced equation:

$$\begin{array}{cccc} CH_4 & + \; 2O_2 & \longrightarrow \; CO_2 & + \; 2H_2O \\ 1\text{ mol} & 2\text{ mol} & 1\text{ mol} & 2\text{ mol} \\ 16\text{ g} & 2(32\text{ g}) & 44\text{ g} & 2(18\text{ g}) \end{array}$$

This tells us that *one* mole of CH_4 reacts with *two* moles of O_2. We are given masses of both CH_4 and O_2, so we calculate the number of moles of each reactant, and then determine the number of moles of each reactant required to react with the other. From these calculations we can identify the limiting reactant. We base the calculation on it.

Solution

$$\underline{?}\text{ mol } CH_4 = 8.0\text{ g } CH_4 \times \frac{1\text{ mol } CH_4}{16\text{ g } CH_4} = \underline{0.50\text{ mol } CH_4}$$

$$\underline{?}\text{ mol } O_2 = 48\text{ g } O_2 \times \frac{1\text{ mol } O_2}{32\text{ g } O_2} = \underline{1.5\text{ mol } O_2}$$

Now we return to the balanced equation. First we calculate the number of moles of O_2 required to react with 0.50 mole of CH_4.

$$\underline{?}\text{ mol } O_2 = 0.50\text{ mol } CH_4 \times \frac{2\text{ mol } O_2}{1\text{ mol } CH_4} = 1\text{ mol } O_2$$

We have 1.5 moles of O_2, but only 1 mole of O_2 is required, so CH_4 is the limiting reactant. Or we can calculate the number of moles of CH_4 required to react with 1.5 moles of O_2.

$$\underline{?}\text{ mol } CH_4 = 1.5\text{ mol } O_2 \times \frac{1\text{ mol } CH_4}{2\text{ mol } O_2} = 0.75\text{ mol } CH_4$$

This tells us that 0.75 mole of CH_4 would be required to react with 1.5 moles of O_2. But we have only 0.50 mole of CH_4, so we see again that CH_4 is the limiting reactant. The reaction must stop when the limiting reactant, CH_4, is used up; we base the calculation on CH_4.

$$\text{g of } CH_4 \longrightarrow \text{mol of } CH_4 \longrightarrow \text{mol of } CO_2 \longrightarrow \text{g of } CO_2$$

$$\underline{?}\text{ g } CO_2 = 8.0\text{ g } CH_4 \times \frac{1\text{ mol } CH_4}{16\text{ g } CH_4} \times \frac{1\text{ mol } CO_2}{1\text{ mol } CH_4} \times \frac{44\text{ g } CO_2}{1\text{ mol } CO_2} = 22\text{ g } CO_2$$

Thus, 22 grams of CO_2 is the most CO_2 that can be produced from 8.0 grams of CH_4 and 48 grams of O_2. If the calculation had been based on O_2 rather than CH_4, the answer would be too big and *wrong*. This would require more CH_4 than we had.

Another approach to problems like Example 3-8 is to calculate the number of moles of each reactant:

$$\underline{?}\text{ mol } CH_4 = 8.0\text{ g } CH_4 \times \frac{1\text{ mol } CH_4}{16\text{ g } CH_4} = \underline{0.50\text{ mol } CH_4}$$

$$\underline{?}\text{ mol } O_2 = 48\text{ g } O_2 \times \frac{1\text{ mol } O_2}{32\text{ g } O_2} = \underline{1.5\text{ mol } O_2}$$

A precipitate of solid $Ni(OH)_2$ forms when colorless NaOH solution is added to green $NiCl_2$ solution.

Then we return to the balanced equation. We first calculate the *required ratio* of reactants as indicated by the balanced chemical equation. We then calculate the *available ratio* of reactants and compare the two:

Required Ratio	**Available Ratio**
$\frac{1 \text{ mol } CH_4}{2 \text{ mol } O_2} = \frac{0.50 \text{ mol } CH_4}{1.00 \text{ mol } O_2}$	$\frac{0.50 \text{ mol } CH_4}{1.50 \text{ mol } O_2} = \frac{0.33 \text{ mol } CH_4}{1.00 \text{ mol } O_2}$

We see that each mole of O_2 would require exactly 0.50 mole of CH_4 to be completely used up. But we have only 0.33 mole of CH_4 for each mole of O_2, so there is *insufficient* CH_4 to react with all of the available O_2. The reaction must stop when the CH_4 is gone; CH_4 is the "limiting reactant" and we must base the calculation on it.

Example 3-9

What is the maximum mass of $Ni(OH)_2$ that could be prepared by mixing two solutions that contain 26.0 grams of $NiCl_2$ and 10.0 grams of NaOH, respectively?

$$NiCl_2 + 2NaOH \longrightarrow Ni(OH)_2 + 2NaCl$$

Plan

Interpreting the balanced equation as usual, we have

$NiCl_2$	+ $2NaOH$	$\longrightarrow$ $Ni(OH)_2$	+ $2NaCl$
1 mol	2 mol	1 mol	2 mol
129.7 g	2(40.0 g)	92.7 g	2(58.4 g)

We determine the number of moles of $NiCl_2$ and NaOH present. Then we find the number of moles of each reactant required to react with the other reactant. These calculations identify the limiting reactant. We base the calculation on it.

Note that even though the reaction occurs in aqueous solution, this calculation is similar to earlier examples because we are given the amounts of both reactants.

Solution

$$\underline{?} \text{ mol } NiCl_2 = 26.0 \text{ g } NiCl_2 \times \frac{1 \text{ mol } NiCl_2}{129.7 \text{ g } NiCl_2} = 0.200 \text{ mol } NiCl_2$$

$$\underline{?} \text{ mol NaOH} = 10.0 \text{ g NaOH} \times \frac{1 \text{ mol NaOH}}{40.0 \text{ g NaOH}} = 0.250 \text{ mol NaOH}$$

We return to the balanced equation and calculate the number of moles of NaOH required to react with 0.200 mole of $NiCl_2$.

$$\underline{?} \text{ mol NaOH} = 0.200 \text{ mol } NiCl_2 \times \frac{2 \text{ mol NaOH}}{1 \text{ mol } NiCl_2} = 0.400 \text{ mol NaOH}$$

But we have only 0.250 mole of NaOH, so NaOH is the limiting reactant.

If we calculate the number of moles of $NiCl_2$ required to react with 0.250 mole of NaOH, we get

$$\underline{?} \text{ mol } NiCl_2 = 0.250 \text{ mol NaOH} \times \frac{1 \text{ mol } NiCl_2}{2 \text{ mol NaOH}} = 0.125 \text{ mol } NiCl_2$$

We see that 0.250 mole of NaOH can react with only 0.125 mole of $NiCl_2$. But we have 0.200 mole of $NiCl_2$. This also tells us that NaOH is the limiting reactant, and so the calculation must be based on NaOH. The reaction must stop when all of the NaOH has been used up.

g of NaOH $\longrightarrow$ mol of NaOH $\longrightarrow$ mol $Ni(OH)_2$ $\longrightarrow$ g of $Ni(OH)_2$

$$\underline{?}\text{ g } Ni(OH)_2 = 10.0\text{ g NaOH} \times \frac{1\text{ mol NaOH}}{40.0\text{ g NaOH}} \times \frac{1\text{ mol } Ni(OH)_2}{2\text{ mol NaOH}} \times \frac{92.7\text{ g } Ni(OH)_2}{1\text{ mol } Ni(OH)_2}$$

$$= 11.6\text{ g } Ni(OH)_2$$

EOC 32, 36

3-4 Percent Yields from Chemical Reactions

The **theoretical yield** from a chemical reaction is the yield calculated by assuming that the chemical reaction goes to completion. In practice we often do not isolate as much product from a reaction mixture as is theoretically possible. There are several reasons. (1) Many reactions do not go to completion; i.e., the reactants are not completely converted to products. (2) In some cases, a particular set of reactants undergoes two or more reactions simultaneously, forming undesired products as well as desired products. Reactions other than the desired one are called "side reactions." (3) In some cases, separation of the desired product from the reaction mixture is so difficult that not all of the product formed is successfully isolated.

In the examples we have worked to this point, the amounts of products that we calculated were theoretical yields.

The term **percent yield** is used to indicate how much of a desired product is obtained from a reaction.

$$\text{percent yield} = \frac{\text{actual yield of product}}{\text{theoretical yield of product}} \times 100\%$$

Consider the preparation of nitrobenzene, $C_5H_6NO_2$, by the reaction of a limited amount of benzene, C_6H_6, with excess nitric acid, HNO_3. The balanced equation for the reaction may be written

C_6H_6 +	HNO_3 $\longrightarrow$	$C_6H_5NO_2$ +	H_2O
1 mol	1 mol	1 mol	1 mol
78.1 g	63.0 g	123.1 g	18.0 g

Example 3-10

A 15.6-gram sample of C_6H_6 is mixed with excess HNO_3. We isolate 18.0 grams of $C_6H_5NO_2$. What is the percent yield of $C_6H_5NO_2$ in this reaction?

Plan

First we interpret the balanced chemical equation to calculate the theoretical yield of $C_6H_5NO_2$. Then we use the actual (isolated) yield with the definition given above to calculate the percent yield.

Solution

We calculate the theoretical yield of $C_6H_5NO_2$.

It is not necessary to know the mass of one mole of HNO_3 to solve this problem.

$$\underline{?}\text{ g } C_6H_5NO_2 = 15.6\text{ g } C_6H_6 \times \frac{1\text{ mol } C_6H_6}{78.1\text{ g } C_6H_6} \times \frac{1\text{ mol } C_6H_5NO_2}{1\text{ mol } C_6H_6} \times \frac{123.1\text{ g } C_6H_5NO_2}{1\text{ mol } C_6H_5NO_2}$$

$$= 24.6\text{ g } C_6H_5NO_2 \leftarrow \text{theoretical yield}$$

This tells us that if *all* the C_6H_6 were converted to $C_6H_5NO_2$ and isolated, we should obtain 24.6 grams of $C_6H_5NO_2$ (100% yield). However, we isolate only 18.0 grams of $C_6H_5NO_2$.

$$\text{percent yield} = \frac{\text{actual yield of product}}{\text{theoretical yield of product}} \times 100\% = \frac{18.0\text{ g}}{24.6\text{ g}} \times 100\%$$

$$= 73.2\text{ percent yield}$$

EOC 44

The amount of nitrobenzene obtained *in this experiment* is 73.2% of the amount that would be expected *if* the reaction had gone to completion, *if* there were no side reactions, and *if* we could have recovered all of the product.

In many important chemical processes, especially those encountered in the chemical industry, several equations are required to describe the chemical change. An analysis of the products often lets us describe the fraction of the change that occurs by each reaction.

3-5 Sequential Reactions

Many chemical reactions occur in a series of steps. They are called **sequential reactions**. Often more than one step (reaction) is required to change starting materials into the desired product. This is true for many reactions that we carry out in the laboratory and for many industrial processes. The amount of desired product from the first reaction is taken as the starting material for the second reaction.

Example 3-11

At high temperatures carbon reacts with water to produce a mixture of carbon monoxide, CO, and hydrogen, H_2.

$$C + H_2O \xrightarrow{\text{red heat}} CO + H_2$$

Carbon monoxide is separated from H_2 and then used to separate nickel from cobalt by forming a volatile compound, nickel tetracarbonyl, $Ni(CO)_4$.

$$Ni + 4CO \longrightarrow Ni(CO)_4$$

What mass of $Ni(CO)_4$ could be obtained from the CO produced by the reaction of 75.0 grams of carbon? Assume 100% reaction and 100% recovery in both steps.

Plan

We interpret both chemical equations in the usual way, and solve the problem in two steps. They tell us that one mole of C produces one mole of CO and that four moles of CO are required to produce one mole of $Ni(CO)_4$.

1. We determine the amount of CO formed in the first reaction. It is most conveniently expressed in moles of CO.
2. We determine the number of grams of $Ni(CO)_4$ that would be formed, in the second reaction, from the number of moles of CO produced in Step 1.

Solution

1.

$$\underset{\substack{1\text{ mol}\\12.0\text{ g}}}{C} + \underset{1\text{ mol}}{H_2O} \longrightarrow \underset{1\text{ mol}}{CO} + \underset{1\text{ mol}}{H_2}$$

$$\underline{?}\text{ mol CO} = 75.0\text{ g} \times \frac{1\text{ mol C}}{12.0\text{ g C}} \times \frac{1\text{ mol CO}}{1\text{ mol C}} = 6.25\text{ mol CO}$$

2.

$$\underset{1\text{ mol}}{Ni} + \underset{4\text{ mol}}{4CO} \longrightarrow \underset{\substack{1\text{ mol}\\96.0\text{ g}}}{Ni(CO)_4}$$

$$\underline{?}\text{ g Ni(CO)}_4 = 6.25\text{ mol CO} \times \frac{1\text{ mol Ni(CO)}_4}{4\text{ mol CO}} \times \frac{96.0\text{ g Ni(CO)}_4}{1\text{ mol Ni(CO)}_4}$$

$$= 150\text{ g Ni(CO)}_4$$

Alternatively, we can set up a series of unit factors based on the conversions in the reaction sequence and solve the problem in one step.

$$\text{g C} \longrightarrow \text{mol C} \longrightarrow \text{mol CO} \longrightarrow \text{mol Ni(CO)}_4 \longrightarrow \text{g Ni(CO)}_4$$

$$\underline{?}\text{ g Ni(CO)}_4 = 75.0\text{ g C} \times \frac{1\text{ mol C}}{12.0\text{ g C}} \times \frac{1\text{ mol CO}}{1\text{ mol C}} \times \frac{1\text{ mol Ni(CO)}_4}{4\text{ mol CO}} \times \frac{96.0\text{ g Ni(CO)}_4}{1\text{ mol Ni(CO)}_4}$$

$$= 150\text{ g Ni(CO)}_4$$

EOC 50, 52

Example 3-12

The Grignard reaction is a two-step reaction that is used to prepare pure hydrocarbons. Consider the preparation of pure ethane, C_2H_6, from ethyl chloride, CH_3CH_2Cl.

$$\text{Step 1:}\quad CH_3CH_2Cl + Mg \longrightarrow CH_3CH_2MgCl$$

$$\text{Step 2:}\quad CH_3CH_2MgCl + HOH \longrightarrow CH_3CH_3 + Mg(OH)Cl$$

We allow 27.2 grams of ethyl chloride to react with excess magnesium. From the first step reaction, CH_3CH_2MgCl is obtained in 79.5% yield. In the second step reaction, a 78.8% yield of CH_3CH_3 is obtained. What mass of CH_3CH_3 is obtained?

Plan

1. We interpret the first step equation as usual and calculate the amount of CH_3CH_2MgCl *obtained.*

$$\underset{\substack{1\text{ mol}\\64.4\text{ g}}}{CH_3CH_2Cl} + \underset{\substack{1\text{ mol}\\24.3\text{ g}}}{Mg} \longrightarrow \underset{\substack{1\text{ mol}\\88.7\text{ g}}}{CH_3CH_2MgCl}$$

$$\text{g } CH_3CH_2Cl \rightarrow \text{mol } CH_3CH_2Cl \rightarrow \text{mol } CH_3CH_2MgCl \rightarrow \text{g } CH_3CH_2MgCl$$

2. Then we interpret the second step equation and calculate the amount of CH_3CH_3 obtained.

$$CH_3CH_2MgCl + HOH \longrightarrow CH_3CH_3 + Mg(OH)Cl$$

CH_3CH_2MgCl	HOH	CH_3CH_3	$Mg(OH)Cl$
1 mol	1 mol	1mol	1 mol
88.7 g		30.0 g	

g CH_3CH_2MgCl → mol CH_3CH_2MgCl → mol CH_3CH_3 → g CH_3CH_3

Solution

The first step reaction gives a 79.5% yield. This gives the unit factor

$$\frac{79.5 \text{ g } CH_3CH_2MgCl \text{ actual}}{100 \text{ g } CH_3CH_2MgCl \text{ theor.}}$$

1. $$\underline{?} \text{ g } CH_3CH_2MgCl = 27.2 \text{ g } CH_3CH_2Cl \times \frac{1 \text{ mol } CH_3CH_2Cl}{64.4 \text{ g } CH_3CH_2Cl} \times \frac{1 \text{ mol } CH_3CH_2MgCl \text{ theor.}}{1 \text{ mol } CH_3CH_2Cl} \times \frac{88.7 \text{ g } CH_3CH_2MgCl \text{ theor.}}{1 \text{ mol } CH_3CH_2MgCl \text{ theor.}} \times \frac{79.5 \text{ g } CH_3CH_2MgCl \text{ actual}}{100 \text{ g } CH_3CH_2MgCl \text{ theor.}} = 29.8 \text{ g } CH_3CH_2MgCl$$

2. $$\underline{?} \text{ g } CH_3CH_3 = 29.8 \text{ g } CH_3CH_2MgCl \times \frac{1 \text{ mol } CH_3CH_2MgCl}{88.7 \text{ g } CH_3CH_2Cl} \times \frac{1 \text{ mol } CH_3CH_3 \text{ theor.}}{1 \text{ mol } CH_3CH_2MgCl} \times \frac{30.0 \text{ g } CH_3CH_3 \text{ theor.}}{1 \text{ mol } CH_3CH_3 \text{ theor.}} \times \frac{78.8 \text{ g } CH_3CH_3 \text{ actual}}{100 \text{ g } CH_3CH_3 \text{ theor.}} = 7.94 \text{ g } CH_3CH_3$$

Alternatively, we could set the calculation up in a single step.

2. $$\underline{?} \text{ g } CH_3CH_3 = 27.2 \text{ g } CH_3CH_2Cl \times \frac{1 \text{ mol } CH_3CH_2Cl}{64.4 \text{ g } CH_3CH_2Cl} \times \frac{1 \text{ mol } CH_3CH_2MgCl}{1 \text{ mol } CH_3CH_2Cl} \times \frac{79.5 \text{ g } CH_3CH_2MgCl \text{ actual}}{100 \text{ g } CH_3CH_2MgCl \text{ theor.}} \times \frac{1 \text{ mol } CH_3CH_3}{1 \text{ mol } CH_3CH_2MgCl} \times \frac{30.0 \text{ g } CH_3CH_3 \text{ theor.}}{1 \text{ mol } CH_3CH_3 \text{ theor.}} \times \frac{78.8 \text{ g } CH_3CH_3 \text{ actual}}{100 \text{ g } CH_3CH_3 \text{ theor.}}$$

$$= 7.94 \text{ g } CH_3CH_3$$

EOC 54, 55

The sodium hydroxide and aluminum in some drain cleaners do not react while they are stored in solid form. When water is added, the NaOH dissolves and begins to act on trapped grease. At the same time, NaOH and Al react to produce H_2 gas; the resulting turbulence helps to dislodge the blockage. Do you see why the container should be kept tightly closed?

Chemists have determined the structures of many naturally occurring compounds. One way of proving the structure involves the synthesis of the natural product (compound) from available starting materials. Professor Grieco, now at Indiana University, was assisted by Majetich and Ohfune in the synthesis of helenalin, a powerful cancer drug, in a forty-step process. This forty-step synthesis gave a remarkable average yield of about 90% for each step, which resulted in an overall yield of about 1.5%.

3-6 Concentrations of Solutions

Many chemical reactions are more conveniently carried out with the reactants in solution rather than as pure solids, liquids, or gases. A **solution** is a homogeneous mixture, at the molecular level, of two or more substances. Simple solutions usually consist of one substance, the **solute**, dissolved in another substance, the **solvent**. The solutions used in the laboratory are usually liquids, and the solvent is often water. These are called **aqueous solutions**. For example, solutions of hydrochloric acid are prepared by dissolving hydrogen chloride (HCl, a gas at room temperature and atmospheric pressure) in water. Solutions of sodium hydroxide are prepared by dissolving solid NaOH in water.

In some solutions, such as a nearly equal mixture of ethyl alcohol and water, the distinction between *solute* and *solvent* is arbitrary.

We often use solutions to supply the reactants for chemical reactions. Solutions allow the most intimate mixing of the reacting substances at the molecular level, much more than would be possible in solid form. (A practical example is drain cleaner, shown in the photo.) Furthermore, the rate of the reaction can often be controlled by adjusting the concentrations of the solutions. In this section we shall study methods for expressing the quantities of the various components present in a given amount of solution.

Concentrations of solutions are expressed in terms of *either* the amount of solute present in a given mass or volume of *solution*, or the amount of solute dissolved in a given mass or volume of *solvent*.

Percent by Mass

Concentrations of solutions may be expressed in terms of percent by mass of solute, which gives the mass of solute per 100 mass units of solution. The gram is the usual mass unit.

$$\% \text{ solute} = \frac{\text{mass of solute}}{\text{mass of solution}} \times 100\%$$

Thus, a solution that is 10.0% calcium gluconate, $Ca(C_6H_{11}O_7)_2$, by mass contains 10.0 grams of calcium gluconate in 100.0 grams of *solution*. This could be described as 10.0 grams of calcium gluconate in 90.0 grams of water. The density of a 10.0% solution of calcium gluconate is 1.07 g/mL, so 100 mL of a 10.0% solution of calcium gluconate has a mass of 107 grams. Observe that 100 grams of a solution usually does *not* occupy 100 mL. Unless otherwise specified, percent means percent *by mass*, and water is the solvent.

A 10.0% solution of $Ca(C_6H_{11}O_7)_2$ is sometimes administered intravenously in emergency treatment for black widow spider bites.

Example 3-13

Calculate the mass of nickel(II) sulfate, $NiSO_4$, contained in 200 grams of a 6.00% solution of $NiSO_4$.

Plan

The percentage information tells us that the solution contains 6.00 grams of $NiSO_4$ per 100 grams of solution. The desired information is the mass of $NiSO_4$ in 200 grams of solution. A unit factor is constructed by placing 6.00 grams of $NiSO_4$ over 100 grams of solution. Multiplication of the mass of the solution, 200 grams, by the unit factor gives the mass of $NiSO_4$ in the solution.

Solution

$$\underline{?} \text{ g } NiSO_4 = 200 \text{ g soln} \times \frac{6.00 \text{ g } NiSO_4}{100 \text{ g soln}} = 12.0 \text{ g } NiSO_4$$

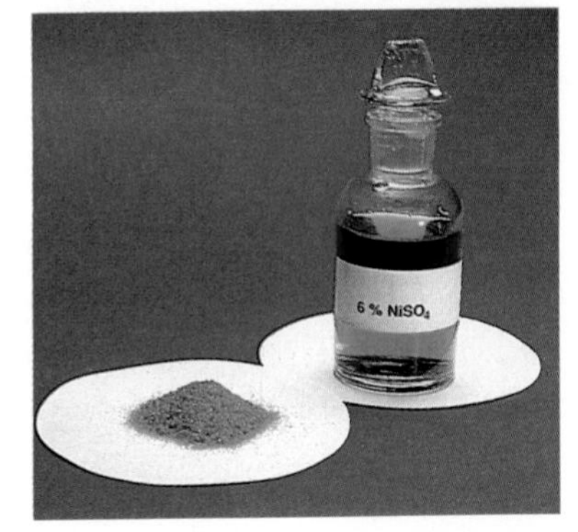

Example 3-14

Calculate the mass of 6.00% $NiSO_4$ solution that contains 40.0 grams of $NiSO_4$.

Plan

Placing 100 grams of solution over 6.00 grams of $NiSO_4$ gives another unit factor.

Solution

$$\underline{?}\text{ g soln} = 40.0\text{ g NiSO}_4 \times \frac{100\text{ g soln}}{6.00\text{ g NiSO}_4} = 667\text{ g soln}$$

EOC 57, 58

Example 3-15

Calculate the mass of $NiSO_4$ contained in 200 mL of a 6.00% solution of $NiSO_4$. The density of the solution is 1.06 g/mL at 25°C.

Plan

The volume of a solution multiplied by its density gives the mass of solution (Section 1-11). The mass of solution is then multiplied by the fraction of that mass due to $NiSO_4$ (6.00 g $NiSO_4$/100 g soln) to give the mass of $NiSO_4$ in 200 mL of solution.

Solution

Volume of solution × density of solution = mass of solution

$$\underline{?}\text{ g NiSO}_4 = \underbrace{200\text{ mL soln} \times \frac{1.06\text{ g soln}}{1.00\text{ mL soln}}}_{212\text{ g soln}} \times \frac{6.00\text{ g NiSO}_4}{100\text{ g soln}} = 12.7\text{ g NiSO}_4$$

EOC 59, 60

Example 3-16

What volume of a solution that is 15.0% iron(III) nitrate contains 30.0 grams of $Fe(NO_3)_3$? The density of the solution is 1.16 g/mL at 25°C.

Plan

Two unit factors relate mass of $Fe(NO_3)_3$ and mass of solution, 15.0 g $Fe(NO_3)_3$/100 g soln and 100 g soln/15.0 g $Fe(NO_3)_3$. The second factor converts grams of $Fe(NO_3)_3$ to grams of solution.

Solution

$$\underline{?}\text{ mL soln} = \underbrace{30.0\text{ g Fe(NO}_3)_3 \times \frac{100\text{ g soln}}{15.0\text{ g Fe(NO}_3)_3}}_{200\text{ g soln}} \times \frac{1.00\text{ mL soln}}{1.16\text{ g soln}} = 172\text{ mL}$$

Note that the answer is not 200 mL but considerably less because 1.00 mL of solution has a mass of 1.16 grams. However, 172 mL of the solution has a mass of 200 grams.

EOC 62, 63

Molarity (molar concentration)

Molarity (M), or molar concentration, is a common unit for expressing the concentrations of solutions. **Molarity** is defined as the number of moles of solute per liter of solution:

The definition of molarity specifies the amount of solute *per unit volume of solution,* whereas percent specifies the amount of solute *per unit mass of solution.* Therefore, molarity depends on temperature and pressure, whereas percent by mass does not.

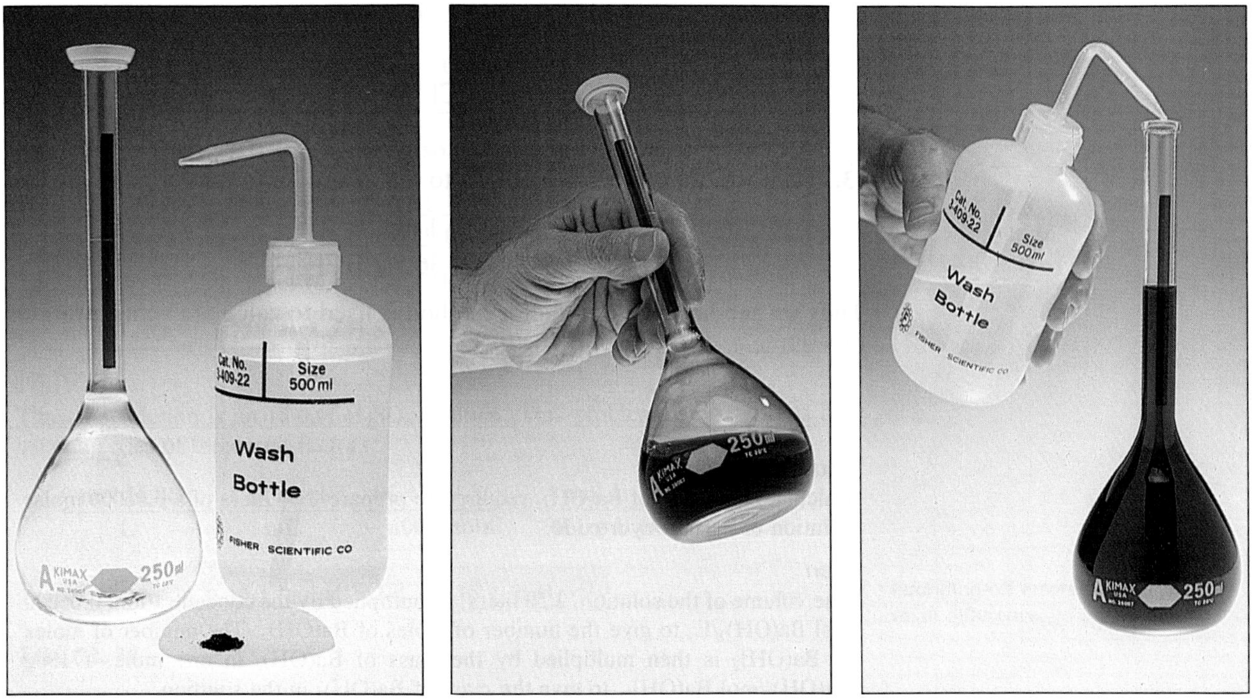

$$\text{molarity} = \frac{\text{number of moles of solute}}{\text{number of liters of solution}}$$

To prepare one liter of a one molar solution, one mole of solute is placed in a one-liter volumetric flask, enough solvent is added to dissolve the solute, and solvent is then added until the volume of the solution is exactly one liter. Students sometimes make the mistake of assuming that a one molar solution contains one mole of solute in a liter of solvent. This is *not* the case; one liter of solvent *plus* one mole of solute usually has a total volume of more than one liter. A 0.100 *M* solution contains 0.100 mole of solute per liter, and a 0.0100 *M* solution contains 0.0100 mole of solute per liter (Figure 3-2).

We often express the volume of a solution in milliliters rather than in liters. Likewise, we may express the amount of solute in millimoles (mmol) rather than in moles. Because one milliliter is 1/1000 of a liter and one millimole is 1/1000 of a mole, molarity also may be expressed as the number of millimoles of solute per milliliter of solution:

$$\text{molarity} = \frac{\text{number of millimoles of solute}}{\text{number of milliliters of solution}}$$

Water is the solvent in *most* of the solutions that we encounter. Unless otherwise indicated, we assume that water is the solvent. When the solvent is other than water, we state this explicitly.

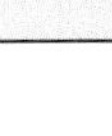

(a) (b) (c)

Figure 3-2
Preparation of 0.0100 *M* solution of $KMnO_4$, potassium permanganate. 250 mL of 0.0100 *M* $KMnO_4$ solution contains 0.395 g of $KMnO_4$ (1 mol = 158 g). (a) 0.395 g of $KMnO_4$ (0.00250 mole) is weighed out carefully and transferred into a 250-mL volumetric flask. (b) The $KMnO_4$ is dissolved in water. (c) Distilled H_2O is added to the volumetric flask until the volume of solution is 250 mL. The flask is then stoppered, and its contents are mixed thoroughly to give a homogeneous solution.

(a) (b) (c)

Figure 3-3
Dilution of solution. (a) A 100-mL volumetric flask is filled to the calibration line with 0.100 *M* potassium dichromate, $K_2Cr_2O_7$, solution. (b) The 0.100 *M* $K_2Cr_2O_7$ solution is transferred into a 1.00-L volumetric flask. The small flask is rinsed with distilled H_2O several times, and the rinse solutions are added to the larger flask. (c) Distilled water is added until the 1.00-L flask contains 1.00 L of solution. The flask is stoppered and its contents are mixed thoroughly. The new solution is 0.0100 *M* $K_2Cr_2O_7$. (100 mL of 0.100 *M* $K_2Cr_2O_7$ solution has been diluted to 1000 mL.)

3-7 Dilution of Solutions

Recall that the definition of molarity is the number of moles of solute divided by the volume of the solution in liters:

$$\text{molarity} = \frac{\text{number of moles of solute}}{\text{number of liters of solution}}$$

Multiplying both sides of the equation by the volume, we obtain

$$\text{volume (in L)} \times \text{molarity} = \text{number of moles of solute}$$

or

$$\text{volume (in mL)} \times \text{molarity} = \text{number of mmol of solute}$$

Multiplication of the volume of a solution by its molar concentration gives the amount of solute in the solution.

When we dilute a solution by mixing it with more solvent, the number of moles of solute present does not change. But the volume and the concentration of the solution *do* change. Because the same number of moles of solute is divided by a larger number of liters of solution, the molarity decreases. Using a subscript 1 to represent the original concentrated solution and a subscript 2 to represent the dilute solution, we obtain

$$\text{volume}_1 \times \text{molarity}_1 = \text{number of moles of solute} = \text{volume}_2 \times \text{molarity}_2$$

or

This relationship also applies when the concentration is changed by evaporating some solvent.

$$V_1 \times M_1 = V_2 \times M_2 \quad \text{(for dilution only)}$$

This expression can be used to calculate any one of four quantities when the other three are known (Figure 3-3). Suppose a certain volume of dilute solution of a given molarity is required for use in the laboratory, and we know the concentration of the stock solution available. Then we can calculate the amount of stock solution that must be used to make the dilute solution.

Caution

Dilution of a concentrated solution, especially of a strong acid or base, frequently liberates a great deal of heat. This can vaporize drops of water as they hit the concentrated solution and can cause dangerous spattering. As a safety precaution, *concentrated solutions of acids or bases are always poured slowly into water*, allowing the heat to be absorbed by the larger quantity of water. Calculations are usually simpler to visualize by assuming that water is added to the concentrated solution.

Example 3-20

Calculate the volume of 18.0 *M* H_2SO_4 required to prepare 1.00 liter of a 0.900 *M* solution of H_2SO_4.

Plan

The volume (1.00 L) and molarity (0.900 *M*) of the final solution, as well as the molarity (18.0 *M*) of the original solution, are given. Therefore, the relation $V_1 \times M_1 = V_2 \times M_2$ can be used, with subscript 1 for the commercial acid solution and subscript 2 for the dilute solution. We solve

$$V_1 \times M_1 = V_2 \times M_2 \qquad \text{for } V_1$$

Solution

$$V_1 = \frac{V_2 \times M_2}{M_1} = \frac{1.00 \text{ L} \times 0.900\ M}{18.0\ M} = 0.0500 \text{ L} = 50.0 \text{ mL}$$

The dilute solution contains 1.00 L × 0.900 *M* = 0.900 mol of H_2SO_4, so 0.900 mole of H_2SO_4 must be present in the original concentrated solution. Indeed, 0.0500 L × 18.0 *M* = 0.900 mol of H_2SO_4.

EOC 79, 80

3-8 Using Solutions in Chemical Reactions

If we plan to carry out a reaction in a solution, we must calculate the amounts of solutions required. If we know the molarity of a solution, we can calculate the amount of solute contained in a specified volume of that solution. This is illustrated in Example 3-21.

Example 3-21

Calculate (a) the number of moles of H_2SO_4, (b) the number of millimoles of H_2SO_4, and (c) the mass of H_2SO_4 in 500 mL of 0.324 *M* H_2SO_4 solution.

Plan

Because we have two parallel calculations in this example, we shall state the plan for each step just before the calculation is done.

Solution

(a) The volume of a solution in liters multiplied by its molarity gives the number of moles of solute, H_2SO_4 in this case.

500 mL is more conveniently expressed as 0.500 L in this problem. By now, you should be able to convert mL to L (and the reverse) without writing out the conversion.

$$\underline{?}\text{ mol } H_2SO_4 = 0.500 \text{ L soln} \times \frac{0.324 \text{ mol } H_2SO_4}{\text{L soln}} = \boxed{0.162 \text{ mol } H_2SO_4}$$

(b) The volume of a solution in milliliters multiplied by its molarity gives the number of millimoles of solute, H_2SO_4:

$$\underline{?}\text{ mmol } H_2SO_4 = 500 \text{ mL soln} \times \frac{0.324 \text{ mmol } H_2SO_4}{\text{mL soln}} = \boxed{162 \text{ mmol } H_2SO_4}$$

(c) We may use the results of *either* (a) or (b) to calculate the mass of H_2SO_4 in the solution:

A mole of H_2SO_4 is 98.1 g. A millimole is 98.1 mg.

$$\underline{?}\text{ g } H_2SO_4 = 0.162 \text{ mol } H_2SO_4 \times \frac{98.1 \text{ g } H_2SO_4}{1 \text{ mol } H_2SO_4} = \boxed{15.9 \text{ g } H_2SO_4}$$

or

$$\underline{?}\text{ mg } H_2SO_4 = 162 \text{ mmol } H_2SO_4 \times \frac{98.1 \text{ mg } H_2SO_4}{1 \text{ mmol } H_2SO_4} = \boxed{1.59 \times 10^4 \text{ mg } H_2SO_4}$$

The mass of H_2SO_4 in the solution can be calculated without solving explicitly for the number of moles (or millimoles) of H_2SO_4:

Volume in liters times molarity gives moles of H_2SO_4. Molarity is a unit factor; i.e.,

$$\frac{\text{mol solute}}{\text{L soln}}$$

$$\underline{?}\text{ g } H_2SO_4 = 0.500 \text{ L soln} \times \frac{0.324 \text{ mol } H_2SO_4}{\text{L soln}} \times \frac{98.1 \text{ g } H_2SO_4}{1 \text{ mol } H_2SO_4} = \boxed{15.9 \text{ g } H_2SO_4}$$

Example 3-22 demonstrates how we can relate the volume of a solution of known concentration of one reactant to the mass of the other reactant.

Example 3-22

Calculate the volume of a 0.324 *M* solution of sulfuric acid required to react completely with 2.792 grams of Na_2CO_3 according to the equation

$$H_2SO_4 + Na_2CO_3 \longrightarrow Na_2SO_4 + CO_2 + H_2O$$

Plan

The balanced equation tells us that one mole of H_2SO_4 reacts with one mole of Na_2CO_3, and we can write

$$\underset{1\text{ mol}}{H_2SO_4} + \underset{\substack{1\text{ mol}\\106.0\text{ g}}}{Na_2CO_3} \longrightarrow \underset{1\text{ mol}}{Na_2SO_4} + \underset{1\text{ mol}}{CO_2} + \underset{1\text{ mol}}{H_2O}$$

We convert (1) grams of Na_2CO_3 to moles of Na_2CO_3, (2) moles of Na_2CO_3 to moles of H_2SO_4, and (3) moles of H_2SO_4 to liters of H_2SO_4 solution.

$$\text{g } Na_2CO_3 \longrightarrow \text{mol } Na_2CO_3 \longrightarrow \text{mol } H_2SO_4 \longrightarrow \text{L } H_2SO_4 \text{ soln}$$

Solution

$$\underline{?}\text{ L } H_2SO_4 = 2.792 \text{ g } Na_2CO_3 \times \frac{1 \text{ mol } Na_2CO_3}{106.0 \text{ g } Na_2CO_3} \times \frac{1 \text{ mol } H_2SO_4}{1 \text{ mol } Na_2CO_3} \times \frac{1 \text{ L } H_2SO_4 \text{ soln}}{0.324 \text{ mol } H_2SO_4}$$

$$= \boxed{0.0813 \text{ L } H_2SO_4 \text{ soln}} \quad \text{or} \quad \boxed{81.3 \text{ mL } H_2SO_4 \text{ soln}}$$

EOC 85

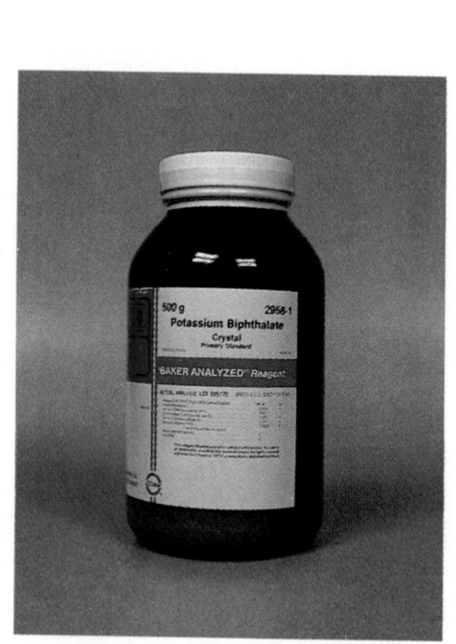

The indicator methyl orange changes from yellow, its color in basic solutions, to orange, its color in acidic solutions, when the reaction in Example 3-22 reaches completion.

Often we must calculate the volume of solution of known molarity that is required to react with a specified volume of another solution. We always examine the balanced chemical equation for the reaction to determine the **reaction ratio**, i.e., the relative numbers of moles (or millimoles) of reactants.

Example 3-23

Find the volume of 0.505 *M* NaOH solution required to react with 40.0 mL of 0.505 *M* H_2SO_4 solution according to the reaction

$$H_2SO_4 + 2NaOH \longrightarrow Na_2SO_4 + 2H_2O$$

Plan

We shall work this example in several steps, stating the "plan," or reasoning, just before each step in the calculation. Then we shall use a single setup to solve the problem.

Solution

The balanced equation tells us that the reaction ratio is one mole of H_2SO_4 to two moles of NaOH:

$$\underset{1\text{ mol}}{H_2SO_4} + \underset{2\text{ mol}}{2NaOH} \longrightarrow \underset{1\text{ mol}}{Na_2SO_4} + \underset{2\text{ mol}}{2H_2O}$$

From the volume and the molarity of the H_2SO_4 solution, we can calculate the number of moles of H_2SO_4:

$$\underline{?}\text{ mol } H_2SO_4 = 0.0400\text{ L } H_2SO_4\text{ soln} \times \frac{0.505\text{ mol } H_2SO_4}{\text{L soln}} = 0.0202\text{ mol } H_2SO_4$$

The volume of H_2SO_4 solution is expressed as 0.0400 L rather than 40.0 mL.

The number of moles of H_2SO_4 is related to the number of moles of NaOH by the reaction ratio, 1 mol H_2SO_4/2 mol NaOH:

$$\underline{?} \text{ mol NaOH} = 0.0202 \text{ mol } H_2SO_4 \times \frac{2 \text{ mol NaOH}}{1 \text{ mol } H_2SO_4} = 0.0404 \text{ mol NaOH}$$

Now we can calculate the volume of 0.505 *M* NaOH solution that contains 0.0404 mole of NaOH:

Again we see that molarity is a unit factor. In this case,

$$\frac{1.00 \text{ L NaOH soln}}{0.505 \text{ mol NaOH}}$$

$$\underline{?} \text{ L NaOH soln} = 0.0404 \text{ mol NaOH} \times \frac{1.00 \text{ L NaOH soln}}{0.505 \text{ mol NaOH}}$$

$$= 0.0800 \text{ L NaOH soln}$$

which we usually call 80.0 mL of NaOH solution.

We have worked through the problem stepwise; let us solve it in a single setup.

L H_2SO_4 soln available ⟶ mol H_2SO_4 available ⟶ mol NaOH soln needed ⟶ L NaOH soln needed

$$\underline{?} \text{ L NaOH soln} = 0.0400 \text{ L } H_2SO_4 \text{ soln} \times \frac{0.505 \text{ mol } H_2SO_4}{\text{L } H_2SO_4 \text{ soln}} \times \frac{2 \text{ mol NaOH}}{1 \text{ mol } H_2SO_4}$$

$$\times \frac{1.00 \text{ L NaOH soln}}{0.505 \text{ mol NaOH}}$$

$$= 0.0800 \text{ L NaOH soln or } 80.0 \text{ mL NaOH soln}$$

We could have retained volumes in mL and expressed molarities as mmol/mL. Try working this example in those terms.

EOC 86

3-9 Titrations

In Example 3-23, we calculated the volume of one solution that is required to react with a given volume of another solution, with the concentrations of *both* solutions given. In the laboratory we often measure the volume of one solution that is required to react with a given volume of another solution of known concentration. Then we calculate the concentration of the first solution. The process is called **titration** (Figure 3-4).

Titration is the process in which a solution of one reactant, the titrant, is carefully added to a solution of another reactant, and the volume of titrant required for complete reaction is measured.

Solutions of accurately known concentrations are called **standard solutions**. We can prepare standard solutions of some substances by dissolving a known weight of solid in enough water to give an accurately known volume of solution. Other substances cannot be weighed out accurately and conveniently because they react with the atmosphere. Instead, we prepare solutions of such substances and then determine their concentrations by titration with a standard solution.

CO_2, H_2O, and O_2 are present in the atmosphere. They react with many substances.

(a)

(b)

(c)

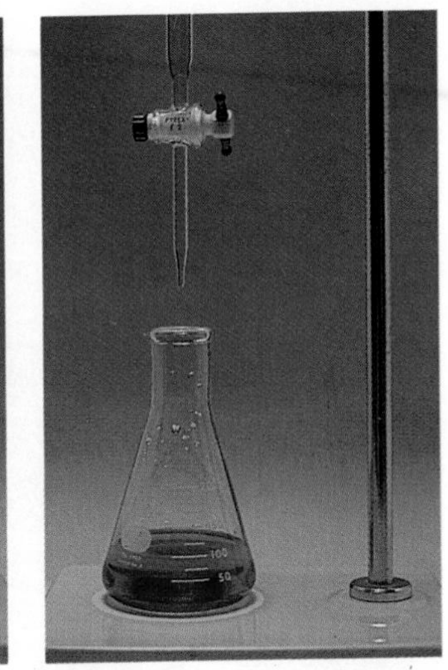

(d)

Figure 3-4
The titration process. (a) A typical setup for titration in a teaching laboratory. The solution to be titrated is placed in an Erlenmeyer flask, and a few drops of indicator are added. The buret is filled with a standard solution (or the solution to be standardized). The volume of solution in the buret is read carefully. (b) The meniscus describes the surface of the liquid in the buret. Aqueous solutions wet glass, so the meniscus of an aqueous solution is always concave. The position of the *bottom* of the meniscus is read and recorded. (c) The solution in the buret is added (dropwise near the end point), with stirring, to the Erlenmeyer flask until the end point is reached. (d) The end point is signaled by the appearance (or change) of color *throughout* the solution being titrated. (A very large excess of indicator was used to make this photograph.) The volume of the liquid is read again—the difference between the final and initial buret readings is the volume of the solution used.

How does one know when to stop a titration—that is, when is the chemical reaction just complete? A few drops of an *indicator* solution are added to the solution to be titrated. An **indicator** is a substance that can exist in different forms, with different colors that depend upon the concentration of H^+ in the solution. At least one of these forms must be very intensely colored so that even very small amounts of it can be seen.

Suppose we titrate an acid solution of unknown concentration by adding a standardized solution of sodium hydroxide dropwise from a **buret** (Figure 3-4). A common buret is graduated in large intervals of 1 mL and in smaller intervals of 0.1 mL so that it is possible to estimate the volume of a solution dispensed to within at least ± 0.02 mL. (Experienced individuals can often read a buret to ± 0.01 mL.) The analyst tries to choose an indicator that changes color clearly at the point at which stoichiometrically equivalent amounts of acid and base have reacted, the **equivalence point**. The point at which the indicator changes color and the titration is stopped is called the **end point**. Ideally, the end point should coincide with the equivalence point. Phenolphthalein is colorless in acidic solution and reddish violet in basic solution. In a titration in which a base is added to an acid, phenolphthalein is often used as the indicator. The end point is signaled by the first appearance of a faint pink coloration that persists for at least 15 seconds as the solution is swirled.

Example 3-24

What is the molarity of a hydrochloric acid solution if 36.7 mL of the HCl solution is required to react with 43.2 mL of 0.236 *M* sodium hydroxide solution?

$$HCl + NaOH \longrightarrow NaCl + H_2O$$

The indicator phenolphthalein changes from colorless, its color in acidic solutions, to pink, its color in basic solutions, when the reaction in Example 3-23 reaches completion. Note the first appearance of a faint pink coloration in the middle beaker; this signals that the end point is near.

Plan

The balanced equation tells us that the reaction ratio is one mole of HCl to one mole of NaOH, which gives the unit factor, 1 mol HCl/1 mol NaOH.

$$\underset{1\text{ mol}}{\text{HCl}} + \underset{1\text{ mol}}{\text{NaOH}} \longrightarrow \underset{1\text{ mol}}{\text{NaCl}} + \underset{1\text{ mol}}{\text{H}_2\text{O}}$$

First we find the number of moles of NaOH. The reaction ratio is one mole of HCl to one mole of NaOH, so the HCl solution must contain the same number of moles of HCl. Then we can calculate the molarity of the HCl solution because we know its volume.

Solution

The volume of a solution (in liters) multiplied by its molarity gives the number of moles of solute.

$$\underline{?}\text{ mol NaOH} = 0.0432\text{ L NaOH soln} \times \frac{0.236\text{ mol NaOH}}{1\text{ L NaOH soln}} = 0.0102\text{ mol NaOH}$$

Because the reaction ratio is one mole of NaOH to one mole of HCl, the HCl solution must contain 0.0102 mole of HCl.

$$\underline{?}\text{ mol HCl} = 0.0102\text{ mol NaOH} \times \frac{1\text{ mol HCl}}{1\text{ mol NaOH}} = 0.0102\text{ mol HCl}$$

We know the volume of the HCl solution, so we can calculate its molarity.

$$\frac{\underline{?}\text{ mol HCl}}{\text{L HCl soln}} = \frac{0.0102\text{ mol HCl}}{0.0367\text{ L HCl soln}} = 0.278\ M\text{ HCl}$$

EOC 95

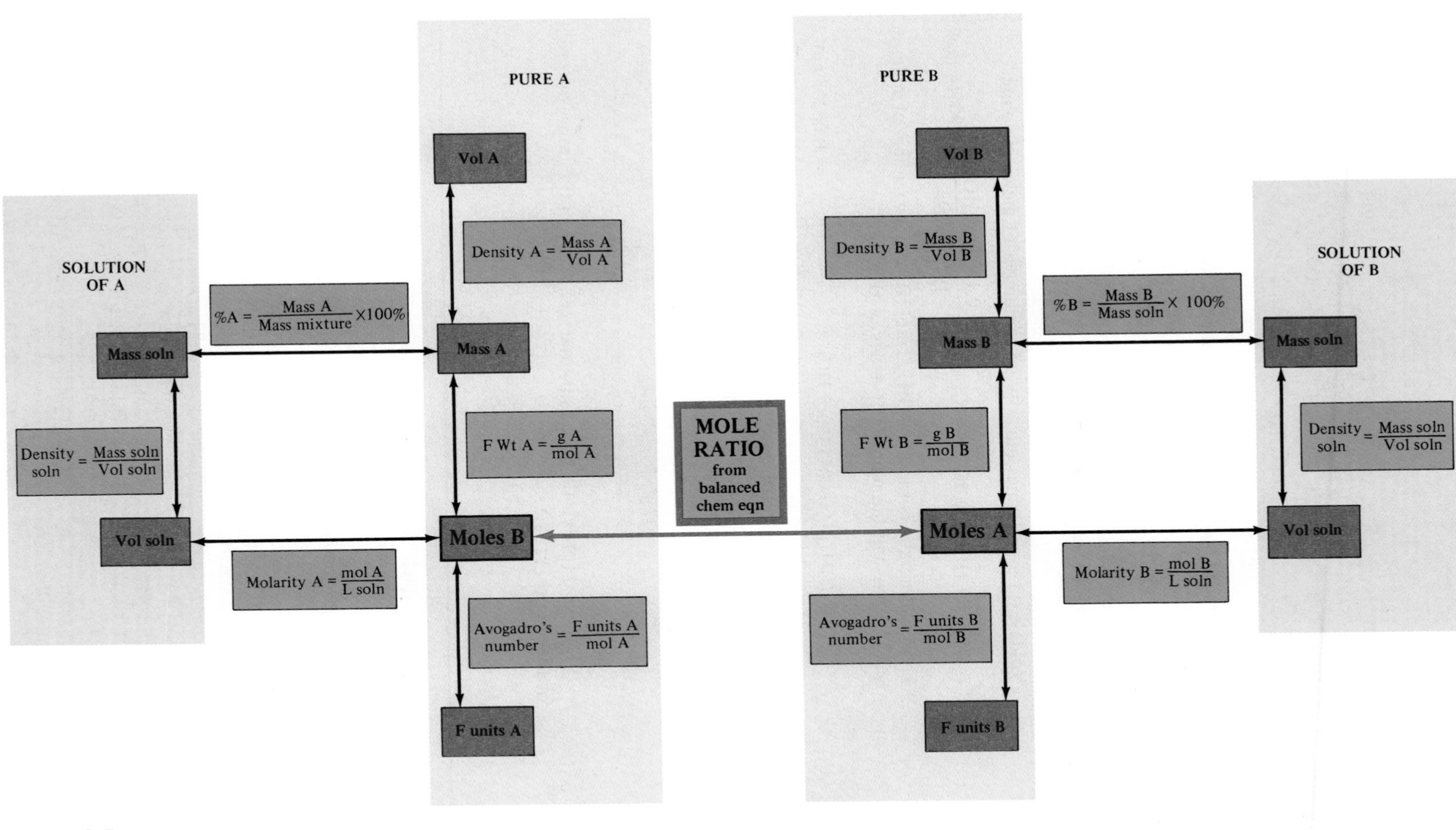

Figure 3-5
Some important relationships in reaction stoichiometry, Chapter 3. The left-hand portion of this diagram relates various ways of describing a single substance, A. It is similar to Figure 2-12, with the addition of the molarity calculation. Similarly, the right-hand portion applies to substance B. The mole concept (red boxes and arrows) relates the amounts of A and B involved in a chemical reaction.

Example 3-25

What is the molarity of a barium hydroxide solution if 40.9 mL of the $Ba(OH)_2$ solution is required to react with 38.3 mL of 0.108 *M* hydrochloric acid solution?

$$Ba(OH)_2 + 2HCl \longrightarrow BaCl_2 + 2H_2O$$

Plan

The balanced equation tells us that the reaction ratio is 1 mol $Ba(OH)_2$/2 mol HCl.

$$\underset{1\text{ mol}}{Ba(OH)_2} + \underset{2\text{ mol}}{2HCl} \longrightarrow \underset{1\text{ mol}}{BaCl_2} + \underset{2\text{ mol}}{2H_2O}$$

We use the same reasoning as in Example 3-24, except that the reaction ratio is 1 mol $Ba(OH)_2$/2 mol HCl.

Solution

Volume (in liters) times molarity gives number of moles of solute.

$$\underline{?}\text{ mol HCl} = 0.0383\text{ L HCl soln} \times \frac{0.108\text{ mol HCl}}{\text{L HCl soln}} = 0.00414\text{ mol HCl}$$

The reaction ratio is 1 mol $Ba(OH)_2$/2 mol HCl, so the number of moles of $Ba(OH)_2$ must be one-half the number of moles of HCl.

$$\underline{?}\text{ mol Ba(OH)}_2 = 0.00414\text{ mol HCl} \times \frac{1\text{ mol Ba(OH)}_2}{2\text{ mol HCl}} = 0.00207\text{ mol Ba(OH)}_2$$

We know the volume of the $Ba(OH)_2$ solution, and so we can calculate its molarity.

$$\frac{\underline{?}\text{ mol Ba(OH)}_2}{\text{L Ba(OH)}_2\text{ soln}} = \frac{0.00207\text{ mol Ba(OH)}_2}{0.0409\text{ L Ba(OH)}_2\text{ soln}} = 0.0506\ M\text{ Ba(OH)}_2$$

EOC 96

Figure 3-5 (previous page) summarizes some of the important relationships in this chapter.

Key Terms

Actual yield The amount of a specified pure product actually obtained from a given reaction. Compare with *Theoretical yield.*

Buret A piece of volumetric glassware, usually graduated in 0.1-mL intervals, that is used in titrations to deliver solutions in a quantitative (dropwise) manner.

Chemical equation Description of a chemical reaction by placing the formulas of reactants on the left and the formulas of products on the right of an arrow. A chemical equation must be balanced; i.e., it must have the same number of each kind of atom on both sides.

Concentration The amount of solute per unit volume or mass of solvent or of solution.

Dilution The process of reducing the concentration of a solute in solution, usually simply by mixing with more solvent.

End point The point at which an indicator changes color and a titration is stopped.

Equivalence point The point at which chemically equivalent amounts of reactants have reacted.

Indicators For acid–base titrations, organic compounds that exhibit different colors in solutions of different acidities; used to determine the point at which reaction between two solutes is complete.

Limiting reactant A substance that stoichiometrically limits the amount of product(s) that can be formed.

Molarity (*M*) The number of moles of solute per liter of solution.

Percent by mass 100% multiplied by the mass of a solute divided by the mass of the solution in which it is contained.

Percent yield 100% times actual yield divided by theoretical yield.

Products Substances produced in a chemical reaction.

Reactants Substances consumed in a chemical reaction.

Reaction ratio The relative amounts of reactants and products involved in a reaction; may be the ratio of moles, millimoles, or masses.

Reaction stoichiometry Description of the quantitative relationships among substances as they participate in chemical reactions.

Sequential reaction A chemical reaction in which several steps are required to convert starting materials into products.

Solute The dispersed (dissolved) phase of a solution.

Solution A homogeneous mixture of two or more substances.

Solvent The dispersing medium of a solution.

Standard solution A solution of accurately known concentration.

Stoichiometry Description of the quantitative relationships among elements and compounds as they undergo chemical changes.

Theoretical yield The maximum amount of a specified product that could be obtained from specified amounts of reactants, assuming complete consumption of the limiting reactant according to only one reaction and complete recovery of the product. Compare with *Actual yield*.

Titration The process by which the volume of a standard solution required to react with a specific amount of a substance is determined.

Exercises

1. What is a chemical equation? What information does it contain?
2. What fundamental law is the basis for balancing a chemical equation?
3. Use words to state explicitly the relationships among numbers of molecules of reactants and products in the equation for the combustion of heptane, C_7H_{16}.

$$C_7H_{16} + 11O_2 \longrightarrow 7CO_2 + 8H_2O$$

Balance each "equation" in Exercises 4–7.

4. (a) $Al + O_2 \rightarrow Al_2O_3$
   (b) $Mn + O_2 \rightarrow Mn_3O_4$
   (c) $Al + H_2SO_4 \rightarrow Al_2(SO_4)_3 + H_2$
   (d) $Ba + H_2O \rightarrow Ba(OH)_2 + H_2$
   (e) $Mg + H_3PO_4 \rightarrow Mg_3(PO_4)_2 + H_2$
5. (a) $P_4 + Cl_2 \rightarrow PCl_3$
   (b) $P_4 + Cl_2 \rightarrow PCl_5$
   (c) $Ca(CH_3COO)_2 + H_2SO_4 \rightarrow CaSO_4 + CH_3COOH$
   (d) $KOH + CO_2 \rightarrow K_2CO_3 + H_2O$
   (e) $TiCl_4 + H_2O \rightarrow TiO_2 + HCl$
6. (a) $KOH + H_2SO_4 \rightarrow K_2SO_4 + H_2O$
   (b) $Ca(OH)_2 + H_3PO_4 \rightarrow Ca_3(PO_4)_2 + H_2O$
   (c) $Ba(OH)_2 + P_4O_{10} \rightarrow Ba_3(PO_4)_2 + H_2O$
   (d) $Al_2O_3 + H_2SeO_4 \rightarrow Al_2(SeO_4)_3 + H_2O$
   (e) $Fe_2O_3 + HClO_4 \rightarrow Fe(ClO_4)_3 + H_2O$
7. (a) $BiCl_3 + H_2S \rightarrow Bi_2S_3 + HCl$
   (b) $CrCl_3 + NH_3 + H_2O \rightarrow Cr(OH)_3 + NH_4Cl$
   (c) $H_3AsO_3 + H_2S \rightarrow As_2S_3 + H_2O$
   (d) $C_2H_6 + O_2 \rightarrow CO_2 + H_2O$
   (e) $C_2H_5OH + O_2 \rightarrow CO_2 + H_2O$

Calculations Based on Chemical Equations

In Exercises 8–15, (a) write the balanced chemical equation that represents the reaction described by words, and then perform calculations to answer parts (b) and (c).

8. (a) Nitrogen, N_2, combines with hydrogen, H_2, to form ammonia, NH_3.
   (b) How many hydrogen molecules would be required to react with 100 nitrogen molecules?
   (c) How many ammonia molecules would be formed in part (b)?
9. (a) Sulfur, S_8, combines with oxygen at elevated temperatures to form sulfur dioxide.
   (b) If 96 oxygen molecules are used up in this reaction, how many sulfur molecules reacted?
   (c) How many sulfur dioxide molecules were formed in part (b)?
10. (a) Pentane, C_5H_{12}, burns in excess oxygen to form carbon dioxide and water.
    (b) How many oxygen molecules would be required to react with 33 pentane molecules?
    (c) How many water molecules would be formed in part (b)?
11. (a) Ethylamine, $C_2H_5NH_2$, burns in excess oxygen to form carbon dioxide, nitrogen dioxide, and water.
    (b) How many oxygen molecules would be required to react with 25 ethylamine molecules?
    (c) How many water molecules would be formed in part (b)?
12. (a) Lime, CaO, dissolves in muriatic acid, HCl, to form calcium chloride, $CaCl_2$, and water.
    (b) How many moles of HCl would be required to dissolve 7.5 mol of CaO?
    (c) How many moles of water would be formed in part (b)?
13. (a) Aluminum building materials have a hard, transparent, protective coating of aluminum oxide, Al_2O_3, formed by reaction with oxygen in the air. The sulfuric acid, H_2SO_4, in acid rain dissolves this protective coating and forms aluminum sulfate, $Al_2(SO_4)_3$, and water.

(b) How many moles of H_2SO_4, are required to react with 3.0 mol of Al_2O_3?
(c) How many moles of $Al_2(SO_4)_3$ are formed in part (b)?

14. (a) Nitromethane, CH_3NO_2, often called "nitro," is used as a fuel additive in some racing vehicles. When CH_3NO_2 burns in excess oxygen, it forms carbon dioxide, nitrogen dioxide, and water.
(b) How many moles of oxygen are required to burn 5.0 mol of CH_3NO_2?
(c) How many moles of water are formed in part (b)?

15. (a) Butane, C_4H_{10}, burns in excess oxygen to form carbon dioxide and water.
(b) If 3.9 mol of oxygen are used up in this reaction, how many moles of butane were burned?
(c) How many moles of water were formed in part (b)?

16. The equation that describes the commercial "roasting" of zinc sulfide is

$$2ZnS + 3O_2 \xrightarrow{\Delta} 2ZnO + 2SO_2$$

What is the mole ratio of (a) O_2 to ZnS, (b) ZnO to ZnS, and (c) SO_2 to ZnS?

17. The reaction between dilute nitric acid and copper is given by the equation

$$3Cu + 8HNO_3 \longrightarrow 3Cu(NO_3)_2 + 2NO + 4H_2O$$

What is the mole ratio of (a) HNO_3 to Cu, (b) NO to Cu, and (c) $Cu(NO_3)_2$ to Cu?

18. How many moles of oxygen can be obtained by the decomposition of 1.00 mol of reactant in each of the following reactions?
(a) $2KClO_3 \rightarrow 2KCl + 3O_2$
(b) $2H_2O_2 \rightarrow 2H_2O + O_2$
(c) $2HgO \rightarrow 2Hg + O_2$
(d) $2NaNO_3 \rightarrow 2NaNO_2 + O_2$
(e) $KClO_4 \rightarrow KCl + 2O_2$

19. For the formation of 1.00 mol of water, which reaction uses the most nitric acid?
(a) $3Cu + 8HNO_3 \rightarrow 3Cu(NO_3)_2 + 2NO + 4H_2O$
(b) $Al_2O_3 + 6HNO_3 \rightarrow 2Al(NO_3)_3 + 3H_2O$
(c) $4Zn + 10HNO_3 \rightarrow 4Zn(NO_3)_2 + NH_4NO_3 + 3H_2O$

20. Consider the reaction

$$NH_3 + O_2 \xrightarrow{\text{not balanced}} NO + H_2O$$

For every 1.50 mol of NH_3, (a) how many moles of O_2 are required, (b) how many moles of NO are produced, and (c) how many moles of H_2O are produced?

21. Consider the reaction

$$2NO + Br_2 \longrightarrow 2NOBr$$

For every 3.00 mol of bromine that reacts, how many moles of (a) NO react and (b) NOBr are produced?

22. Find the mass of chlorine that will combine with 3.18 g of hydrogen to form hydrogen chloride:

$$H_2 + Cl_2 \longrightarrow 2HCl$$

23. What mass of solid AgCl will precipitate from a solution containing 1.50 g of $CaCl_2$ if an excess amount of $AgNO_3$ is added?

$$CaCl_2 + 2AgNO_3 \longrightarrow 2AgCl + Ca(NO_3)_2$$

24. A sample of magnetic iron oxide, Fe_3O_4, reacted completely with hydrogen at red heat. The water vapor formed by the reaction

$$Fe_3O_4 + 4H_2 \xrightarrow{\Delta} 3Fe + 4H_2O$$

was condensed and found to weigh 7.5 g. Calculate the mass of Fe_3O_4 that reacted.

25. What masses of cobalt(II) chloride and of hydrogen fluoride are needed to prepare 10.0 moles of cobalt(II) fluoride by the following reaction?

$$CoCl_2 + 2HF \longrightarrow CoF_2 + 2HCl$$

26. Gaseous chlorine and gaseous fluorine undergo a combination reaction to form the interhalogen compound ClF. Write the chemical equation for this reaction and calculate the mass of fluorine needed to react with 3.27 g of Cl_2.

27. Dinitrogen pentoxide, N_2O_5, undergoes a decomposition reaction to form nitrogen dioxide, NO_2, and oxygen. Write the chemical equation for this reaction. A 0.165-g sample of O_2 was produced by the reaction. What mass of NO_2 was produced?

28. Gaseous chlorine will displace bromide ion from an aqueous solution of potassium bromide to form aqueous potassium chloride and aqueous bromine. Write the chemical equation for this reaction. What mass of bromine will be produced if 0.289 g of chlorine undergoes reaction?

29. Solid zinc sulfide reacts with hydrochloric acid to form a mixture of aqueous zinc chloride and hydrogen sulfide, H_2S. Write the chemical equation for this reaction. What mass of zinc sulfide is needed to react with 10.65 g of HCl?

***30.** An impure sample of $CuSO_4$ weighing 5.52 g was dissolved in water and allowed to react with excess zinc.

$$CuSO_4 + Zn \longrightarrow ZnSO_4 + Cu$$

What was the percent $CuSO_4$ in the sample if 1.49 g of Cu was produced?

31. You are designing an experiment for the preparation of hydrogen. For the production of equal amounts of hydrogen, which metal, Zn or Al, is less expensive if Zn costs about half as much as Al on a mass basis?

$$Zn + 2HCl \longrightarrow ZnCl_2 + H_2$$
$$2Al + 6HCl \longrightarrow 2AlCl_3 + 3H_2$$

Limiting Reactant

32. How many grams of NH_3 can be prepared from 77.3 grams of N_2 and 14.2 grams of H_2?

$$N_2 + 3H_2 \longrightarrow 2NH_3$$

33. Silver nitrate solution reacts with barium chloride solution according to the equation

$$2AgNO_3 + BaCl_2 \longrightarrow Ba(NO_3)_2 + 2AgCl$$

All of the substances involved in this reaction are soluble in water except silver chloride, AgCl, which forms a solid (precipitate) at the bottom of the flask. Suppose we mix together a solution containing 12.6 g of $AgNO_3$ and 8.4 g of $BaCl_2$. What mass of AgCl would be formed?

***34.** "Superphosphate," a water-soluble fertilizer, is a mixture of $Ca(H_2PO_4)_2$ and $CaSO_4$ on a 1:2 *mole* basis. It is formed by the reaction

$$Ca_3(PO_4)_2 + 2H_2SO_4 \longrightarrow Ca(H_2PO_4)_2 + 2CaSO_4$$

We treat 250 g of $Ca_3(PO_4)_2$ with 150 g of H_2SO_4. How many grams of superphosphate could be formed?

35. Silicon carbide, an abrasive, is made by the reaction of silicon dioxide with graphite:

$$SiO_2 + C \xrightarrow{\Delta} SiC + CO \qquad \text{(balanced?)}$$

We mix 377 g of SiO_2 and 44.6 g of C. If the reaction proceeds as far as possible, which reactant will be left over? How much of this reactant will remain?

36. What mass of potassium can be produced by the reaction of 100.0 g of Na with 100.0 g of KCl?

$$Na + KCl \xrightarrow{\Delta} NaCl + K$$

37. A reaction mixture contains 25.0 g of PCl_3 and 45.0 g of PbF_2. What mass of $PbCl_2$ can be obtained from the following reaction?

$$3PbF_2 + 2PCl_3 \longrightarrow 2PF_3 + 3PbCl_2$$

How much of which reactant will be left unchanged?

38. What mass of $BaSO_4$ will be produced by the reaction of 33.2 g of Na_2SO_4 with 43.5 g of $Ba(NO_3)_2$?

$$Ba(NO_3)_2 + Na_2SO_4 \longrightarrow BaSO_4 + 2NaNO_3$$

39. Consider the reaction

$$3HCl + 3HNF_2 \longrightarrow 2ClNF_2 + NH_4Cl + 2HF$$

A mixture of 8.00 g of HCl and 10.00 g of HNF_2 is allowed to react. If only 15% of the limiting reactant does react, what is the composition of the final mixture?

Percent Yield from Chemical Reactions

40. What mass of chromium is present in 150 grams of an ore of chromium that is 67.0% chromite, $FeCr_2O_4$, and 33.0% impurities by mass? If 87.5% of the chromium can be recovered from 125 grams of the ore, what mass of pure chromium is obtained?

41. A particular ore of lead, galena, is 10% lead sulfide, PbS, and 90% impurities by weight. What mass of lead is contained in 75 grams of this ore?

42. The percent yield for the reaction

$$PCl_3 + Cl_2 \longrightarrow PCl_5$$

is 85.0%. What mass of PCl_5 would be expected from the reaction of 38.5 g of PCl_3 with excess chlorine?

43. The percent yield for the following reaction carried out in carbon tetrachloride solution

$$Br_2 + Cl_2 \longrightarrow 2BrCl$$

is 57.0%. (a) What amount of BrCl would be formed from the reaction of 0.0100 mol Br_2 with 0.0100 mol Cl_2? (b) What amount of Br_2 is left unchanged?

44. Solid silver nitrate undergoes thermal decomposition to form silver metal, nitrogen dioxide, and oxygen. Write the chemical equation for this reaction. A 0.362-g sample of silver metal was obtained from the decomposition of a 0.575-g sample of $AgNO_3$. What is the percent yield of the reaction?

45. Gaseous nitrogen and hydrogen undergo a reaction to form gaseous ammonia (the Haber process). Write the chemical equation for this reaction. At a temperature of 400°C and a total pressure of 250 atm, 0.720 g of NH_3 was produced by the reaction of 2.80 g of N_2 with excess H_2. What is the percent yield of the reaction?

46. Ethylene oxide, C_2H_4O, a fumigant sometimes used by exterminators, is synthesized in 89% yield by reaction of ethylene bromohydrin, C_2H_5OBr, with sodium hydroxide:

$$C_2H_5OBr + NaOH \longrightarrow C_2H_4O + NaBr + H_2O$$

How many grams of ethylene bromohydrin would be consumed in the production of 255 g of ethylene oxide, at 89% yield?

***47.** How much 68% Na_2SO_4 could be produced from 375 g of 88% pure NaCl?

$$2NaCl + H_2SO_4 \longrightarrow Na_2SO_4 + 2HCl$$

***48.** Calcium carbide is made in an electric furnace by the reaction

$$CaO + 3C \longrightarrow CaC_2 + CO$$

The crude product is usually 85% CaC_2 and 15% unreacted CaO. (a) How much CaO should we start with to produce 250 kg of crude product? (b) How much CaC_2 would this crude product contain?

49. Ethylene glycol, $C_2H_6O_2$, is used as antifreeze in automobile radiators. A method of producing small amounts of ethylene glycol in the laboratory is by reaction of 1,2-dichloroethane with sodium carbonate in a water solution, followed by distillation of the reaction mixture to purify the ethylene glycol.

$$C_2H_4Cl_2 + Na_2CO_3 + H_2O \longrightarrow C_2H_6O_2 + 2NaCl + CO_2$$

When 27.4 g of 1,2-dichloroethane is used in this reaction, 10.3 g of ethylene glycol is obtained. (a) Calculate the theoretical yield of ethylene glycol. (b) What is the percent yield of ethylene glycol in this process? (c) What mass of Na_2CO_3 is consumed?

Sequential Reactions

50. Consider the two-step process for the formation of tellurous acid described by the following equations:

$$TeO_2 + 2OH^- \longrightarrow TeO_3^{2-} + H_2O$$
$$TeO_3^{2-} + 2H^+ \longrightarrow H_2TeO_3$$

What mass of H_2TeO_3 would be formed from 62.1 g of TeO_2, assuming 100% yield?

51. Consider the formation of cyanogen, C_2N_2, and its subsequent decomposition in water given by the equations

$$2Cu^{2+} + 6CN^- \longrightarrow 2[Cu(CN)_2]^- + C_2N_2$$
$$C_2N_2 + H_2O \longrightarrow HCN + HOCN$$

How much hydrocyanic acid, HCN, can be produced from 10.00 g of KCN, assuming 100% yield?

52. What mass of potassium chlorate would be required to supply the proper amount of oxygen needed to burn 35.0 g of methane, CH_4?

$$2KClO_3 \longrightarrow 2KCl + 3O_2$$
$$CH_4 + 2O_2 \longrightarrow CO_2 + 2H_2O$$

53. Hydrogen, obtained by the electrical decomposition of water, was combined with chlorine to produce 51.0 g of hydrogen chloride. Calculate the mass of water decomposed.

$$2H_2O \longrightarrow 2H_2 + O_2$$
$$H_2 + Cl_2 \longrightarrow 2HCl$$

***54.** About half of the world's production of pigments for paints involves the formation of white TiO_2. In the United States, it is made on a large scale by the *chloride process*, starting with ores containing only small amounts of rutile, TiO_2. The ore is treated with chlorine and carbon (coke). This produces $TiCl_4$ and gaseous products:

$$2TiO_2 + 3C + 4Cl_2 \longrightarrow 2TiCl_4 + CO_2 + 2CO$$

The $TiCl_4$ is then converted into TiO_2 of high purity:

$$TiCl_4 + O_2 \longrightarrow TiO_2 + 2Cl_2$$

Suppose the first process can be carried out with 65.0% yield and the second with 92.0% yield. How many kg of TiO_2 could be produced starting with 1.00 metric ton (1.00×10^6 g) of an ore that is 0.25% rutile?

55. When sulfuric acid dissolves in water, the following reactions take place:

$$H_2SO_4 \longrightarrow H^+ + HSO_4^-$$
$$HSO_4^- \longrightarrow H^+ + SO_4^{2-}$$

The first reaction is 100.0% complete and the second reaction is 10.0% complete. Calculate the concentrations of the various ions in a 0.100 *M* aqueous solution of H_2SO_4.

***56.** The chief ore of zinc is the sulfide, ZnS. The ore is concentrated by flotation and then heated in air, which converts the ZnS to ZnO.

$$2ZnS + 3O_2 \longrightarrow 2ZnO + 2SO_2$$

The ZnO is then treated with dilute H_2SO_4

$$ZnO + H_2SO_4 \longrightarrow ZnSO_4 + H_2O$$

to produce an aqueous solution containing the zinc as $ZnSO_4$. An electrical current is passed through the solution to produce the metal.

$$2ZnSO_4 + 2H_2O \longrightarrow 2Zn + 2H_2SO_4 + O_2$$

What mass of Zn will be obtained from an ore containing 100. kg of ZnS? Assume the flotation process to be 91% efficient, the electrolysis step to be 98% efficient, and the other steps to be 100% efficient.

Concentrations of Solutions—Percent by Mass

57. What mass of an 8.65% solution of potassium dichromate contains 60.0 g of $K_2Cr_2O_7$? What mass of water does this amount of solution contain?

58. Calculate the mass of an 8.30% solution of ammonium chloride, NH_4Cl, that contains 100 g of water. What mass of NH_4Cl does this amount of solution contain?

59. The density of an 18.0% solution of ammonium chloride, NH_4Cl, solution is 1.05 g/mL. What mass of NH_4Cl does 350 mL of this solution contain?

60. The density of an 18.0% solution of ammonium sulfate, $(NH_4)_2SO_4$, is 1.10 g/mL. What mass of $(NH_4)_2SO_4$ would be required to prepare 350 mL of this solution?

61. What volume of the solution of NH_4Cl described in Exercise 59 contains 80.0 g of NH_4Cl?

62. What volume of the solution of $(NH_4)_2SO_4$ described in Exercise 60 contains 80.0 g of $(NH_4)_2SO_4$?

*63. A reaction requires 37.8 g of NH_4Cl. What volume of the solution described in Exercise 59 would you use if you wished to use a 20.0% excess of NH_4Cl?

Concentrations of Solutions—Molarity

64. What is the molarity of a solution that contains 490 g of phosphoric acid, H_3PO_4, in 2.00 L of solution?
65. What is the molarity of a solution that contains 1.37 g of sodium chloride in 25.0 mL of solution?
66. What is the molarity of a solution containing 0.155 mol H_3PO_4 in 200 mL of solution?
67. A solution contains 0.100 mole per liter of each of the following acids: HCl, H_2SO_4, H_3PO_4.
(a) Is the molarity the same for each acid?
(b) Is the number of molecules per liter the same for each acid?
(c) Is the mass per liter the same for each acid?
68. A solution contains 1.05 g of a rubbing alcohol, $(CH_3)_2CHOH$, in 100 mL of solution. Find (a) the molarity of the solution and (b) the number of moles in 1.00 mL of solution.
69. (a) Calculate the molarity of caffeine in a 12-oz cola drink containing 50 mg caffeine, $C_8H_{10}N_4O_2$.
(b) Cola drinks are usually 5.06×10^{-3} *M* with respect to H_3PO_4. How much of this acid is in a 250-mL drink? (1 oz = 29.6 mL)
70. How many grams of the cleansing agent Na_3PO_4 (a) are needed to prepare 200 mL of 0.25 *M* solution, and (b) are in 200 mL of 0.25 *M* solution?
71. How many kg of ethylene glycol, $C_2H_6O_2$, are needed to prepare a 9.00 *M* solution to protect a 15.0-L car radiator against freezing? What is the mass of $C_2H_6O_2$ in 15.0 L of 9.00 *M* solution?
72. A solution made by dissolving 18.0 g of $CaCl_2$ in 72.0 g of water has a density of 1.180 g/mL at 20°C.
(a) What is the percent by mass of $CaCl_2$ in the solution?
(b) What is the molarity of $CaCl_2$ in the solution?
73. Stock phosphoric acid solution is 85.0% H_3PO_4 and has a specific gravity of 1.70. What is the molarity of the solution?
74. Stock hydrofluoric acid solution is 49.0% HF and has a specific gravity of 1.17. What is the molarity of the solution?
75. What mass of sodium sulfate, Na_2SO_4, is contained in 750 mL of a 2.00 molar solution?
76. What is the molarity of a barium chloride solution prepared by dissolving 3.50 g of $BaCl_2 \cdot 2H_2O$ in enough water to make 500 mL of solution?
77. What mass of potassium benzoate trihydrate, $KC_7H_5O_2 \cdot 3H_2O$, is needed to prepare one liter of a 0.125 molar solution of potassium benzoate?
78. What volume of 0.850 *M* $CuSO_4$ solution can be prepared from 75.0 g of copper(II) sulfate pentahydrate, $CuSO_4 \cdot 5H_2O$?

Dilution of Solutions

79. Commercially available concentrated sulfuric acid is 18.0 *M* H_2SO_4. Calculate the volume of concentrated sulfuric acid required to prepare 2.50 L of 0.150 *M* H_2SO_4 solution.
80. Commercial concentrated hydrochloric acid is 12.0 *M* HCl. What volume of concentrated hydrochloric acid is required to prepare 3.50 L of 2.40 *M* HCl solution?
81. Calculate the volume of 2.00 *M* NaOH solution required to prepare 100 mL of a 0.500 *M* solution of NaOH.
82. Calculate the volume of 0.0500 *M* $Ba(OH)_2$ solution that contains the same number of moles of $Ba(OH)_2$ as 120 mL of 0.0800 *M* $Ba(OH)_2$ solution.

*83. Calculate the resulting molarity when 50.0 mL of 2.30 *M* NaCl solution is mixed with 80.0 mL of 1.40 *M* NaCl.

*84. Calculate the resulting molarity when 125 mL of 6.00 *M* H_2SO_4 solution is mixed with 225 mL of 3.00 *M* H_2SO_4.

Using Solutions in Chemical Reactions

85. What volume of 0.50 *M* HBr is required to react completely with 0.75 mol of $Ca(OH)_2$?

$$2HBr + Ca(OH)_2 \longrightarrow CaBr_2 + 2H_2O$$

86. What volume of 0.324 *M* HNO_3 solution is required to react completely with 22.0 mL of 0.0612 *M* $Ba(OH)_2$?

$$Ba(OH)_2 + 2HNO_3 \longrightarrow Ba(NO_3)_2 + 2H_2O$$

87. What is the concentration, in mol/L, of an HCl solution if 23.65 mL reacts completely with (neutralizes) 25.00 mL of a 0.1037 *M* solution of NaOH?

$$HCl + NaOH \longrightarrow NaCl + H_2O$$

88. An excess of $AgNO_3$ reacts with 100.0 mL of an $AlCl_3$ solution to give 0.275 g of AgCl. What is the concentration, in mol/L, of the $AlCl_3$ solution?

$$AlCl_3 + 3AgNO_3 \longrightarrow 3AgCl + Al(NO_3)_3$$

89. An impure sample of solid Na_2CO_3 was allowed to react with 0.1026 *M* HCl.

$$Na_2CO_3 + 2HCl \longrightarrow 2NaCl + CO_2 + H_2O$$

A 0.1247-g sample of sodium carbonate required 14.78 mL of HCl. What is the purity of the sodium carbonate?

90. Calculate the theoretical yield of AgCl formed from the reaction of an aqueous solution containing excess $ZnCl_2$ with 35.0 mL of 0.325 *M* $AgNO_3$.

$$ZnCl_2 + 2AgNO_3 \longrightarrow Zn(NO_3)_2 + 2AgCl$$

Titrations

91. Define and illustrate the following terms clearly and concisely: (a) standard solution; (b) titration.
92. Distinguish between the *equivalence point* and *end point* of a titration.

93. What volume of 0.275 molar hydrochloric acid solution reacts with 36.4 mL of 0.150 molar sodium hydroxide solution? (See Example 3-24.)

94. What volume of 0.112 molar sodium hydroxide solution would be required to react with 25.3 mL of 0.400 molar sulfuric acid solution? (See Example 3-23.)

95. A 0.08964 *M* solution of NaOH was used to titrate a solution of unknown concentration of HCl. A 30.00-mL sample of the HCl solution required 24.21 mL of the NaOH solution for complete reaction. What is the molarity of the HCl solution? (See Example 3-24.)

96. A 34.53-mL sample of a solution of sulfuric acid, H_2SO_4, reacts with 27.86 mL of 0.08964 *M* NaOH solution. Calculate the molarity of the sulfuric acid solution. (See Example 3-23.)

97. Benzoic acid, C_6H_5COOH, is sometimes used for the standardization of solutions of bases. A 1.862-g sample of the acid reacts with 31.62 mL of an NaOH solution. What is the molarity of the base solution?

$$C_6H_5COOH + NaOH \longrightarrow C_6H_5COONa + H_2O$$

98. An antacid tablet containing calcium carbonate as an active ingredient requires 22.6 mL of 0.0932 *M* HCl for complete reaction. What mass of $CaCO_3$ did the tablet contain?

$$2HCl + CaCO_3 \longrightarrow CaCl_2 + H_2O + CO_2$$

Mixed Exercises

*99. What mass of sulfuric acid can be obtained from 1.00 kg of sulfur by the following series of reactions?

$$S + O_2 \xrightarrow{98\%\ \text{yield}} SO_2$$

$$2SO_2 + O_2 \xrightarrow{96\%\ \text{yield}} 2SO_3$$

$$SO_3 + H_2SO_4 \xrightarrow{100\%\ \text{yield}} H_2S_2O_7$$

$$H_2S_2O_7 + H_2O \xrightarrow{97\%\ \text{yield}} 2H_2SO_4$$

*100. What is the total mass of products formed when 33.8 g of carbon disulfide is burned in air? What mass of carbon disulfide would have to be burned to produce a mixture of carbon dioxide and sulfur dioxide that has a mass of 54.2 g?

$$CS_2 + 3O_2 \xrightarrow{\Delta} CO_2 + 2SO_2$$

*101. A mixture of calcium oxide, CaO, and calcium carbonate, $CaCO_3$, that had a mass of 1.844 g was heated until all the calcium carbonate was decomposed according to the following equation. After heating, the sample weighed 1.462 g. Calculate the masses of CaO and $CaCO_3$ present in the original sample.

$$CaCO_3 \xrightarrow{\Delta} CaO + CO_2$$

Some Types of Chemical Reactions

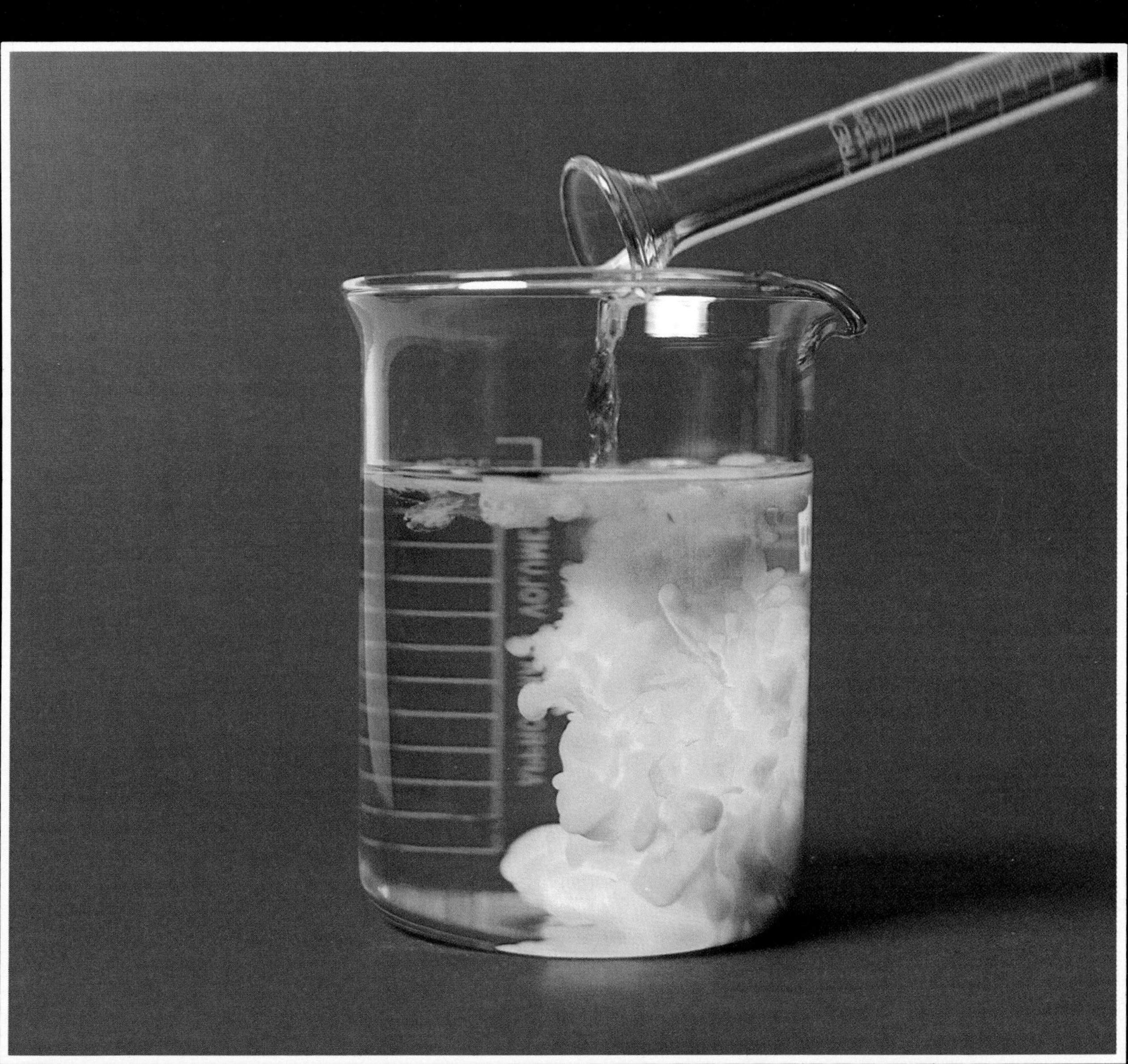

Pouring ammonium sulfide solution into a solution of cadmium nitrate gives a precipitate of cadmium sulfide.

$(NH_4)_2S + Cd(NO_3)_2 \rightarrow CdS(s) + 2NH_4NO_3$

Cadmium sulfide is used as a pigment in artists' oil-based paints.

Outline

Objectives

As you study this chapter, you should learn

- ☐ About the periodic table and the classification of elements
- ☐ About reactions of solutes in aqueous solutions
- ☐ To recognize nonelectrolytes, strong electrolytes, and weak electrolytes
- ☐ The classification of acids, bases, and salts
- ☐ Which kinds of compounds are soluble and which kinds are insoluble in water
- ☐ How to describe reactions in aqueous solutions by writing formula unit equations as well as total ionic and net ionic equations
- ☐ About displacement reactions and the activity series
- ☐ About precipitation reactions
- ☐ About oxidation numbers
- ☐ To balance equations for oxidation–reduction reactions

We observe that some elements form compounds with only a few (or no) other elements, while others combine with nearly every other element (oxygen) or form millions of compounds (carbon). Some elements are metals, and others obviously lack metallic properties. Some substances are gases, others are liquids, and still others are solids. Some solids are soft (paraffin), others are quite hard (diamond), and others are quite strong (steel). Clearly, there are significant differences among the chemical bonds and other attractive forces that hold atoms together in different substances.

Compounds are formed when atoms of different elements are joined together by chemical bonds. Chemical bonds are broken when atoms are separated and the original compounds cease to exist. The attractive forces between atoms are electrical in nature, and chemical reactions between atoms involve *changes* in their electronic structures.

As we shall see in the next few chapters,

> the positions of elements in the periodic table are related to the arrangements of their electrons; these determine the chemical and physical properties of the elements.

Let us now turn our attention to the periodic table and to **chemical periodicity**, the variation in properties of elements with their positions in the periodic table.

4-1 The Periodic Table: Metals, Nonmetals, and Metalloids

In 1869 the Russian chemist Dimitri Mendeleev and the German chemist Lothar Meyer independently published arrangements of known elements that are much like the periodic table in use today. Mendeleev's classification was based primarily on chemical properties of the elements, whereas Meyer's classification was based largely on physical properties. The tabulations were surprisingly similar. Both emphasized the *periodicity*, or regular periodic repetition, of properties with increasing atomic weight.

Pronounced "men-del-*lay*-ev."

Mendeleev arranged the known elements in order of increasing atomic weight in successive sequences so that elements with similar chemical properties fell in the same column. He noted that both physical and chemical properties of the elements vary in a periodic fashion with atomic weight. His periodic table of 1872 contained the 62 elements that were known then (Figure 4-1).

Consider H, Li, Na, and K, all of which appear in "Gruppe I" of Mendeleev's table. All were known to combine with F, Cl, Br, and I of "Gruppe VII" to produce compounds that have similar formulas such as HF, LiCl, NaCl, and KI. All these compounds dissolve in water to produce solutions that conduct electricity. The "Gruppe II" elements form compounds such as $BeCl_2$, $MgBr_2$, and $CaCl_2$, as well as compounds with O and S from "Gruppe VI" such as MgO, CaO, MgS, and CaS.

In most areas of human endeavor progress is slow and faltering. However, there is an occasional individual who develops concepts and techniques that clarify confused situations. Mendeleev was such an individual. One of the brilliant successes of his periodic table was that it provided for elements that were unknown at the time. When he encountered "missing" elements, Mendeleev left blank spaces. Some appreciation of his genius in constructing the table as he did can be gained by comparing the predicted (1871) and observed properties of germanium, which was not discovered until 1886. Mendeleev called the undiscovered element eka-silicon because it fell below

REIHEN	GRUPPE I — R^2O	GRUPPE II — RO	GRUPPE III — R^2O^3	GRUPPE IV RH^4 RO^2	GRUPPE V RH^3 R^2O^5	GRUPPE VI RH^2 RO^3	GRUPPE VII RH R^2O^7	GRUPPE VIII — RO^4
1	H = 1							
2	Li = 7	Be = 9,4	B = 11	C = 12	N = 14	O = 16	F = 19	
3	Na = 23	Mg = 24	Al = 27,3	Si = 28	P = 31	S = 32	Cl = 35,5	
4	K = 39	Ca = 40	— = 44	Ti = 48	V = 51	Cr = 52	Mn = 55	Fe = 56, Co = 59, Ni = 59, Cu = 63.
5	(Cu = 63)	Zn = 65	— = 68	— = 72	As = 75	Se = 78	Br = 80	
6	Rb = 85	Sr = 87	?Yt = 88	Zr = 90	Nb = 94	Mo = 96	— = 100	Ru = 104, Rh = 104, Pd = 106, Ag = 108.
7	(Ag = 108)	Cd = 112	In = 113	Sn = 118	Sb = 122	Te = 125	J = 127	
8	Cs = 133	Ba = 137	?Di = 138	?Ce = 140	—	—	—	— — — —
9	(—)	—	—	—	—	—	—	
10	—	—	?Er = 178	?La = 180	Ta = 182	W = 184	—	Os = 195, Ir = 197, Pt = 198, Au = 199.
11	(Au = 199)	Hg = 200	Tl = 204	Pb = 207	Bi = 208	—	—	
12	—	—	—	Th = 231	—	U = 240	—	— — — —

Figure 4-1
Mendeleev's early periodic table (1872). "J" is the German symbol for iodine.

silicon in his table. He was familiar with the properties of germanium's neighboring elements. They served as the basis for his predictions of properties of germanium (Table 4-1). Some modern values for properties of germanium differ significantly from those reported in 1886. But many of the values upon which Mendeleev based his predictions were inaccurate, as were most of the 1886 values for Ge.

Because Mendeleev's arrangement of the elements was based on increasing *atomic weights*, several elements appeared to be out of place in his table. Mendeleev put the controversial elements (Te and I, Co and Ni) in locations consistent with their properties. He thought the apparent reversal of atomic weights was due to inaccurate values for those weights. Careful redetermination showed that the values were correct. Resolution of the problem of these "out-of-place" elements had to await the development of the concept of *atomic number*, approximately 50 years after Mendeleev's work. The **atomic number** of an element is the number of protons in the nucleus of its atoms. (It is also the number of electrons in an atom of an element.) This quantity is fundamental to the identity of each element because it is related to the electrical make-up of atoms. Elements are arranged in the periodic table in order of increasing atomic number. With the development of this concept, the **periodic law** attained essentially its present form:

Silicon (top), germanium, and tin (bottom).

The properties of the elements are periodic functions of their atomic numbers.

The periodic law tells us that if we arrange the elements in order of increasing atomic number, we periodically encounter elements that have similar chemical and physical properties. The presently used "long form" of the period table (Table 4-2) is such an arrangement. The vertical columns are referred to as **groups** or **families**, and the horizontal rows are called

Table 4-1
Predicted and Observed Properties of Germanium

Property	Eka-Silicon Predicted, 1871	Germanium Reported, 1886	Modern Values
Atomic weight	72	72.32	72.61
Atomic volume	13 cm^3	13.22 cm^3	13.5 cm^3
Specific gravity	5.5	5.47	5.35
Specific heat	0.073 cal/g°C	0.076 cal/g°C	0.074 cal/g°C
Maximum valence*	4	4	4
Color	Dark gray	Grayish white	Grayish white
Reaction with water	Will decompose steam with difficulty	Does not decompose water	Does not decompose water
Reactions with acids and alkalis	Slight with acids; more pronounced with alkalis	Not attacked by HCl or dilute aqueous NaOH; reacts vigorously with molten NaOH	Not dissolved by HCl or H_2SO_4 or dilute NaOH; dissolved by concentrated NaOH
Formula of oxide	EsO_2	GeO_2	GeO_2
Specific gravity of oxide	4.7	4.703	4.228
Specific gravity of tetrachloride	1.9 at 0°C	1.887 at 18°C	1.8443 at 30°C
Boiling point of tetrachloride	100°C	86°C	84°C
Boiling point of tetraethyl derivative	160°C	160°C	186°C

* ''Valence'' refers to the combining power of a specific element.

periods. Elements in a *group* have similar chemical and physical properties, while those within a *period* have properties that change progressively across the table. Several groups of elements have common names that are used so frequently they should be learned. The Group IA elements, except H, are referred to as **alkali metals**, and the Group IIA elements are called the **alkaline earth metals**. The Group VIIA elements are called **halogens**, which means ''salt formers,'' and the Group 0 elements are called **noble** (or **rare**) **gases**.

Alkaline means basic. The character of basic compounds is described in Section 10-4.

Three of the halogens: (left to right) chlorine, bromine, iodine.

Table 4-2
The Periodic Table*

Metals / Nonmetals / Metalloids

	IA	IIA	IIIB	IVB	VB	VIB	VIIB	VIIIB	VIIIB	VIIIB	IB	IIB	IIIA	IVA	VA	VIA	VIIA	0
1	1 H																	2 He
2	3 Li	4 Be											5 B	6 C	7 N	8 O	9 F	10 Ne
3	11 Na	12 Mg											13 Al	14 Si	15 P	16 S	17 Cl	18 Ar
4	19 K	20 Ca	21 Sc	22 Ti	23 V	24 Cr	25 Mn	26 Fe	27 Co	28 Ni	29 Cu	30 Zn	31 Ga	32 Ge	33 As	34 Se	35 Br	36 Kr
5	37 Rb	38 Sr	39 Y	40 Zr	41 Nb	42 Mo	43 Tc	44 Ru	45 Rh	46 Pd	47 Ag	48 Cd	49 In	50 Sn	51 Sb	52 Te	53 I	54 Xe
6	55 Cs	56 Ba	57 La *	72 Hf	73 Ta	74 W	75 Re	76 Os	77 Ir	78 Pt	79 Au	80 Hg	81 Tl	82 Pb	83 Bi	84 Po	85 At	86 Rn
7	87 Fr	88 Ra	89 Ac †															

*	58 Ce	59 Pr	60 Nd	61 Pm	62 Sm	63 Eu	64 Gd	65 Tb	66 Dy	67 Ho	68 Er	69 Tm	70 Yb	71 Lu
†	90 Th	91 Pa	92 U	93 Np	94 Pu	95 Am	96 Cm	97 Bk	98 Cf	99 Es	100 Fm	101 Md	102 No	103 Lr

* There are other systems for numbering the groups in the periodic table. We use the standard American system.

About 80% of the elements are metals.

The physical and chemical properties that distinguish metals from nonmetals are summarized in Tables 4-3 and 4-4. The general properties of metals and nonmetals are opposite. Not all metals and nonmetals possess all these properties, but they share most of them to varying degrees. The physical properties of metals can be explained on the basis of metallic bonding in solids (Section 13-17).

Table 4-2, the periodic table, shows how we divide the known elements into *metals* (shown in blue), *nonmetals* (yellow), and *metalloids* (green). The elements to the left of those touching the heavy stairstep line are *metals* (except hydrogen), while those to the right are *nonmetals*. Such a classification is somewhat arbitrary, and several elements do not fit neatly into either class. The elements adjacent to the heavy line are often called *metalloids* (or semimetals), because they are metallic (or nonmetallic) only to a limited degree.

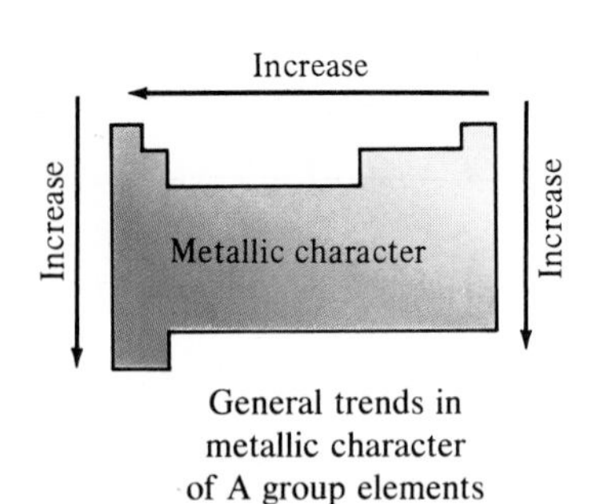

General trends in metallic character of A group elements with position in the periodic table

Metallic character increases from top to bottom and from right to left with respect to position in the periodic table.

Cesium, atomic number 55, is the most active naturally occurring metal.

Copper is rolled into sheets.

Table 4-3
Some Physical Properties of Metals and Nonmetals

Metals	Nonmetals
1. High electrical conductivity that decreases with increasing temperature	**1.** Poor electrical conductivity (except carbon in the form of graphite)
2. High thermal conductivity	**2.** Good heat insulators (except carbon in the form of diamond)
3. Metallic gray or silver luster*	**3.** No metallic luster
4. Almost all are solids†	**4.** Solids, liquids, or gases
5. Malleable (can be hammered into sheets)	**5.** Brittle in solid state
6. Ductile (can be drawn into wires)	**6.** Nonductile
7. Solid state characterized by metallic bonding	**7.** Covalently bonded molecules; noble gases are monatomic

* Except copper and gold.
† Except mercury; cesium and gallium melt in protected hand.

Nonmetallic character increases from bottom to top and from left to right in the periodic table.

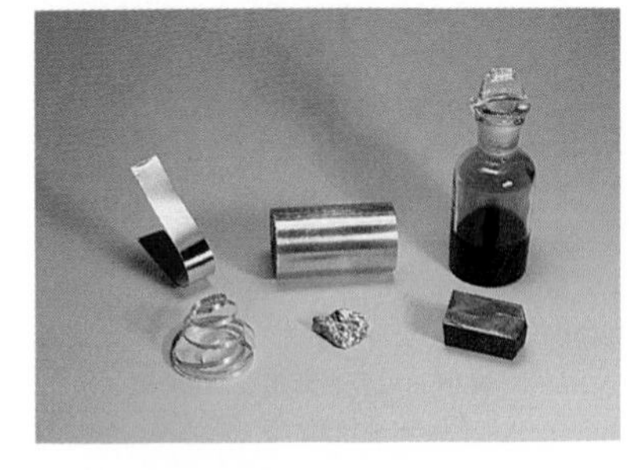

Two metals (left), copper and magnesium; two metalloids (center), aluminum and silicon; two nonmetals (right), bromine and carbon.

Fluorine, atomic number 9, is the most active nonmetal.

Metalloids show some properties that are characteristic of both metals and nonmetals. Many of the metalloids, such as silicon, germanium, and antimony, act as semiconductors, which are important in solid-state electronic circuits. **Semiconductors** are insulators at lower temperatures, but become conductors at higher temperatures (Section 13-17).

Aluminum is the most metallic of the metalloids and is sometimes classified as a metal. It is metallic in appearance, and an excellent conductor of electricity, but its electrical conductivity *increases with increasing temperature*. The conductivities of metals, by contrast, decrease with increasing temperature.

Table 4-4
Some Chemical Properties of Metals and Nonmetals

Metals	Nonmetals
1. Outer shells contain few electrons—usually three or fewer	**1.** Outer shells contain four or more electrons*
2. Form cations by losing electrons	**2.** Form anions by gaining electrons†
3. Form ionic compounds with nonmetals	**3.** Form ionic compounds with metals† and molecular (covalent) compounds with other nonmetals

* Except hydrogen and helium.
† Except the noble gases.

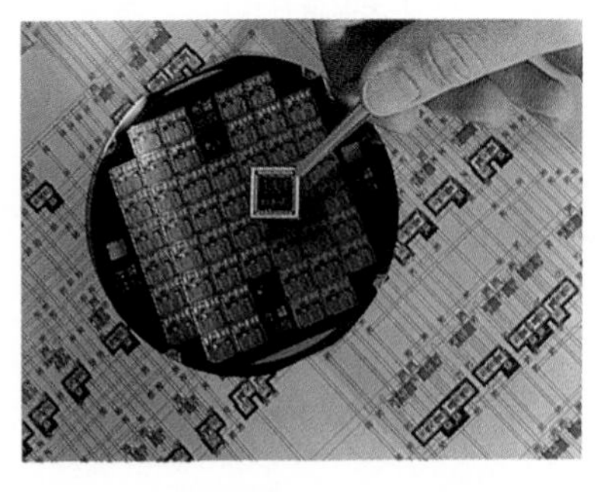

Silicon, a metalloid, is widely used in the manufacture of electronic chips.

Chemistry in Use. . .
The Periodic Table

The periodic table is almost the first thing a student of chemistry encounters. It appears invariably in textbooks, in lecture halls, and in laboratories. The most knowledgeable scientists consider it an indispensable reference. And yet, less than 150 years ago, the idea of arranging the elements by atomic weight or number was considered absurd. At an 1866 meeting of the Chemical Society at Burlington House, England, J. A. R. Newlands presented a theory he called the Law of Octaves. It stated that when the known elements were listed by increasing atomic weights, those that were eight places apart would be similar, much like notes on a piano keyboard. His colleagues' reactions are probably summed up best by the remark of a Professor Foster: "Have you thought of arranging the elements according to their initial letters? Maybe some better connections would come to light that way."

It is not surprising that poor Newlands was not taken seriously. In the 1860s, little information was available to illustrate relationships among the elements. Only 62 of them had been distinguished from more complex substances when Mendeleev first announced his discovery of the Periodic Law in 1869. However, as advances in atomic theory were made and as new experiments contributed to the understanding of chemical behavior, some scientists had begun to see similarities and patterns among the elements. In 1829 Johann Dobereiner first noticed *triads,* series of three elements in which the characteristics of the middle element are an average of the characteristics of the others. One such triad consists of Li, Na, and K (atomic weights 6.9, 23.0, and 39.1 g/mol). Beguyer de Chantcourtois, in 1862, assigned them to a repeating pattern known as a Telluric Screw, based on the observation that elements whose atomic weights differ by 16 g/mol are often similar. And in 1869 Lothar Meyer and Dmitri Mendeleev independently published similar versions of the now-famous periodic table.

Mendeleev's discovery was the result of many years of hard work. He gathered information on the elements from all corners of the earth—by corresponding with colleagues, studying books and papers, and redoing experiments to confirm data. He put the statistics of each element on a small card and pinned the cards to his laboratory wall, where he arranged and rearranged them many times until he was sure that they were in the right order. One especially farsighted feature of Mendeleev's accomplishment was his realization that some elements were missing from the table. His predictions for the properties of these substances (gallium, scandium, and germanium) were found to be remarkably accurate considering the data available to him. (It is important to remember that Mendeleev's periodic table organization was devised more than 50 years before the discovery and characterization of subatomic particles.)

Since its birth in 1869, the periodic table has been discussed and

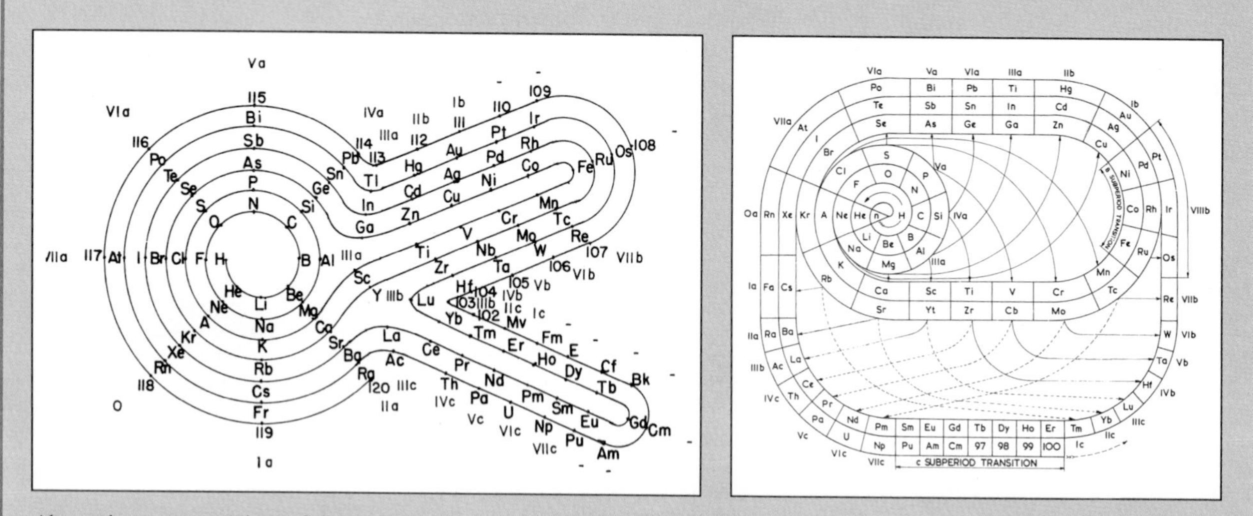

Alternative representations of the periodic table, as proposed by Charles Janet, 1928 (left), and John D. Clark, 1950 (right).

revised many times. Spectroscopic and other discoveries have filled in the blanks left by Mendeleev and added a new column consisting of the noble gases. As scientists learned more about atomic structure, the basis for ordering was changed from atomic weight to atomic number. The perplexing rare earths were sorted out and given a special place, along with many of the elements created by atomic bombardment. Even the form of the table has been experimented with, resulting in everything from spiral and circular tables to exotic shapes such as those suggested by Janet and by Clark. Recently a three-dimensional periodic table that takes into account valence-shell energies has been proposed by Professor Leland C. Allen of Princeton University.

It is certain that the future will disclose new and exciting information on the elements that make up our universe. During the past century, chemistry has become a fast-moving science in which methods and instruments are often outdated within a few years. But it is doubtful that our old friend, the periodic table, will ever become obsolete. It may be modified, but it will always stand as a statement of basic relationships in chemistry and as a monument to the wisdom and insight of its creator, Dmitri Mendeleev.

Lisa L. Saunders
Chemistry major
University of Texas at Austin

4-2 Aqueous Solutions—An Introduction

Approximately three fourths of the earth's surface is covered with water. The body fluids of all plants and animals are mainly water. Thus we can see that many important chemical reactions occur in aqueous (water) solutions, or in contact with water. In Chapter 3, we introduced solutions and methods of expressing concentrations of solutions. It is useful to know the kinds of substances that are soluble in water, and the forms in which they exist, before we begin our systematic study of reactions.

1 Electrolytes and Extent of Ionization

Solutes that are water-soluble can be classified as either electrolytes or nonelectrolytes. **Electrolytes** are substances whose aqueous solutions conduct electrical current. **Strong electrolytes** are substances that conduct electricity well in dilute aqueous solution. **Weak electrolytes** conduct electricity poorly in dilute aqueous solution. Aqueous solutions of **nonelectrolytes** do not conduct electricity. Electrical current is carried through aqueous solution by the movement of ions. The strength of an electrolyte depends upon the number of ions in solution and also on the charges on these ions (see Figure 4-2).

Recall that *ions* are charged particles. The movement of charged particles conducts electricity.

Dissociation refers to the process in which a solid *ionic compound*, such as NaCl, separates into its ions in solution. **Ionization** refers to the process in which a *molecular compound* separates to form ions in solution. Molecular compounds, for example *pure* HCl, exist as discrete molecules and do not contain ions. However, many such compounds ionize in solution to form ions.

Three major classes of solutes are strong electrolytes: (1) strong acids, (2) strong soluble bases, and (3) most soluble salts. *These compounds are completely or nearly completely ionized (or dissociated) in dilute aqueous solutions*, and therefore are strong electrolytes.

Acids and bases are further identified in Subsections 2, 3, and 4.

(a)

(b)

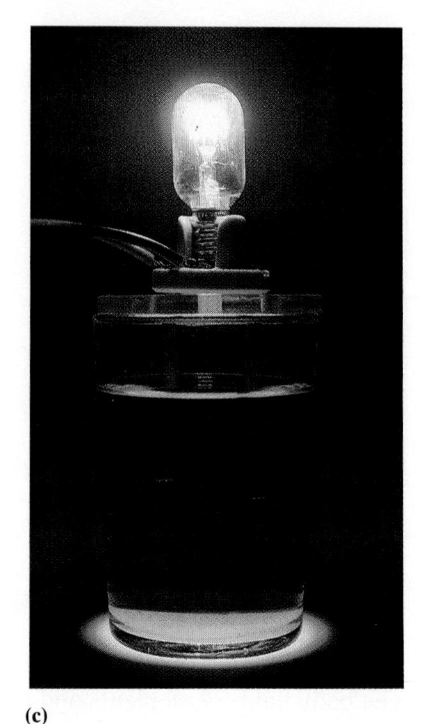

(c)

Figure 4-2
An experiment to demonstrate the presence of ions in solution. Two copper electrodes dip into a liquid in a beaker. When the liquid contains significant concentrations of ions, the ions move between the electrodes to complete the circuit (which includes a light bulb). (a) Pure water is a nonelectrolyte. (b) A solution of a weak electrolyte, acetic acid (CH_3COOH); it contains low concentrations of ions, and so the bulb glows dimly. (c) A solution of a strong electrolyte, potassium chromate (K_2CrO_4); it contains a high concentration of ions, and so the bulb glows brightly.

Recall that positively charged ions are called *cations* while negatively charged ions are called *anions*. Even though the formula for a salt may include H or OH, it must contain another cation *and* another anion. For example, $NaHSO_4$ and $Al(OH)_2Cl$ are salts.

An **acid** can be defined as a substance that produces hydrogen ions, H^+, in aqueous solutions. A **base** is a substance that produces hydroxide ions, OH^-, in aqueous solutions. A **salt** is a compound that contains a cation other than H^+ and an anion other than hydroxide ion, OH^-, or oxide ion, O^{2-}. As we shall see later in this chapter, salts are formed when acids react with bases.

2 Strong and Weak Acids

Many properties of aqueous solutions of acids are due to $H^+(aq)$ ions. These are described in Section 10-4.

As a matter of convenience we place acids into two classes: strong acids and weak acids. **Strong acids** ionize (separate into ions) completely, or very nearly completely, in dilute aqueous solution. The seven common strong acids and their anions are listed in Table 4-5. Please learn this list because it is short. (The list of common weak acids is long.)

Because strong acids ionize completely or very nearly completely in dilute solutions, their solutions contain (predominantly) the ions of the acid rather than acid molecules. Consider the ionization of hydrochloric acid. Pure hydrogen chloride, HCl, is a molecular compound that is a gas at room

Table 4-5
Common Strong Acids and Their Anions

Common Strong Acids		Anions of These Strong Acids	
Formula	Name	Formula	Name
HCl	hydrochloric acid	Cl^-	chloride ion
HBr	hydrobromic acid	Br^-	bromide ion
HI	hydroiodic acid	I^-	iodide ion
HNO_3	nitric acid	NO_3^-	nitrate ion
$HClO_4$	perchloric acid	ClO_4^-	perchlorate ion
$HClO_3$	chloric acid	ClO_3^-	chlorate ion
H_2SO_4	sulfuric acid	HSO_4^-	hydrogen sulfate ion
		SO_4^{2-}	sulfate ion

temperature and atmospheric pressure. When it dissolves in water, it reacts nearly 100% to produce a solution that contains hydrogen ions and chloride ions:

$$HCl(g) \xrightarrow{H_2O} H^+(aq) + Cl^-(aq) \qquad \text{(to completion)}$$

Similar equations can be written for all strong acids.

The species we have shown as $H^+(aq)$ is sometimes shown as H_3O^+ or $[H(H_2O)^+]$ to emphasize hydration. However, it really exists in varying degrees of hydration, such as $H(H_2O)^+$, $H(H_2O)_2^+$, and $H(H_2O)_3^+$.

Weak acids ionize only slightly (usually less than 5%) in dilute aqueous solution; the list of weak acids is very long. Many common ones are listed in Appendix F, and a few of them and their anions are given in Table 4-6.

To give a more complete description of reactions, we indicate the physical states of reactants and products: (g) for gases, (ℓ) for liquids, and (s) for solids. The notation (aq) following ions indicates that they are hydrated in aqueous solution; that is, they interact with water molecules in solution. The complete ionization of a strong electrolyte is indicated by a single arrow (→).

Because there are so many weak acids, please learn the list of common strong acids (Table 4-5). Then assume that the other acids you encounter are weak.

Table 4-6
Some Common Weak Acids and Their Anions

Common Weak Acids		Anions of These Weak Acids	
Formula	Name	Formula	Name
HF*	hydrofluoric acid	F^-	fluoride ion
CH_3COOH	acetic acid	CH_3COO^-	acetate ion
HCN	hydrocyanic acid	CN^-	cyanide ion
HNO_2†	nitrous acid	NO_2^-	nitrite ion
H_2CO_3†	carbonic acid	HCO_3^-	hydrogen carbonate ion
		CO_3^{2-}	carbonate ion
H_2SO_3†	sulfurous acid	HSO_3^-	hydrogen sulfite ion
		SO_3^{2-}	sulfite ion
H_3PO_4	phosphoric acid	$H_2PO_4^-$	dihydrogen phosphate ion
		HPO_4^{2-}	hydrogen phosphate ion
		PO_4^{3-}	phosphate ion
$(COOH)_2$	oxalic acid	$H(COO)_2^-$	hydrogen oxalate ion
		$(COO)_2^{2-}$	oxalate ion

* HF is a weak acid, whereas HCl, HBr, and HI are strong acids.
† Free acid molecules exist only in dilute aqueous solution or not at all. However, many salts of these acids are common, stable compounds.

Citrus fruits contain citric acid, and so their juices are acidic. This is shown here by the color changes on the indicator paper.

We usually write the formulas of inorganic acids with hydrogen written first. Organic acids can often be recognized by the presence of the COOH group in the formula.

The equation for the ionization of acetic acid, CH_3COOH, in water is typical of weak acids:

$$CH_3COOH(aq) \rightleftharpoons H^+(aq) + CH_3COO^-(aq) \qquad \text{(reversible)}$$

The double arrow ($\rightleftharpoons$) generally signifies that the reaction occurs in *both* directions and that the forward reaction does not go to completion. All of us are familiar with solutions of acetic acid. Vinegar is 5% acetic acid by mass. Our use of oil and vinegar as a salad dressing tells us that acetic acid is a weak acid. To be specific, acetic acid is 0.5% ionized (and 99.5% non-ionized) in 5% solution.

Our stomachs have linings that are much more resistant to attack by acids than are our other tissues.

Acetic acid is one of the organic acids, most of which are weak. A multitude of organic acids occur in living systems. Organic acids contain the carboxylate grouping of atoms, —COOH. They can ionize slightly by breakage of the O—H bond, as shown here for acetic acid:

The carboxylate group —COOH is

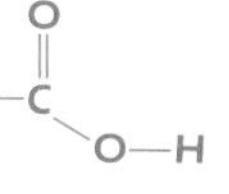

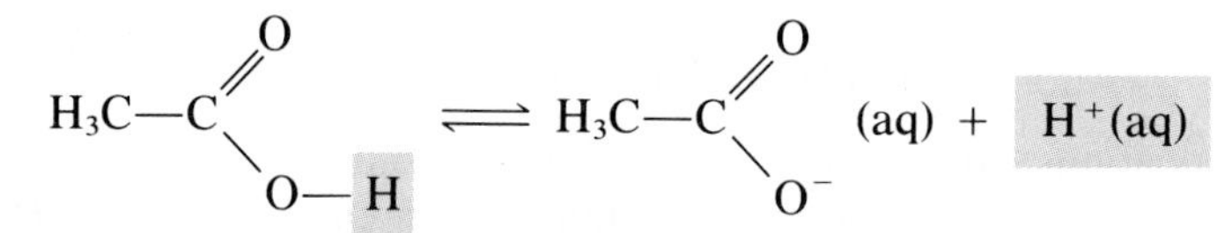

Other organic acids have other groups in the position of the H_3C— group in acetic acid.

Organic acids are discussed in Chapter 31. Carbonic acid, H_2CO_3, and hydrocyanic acid, $HCN(aq)$, are two common acids that contain carbon but that are considered to be *inorganic* acids. Inorganic acids are often called **mineral acids** because they are obtained primarily from mineral (nonliving) sources.

Inorganic acids may be strong or weak; most organic acids are weak.

Example 4-1

In the following lists of common acids, which are strong and which are weak?

(a) H_3PO_4, HCl, H_2CO_3, HNO_3; (b) $HClO_4$, H_2SO_4, $HClO$, HF

Plan

We recall that Table 4-5 lists the common strong acids. Other *common* acids are assumed to be weak.

Solution

(a) HCl and HNO_3 are strong acids; H_3PO_4 and H_2CO_3 are weak acids.
(b) $HClO_4$ and H_2SO_4 are strong acids; $HClO$ and HF are weak acids.

3 Reversible Reactions

Reactions that can occur in both directions are **reversible reactions**. We use a double arrow ($\rightleftharpoons$) to indicate that a reaction is *reversible*. What is the fundamental difference between reactions that go to completion and those that are reversible? We have seen that the ionization of HCl in water is nearly complete. Suppose we dissolve some table salt, NaCl, in water and then add some dilute nitric acid to it. The resulting solution contains hydrogen ions and chloride ions, the products of the ionization of HCl, as well as sodium ions and nitrate ions. The H^+ and Cl^- ions do *not* react significantly to form nonionized molecules of HCl, the reverse of the ionization of HCl.

$$H^+(aq) + Cl^-(aq) \longrightarrow \text{no reaction}$$

In contrast, when a sample of sodium acetate, $NaCH_3COO$, is dissolved in H_2O and mixed with nitric acid, the resulting solution initially contains Na^+, CH_3COO^-, H^+, and NO_3^- ions. But most of the H^+ and CH_3COO^- ions combine to produce nonionized molecules of acetic acid, the reverse of the ionization of the acid. Thus, the ionization of acetic acid, like that of any other weak electrolyte, is reversible.

Na^+ and NO_3^- ions do not combine because $NaNO_3$ is soluble in water.

$$H^+(aq) + CH_3COO^-(aq) \rightleftharpoons CH_3COOH(aq) \qquad \text{(reversible)}$$

4 Strong Soluble Bases, Insoluble Bases, and Weak Bases

Most common bases are ionic metal hydroxides. Most are insoluble in water. **Strong soluble bases** are soluble in water and are dissociated completely in dilute aqueous solution. The common strong soluble bases are listed in Table 4-7. They are the hydroxides of the Group IA metals and the heavier members of Group IIA. The equation for the dissociation of sodium hydroxide in water is typical. Similar equations can be written for other strong soluble bases.

Solutions of bases have a set of common properties due to the OH^- ion. These are described in Section 10-4.

$$NaOH(s) \xrightarrow{H_2O} Na^+(aq) + OH^-(aq) \qquad \text{(to completion)}$$

Strong soluble bases are ionic compounds in the solid state.

Other metals form ionic hydroxides, but these are so sparingly soluble in water that they cannot produce strongly basic solutions. They are called **insoluble bases**. Typical examples include $Cu(OH)_2$, $Zn(OH)_2$, $Fe(OH)_2$, and $Fe(OH)_3$.

Many **weak bases**, such as ammonia, NH_3, are very soluble in water but ionize only slightly in solution.

$$NH_3(aq) + H_2O(\ell) \rightleftharpoons NH_4^+(aq) + OH^-(aq) \qquad \text{(reversible)}$$

The weak bases are *molecular* substances that ionize only slightly in water; they are sometimes called molecular bases.

Table 4-7
Common Strong Soluble Bases

Group IA		Group IIA	
LiOH	lithium hydroxide		
NaOH	sodium hydroxide		
KOH	potassium hydroxide	$Ca(OH)_2$	calcium hydroxide
RbOH	rubidium hydroxide	$Sr(OH)_2$	strontium hydroxide
CsOH	cesium hydroxide	$Ba(OH)_2$	barium hydroxide

Example 4-2

From the following lists, choose (i) the strong soluble bases, (ii) the insoluble bases, and (iii) the weak bases. (a) $NaOH$, $Cu(OH)_2$, $Pb(OH)_2$, $Ba(OH)_2$; (b) $Fe(OH)_3$, KOH, $Mg(OH)_2$, $Sr(OH)_2$, NH_3

Plan

(i) We recall that Table 4-7 lists the *common strong soluble bases.* (ii) Other common metal hydroxides are assumed to be *insoluble* bases. (iii) Ammonia and its derivatives, the amines, are the common *weak bases.*

Solution

(a) (i) The strong soluble bases are $NaOH$ and $Ba(OH)_2$, so
(ii) the insoluble bases are $Cu(OH)_2$ and $Pb(OH)_2$.
(b) (i)The strong soluble bases are KOH and $Sr(OH)_2$, so
(ii) the insoluble bases are $Fe(OH)_3$ and $Mg(OH)_2$, and
(iii) the weak base is NH_3.

EOC 30–32

5 Solubility Rules for Compounds in Aqueous Solution

There is no sharp dividing line between "soluble" and "insoluble" compounds. Compounds whose solubilities fall near the arbitrary dividing line are called "moderately soluble" compounds.

Solubility is a complex phenomenon, and it is not possible to state simple rules that cover all cases. The following rules for solutes in aqueous solutions will be very useful for nearly all acids, bases, and salts encountered in general chemistry. Compounds whose solubility in water is less than about 0.02 mole per liter are usually classified as insoluble compounds, whereas those that are more soluble are classified as soluble compounds. No gaseous or solid substances are infinitely soluble in water. You may wish to review Table 2-2 on page 52, a list of some common ions. Table 7-4 on page 285 contains a more comprehensive list.

1. The common inorganic acids are soluble in water. Low molecular weight organic acids are soluble.
2. The common compounds of the Group IA metals (Li, Na, K, Rb, Cs) and the ammonium ion, NH_4^+, are soluble in water.
3. The common nitrates, NO_3^-; acetates, CH_3COO^-; chlorates, ClO_3^-; and perchlorates, ClO_4^-, are soluble in water.
4. (a) The common chlorides, Cl^-, are soluble in water except $AgCl$, Hg_2Cl_2, and $PbCl_2$.
   (b) The common bromides, Br^-, and iodides, I^-, show approximately the same solubility behavior as chlorides, but there are some exceptions. As these halide ions (Cl^-, Br^-, I^-) increase in size, the solubilities of their slightly soluble compounds decrease.
5. The common sulfates, SO_4^{2-}, are soluble in water except $PbSO_4$, $BaSO_4$, and $HgSO_4$; $CaSO_4$ and Ag_2SO_4 are moderately soluble.
6. The common metal hydroxides, OH^-, are *insoluble* in water except those of the Group IA metals and the heavier members of the Group IIA metals, beginning with $Ca(OH)_2$.

7. The common carbonates, CO_3^{2-}, phosphates, PO_4^{3-}, and arsenates, AsO_4^{3-}, are *insoluble* in water except those of the Group IA metals and NH_4^+. $MgCO_3$ is moderately soluble.
8. The common sulfides, S^{2-}, are *insoluble* in water except those of the Group IA and Group IIA metals and the ammonium ion.

The fact that a substance is insoluble in water does not mean that it cannot take part in a reaction in contact with water.

Table 4-8 summarizes much of the information about the solubility rules.

We shall now discuss chemical reactions in some detail. Because millions of reactions are known, it is useful to group them into classes, or types, so that we can deal with such massive amounts of information systematically. We shall classify them as (1) precipitation reactions, (2) acid–base reactions, (3) displacement reactions, and (4) oxidation–reduction reactions. We shall also distinguish among (1) formula unit, (2) total ionic, and (3) net ionic equations for chemical reactions and indicate the advantages and disadvantages of these methods for representing chemical reactions. As we study different kinds of chemical reactions, we shall learn to predict the products of other similar reactions.

In Chapter 6 we shall describe typical reactions of hydrogen, oxygen, and their compounds. These reactions will illustrate periodic relationships with

Table 4-8
Solubility of Common Ionic Compounds in Water

Generally Soluble	Exceptions
Na^+, K^+, NH_4^+ compounds	No common exceptions
chlorides (Cl^-)	Insoluble: $AgCl$, Hg_2Cl_2 Soluble in hot water: $PbCl_2$
bromides (Br^-)	Insoluble: $AgBr$, Hg_2Br_2, $PbBr_2$ Moderately soluble: $HgBr_2$
iodides (I^-)	Insoluble: many heavy metal iodides
sulfates (SO_4^{2-})	Insoluble: $BaSO_4$, $PbSO_4$, $HgSO_4$ Moderately soluble: $CaSO_4$, $SrSO_4$, Ag_2SO_4
nitrates (NO_3^-), nitrites (NO_2^-)	Moderately soluble: $AgNO_2$
chlorates (ClO_3^-), perchlorates (ClO_4^-), permanganates (MnO_4^-)	Moderately soluble: $KClO_4$
acetates (CH_3COO^-)	Moderately soluble: $AgCH_3COO$
Generally Insoluble	**Exceptions**
sulfides (S^{2-})	Soluble: those of NH_4^+, Na^+, K^+, Mg^{2+}, Ca^{2+}
oxides (O^{2-}), hydroxides (OH^-)	Soluble: Li_2O*, $LiOH$, Na_2O*, $NaOH$, K_2O*, KOH, BaO*, $Ba(OH)_2$ Moderately soluble: CaO*, $Ca(OH)_2$, SrO*, $Sr(OH)_2$
carbonates (CO_3^{2-}), phosphates (PO_4^{3-}), arsenates (AsO_4^{3-})	Soluble: those of NH_4^+, Na^+, K^+

* Dissolves with evolution of heat and formation of hydroxides.

respect to chemical properties. Our system is not an attempt to transform nature so that it fits into small categories; rather it is an attempt to give some order to our observations of nature. We shall see that many oxidation–reduction reactions also fit into more than one category, and that some reactions do not fit neatly into any of these categories.

We have distinguished between strong and weak electrolytes and between soluble and insoluble compounds. Let us now see how we can describe chemical reactions in aqueous solutions.

4-3 Describing Reactions in Aqueous Solutions

Many important chemical reactions occur in aqueous solutions. Let us consider some of these kinds of reactions and show how we write the chemical equations that describe them.

We use three kinds of chemical equations to describe reactions in aqueous solutions. Table 4-9 shows the kinds of information about each substance that we use in writing equations for reactions in aqueous solutions. Some typical examples are included. Please study Table 4-9 carefully.

1. In **formula unit (molecular) equations**, we show complete formulas for all compounds. When metallic copper is added to a solution of (colorless) silver nitrate, the more active metal—copper—displaces silver ions from the solution. The resulting solution contains blue copper(II) nitrate, and metallic silver forms as a finely divided solid (Figure 4-3):

Because we have not studied periodic trends in properties of transition metals, it would be difficult for you to predict that Cu is more active than Ag. The fact that this reaction occurs (see Figure 4-3) shows that it is.

$$2AgNO_3(aq) + Cu(s) \longrightarrow 2Ag(s) + Cu(NO_3)_2(aq)$$

Both silver nitrate and copper(II) nitrate are soluble ionic compounds (for solubility rules see page 132 and Table 4-8).

Table 4-9
Bonding, Solubility, Electrolyte Characteristics, and Predominant Forms of Solutes in Contact with Water

	Acids		Bases			Salts	
	Strong acids	Weak acids	Strong soluble bases	Insoluble bases	Weak bases	Soluble salts	Insoluble salts
Examples	HCl, HNO_3	CH_3COOH, HF	NaOH, $Ca(OH)_2$	$Mg(OH)_2$, $Al(OH)_3$	NH_3, CH_3NH_2	KCl, $NaNO_3$, NH_4Br	$BaSO_4$, AgCl, $Ca_3(PO_4)_2$
Pure compound ionic or covalent?	Covalent	Covalent	Ionic	Ionic	Covalent	Ionic	Ionic
Water soluble or insoluble?	Soluble*	Soluble*	Soluble	Insoluble	Soluble†	Soluble	Insoluble
~100% ionized or dissociated in dilute aqueous solution?	Yes	No	Yes	(footnote ‡)	No	Yes§	(footnote ‡)
Written in ionic equations as	Separate ions	Molecules	Separate ions	Complete formulas	Molecules	Separate ions	Complete formulas

* Most common inorganic acids and the low-molecular-weight organic acids (—COOH) are water soluble.
† The low-molecular-weight amines are water-soluble.
‡ The *very small concentrations* of "insoluble" metal hydroxides and insoluble salts in saturated aqueous solutions are nearly completely dissociated.
§ There are a few exceptions. A few soluble salts are molecular (and not ionic) compounds.

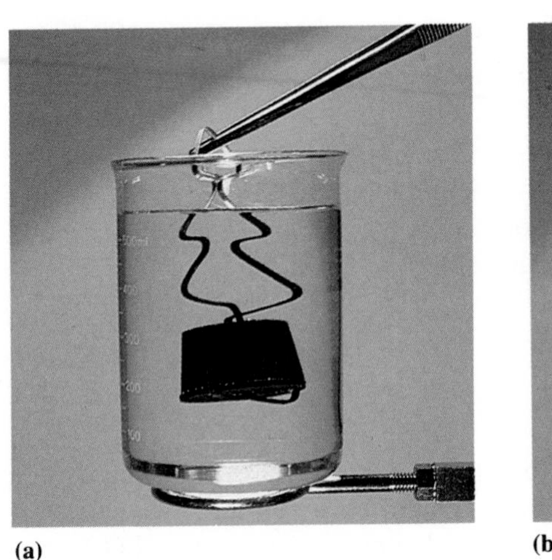
(a)

(b)

Figure 4-3
(a) Copper wire and a silver nitrate solution. (b) The copper wire has been placed in the solution and some finely divided silver has deposited on the wire. The solution is blue because it contains copper(II) nitrate.

2. In **total ionic equations**, formulas are written to show the (predominant) form in which each substance exists when it is in contact with aqueous solution. We often use brackets in total ionic equations to show ions that have a common source or that remain in solution after reaction is complete. The total ionic equation for this reaction is

$$2[Ag^+(aq) + NO_3^-(aq)] + Cu(s) \longrightarrow 2Ag(s) + [Cu^{2+}(aq) + 2NO_3^-(aq)]$$

Examination of the total ionic equation shows that NO_3^- ions do not participate in the reaction. They are sometimes called "spectator" ions.

3. The **net ionic equation** shows only the species that react. The net ionic equation is obtained by eliminating the spectator ions and the brackets from the total ionic equation.

$$2Ag^+(aq) + Cu(s) \longrightarrow 2Ag(s) + Cu^{2+}(aq)$$

Brackets are not used in net ionic equations.

Net ionic equations allow us to focus on the *essence* of a chemical reaction in aqueous solutions—that is, the changes that *really* occur. On the other hand, if we are dealing with stoichiometric calculations we frequently must deal with formula weights and therefore with the *complete* formulas of all species. In such cases, formula unit equations are more useful. Total ionic equations provide the bridge between the two.

Here is why it is important to know how and when to construct net ionic equations from molecular equations.

In general, you must answer two questions about a substance to determine whether it should be written in ionic form in an ionic equation:

1. Is it soluble in water?
2. If it is soluble, is it mostly ionized or dissociated in water?

If *both* answers are yes, the substance is a soluble strong electrolyte, and its formula is written in ionic form. If *either* answer is no, the full formula is written. To answer these questions it is necessary to know the lists of strong acids (Table 4-5) and strong soluble bases (Table 4-7). These acids and bases are completely or almost completely ionized in dilute aqueous solutions. Other common acids and bases are either insoluble or only slightly ionized. In addition, the solubility rules (page 132 and Table 4-8) allow you to determine which salts are soluble in water. Most salts that are soluble in water are also strong electrolytes. Exceptions such as lead acetate, $Pb(CH_3COO)_2$, which is soluble but predominantly nonionized, will be noted as they are encountered.

> The only common substances that should be written in ionized or dissociated form in ionic equations are (1) strong acids, (2) strong soluble bases, and (3) soluble ionic salts.

4-4 Precipitation Reactions

To understand the discussion of precipitation reactions, you *must* know the solubility rules (page 132) and Table 4-8.

In **precipitation reactions** an insoluble solid, a **precipitate**, forms and then settles out of solution. Our teeth and bones were formed by very slow precipitation reactions in which mostly calcium phosphate $Ca_3(PO_4)_2$ was deposited in the correct geometric arrangements.

An example of a precipitation reaction is the formation of bright yellow insoluble lead(II) chromate as a result of mixing solutions of the soluble ionic compounds lead(II) nitrate and potassium chromate (Figure 4-4). The other product of the reaction is KNO_3, a soluble ionic salt.

The balanced formula unit, total ionic, and net ionic equations for this reaction follow.

$$Pb(NO_3)_2(aq) + K_2CrO_4(aq) \longrightarrow PbCrO_4(s) + 2KNO_3(aq)$$

$$[Pb^{2+}(aq) + 2\ NO_3^-(aq)] + [2K^+(aq) + CrO_4^{2-}(aq)] \longrightarrow PbCrO_4(s) + 2[K^+(aq) + NO_3^-(aq)]$$

$$Pb^{2+}(aq) + CrO_4^{2-}(aq) \longrightarrow PbCrO_4(s)$$

Figure 4-4
A precipitation reaction. When K_2CrO_4 solution is added to aqueous $Pb(NO_3)_2$ solution, the yellow compound $PbCrO_4$ precipitates. The resulting solution contains K^+ and NO_3^- ions, the ions of KNO_3.

Another important precipitation reaction involves the formation of insoluble carbonates (solubility rule 7). Limestone deposits are mostly calcium carbonate, $CaCO_3$, although many also contain significant amounts of magnesium carbonate, $MgCO_3$.

Suppose we mix together aqueous solutions of sodium carbonate, Na_2CO_3, and calcium chloride, $CaCl_2$. We recognize that *both* Na_2CO_3 and $CaCl_2$ (solubility rules 2, 4a, and 7) are soluble ionic compounds. At the instant of mixing, the resulting solution contains four ions:

$$Na^+(aq),\quad CO_3^{2-}(aq),\quad Ca^{2+}(aq),\quad Cl^-(aq)$$

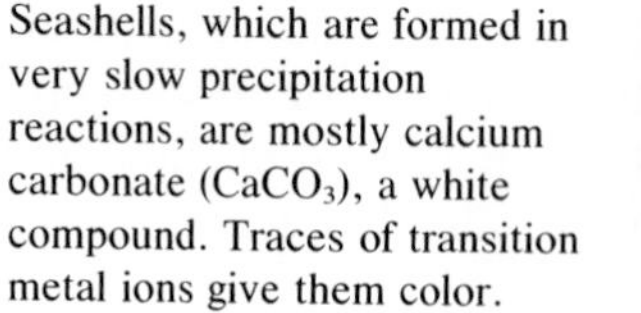

Seashells, which are formed in very slow precipitation reactions, are mostly calcium carbonate ($CaCO_3$), a white compound. Traces of transition metal ions give them color.

One pair of ions, Na^+ and Cl^-, *cannot* form an insoluble compound (solubility rules 2 and 4). We look for a pair of ions that would form an insoluble compound. Ca^{2+} ions and CO_3^{2-} ions are such a combination; they form insoluble $CaCO_3$ (solubility rule 7). The equations for the reaction follow.

$$CaCl_2(aq) + Na_2CO_3(aq) \longrightarrow CaCO_3(s) + 2\,NaCl(aq)$$

$$[Ca^{2+}(aq) + 2\,Cl^-(aq)] + [2Na^+(aq) + CO_3^{2-}(aq)] \longrightarrow CaCO_3(s) + 2[Na^+(aq) + Cl^-(aq)]$$

$$Ca^{2+} + CO_3^{2-}(aq) \longrightarrow CaCO_3(s)$$

Example 4-3

Will a precipitate form when aqueous solutions of $Ca(NO_3)_2$ and NaCl are mixed in reasonable concentrations?

Plan

We recognize that both $Ca(NO_3)_2$ (solubility rule 3) and NaCl (solubility rules 2 and 4) are soluble compounds. At the instant of mixing, the resulting solution contains four ions:

$$Ca^{2+}(aq), \quad NO_3^-(aq), \quad Na^+(aq), \quad Cl^-(aq)$$

New combinations of ions *could* be $CaCl_2$ and $NaNO_3$. Solubility rule 4 tells us that $CaCl_2$ is a soluble compound, while solubility rules 2 and 3 tell us that $NaNO_3$ is a soluble compound.

Solution

Therefore no precipitate forms in this solution.

Example 4-4

Will a precipitate form when aqueous solutions of $CaCl_2$ and K_3PO_4 are mixed in reasonable concentrations? Write the appropriate equations for any reaction.

Plan

Both $CaCl_2$ (solubility rule 4) and K_3PO_4 (solubility rule 2) are soluble compounds. At the instant of mixing, four ions are present in the solution:

$$Ca^{2+}(aq), \quad Cl^-(aq), \quad K^+(aq), \quad PO_4^{3-}(aq)$$

New combinations of these ions *could* be KCl and $Ca_3(PO_4)_2$. Solubility rules 2 and 4 tell us that potassium chloride, KCl, is a soluble compound.

Solution

Solubility rule 7 tells us that calcium phosphate, $Ca_3(PO_4)_2$, is an insoluble compound and so it forms a precipitate.

The equations for the formation of calcium phosphate follow.

$$3CaCl_2(aq) + 2K_3PO_4(aq) \longrightarrow Ca_3(PO_4)_2(s) + 6KCl(aq)$$

$$3[Ca^{2+}(aq) + 2\,Cl^-(aq)] + 2[3K^+(aq) + PO_4^{3-}(aq)] \longrightarrow Ca_3(PO_4)_2(s) + 6[K^+(aq) + Cl^-(aq)]$$

$$3Ca^{2+}(aq) + 2PO_4^{3-}(aq) \longrightarrow Ca_3(PO_4)_2(s)$$

EOC 40, 42

4-5 Acid–Base Reactions

The manufacture of fertilizers consumes more H_2SO_4 *and* more NH_3 than any other single use.

Acid–base reactions are among the most important kinds of chemical reactions. Many acid–base reactions occur in nature in both plants and animals. Many acids and bases are essential compounds in an industrialized society (see Table 4-10). For example, approximately 350 pounds of sulfuric acid, H_2SO_4, and approximately 135 pounds of ammonia, NH_3, are required to support the lifestyle of an average American for one year.

The reaction of an acid with a metal hydroxide base produces a salt and water. Such reactions are called **neutralization reactions** because the typical properties of acids and bases are neutralized.

In a *neutralization reaction* H^+ ions and OH^- combine to form water molecules.

Consider the reaction of hydrochloric acid, HCl, with aqueous sodium hydroxide, NaOH. Table 4-5 tells us that HCl is a strong acid, and Table 4-7 tells us that NaOH is a strong soluble base. The salt sodium chloride, NaCl, is formed in this reaction. It contains the cation of its parent base,

Table 4-10
1990 Production of Acids, Bases, and Salts in the United States

Formula	Name	Billions of Pounds	Major Uses
H_2SO_4	sulfuric acid	88.56	Manufacture of fertilizers and other chemicals
CaO, $Ca(OH)_2$	lime (calcium oxide and calcium hydroxide)	34.80	Manufacture of other chemicals, steelmaking, water treatment
NH_3	ammonia	33.92	Fertilizer; manufacture of fertilizers and other chemicals
H_3PO_4	phosphoric acid	24.35	Manufacture of fertilizers and detergents
$NaOH$	sodium hydroxide	23.28	Manufacture of other chemicals, pulp and paper, soap and detergents, aluminum, textiles
Na_2CO_3	sodium carbonate (soda ash)	19.85	Manufacture of glass, other chemicals, detergents, pulp, and paper
HNO_3	nitric acid	15.50	Manufacture of fertilizers, explosives, plastics, and lacquers
NH_4NO_3	ammonium nitrate	14.21	Fertilizer and explosive
$C_6H_4(COOH)_2$*	terephthalic acid	7.69	Manufacture of fibers (polyesters), films, and bottles
$(NH_4)_2SO_4$	ammonium sulfate	4.99	Fertilizer
HCl	hydrochloric acid	4.68	Manufacture of other chemicals and rubber; metal cleaning
CH_3COOH*	acetic acid	3.76	Manufacture of acetate esters
KOH, K_2CO_3	potash	3.62	Manufacture of fertilizers
$Al_2(SO_4)_2$	aluminum sulfate	2.42	Water treatment; dyeing textiles
Na_2SiO_3	sodium silicate	1.76	Manufacture of detergents, cleaning agents, and adhesives
$C_4H_8(COOH)_2$*	adipic acid	1.64	Manufacture of Nylon 66
Na_2SO_4	sodium sulfate	1.47	Manufacture of paper, glass, and detergents
$CaCl_2$	calcium chloride	1.38	De-icing roads in winter, controlling dust in summer, concrete additive

* Organic compound.

Na^+, and the anion of its parent acid, Cl^-. Solubility rules 2 and 4 tell us that NaCl is a soluble salt.

$$HCl(aq) + NaOH(aq) \longrightarrow H_2O(\ell) + NaCl(aq)$$

$$[H^+(aq) + Cl^-(aq)] + [Na^+(aq) + OH^-(aq)] \longrightarrow H_2O(\ell) + [Na^+(aq) + Cl^-(aq)]$$

$$H^+(aq) + OH^-(aq) \longrightarrow H_2O(\ell)$$

The net ionic equation for *all* reactions of strong acids with strong soluble bases that form soluble salts and water is

$$H^+(aq) + OH^-(aq) \longrightarrow H_2O(\ell)$$

Example 4-5

Predict the products of the reaction between HI(aq) and $Ca(OH)_2$(aq). Write balanced formula unit, total ionic, and net ionic equations.

Plan

This is an acid–base neutralization reaction; the products are H_2O and the salt that contains the cation of the base, Ca^{2+}, and the anion of the acid, I^-; CaI_2 is a soluble salt (solubility rule 4). HI is a strong acid (Table 4-5), $Ca(OH)_2$ is a strong soluble base (Table 4-7), and CaI_2 is a soluble ionic salt, so all are written in ionic form.

Solution

$$2HI(aq) + Ca(OH)_2(aq) \longrightarrow CaI_2(aq) + 2H_2O(\ell)$$

$$2[H^+(aq) + I^-(aq)] + [Ca^{2+}(aq) + 2OH^-(aq)] \longrightarrow [Ca^{2+}(aq) + 2I^-(aq)] + 2H_2O(\ell)$$

We cancel the spectator ions.

$$2H^+(aq) + 2OH^-(aq) \longrightarrow 2H_2O(\ell)$$

Dividing by 2 gives the net ionic equation

$$H^+(aq) + OH^-(aq) \longrightarrow H_2O(\ell)$$

Recall that in balanced equations we show the smallest whole-number coefficients possible.

EOC 45

Reactions of *weak* acids with strong soluble bases also produce salts and water, but there is a significant difference in the balanced ionic equations because weak acids are only *slightly* ionized.

Example 4-6

Write balanced formula unit, total ionic, and net ionic equations for the reaction of acetic acid with potassium hydroxide.

Plan

Neutralization reactions produce a salt and water. CH_3COOH is a weak acid (Table 4-6) and so it is written in molecular form. KOH is a strong soluble base

(Table 4-7) and KCH_3COO is a soluble salt (solubility rules 2 and 3), and so both are written in ionic form.

Solution

$$CH_3COOH(aq) + KOH(aq) \longrightarrow KCH_3COO(aq) + H_2O(\ell)$$

$$CH_3COOH(aq) + [K^+(aq) + OH^-(aq)] \longrightarrow [K^+(aq) + CH_3COO^-(aq)] + H_2O(\ell)$$

The spectator ion is K^+, the cation of the strong soluble base, KOH.

$$CH_3COOH(aq) + OH^-(aq) \longrightarrow CH_3COO^-(aq) + H_2O(\ell)$$

Thus, we see that *this* net ionic equation includes *molecules* of the weak acid and *anions* of the weak acid.

EOC 46, 47

A monoprotic acid contains one acidic H per formula unit.

The reactions of *weak monoprotic acids* with *strong soluble bases* that form *soluble salts* can be represented in general terms as

$$HA(aq) + OH^-(aq) \longrightarrow A^-(aq) + H_2O(\ell)$$

where HA represents the weak acid and A^- represents its anion.

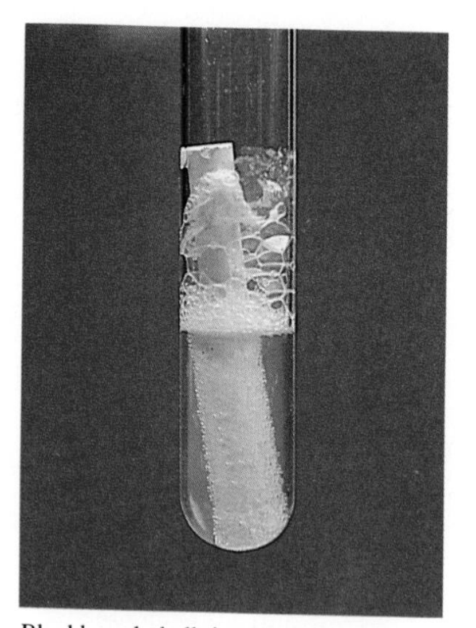

Blackboard chalk is mostly calcium carbonate, $CaCO_3$. Bubbles of carbon dioxide, CO_2, are clearly visible in this photograph of $CaCO_3$ dissolving in HCl.

Removal of Ions from Aqueous Solutions

Many reactions that occur in aqueous solution result in the removal of ions from the solution. This happens in one of three ways: (1) formation of predominantly nonionized molecules (weak or nonelectrolyte) in solution, (2) formation of a precipitate, or (3) formation of a gas that escapes. We have discussed examples of the first two in acid–base neutralization and precipitation reactions.

Let us illustrate the third case. When an acid—for example, hydrochloric acid—is added to solid calcium carbonate, a reaction occurs in which carbonic acid, a weak acid, is produced.

$$2HCl(aq) + CaCO_3(s) \longrightarrow H_2CO_3(aq) + CaCl_2(aq)$$

$$2[H^+(aq) + Cl^-(aq)] + CaCO_3(s) \longrightarrow H_2CO_3(aq) + [Ca^{2+}(aq) + 2Cl^-(aq)]$$

$$2H^+(aq) + CaCO_3(s) \longrightarrow H_2CO_3(aq) + Ca^{2+}(aq)$$

The heat generated in the reaction causes thermal decomposition of carbonic acid to gaseous carbon dioxide and water:

$$H_2CO_3(aq) \longrightarrow CO_2(g) + H_2O(\ell)$$

Most of the CO_2 bubbles off and the reaction goes to completion (with respect to the limiting reactant). The net effect is the conversion of ionic species into nonionized molecules of a gas (CO_2) and water.

4-6 Displacement Reactions

Reactions in which one element displaces another from a compound are called **displacement reactions**. Active metals displace less active metals or hydrogen from their compounds in aqueous solution. Active metals are those that readily lose electrons to form cations (see Table 4-11).

1 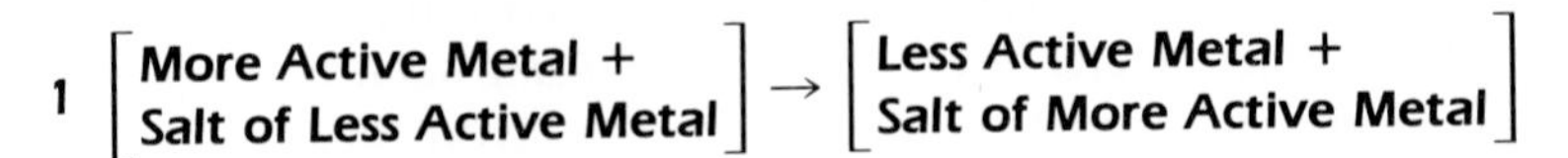

The reaction of copper with silver nitrate that was described in detail in Section 4-3 is typical. Please refer to it.

Table 4-11
Activity Series of the Metals

Metal	Displace hydrogen from nonoxidizing acids	Displace hydrogen from steam	Displace hydrogen from cold water
Li	Displace hydrogen from nonoxidizing acids	Displace hydrogen from steam	Displace hydrogen from cold water
K			
Ca			
Na			
Mg			
Al			
Mn			
Zn			
Cr			
Fe			
Cd			
Co			
Ni			
Sn			
Pb			
H (a nonmetal)			
Sb (a metalloid)			
Cu			
Hg			
Ag			
Pt			
Au			

(Li through Na: displace hydrogen from cold water; Li through Fe: displace hydrogen from steam; Li through Pb: displace hydrogen from nonoxidizing acids.)

Example 4-7

A large piece of zinc metal is placed in a copper(II) sulfate, $CuSO_4$, solution. The blue solution becomes colorless as copper metal falls to the bottom of the container. The resulting solution contains zinc sulfate, $ZnSO_4$. Write balanced formula unit, total ionic, and net ionic equations for the reaction.

Plan

The metals zinc and copper are *not* ionized or dissociated in contact with H_2O. Both $CuSO_4$ and $ZnSO_4$ are soluble salts (solubility rule 5), and so they are written in ionic form.

Solution

$$CuSO_4(aq) + Zn(s) \longrightarrow Cu(s) + ZnSO_4(aq)$$

$$[Cu^{2+}(aq) + SO_4^{2-}(aq)] + Zn(s) \longrightarrow Cu(s) + [Zn^{2+}(aq) + SO_4^{2-}(aq)]$$

$$Cu^{2+}(aq) + Zn(s) \longrightarrow Cu(s) + Zn^{2+}(aq)$$

In this *displacement reaction*, the more active metal, zinc, displaces the ions of the less active metal, copper, from aqueous solution.

A strip of zinc was placed in a blue solution of copper(II) sulfate, $CuSO_4$. The copper has been displaced from solution and has fallen to the bottom of the beaker. The resulting zinc sulfate solution is colorless.

2 [Active Metal + Nonoxidizing Acid] → [Hydrogen + Salt of Acid]

A common method for the preparation of small amounts of hydrogen involves the reaction of active metals with nonoxidizing acids, such as HCl and H_2SO_4. For example, when zinc is dissolved in H_2SO_4, the reaction produces zinc sulfate; hydrogen is displaced from the acid, and it bubbles off as gaseous H_2. The formula unit equation for this reaction is

$$Zn(s) + \underset{\text{strong acid}}{H_2SO_4(aq)} \longrightarrow \underset{\text{soluble salt}}{ZnSO_4(aq)} + H_2(g)$$

Both sulfuric acid (in dilute solution) and zinc sulfate exist primarily as ions, so the total ionic equation is

$$Zn(s) + [2H^+(aq) + SO_4^{2-}(aq)] \longrightarrow [Zn^{2+}(aq) + SO_4^{2-}(aq)] + H_2(g)$$

Zinc dissolves in dilute H_2SO_4 to produce H_2 and a solution that contains $ZnSO_4$.

Elimination of unreacting species common to both sides of the total ionic equation gives the net ionic equation:

$$Zn(s) + 2H^+(aq) \longrightarrow Zn^{2+}(aq) + H_2(g)$$

Table 4-11 lists the **activity series**. When any metal listed above hydrogen in this series is added to solutions of *nonoxidizing* acids such as hydrochloric acid, HCl, and sulfuric acid, H_2SO_4, the metal dissolves to produce hydrogen, and a salt is formed. HNO_3 is the common *oxidizing acid*. It reacts with active metals to produce oxides of nitrogen, but *not* hydrogen, H_2.

Example 4-8

Which of the following metals can displace hydrogen from hydrochloric acid solution? Write appropriate equations for any reactions that can occur.

Al, Cu, Ag

Plan

The activity series of the metals, Table 4-11, tells us that copper and silver *do not* displace hydrogen from solutions of nonoxidizing acids. Aluminum is an active metal that can displace H_2 from HCl and form aluminum chloride (photo on page 150).

Solution

$$6HCl(aq) + 2Al(s) \longrightarrow 3H_2(g) + 2AlCl_3(aq)$$

$$6[H^+(aq) + Cl^-(aq)] + 2Al(s) \longrightarrow 3H_2(g) + 2[Al^{3+}(aq) + 3Cl^-(aq)]$$

$$6H^+(aq) + 2Al(s) \longrightarrow 3H_2(g) + 2Al^{3+}(aq)$$

EOC 55, 56

Very active metals can even displace hydrogen from water. However, such reactions of very active metals of Group IA are dangerous because they generate enough heat to cause explosive ignition of the hydrogen (Figure 4-5). The reaction of potassium, or another metal of Group IA, with water is also a *displacement reaction*:

$$2K(s) + 2H_2O(\ell) \longrightarrow 2[K^+(aq) + OH^-(aq)] + H_2(g)$$

Figure 4-5
Potassium, like other Group IA metals, reacts vigorously with water. The room was completely dark, and all the light for this photograph was produced by dropping a small piece of potassium into a beaker of water.

Example 4-9

Which of the following metals can displace hydrogen from water at room temperature? Write appropriate equations for any reactions that can occur.

Sn, Ca, Hg

Plan

The activity series, Table 4-11, tells us that tin and mercury *cannot* displace hydrogen from water. Calcium is a very active metal (Table 4-11) that displaces hydrogen from cold water and forms calcium hydroxide, a strong soluble base.

Solution

$$Ca(s) + 2H_2O(\ell) \longrightarrow H_2(g) + Ca(OH)_2(aq)$$

$$Ca(s) + 2H_2O(\ell) \longrightarrow H_2(g) + [Ca^{2+}(aq) + 2OH^-(aq)]$$

$$Ca(s) + 2H_2O(\ell) \longrightarrow H_2(g) + Ca^{2+}(aq) + 2OH^-(aq)$$

EOC 59, 60

The reaction of calcium with water at room temperature produces a lazy stream of bubbles of hydrogen.

3 [Active Nonmetal + Salt of Less Active Nonmetal] → [Less Active Nonmetal + Salt of Active Nonmetal]

Many *nonmetals* displace less active nonmetals from combination with a metal or other cation. For example, when chlorine is bubbled through a solution containing bromide ions (derived from a soluble ionic salt such as sodium bromide, NaBr), chlorine displaces bromide ions to form elemental bromine and chloride ions (as aqueous sodium chloride):

$$\underset{\text{chlorine}}{Cl_2(g)} + \underset{\text{sodium bromide}}{2[Na^+(aq) + Br^-(aq)]} \longrightarrow \underset{\text{sodium chloride}}{2[Na^+(aq) + Cl^-(aq)]} + \underset{\text{bromine}}{Br_2(\ell)}$$

Similarly, when bromine is added to a solution containing iodide ions, the iodide ions are displaced by bromine to form iodine and bromide ions:

$$\underset{\text{bromine}}{Br_2(\ell)} + \underset{\text{sodium iodide}}{2[Na^+(aq) + I^-(aq)]} \longrightarrow \underset{\text{sodium bromide}}{2[Na^+(aq) + Br^-(aq)]} + \underset{\text{iodine}}{I_2(s)}$$

Each halogen will displace less active (heavier) halogens from their binary salts; i.e., the order of increasing activities is

$$I_2 < Br_2 < Cl_2 < F_2$$

Activity of the halogens decreases as the group is descended.

Conversely, a halogen will *not* displace more active (lighter) members from their salts:

$$I_2(s) + 2F^- \longrightarrow \text{no reaction}$$

Example 4-10

Which of the following combinations would result in a displacement reaction? Write appropriate equations for any reactions that occur.

(a) $I_2(s) + NaBr(aq) \longrightarrow$

(b) $Cl_2(g) + NaI(aq) \longrightarrow$

(c) $Br_2(\ell) + NaCl(aq) \longrightarrow$

Plan

The activity of the halogens decreases from top to bottom in the periodic table. We see (a) that Br is above I and (c) that Cl is above Br in the periodic table. Therefore neither combination (a) nor combination (c) could result in reaction. Cl is above I in the periodic table, and so combination (b) results in a displacement reaction.

Solution

The more active halogen, Cl_2, displaces the less active halogen, I_2, from its compounds.

$$2NaI(aq) + Cl_2(g) \longrightarrow I_2(s) + 2NaCl(aq)$$

$$2[Na^+(aq) + I^-(aq)] + Cl_2(g) \longrightarrow I_2(s) + 2[Na^+(aq) + Cl^-(aq)]$$

$$2I^-(aq) + Cl_2(g) \longrightarrow I_2(s) + 2Cl^-(aq)$$

EOC 61, 62

4-7 Oxidation Numbers

Many reactions involve the transfer of electrons from one species to another. They are called **oxidation–reduction reactions** or simply **redox reactions**. Some of the reactions discussed earlier in this chapter are also redox reactions, although they were not identified as such. We use oxidation numbers to keep track of electron transfers.

Table 4-12
Common Oxidation Numbers for Group A Elements in Compounds and Ions

Element(s)	Common Ox. Nos.	Examples	Other Ox. Nos.
H	+1	H_2O, CH_4, NH_4Cl	−1 in metal hydrides, e.g., NaH, CaH_2
Group IA	+1	KCl, NaH, $RbNO_3$, K_2SO_4	None
Group IIA	+2	$CaCl_2$, MgH_2, $Ba(NO_3)_2$, $SrSO_4$	None
Group IIIA	+3	$AlCl_3$, BF_3, $Al(NO_3)_3$, GaI_3	None in common compounds
Group IVA	+2 +4	CO, PbO, $SnCl_2$, $Pb(NO_3)_2$ CCl_4, SiO_2, SiO_3^{2-}, $SnCl_4$	Many others are also seen, including −4, −3, −2, −1, +1, +3
Group VA	−3 in binary compounds with metals −3 in NH_4^+, binary compounds with H	Mg_3N_2, Na_3P, Cs_3As NH_3, PH_3, AsH_3, NH_4^+	+3, e.g., NO_2^-, PCl_3 +5, e.g., NO_3^-, PO_4^{3-}, AsF_5, P_4O_{10}
O	−2	H_2O, P_4O_{10}, Fe_2O_3, CaO, ClO_3^-	+2 in OF_2 −1 in peroxides, e.g., H_2O_2, Na_2O_2 $-\frac{1}{2}$ in superoxides, e.g., KO_2, RbO_2
Group VIA (other than O)	−2 in binary compounds with metals and H −2 in binary compounds with NH_4^+	H_2S, CaS, Fe_2S_3, Na_2Se $(NH_4)_2S$, $(NH_4)_2Se$	+4 with O and the lighter halogens, e.g., SO_2, SeO_2, Na_2SO_3, SO_3^{2-}, SF_4 +6 with O and the lighter halogens, e.g., SO_3, TeO_3, H_2SO_4, SO_4^{2-}, SF_6
Group VIIA	−1 in binary compounds with metals and H −1 in binary compounds with NH_4^+	MgF_2, KI, $ZnCl_2$, $FeBr_3$ NH_4Cl, NH_4Br	Except F, with O and the lighter halogens +1, e.g., BrF, ClO^-, BrO^- +3, e.g., ICl_3, ClO_2^-, BrO_2^- +5, e.g., BrF_5, ClO_3^-, BrO_3^- +7, e.g., IF_7, ClO_4^-, BrO_4^-

The **oxidation number**, or **oxidation state**, of an element in a simple *binary* ionic compound is the number of electrons gained or lost by an atom of that element when it forms the compound. In the case of a single-atom ion, it corresponds to the actual charge on the ion. In molecular compounds, oxidation numbers do not have the same physical significance they have in ionic compounds. However, they are very useful mechanical aids in writing formulas and in balancing equations. In molecular species, the oxidation numbers are assigned according to an arbitrary, but useful, set of rules. The element farther to the right and higher up in the periodic table is assigned a negative oxidation number, and the element farther to the left and lower down in the periodic table is assigned a positive oxidation number.

Binary means two. Binary compounds contain two elements.

The general rules for assigning oxidation numbers follow. These rules are not comprehensive, but they cover most cases.

1. The oxidation number of any free, uncombined element is zero. This includes multiatomic elements such as H_2, O_2, O_3, and S_8.

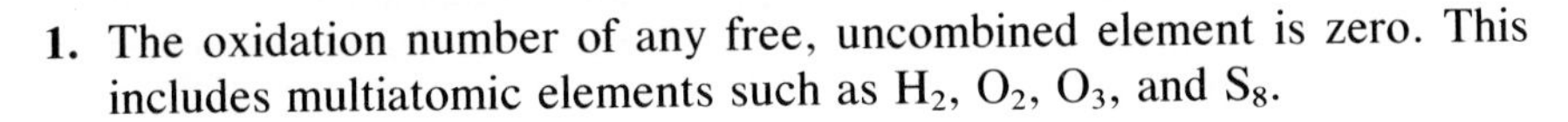

Table 4-13
The Most Common Nonzero Oxidation States (numbers) of the Elements

IA	IIA	IIIB		IVB	VB	VIB	VIIB	VIIIB	VIIIB	VIIIB	IB	IIB	IIIA	IVA	VA	VIA	VIIA	0
1 H +1 +1																	1 H +1 −1	2 He
3 Li +1	4 Be +2												5 B +3	6 C +4 +2 −4	7 N +5 +4 +3 +2 +1 −3	8 O −1 −2	9 F −1	10 Ne
11 Na +1	12 Mg +2												13 Al +3	14 Si +4 −4	15 P +5 +3 −3	16 S +6 +4 +2 −2	17 Cl +7 +5 +3 +1 −1	18 Ar
19 K +1	20 Ca +2	21 Sc +3		22 Ti +4 +3 +2	23 V +5 +4 +3 +2	24 Cr +6 +3 +2	25 Mn +7 +6 +4 +3 +2	26 Fe +3 +2	27 Co +3 +2	28 Ni +2	29 Cu +2 +1	30 Zn +2	31 Ga +3	32 Ge +4 −4	33 As +5 +3 −3	34 Se +6 +4 −2	35 Br +7 +5 +3 +1 −1	36 Kr +4 +2
37 Rb +1	38 Sr +2	39 Y +3		40 Zr +4	41 Nb +5 +4	42 Mo +6 +4 +3	43 Tc +7 +6 +4	44 Ru +8 +6 +4 +3	45 Rh +4 +3 +2	46 Pd +4 +2	47 Ag +1	48 Cd +2	49 In +3	50 Sn +4 +2	51 Sb +5 +3 −3	52 Te +6 +4 −2	53 I +7 +5 +3 +1 −1	54 Xe +6 +4 +2
55 Cs +1	56 Ba +2	57 La +3	58 Ce → 71 Lu +3	72 Hf +4	73 Ta +5	74 W +6 +4	75 Re +7 +6 +4	76 Os +8 +4	77 Ir +4 +3	78 Pt +4 +2	79 Au +3 +1	80 Hg +2 +1	81 Tl +3 +1	82 Pb +4 +2	83 Bi +5 +3	84 Po +2	85 At −1	86 Rn

Groups

The terms "oxidation number" and "oxidation state" are used interchangeably.

2. The oxidation number of an element in a simple (monatomic) ion is the charge on the ion. In a polyatomic ion, the sum of the oxidation numbers of the constituent atoms is equal to the charge on the ion.
3. The sum of the oxidation numbers of all atoms in the compound is zero.

Tables 4-12 and 4-13 (see pp. 144–145) summarize the common nonzero oxidation numbers for the Group A elements.

By convention, *oxidation numbers* are represented as $+n$ and $-n$, while ionic charges are represented as $n+$ and $n-$. We shall circle oxidation numbers associated with formulas and show them in red. Both oxidation numbers and ionic charges can be combined algebraically.

The formula for ammonia, NH_3, is usually written with the more electronegative element first—the only reason is that it has been written this way for many years.

Example 4-11

Determine the oxidation numbers of nitrogen in the following species: (a) N_2O_4, (b) NH_3, (c) NO_3^-, (d) N_2.

Plan

We recall that oxidation numbers are represented *per atom* and that the sum of the oxidation numbers in a molecule is zero, while the sum of the oxidation numbers in an ion equals the charge on the ion.

Solution

(a) The oxidation number of O is -2. The sum of the oxidation numbers for all atoms in a molecule must be zero:

$$\text{ox. no./atom:} \quad \overset{(x)\,(-2)}{N_2O_4}$$

$$\text{total ox. no.:} \quad 2x + 4(-2) = 0 \quad \text{or} \quad x = \boxed{+4}$$

(b) The oxidation number of H is $+1$:

$$\text{ox. no./atom:} \quad \overset{(x)(+1)}{NH_3}$$

$$\text{total ox. no.:} \quad x + 3(1) = 0 \quad \text{or} \quad x = \boxed{-3}$$

(c) The sum of the oxidation numbers for all atoms in an ion equals the charge on the ion:

$$\text{ox. no./atom:} \quad \overset{(x)(-2)}{NO_3^-}$$

$$\text{total ox. no.:} \quad x + 3(-2) = -1 \quad \text{or} \quad x = \boxed{+5}$$

(d) The oxidation number of any free element is zero.

EOC 67, 68

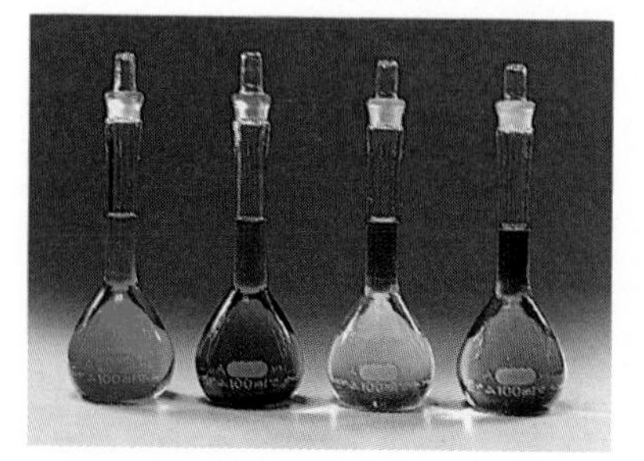

Aqueous solutions of some compounds that contain chromium. Left to right: chromium(II) chloride ($CrCl_2$) is blue; chromium(III) chloride ($CrCl_3$) is green; potassium chromate (K_2CrO_4) is yellow; potassium dichromate ($K_2Cr_2O_7$) is orange.

4-8 Oxidation–Reduction Reactions—An Introduction

Several reactions that we discussed earlier are redox reactions.

Displacement reactions are *always* redox reactions. Acid–base reactions are *never* redox reactions.

We now describe the important terms and illustrate their meanings.

Oxidation–reduction reactions occur in every area of chemistry and biochemistry. We learn to identify oxidizing agents and reducing agents and to balance oxidation–reduction equations. These skills are necessary for the study of electrochemistry in Chapter 21. Electrochemistry involves electron transfer between physically separated oxidizing and reducing agents and interconversions between chemical energy and electrical energy. These skills are also fundamental to biology and biochemistry, because many reactions associated with metabolism are redox reactions.

The term "oxidation" originally referred to the combination of a substance with oxygen. This results in an increase in the oxidation number of an element in that substance. According to the original definition, the following reactions involve oxidation of the substance shown on the far left of each equation. Oxidation numbers are shown for *one* atom of the indicated kind.

1. The formation of rust, Fe_2O_3, iron(III) oxide: — oxidation state of Fe

$$4Fe(s) + 3O_2(g) \longrightarrow 2Fe_2O_3(s) \qquad 0 \longrightarrow +3$$

2. Combustion reactions: — oxidation state of C

$$C(s) + O_2(g) \longrightarrow CO_2(g) \qquad 0 \longrightarrow +4$$

$$2CO(g) + O_2(g) \longrightarrow 2CO_2(g) \qquad +2 \longrightarrow +4$$

$$C_3H_8(g) + 5O_2(g) \longrightarrow 3CO_2(g) + 4H_2O(g) \qquad -8/3 \longrightarrow +4$$

Oxidation number is a formal concept adopted for our convenience. The numbers are determined solely by reliance upon rules. These rules can result in a fractional oxidation number, as shown here. This does not mean that electronic charges are split.

Originally *reduction* described the removal of oxygen from a compound. Oxide ores are reduced to metals (a very real reduction in mass). For example, tungsten for use in light bulb filaments can be prepared by reduction of tungsten(VI) oxide with hydrogen at 1200°C:

oxidation number of W

$$WO_3(s) + 3H_2(g) \longrightarrow W(s) + 3H_2O(g) \qquad +6 \longrightarrow 0$$

Tungsten is reduced and its oxidation state decreases from +6 to zero. Hydrogen is oxidized from zero to the +1 oxidation state. The terms "oxidation" and "reduction" are now applied much more broadly.

In biological systems *oxidation* usually corresponds to the removal of hydrogen.

In biological systems *reduction* usually corresponds to the addition of hydrogen to molecules or polyatomic ions.

Oxidation is an algebraic increase in oxidation number and corresponds to the loss, or apparent loss, of electrons. **Reduction** is an algebraic decrease in oxidation number and corresponds to a gain, or apparent gain, of electrons.

Electrons are neither created nor destroyed in chemical reactions. So oxidation and reduction always occur simultaneously, and to the same extent, in ordinary chemical reactions. In the four equations cited previously as *examples of oxidation*, the oxidation numbers of iron and carbon atoms increase as they are oxidized. In each case oxygen is reduced as its oxidation number decreases from zero to −2.

Oxidizing agents are species that (1) oxidize other substances, (2) are reduced, and (3) gain (or appear to gain) electrons. Reducing agents are species that (1) reduce other substances, (2) are oxidized, and (3) lose (or appear to lose) electrons.

The following abbreviations are widely used:

ox. no. = oxidation number
ox. agt. = oxidizing agent
red. agt. = reducing agent

The equations below represent examples of redox reactions. Oxidation numbers are shown above the formulas, and oxidizing and reducing agents are indicated:

$$\underset{\text{red. agt.}}{2\overset{(0)}{Fe}(s)} + \underset{\text{ox. agt.}}{3\overset{(0)}{Cl_2}(g)} \longrightarrow 2\overset{(+3)}{Fe}\overset{(-1)}{Cl_3}(s)$$

$$\underset{\text{red. agt.}}{2\overset{(+3)}{Fe}\overset{(-1)}{Br_3}(aq)} + \underset{\text{ox. agt.}}{3\overset{(0)}{Cl_2}(g)} \longrightarrow 2\overset{(+3)}{Fe}\overset{(-1)}{Cl_3}(aq) + 3\overset{(0)}{Br_2}(\ell)$$

Iron reacting with chlorine to form iron(III) chloride.

Equations for redox reactions can also be written as total ionic and net ionic equations. For example, the previous equation may also be written as shown below. We distinguish between oxidation numbers and actual charges on ions by denoting oxidation numbers as $+n$ or $-n$ *in red circles just above the symbols of the elements*, and actual charges as $n+$ or $n-$ above and to the right of formulas of ions.

$$2[Fe^{3+}(aq) + 3Br^-(aq)] + 3Cl_2(g) \longrightarrow 2[Fe^{3+}(aq) + 3Cl^-(aq)] + 3Br_2(\ell)$$

The spectator ions, Fe^{3+}, do not participate in electron transfer. Their cancellation allows us to focus on the oxidizing agent, $Cl_2(g)$, and the reducing agent, $Br^-(aq)$.

$$2Br^-(aq) + Cl_2(g) \longrightarrow 2Cl^-(aq) + Br_2(\ell)$$

A **disproportionation reaction** is a redox reaction in which the same species is oxidized and reduced.

Example 4-12

Write each of the following formula unit equations as a net ionic equation if the two differ. Which ones are redox reactions? For the redox reactions, identify the oxidizing agent, the reducing agent, the species oxidized, and the species reduced.

(a) $2AgNO_3(aq) + Cu(s) \longrightarrow Cu(NO_3)_2(aq) + 2Ag(s)$

(b) $2KClO_3(s) \xrightarrow{\Delta} 2KCl(s) + 3O_2(g)$

(c) $3AgNO_3(aq) + K_3PO_4(aq) \longrightarrow Ag_3PO_4(s) + 3KNO_3(aq)$

Plan

To write ionic equations, we must recognize compounds that are (1) soluble in water and (2) ionized or dissociated in aqueous solutions. To determine which are oxidation–reduction reactions, we must assign an oxidation number to each element.

Solution

(a) According to the solubility rules (page 132), both silver nitrate, $AgNO_3$, and copper(II) nitrate, $Cu(NO_3)_2$, are water-soluble ionic compounds. The total ionic equation and oxidation numbers are

$$2[\overset{(+1)}{Ag^+}(aq) + \overset{(+5)}{N}\overset{(-2)}{O_3}{}^-(aq)] + \overset{(0)}{Cu}(s) \longrightarrow [\overset{(+2)}{Cu^{2+}}(aq) + 2\overset{(+5)}{N}\overset{(-2)}{O_3}{}^-(aq)] + 2\overset{(0)}{Ag}(s)$$

(Cu → Cu: +2; Ag → Ag: −1)

Metallic silver formed by immersing a spiral of copper wire in a silver nitrate solution (Example 4-12a).

The nitrate ions, NO_3^-, are spectator ions. Canceling them from both sides gives the net ionic equation:

$$\overset{+1}{2Ag^+(aq)} + \overset{0}{Cu(s)} \longrightarrow \overset{+2}{Cu^{2+}(aq)} + \overset{0}{2Ag(s)}$$

This is a redox equation. The oxidation number of silver decreases from + 1 to zero; silver ion is reduced and is the oxidizing agent. The oxidation number of copper increases from zero to +2; copper is oxidized and is the reducing agent.

(b) This reaction involves two solids and a gas, so the formula unit and net ionic equations are identical. It is a redox reaction:

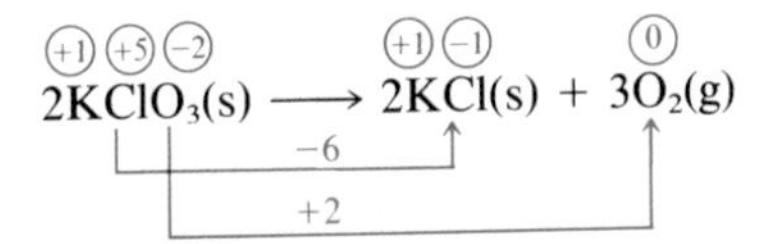

$$2KClO_3(s) \longrightarrow 2KCl(s) + 3O_2(g)$$

(oxidation numbers: K +1, Cl +5, O −2 in $KClO_3$; K +1, Cl −1 in KCl; O 0 in O_2; change for Cl: −6; change for O: +2)

The oxidizing agent is $KClO_3$. The chlorine in $KClO_3$ is reduced from the +5 to the −1 oxidation state. The oxidation number of oxygen increases from −2 to zero; oxygen is oxidized and is the reducing agent. We might also say that $KClO_3$ is both oxidizing agent and reducing agent; this is a disproportionation reaction.

(c) The solubility rules indicate that all these salts are soluble and ionic except for silver phosphate, Ag_3PO_4. The total ionic equation is

$$3[Ag^+(aq) + NO_3^-(aq)] + [3K^+(aq) + PO_4^{3-}(aq)] \longrightarrow Ag_3PO_4(s) + 3[K^+(aq) + NO_3^-(aq)]$$

Eliminating the spectator ions gives the net ionic equation:

$$\overset{+1}{3Ag^+(aq)} + \overset{+5\,-2}{PO_4^{3-}(aq)} \longrightarrow \overset{+1\,+5\,-2}{Ag_3PO_4(s)}$$

There are no changes in oxidation numbers; this is not a redox reaction.

EOC 73, 74

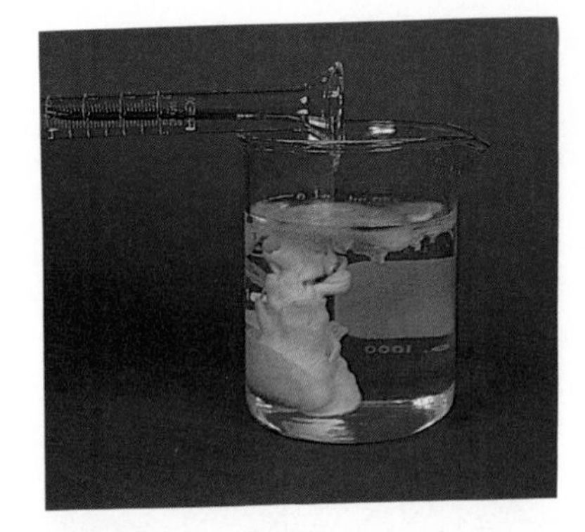

The reaction of $AgNO_3(aq)$ and $K_3PO_4(aq)$ is a precipitation reaction (Example 4-12c).

Balancing Oxidation–Reduction Equations

Our rules for assigning oxidation numbers are constructed so that

the total increase in oxidation numbers must equal the total decrease in oxidation numbers in all redox reactions.

This equivalence provides the basis for balancing redox equations. Although there is no single "best method" for balancing all redox equations, two methods are particularly useful: (1) the change-in-oxidation-number method and (2) the ion–electron method, which is used extensively in electrochemistry (Chapter 21).

Most redox equations can be balanced by both methods, but in some instances one may be easier to use than the other.

4-9 Change-in-Oxidation-Number Method

The next few examples illustrate this method, which is based on *equal total increases and decreases in oxidation numbers*. While many redox equations can be balanced by simple inspection, you should learn the method because it can be used to balance difficult equations. The general procedure follows.

1. Write as much of the overall *unbalanced* equation as possible.
2. Assign oxidation numbers to find the elements that undergo changes in oxidation numbers.
3. a. Draw a bracket to connect atoms of the element that are oxidized. Show the increase in oxidation number *per atom*. Draw a bracket to connect atoms of the element that are reduced. Show the decrease in oxidation number *per atom*.
   b. Determine the factors that will make the *total* increase and decrease in oxidation numbers equal.
4. Insert coefficients into the equation to make the total increase and decrease in oxidation numbers equal.
5. Balance the other atoms by inspection.

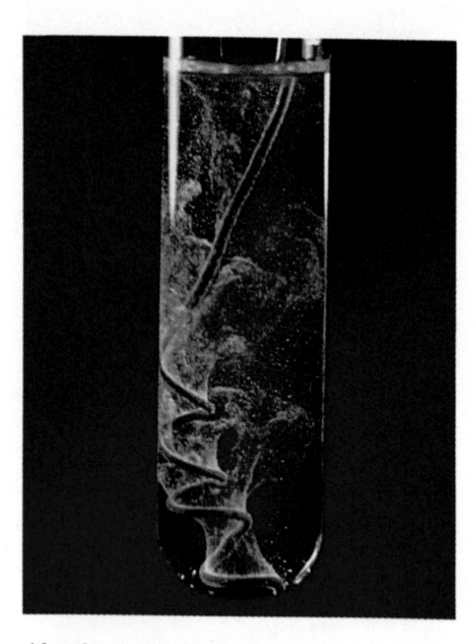

Aluminum wire reacting with hydrochloric acid.

Example 4-13

Aluminum reacts with hydrochloric acid to form aqueous aluminum chloride and gaseous hydrogen. Balance the formula unit equation and identify the oxidizing and reducing agents.

Plan

We follow the five-step procedure, one step at a time.

Solution

The unbalanced formula unit equation and oxidation numbers (Steps 1 and 2) are

$$\overset{+1\,-1}{HCl(aq)} + \overset{0}{Al(s)} \longrightarrow \overset{+3\,-1}{AlCl_3(aq)} + \overset{0}{H_2(g)}$$

The oxidation number of Al increases from 0 to +3. Al is the reducing agent; it is oxidized. The oxidation number of H decreases from + 1 to 0. HCl is the oxidizing agent; it is reduced.

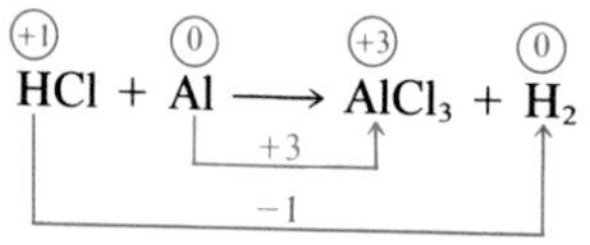

$$\overset{+1}{HCl} + \overset{0}{Al} \longrightarrow \overset{+3}{AlCl_3} + \overset{0}{H_2} \qquad \text{(Step 3a)}$$

(Al → $AlCl_3$: +3; HCl → H_2: −1)

We make the *total* increase and decrease in oxidation numbers equal (Step 3b):

Oxidation Numbers	Change/Atom	Equalizing Changes Gives
Al = 0 ⟶ Al = +3	+3	1(+3) = +3
H = +1 ⟶ H = 0	−1	3(−1) = −3

Each change must be multiplied by two because there are two H's in each H_2.

$$2(+3) = +6 \text{ (total increase)} \qquad 2(-3) = -6 \text{ (total decrease)}$$

We need 2 Al's and 6 H's on each side of the equation (Step 4):

$$6HCl(aq) + 2Al(s) \longrightarrow 2AlCl_3(aq) + 3H_2(g)$$

EOC 77, 78

All balanced equations must satisfy two criteria:

1. There must be mass balance. That is, the same number of atoms of each kind must be shown as reactants and products.
2. There must be charge balance. The sums of actual charges on the left and right sides of the equation must be equal.

Example 4-14

Copper is a widely used metal. Before it is welded (brazed), copper is cleaned by dipping it into nitric acid. HNO_3 oxidizes Cu to Cu^{2+} ions and is reduced to NO. The other product is H_2O. Write the balanced net ionic and formula unit equations for the reaction. Excess HNO_3 is present.

Plan

In writing ionic equations, we recall that strong acids, strong soluble bases, and most soluble salts are strong electrolytes. Then we apply our five-step procedure for redox equations.

Solution

We write the unbalanced net ionic equation and assign oxidation numbers. HNO_3 is a strong acid.

$$H^+(aq) + \overset{+5}{N}O_3^-(aq) + \overset{0}{Cu}(s) \longrightarrow \overset{+2}{Cu}^{2+}(aq) + \overset{+2}{N}O(g) + H_2O(\ell)$$

We see that copper is oxidized; it is the reducing agent. Nitrate ions are reduced; they are the oxidizing agent.

$$H^+(aq) + \overset{+5}{N}O_3^-(aq) + \overset{0}{Cu}(s) \longrightarrow \overset{+2}{Cu}^{2+}(aq) + \overset{+2}{N}O(g) + H_2O(\ell)$$

(Cu → Cu: +2; N → N: −3)

We make the *total* increase and decrease in oxidation numbers equal:

Oxidation Numbers	Change/Atom	Equalizing Changes Gives
$Cu = 0 \longrightarrow Cu = +2$	+2	$3(+2) = +6$
$N = +5 \longrightarrow N = +2$	−3	$2(-3) = -6$

Now we balance the *redox part* of the reaction:

$$H^+ + 2\,NO_3^- + 3\,Cu \longrightarrow 3\,Cu^{2+} + 2\,NO + H_2O$$

There are six O's on the left in NO_3^- ions. A coefficient of 4 before H_2O balances O and gives eight H's on the right. So we need eight H^+ ions on the left to balance the net ionic equation.

$$8H^+(aq) + 2NO_3^-(aq) + 3Cu(s) \longrightarrow 3Cu^{2+}(aq) + 2NO(g) + 4H_2O(\ell)$$

This solution contains excess HNO_3, so NO_3^- is the only anion present in significant concentration. Therefore, we add six more NO_3^- ions on each side to give the balanced formula unit equation:

$$8HNO_3(aq) + 3Cu(s) \longrightarrow 3Cu(NO_3)_2(aq) + 2NO(g) + 4H_2O(\ell)$$

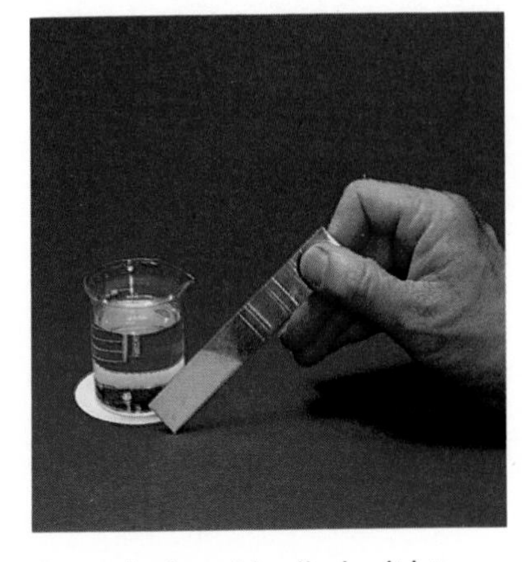

Copper is cleaned by dipping it into nitric acid.

4-10 Adding H^+, OH^-, or H_2O to Balance Oxygen or Hydrogen

Frequently we need more oxygen or hydrogen to complete the mass balance for a reaction in aqueous solution. However, we must be careful not to introduce other changes in oxidation number or to use species that could not actually be present in the solution. (We cannot add H_2 or O_2 to equations because these species are not present in aqueous solutions. Acidic solutions do not contain significant concentrations of OH^- ions. Basic solutions do not contain significant concentrations of H^+ ions.) We accomplish this balance as follows:

In acidic solution: We add only H^+ or H_2O (*not* OH^- in acidic solution)
In basic solution: We add only OH^- or H_2O (*not* H^+ in basic solution)

The following chart shows how to balance hydrogen and oxygen.

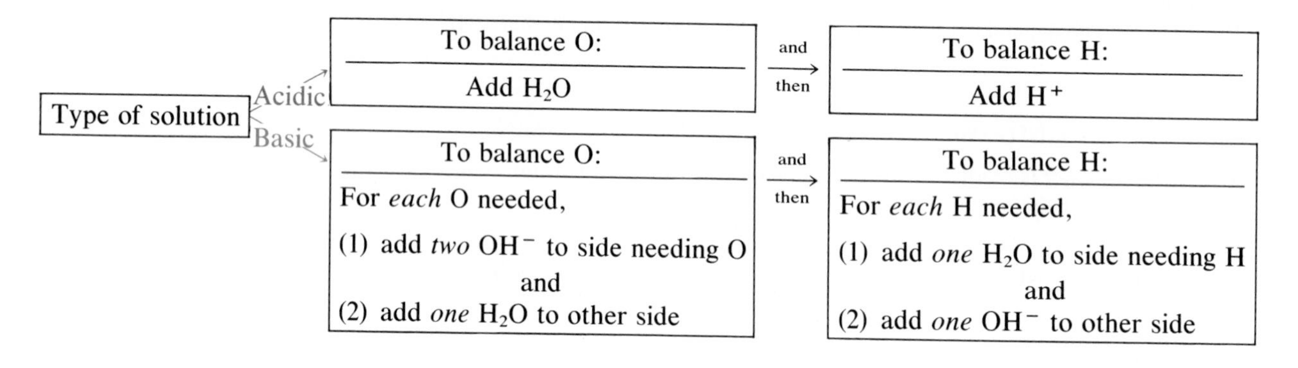

Example 4-15

"Drāno" drain cleaner is solid sodium hydroxide that contains some aluminum turnings. When Drāno is added to water, the NaOH dissolves rapidly with the evolution of a lot of heat. The Al reduces H_2O in the basic solution to produce $[Al(OH)_4]^-$ ions and H_2 gas, which gives the bubbling action. Write the balanced net ionic and formula unit equations for this reaction.

Plan

We are given formulas for reactants and products. Recall that NaOH is a strong soluble base (OH^- and H_2O can be added to either side as needed). We apply our five-step procedure.

Solution

We write the unbalanced net ionic equation and assign oxidation numbers:

$$OH^-(aq) + \overset{(0)}{Al}(s) + \overset{(+1)}{H_2}O(\ell) \longrightarrow [\overset{(+3)}{Al}(OH)_4]^-(aq) + \overset{(0)}{H_2}(g)$$

Aluminum is oxidized; it is the reducing agent. H_2O is reduced; it is the oxidizing agent.

Oxidation numbers (circled above the formulas): H in H_2O = +1; Al(s) = 0; Al in $[Al(OH)_4]^-$ = +3; H in H_2 = 0

$$OH^-(aq) + H_2O(\ell) + Al(s) \longrightarrow [Al(OH)_4]^-(aq) + H_2(g)$$

Change in oxidation number: Al, +3; H, −1

We make the *total* increase and decrease in oxidation numbers equal:

Oxidation Numbers	Change/Atom	Equalizing Changes Gives
$Al = 0 \longrightarrow Al = +3$	$+3$	$1(+3) = +3$
$H = +1 \longrightarrow H = 0$	-1	$3(-1) = -3$

Each change must be multiplied by two because there are 2 H's in each H_2.

$2(+3) = +6$ (total increase) $\quad 2(-3) = -6$ (total decrease)

Now we balance the redox part of the equation. We need 2 Al's on each side. Because only one H in each H_2O molecule is reduced, we show six H_2O on the left and three H_2 on the right:

$$OH^- + 6H_2O + 2Al \longrightarrow 2[Al(OH)_4]^- + 3H_2$$

The net charge on the right is 2−, and so we need two OH^- on the left to balance the net ionic equation:

$$2OH^-(aq) + 6H_2O(\ell) + 2Al(s) \longrightarrow 2[Al(OH)_4]^-(aq) + 3H_2(g)$$

This reaction occurs in excess NaOH solution. We need two $Na^+(aq)$ on each side to balance the negative charges:

$$2NaOH(aq) + 6H_2O(\ell) + 2Al(s) \longrightarrow 2Na[Al(OH)_4](aq) + 3H_2(g)$$

EOC 79

The Drāno reaction.

Example 4-16

The breathalyzer detects the presence of ethanol (ethyl alcohol) in the breath of persons suspected of drunken driving. It utilizes the oxidation of ethanol to acetaldehyde by dichromate ions in acidic solution. The $Cr_2O_7^{2-}(aq)$ ion is orange (see page 391). The $Cr^{3+}(aq)$ ion is green. The appearance of a green color signals alcohol in the breath that exceeds the legal limit. Balance the net ionic equation for this reaction.

$$H^+(aq) + Cr_2O_7^{2-}(aq) + C_2H_5OH(\ell) \longrightarrow Cr^{3+}(aq) + C_2H_4O(\ell) + H_2O(\ell)$$

Plan

We are given the unbalanced equation, which includes H^+. This tells us that the reaction occurs in acidic solution. We apply our five-step procedure.

Solution

We first assign oxidation numbers to the elements that change:

Oxidation numbers (circled above the formulas): C in C_2H_5OH = −2; Cr in $Cr_2O_7^{2-}$ = +6; Cr^{3+} = +3; C in C_2H_4O = −1

$$H^+ + C_2H_5OH + Cr_2O_7^{2-} \longrightarrow Cr^{3+} + C_2H_4O + H_2O$$

We see that ethanol is oxidized; it is the reducing agent. $Cr_2O_7^{2-}$ ions are reduced; they are the oxidizing agent.

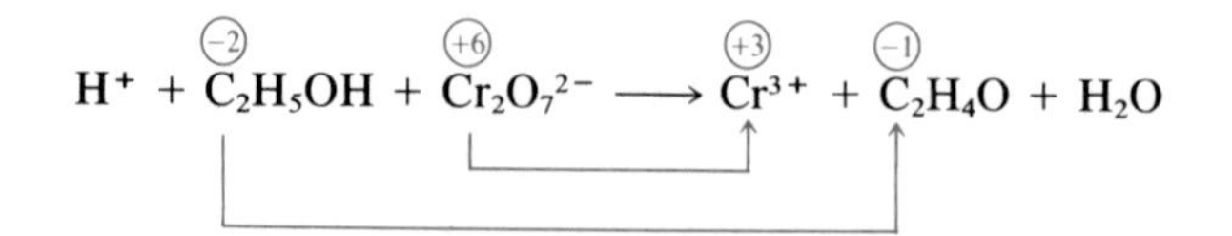

Oxidation Numbers	Change/Atom	Equalizing Changes Gives
$Cr = +6 \longrightarrow Cr = +3$	-3	$1(-3) = -3$
$C = -2 \longrightarrow C = -1$	$+1$	$3(+1) = +3$

Each change must be multiplied by two because there are two Cr's in each $Cr_2O_7^{2-}$ and two C's in C_2H_5OH.

$$2(-3) = -6 \text{ (total decrease)} \qquad 2(+3) = +6 \text{ (total increase)}$$

We need 2 Cr's and 6 C's on each side of the equation to balance the redox part:

$$H^+ + 3C_2H_5OH + Cr_2O_7^{2-} \longrightarrow 2Cr^{3+} + 3C_2H_4O + H_2O$$

Now we balance H and O using our chart. There are 10 O's on the left and only 4 O's on the right. So we add 6 *more* H_2O molecules on the right.

$$H^+ + 3C_2H_5OH + Cr_2O_7^{2-} \longrightarrow 2Cr^{3+} + 3C_2H_4O + 7H_2O$$

Now there are 26 H's on the right and only 19 on the left. So we add 7 *more* H^+ ions on the left to give the balanced net ionic equation.

$$8H^+(aq) + 3C_2H_5OH(\ell) + Cr_2O_7^{2-}(aq) \longrightarrow 2Cr^{3+}(aq) + 3C_2H_4O(\ell) + 7H_2O(\ell)$$

EOC 80

Every balanced equation must have both mass balance and charge balance. Once the redox part of an equation has been balanced, we may count *either* atoms or charges. After we balanced the redox part in Example 4-16, we had

$$H^+ + 3C_2H_5OH + Cr_2O_7^{2-} \longrightarrow 2Cr^{3+} + 3C_2H_4O + H_2O$$

The net charge on the left side is $(1 + 2-) = 1-$. On the right, it is $2(3+) = 6+$. Because H^+ is the *only charged species whose coefficient isn't known*, we add 7 *more* H^+ to give a net charge of 6+ on both sides.

$$8H^+ + 3C_2H_5OH + Cr_2O_7^{2-} \longrightarrow 2Cr^{3+} + 3C_2H_4O + H_2O$$

Now we have 10 O's on the left and only 4 O's on the right. We add six *more* H_2O molecules to give the balanced net ionic equation.

$$8H^+(aq) + 3C_2H_5OH(\ell) + Cr_2O_7^{2-}(aq) \longrightarrow 2Cr^{3+}(aq) + 3C_2H_4O(\ell) + 7H_2O(\ell)$$

How can you tell whether to balance atoms or charges first? Look at the equation *after you have balanced the redox part*. Decide which is simpler, and do that. In the preceding equation, it is easier to balance charges than to balance atoms.

Example 4-19

Potassium permanganate oxidizes iron(II) sulfate to iron(III) sulfate in sulfuric acid solution. Permanganate ions are reduced to manganese(II) ions. Write the balanced net ionic and formula unit equations for this reaction.

Plan

We use the given information to write the skeletal equation. The reaction occurs in H_2SO_4 solution; we can add H^+ and H_2O as needed to construct and balance the appropriate half-reactions. Then we proceed as in earlier examples.

Solution

$$Fe^{2+} + MnO_4^- \longrightarrow Fe^{3+} + Mn^{2+} \quad \text{(skeletal equation)}$$

$$Fe^{2+} \longrightarrow Fe^{3+} \quad \text{(ox. half-reaction)}$$

$$Fe^{2+} \longrightarrow Fe^{3+} + 1e^- \quad \text{(balanced ox. half-reaction)}$$

$$MnO_4^- \longrightarrow Mn^{2+} \quad \text{(red. half-reaction)}$$

$$MnO_4^- + 8H^+ \longrightarrow Mn^{2+} + 4H_2O$$

$$MnO_4^- + 8H^+ + 5e^- \longrightarrow Mn^{2+} + 4H_2O \quad \text{(balanced red. half-reaction)}$$

One half-reaction involves one electron and the other involves five electrons. Now we balance the electron transfer and then add the two equations term by term. This gives the balanced net ionic equation.

$$5(Fe^{2+} \longrightarrow Fe^{3+} + 1e^-)$$

$$1(MnO_4^- + 8H^+ + 5e^- \longrightarrow Mn^{2+} + 4H_2O)$$

$$5Fe^{2+}(aq) + MnO_4^-(aq) + 8H^+(aq) \longrightarrow 5Fe^{3+}(aq) + Mn^{2+}(aq) + 4H_2O(\ell)$$

The reaction occurs in H_2SO_4 solution. The SO_4^{2-} ion is the counter anion in the formula unit equation. Because the Fe^{3+} ion occurs twice in $Fe_2(SO_4)_3$, there must be an even number of Fe. So the net ionic equation is multiplied by two. Then we add $18SO_4^{2-}$ to each side to give complete formulas in the balanced formula unit equation.

$$10FeSO_4(aq) + 2KMnO_4(aq) + 8H_2SO_4(aq) \longrightarrow 5Fe_2(SO_4)_3(aq) + 2MnSO_4(aq) + K_2SO_4(aq) + 8H_2O(\ell)$$

EOC 81, 82

Key Terms

Acid A substance that produces H^+(aq) ions in aqueous solution. Strong acids ionize completely or almost completely in dilute aqueous solution. Weak acids ionize only slightly.

Active metal A metal that loses electrons readily to form cations.

Activity series A listing of metals (and hydrogen) in order of decreasing activity.

Amphoterism The ability to react with both acids and bases.

Atomic number The number of protons in the nucleus of an atom of an element.

Base A substance that produces $OH^-(aq)$ ions in aqueous solution. Strong soluble bases are soluble in water and are completely *dissociated*. Weak bases ionize only slightly.

Chemical periodicity The variation in properties of elements with their positions in the periodic table.

Combustion reaction A highly exothermic reaction of a substance with oxygen, usually with a visible flame.

Displacement reaction A reaction in which one element displaces another from a compound.

Disproportionation reaction A redox reaction in which the oxidizing agent and the reducing agent are the same species.

Dissociation In aqueous solution, the process in which a *solid ionic compound* separates into its ions.

Electrolyte A substance whose aqueous solutions conduct electricity.

Formula unit equation An equation for a chemical reaction in which all formulas are written as complete formulas.

Group (family) The elements in a vertical column of the periodic table.

Ionization In aqueous solution, the process in which a *molecular* compound reacts with water and forms ions.

Metal An element below and to the left of the stepwise division (metalloids) in the upper right corner of the periodic table; about 80% of the known elements are metals.

Metalloids Elements with properties intermediate between metals and nonmetals: B, Al, Si, Ge, As, Sb, Te, Po, and At.

Net ionic equation An equation that results from canceling spectator ions and eliminating brackets from a total ionic equation.

Neutralization The reaction of an acid with a base to form a salt and water. Usually, the reaction of hydrogen ions with hydroxide ions to form water molecules.

Nonelectrolyte A substance whose aqueous solutions do not conduct electricity.

Nonmetals Elements above and to the right of the metalloids in the periodic table.

Oxidation An algebraic increase in oxidation number; may correspond to a loss of electrons.

Oxidation numbers Arbitrary numbers that can be used as mechanical aids in writing formulas and balancing equations; for single-atom ions they correspond to the charge on the ion; more electronegative atoms are assigned negative oxidation numbers.

Oxidation states See *Oxidation numbers*.

Oxidation–reduction reaction A reaction in which oxidation and reduction occur; also called redox reactions.

Oxidizing agent The substance that oxidizes another substance and is reduced.

Period The elements in a horizontal row of the periodic table.

Periodicity Regular periodic variations of properties of elements with atomic number (and position in the periodic table).

Periodic law The properties of the elements are periodic functions of their atomic numbers.

Periodic table An arrangement of elements in order of increasing atomic number that also emphasizes periodicity.

Precipitate An insoluble solid that forms and separates from a solution.

Precipitation reaction A reaction in which a precipitate forms.

Reducing agent The substance that reduces another substance and is oxidized.

Reduction An algebraic decrease in oxidation number; may correspond to a gain of electrons.

Reversible reaction A reaction that occurs in both directions; indicated by double arrows ($\rightleftharpoons$).

Salt A compound that contains a cation other than H^+ and an anion other than OH^- or O^{2-}.

Spectator ions Ions in solution that do not participate in a chemical reaction.

Strong electrolyte A substance that conducts electricity well in dilute aqueous solution.

Semiconductor A substance that does not conduct electricity at low temperatures but does so at higher temperatures.

Ternary acid An acid containing three elements: H, O, and (usually) another nonmetal.

Total ionic equation An equation for a chemical reaction written to show the predominant form of all species in aqueous solution or in contact with water.

Weak electrolyte A substance that conducts electricity poorly in dilute aqueous solution.

Exercises

The Periodic Table

1. State the periodic law. What does it mean?

2. What was Mendeleev's contribution to the construction of the modern periodic table?

3. Consult a handbook of chemistry and look up melting points of the elements of periods 2 and 3. Show that melting point is a property that varies periodically for these elements.

***4.** Mendeleev's periodic table was based on increasing atomic weight, whereas the modern periodic table is based on increasing atomic number. In the modern table argon comes before potassium, yet it has a higher atomic weight. Explain how this can be.

5. Estimate the density of antimony from the following densities (g/cm^3): As, 5.72; Bi, 9.8; Sn, 7.30; Te, 6.24. Show how you arrived at your answer.
6. Estimate the specific heat of antimony from the following specific heats ($J/g \cdot °C$): As, 0.34; Bi, 0.14; Sn, 0.23; Te, 0.20. Show how you arrived at your answer.
7. Estimate the density of selenium from the following densities (g/cm^3): S, 2.07; Te, 6.24; As, 5.72; Br, 3.12. Show how you arrived at your answer.
8. Given the following melting points in °C, estimate the value for CBr_4: CF_4, −184; CCl_4, −23; CI_4, 171 (decomposes).
9. Calcium and magnesium form the following compounds: $CaCl_2$, $MgCl_2$, CaO, MgO, Ca_3N_2, and Mg_3N_2. Predict the formula for a compound of (a) barium and sulfur, (b) strontium and iodine.
10. The formulas of some hydrides of second-period representative elements are as follows: BeH_2, BH_3, CH_4, NH_3, H_2O, HF. A famous test in criminology laboratories for the presence of arsenic (As) involves the formation of arsine, the hydride of arsenic. Predict the formula of arsine.
11. Distinguish between the following terms clearly and concisely, and provide specific examples of each: groups (families) of elements, and periods of elements.
12. Write names and symbols for (a) the alkaline earth metals, (b) the Group IIIA elements, (c) the Group VIB elements.
13. Write names and symbols for (a) the alkali metals, (b) the noble gases, (c) the Group IB elements.
14. Define and illustrate the following terms clearly and concisely: (a) metals, (b) nonmetals, (c) noble gases.

Aqueous Solutions

15. Define and distinguish among (a) strong electrolytes, (b) weak electrolytes, and (c) nonelectrolytes.
16. Three common classes of compounds are electrolytes. Name them and give an example of each.
17. Define (a) acids, (b) bases, and (c) salts.
18. How can a salt be related to a particular acid and a particular base?
19. List the names and formulas of the common strong acids.
20. Write equations for the ionization of the following acids: (a) hydrochloric acid, (b) nitric acid, (c) perchloric acid.
21. List names and formulas of five weak acids.
22. What are reversible reactions? Give some examples.
23. Write equations for the ionization of the following acids. Which ones ionize only slightly? (a) HF, (b) HNO_2, (c) CH_3COOH.
24. List names and formulas of the common strong soluble bases.
25. What is the difference between ionization and dissociation in aqueous solution?
26. The most common weak base is present in a common household chemical. Write the equation for the ionization of this weak base.
27. Summarize the electrical properties of strong electrolytes, weak electrolytes, and nonelectrolytes.
28. Write the formulas of two soluble and two insoluble chlorides, sulfates, and hydroxides.
29. Describe an experiment for classifying each of these compounds as a strong electrolyte, a weak electrolyte, or a nonelectrolyte: K_2CO_3, HCN, CH_3OH, H_2S, H_2SO_4, NH_3.
30. (a) Which of these are acids? HBr, NH_3, H_2SeO_4, BF_3, H_3SbO_4, $Al(OH)_3$, H_2S, C_6H_6, CsOH, H_3BO_3, HCN. (b) Which of these are bases? NaOH, H_2Se, BCl_3, NH_3.

*31. Classify each substance as either an electrolyte or a nonelectrolyte: NH_4Cl, HI, C_6H_6, $Zn(CH_3COO)_2$, $Cu(NO_3)_2$, CH_3COOH, $C_{12}H_{22}O_{11}$, LiOH, $KHCO_3$, CCl_4, $La_2(SO_4)_3$, I_2.

*32. Classify each substance as either a strong or weak electrolyte, and then list (a) the strong acids, (b) the strong bases, (c) the weak acids, and (d) the weak bases. NaCl, $MgSO_4$, HCl, $H_2C_2O_4$, $Ba(NO_3)_2$, H_3PO_4, $Sr(OH)_2$, HNO_3, HI, $Ba(OH)_2$, LiOH, C_2H_5COOH, NH_3, KOH, $MgMoO_4$, HCN, $HClO_4$.

33. Based on the solubility rules given in Table 4-8, how would you write the formulas for the following substances in a net ionic equation? (a) $PbSO_4$, (b) $Na(CH_3COO)$, (c) $(NH_4)_2CO_3$, (d) MnS, (e) $BaCl_2$.
34. Repeat Exercise 33 for the following: (a) $(NH_4)_2SO_4$, (b) NaBr, (c) $Ba(OH)_2$, (d) $Mg(OH)_2$, (e) K_2CO_3.

Precipitation Reactions

Refer to the solubility rules on page 132. Classify the compounds in Exercises 35 through 38 as soluble, moderately soluble, or insoluble in water.

35. (a) $NaClO_4$, (b) AgCl, (c) $Pb(NO_3)_2$, (d) KOH, (e) $MgSO_4$
36. (a) $BaSO_4$, (b) $Al(NO_3)_3$, (c) CuS, (d) Na_2S, (e) $Ca(CH_3COO)_2$
37. (a) $Fe(NO_3)_3$, (b) $Hg(CH_3COO)_2$, (c) $BeCl_2$, (d) $NiSO_4$, (e) $CaCO_3$
38. (a) $KClO_3$, (b) NH_4Br, (c) NH_3, (d) HNO_2, (e) PbS

Exercises 39 and 40 describe precipitation reactions *in aqueous solutions*. For each, write balanced (i) formula unit, (ii) total ionic, and (iii) net ionic equations. Refer to the solubility rules as necessary.

39. (a) Black-and-white photographic film contains some silver bromide, which can be formed by the reaction of sodium bromide with silver nitrate.
(b) Barium sulfate is used when X-rays of the gastrointestinal tract are made. Barium sulfate can be prepared by reacting barium chloride with dilute sulfuric acid.
(c) In water purification small solid particles are often

"trapped" as aluminum hydroxide precipitates and falls to the bottom of the sedimentation pool. Aluminum sulfate reacts with calcium hydroxide (from lime) to form aluminum hydroxide and calcium sulfate.

*40. (a) Our bones are mostly calcium phosphate. Calcium chloride reacts with potassium phosphate to form calcium phosphate and potassium chloride.
(b) Mercury compounds are very poisonous. Mercury(II) nitrate reacts with sodium sulfide to form mercury(II) sulfide, which is very insoluble, and sodium nitrate.
(c) Chromium(III) ions are very poisonous. They can be removed from solution by precipitating very insoluble chromium(III) hydroxide. Chromium(III) chloride reacts with calcium hydroxide to form chromium(III) hydroxide and calcium chloride.

In Exercises 41 and 42, write balanced (i) formula unit, (ii) total ionic, and (iii) net ionic equations for the reactions that occur when *aqueous solutions* of the compounds are mixed.

41. (a) $Ba(NO_3)_2 + K_2CO_3 \rightarrow$
(b) $NaOH + CoCl_2 \rightarrow$
(c) $Al_2(SO_4)_3 + NaOH \rightarrow$

42. (a) $Cu(NO_3)_2 + Na_2S \rightarrow$
(b) $CdSO_4 + H_2S \rightarrow$
(c) $Bi_2(SO_4)_3 + (NH_4)_2S \rightarrow$

43. Use the solubility rules to determine whether or not reactions will occur when aqueous solutions of the following compounds are mixed.
(a) $Hg(NO_3)_2(aq) + Na_2S(aq) \rightarrow$
(b) $Al(NO_3)_3(aq) + LiOH(aq) \rightarrow$
(c) $Li_2SO_3(aq) + NaCl(aq) \rightarrow$
(d) $Fe(OH)_3(s) + KNO_3(aq) \rightarrow$
Write net ionic equations for those reactions that occur.

44. Repeat Exercise 43 for
(a) $Al(OH)_3(s) + NaNO_3(aq) \rightarrow$
(b) $NaBr(aq) + NH_4I(aq) \rightarrow$
(c) $AgNO_3(aq) + HCl(aq) \rightarrow$
(d) $CaCl_2(aq) + Na_2CO_3(aq) \rightarrow$

Acid–Base Reactions

In Exercises 45 through 48, write balanced (i) formula unit, (ii) total ionic, and (iii) net ionic equations for the reactions that occur between the acid and the base. Assume that all reactions occur in water or in contact with water.

45. (a) hydrochloric acid + barium hydroxide
(b) dilute sulfuric acid + potassium hydroxide
(c) perchloric acid + aqueous ammonia

46. (a) acetic acid + calcium hydroxide
(b) sulfurous acid + sodium hydroxide
(c) hydrofluoric acid + lithium hydroxide

***47.** (a) sodium hydroxide + hydrosulfuric acid
(b) barium hydroxide + hydrosulfuric acid
(c) lead(II) hydroxide + hydrosulfuric acid

48. (a) sodium hydroxide + sulfuric acid
(b) calcium hydroxide + phosphoric acid
(c) copper(II) hydroxide + nitric acid

In Exercises 49 through 52, write balanced (i) formula unit, (ii) total ionic, and (iii) net ionic equations for the reaction of an acid and a base that will produce the indicated salts.

49. (a) potassium chloride, (b) sodium phosphate, (c) barium acetate

50. (a) calcium perchlorate, (b) ammonium sulfate, (c) copper(II) sulfide

***51.** (a) sodium carbonate, (b) barium carbonate, (c) nickel(II) nitrate

***52.** (a) sodium sulfide, (b) barium phosphate, (c) lead(II) arsenate

53. Write a balanced equation for the preparation of each of the following salts by a neutralization reaction. SrC_2O_4 is insoluble in water. $Ca(NO_3)_2$, SrC_2O_4, $ZnSO_3$, $(NH_4)_2CO_3$.

54. Write the formulas for the acid and the base that could react to form each of the following *insoluble* salts. (a) $CuCO_3$, (b) Ag_2CrO_4, (c) $Hg_3(PO_4)_2$.

Displacement Reactions

55. Which of the following would displace hydrogen when a piece of the metal is dropped into dilute H_2SO_4 solution? Write balanced net ionic equations for the reactions: Zn, Cu, Fe, Ag.

56. Which of the following metals would displace copper from an aqueous solution of copper(II) sulfate? Write balanced net ionic equations for the reactions: Hg, Zn, Fe, Ag.

57. Arrange the metals listed in Exercise 55 in order of increasing activity.

58. Arrange the metals listed in Exercise 56 in order of increasing activity.

59. Which of the following metals would displace hydrogen from cold water? Write balanced net ionic equations for the reactions: Zn, Na, Ca, Fe.

60. Arrange the metals listed in Exercise 59 in order of increasing activity.

61. What is the order of increasing activity of the halogens?

62. Of the possible displacement reactions shown, which one(s) could occur?
(a) $2Cl^-(aq) + Br_2(\ell) \rightarrow 2Br^-(aq) + Cl_2(g)$
(b) $2Br^-(aq) + F_2(g) \rightarrow 2F^-(aq) + Br_2(\ell)$
(c) $2I^-(aq) + Cl_2(g) \rightarrow 2Cl^-(aq) + I_2(s)$
(d) $2Br^-(aq) + Cl_2(g) \rightarrow 2Cl^-(aq) + Br_2(\ell)$

63. (a) Name two common metals—one that *does not* displace hydrogen from water, and one that *does not* displace hydrogen from water or acid solutions.
(b) Name two common metals—one that *does* displace hydrogen from water, and one that displaces hydrogen from acid solutions but not from water. Write net ionic equations for the reactions that occur.

64. Predict the products of each mixture. If a reaction occurs, write the net ionic equation. If no reaction occurs, write "no reaction."
   (a) $Cd^{2+}(aq) + Al \rightarrow$
   (b) $Ca + H_2O \rightarrow$
   (c) $Ni + H_2O \rightarrow$
   (d) $Hg + HCl(aq) \rightarrow$
   (e) $Ni + H_2SO_4(aq) \rightarrow$
   (f) $Fe + H_2SO_4(aq) \rightarrow$
65. Use the activity series to predict whether or not the following reactions will occur:
   (a) $Fe(s) + Mg^{2+} \rightarrow Mg(s) + Fe^{2+}$
   (b) $Ni(s) + Cu^{2+} \rightarrow Ni^{2+} + Cu(s)$
   (c) $Cu(s) + 2H^+ \rightarrow Cu^{2+} + H_2(g)$
   (d) $Mg(s) + H_2O(g) \rightarrow MgO(s) + H_2(g)$
66. Repeat Exercise 65 for
   (a) $Sn(s) + Ca^{2+} \rightarrow Sn^{2+} + Ca(s)$
   (b) $Al_2O_3(s) + 3H_2(g) \xrightarrow{\Delta} 2Al(s) + 3H_2O(g)$
   (c) $Ca(s) + 2H^+ \rightarrow Ca^{2+} + H_2(g)$
   (d) $Cu(s) + Pb^{2+} \rightarrow Cu^{2+} + Pb(s)$

Oxidation Numbers

67. Assign oxidation numbers to the element specified in each group of compounds.
   (a) N in NO, N_2O_3, N_2O_4, NH_3, N_2H_4, NH_2OH, HNO_3
   (b) C in CO, CO_2, CH_2O, CH_4O, C_2H_6O, $(COOH)_2$, Na_2CO_3
   (c) S in S_8, H_2S, SO_2, SO_3, Na_2SO_3, H_2SO_4, K_2SO_4
68. Assign oxidation numbers to the element specified in each group of compounds.
   (a) P in PCl_3, P_4O_6, P_4O_{10}, HPO_3, H_3PO_4, $POCl_3$, $H_4P_2O_7$, $Mg_3(PO_4)_2$
   (b) Cl in Cl_2, HCl, HClO, $HClO_2$, $KClO_3$, Cl_2O_7, $Ca(ClO_4)_2$
   (c) Mn in MnO, MnO_2, $Mn(OH)_2$, K_2MnO_4, $KMnO_4$, Mn_2O_7
   (d) O in OF_2, Na_2O, Na_2O_2, KO_2
69. Assign oxidation numbers to the element specified in each group of ions.
   (a) S in S^{2-}, SO_3^{2-}, SO_4^{2-}, $S_2O_3^{2-}$, $S_4O_6^{2-}$
   (b) Cr in CrO_2^-, $Cr(OH)_4^-$, CrO_4^{2-}, $Cr_2O_7^{2-}$
   (c) B in BO_2^-, BO_3^{3-}, $B_4O_7^{2-}$
70. Assign oxidation numbers to the element specified in each group of ions.
   (a) N in N^{3-}, NO_2^-, NO_3^-, N_3^-, NH_4^+
   (b) Br in Br^-, BrO^-, BrO_3^-, BrO_4^-

Oxidation–Reduction Reactions

71. Define and illustrate the following terms: (a) oxidation, (b) reduction, (c) oxidizing agent, (d) reducing agent.
72. Why must oxidation and reduction always occur simultaneously in chemical reactions?
73. Determine which of the following are oxidation–reduction reactions. For those that are, identify the oxidizing and reducing agents.
   (a) $3Zn(s) + 2CoCl_3(aq) \rightarrow 3ZnCl_2(aq) + 2Co(s)$
   (b) $ICl(s) + H_2O(\ell) \rightarrow HCl(aq) + HOI(aq)$
   (c) $3HCl(aq) + HNO_3(aq) \rightarrow Cl_2(g) + NOCl(g) + 2H_2O(\ell)$
   (d) $Fe_2O_3(s) + 3CO(g) \xrightarrow{\Delta} 2Fe(s) + 3CO_2(g)$
74. Determine which of the following are oxidation–reduction reactions. For those that are, identify the oxidizing and reducing agents.
   (a) $HgCl_2(aq) + 2KI(aq) \rightarrow HgI_2(s) + 2KCl(aq)$
   (b) $4NH_3(g) + 3O_2(g) \rightarrow 2N_2(g) + 6H_2O(g)$
   (c) $CaCO_3(s) + 2HNO_3(aq) \rightarrow Ca(NO_3)_2(aq) + CO_2(g) + H_2O(\ell)$
   (d) $PCl_3(\ell) + 3H_2O(\ell) \rightarrow 3HCl(aq) + H_3PO_3(aq)$
75. What is oxidized, what is reduced, what is the oxidizing agent, and what is the reducing agent in each reaction?
   (a) $Mg(s) + Sn^{2+}(aq) \rightarrow Sn(s) + Mg^{2+}(aq)$
   (b) $2H_2O_2(\ell) \rightarrow 2H_2O(\ell) + O_2(g)$
   (c) $3H_2SO_3(aq) + HIO_3(aq) \rightarrow 3H_2SO_4(aq) + HI(aq)$
   (d) $CH_4(g) + 4Cl_2(g) \rightarrow CCl_4(\ell) + 4HCl(g)$
76. What mass of Zn is needed to displace 12.5 g of Cu from $CuSO_4 \cdot 5H_2O$?

Change-in-Oxidation-Number Method

In Exercises 77 and 78, write balanced formula unit equations for the reactions described by words.

77. (a) Carbon reacts with hot concentrated nitric acid to form carbon dioxide, nitrogen dioxide, and water.
   (b) Sodium reacts with water to form aqueous sodium hydroxide and gaseous hydrogen.
   (c) Zinc reacts with sodium hydroxide solution to form aqueous sodium tetrahydroxozincate and gaseous hydrogen. (The tetrahydroxozincate ion is $[Zn(OH)_4]^{2-}$.)

*78. (a) Iron reacts with hydrochloric acid to form aqueous iron(II) chloride and gaseous hydrogen.
   (b) Chromium reacts with sulfuric acid to form aqueous chromium(III) sulfate and gaseous hydrogen.
   (c) Tin reacts with concentrated nitric acid to form tin(IV) oxide, nitrogen dioxide, and water.

79. Balance the following ionic equations.
   (a) $Cr(OH)_4^-(aq) + OH^-(aq) + H_2O_2(aq) \rightarrow CrO_4^{2-}(aq) + H_2O(\ell)$
   (b) $MnO_2(s) + H^+(aq) + NO_2^-(aq) \rightarrow NO_3^-(aq) + Mn^{2+}(aq) + H_2O(\ell)$
   (c) $Sn(OH)_3^-(aq) + Bi(OH)_3(s) + OH^-(aq) \rightarrow Sn(OH)_6^{2-}(aq) + Bi(s)$
80. Balance the following ionic equations.
   (a) $MnO_4^-(aq) + H^+(aq) + Br^-(aq) \rightarrow Mn^{2+}(aq) + Br_2(\ell) + H_2O(\ell)$
   (b) $Cr_2O_7^{2-}(aq) + H^+(aq) + I^-(aq) \rightarrow Cr^{3+}(aq) + I_2(s) + H_2O(\ell)$
   (c) $MnO_4^-(aq) + SO_3^{2-}(aq) + H^+(aq) \rightarrow Mn^{2+}(aq) + SO_4^{2-}(aq) + H_2O(\ell)$
   (d) $Cr_2O_7^{2-}(aq) + Fe^{2+}(aq) + H^+(aq) \rightarrow Cr^{3+}(aq) + Fe^{3+}(aq) + H_2O(\ell)$

Ion–Electron Method

81. Balance the following ionic equations.
(a) $CrO_4^{2-}(aq) + H_2O(\ell) + HSnO_2^-(aq) \rightarrow CrO_2^-(aq) + OH^-(aq) + HSnO_3^-(aq)$
(b) $C_2H_4(g) + MnO_4^-(aq) + H^+(aq) \rightarrow CO_2(g) + Mn^{2+}(aq) + H_2O(\ell)$
(c) $H_2S(aq) + H^+(aq) + Cr_2O_7^{2-}(aq) \rightarrow Cr^{3+}(aq) + S(s) + H_2O(\ell)$
(d) $ClO_3^-(aq) + H_2O(\ell) + I_2(s) \rightarrow IO_3^-(aq) + Cl^-(aq) + H^+(aq)$
(e) $Cu(s) + H^+(aq) + SO_4^{2-}(aq) \rightarrow Cu^{2+}(aq) + H_2O(\ell) + SO_2(g)$

82. Balance the following ionic equations.
(a) $Al(s) + NO_3^-(aq) + OH^-(aq) + H_2O \rightarrow Al(OH)_4^-(aq) + NH_3(g)$
(b) $NO_2(g) + OH^-(aq) \rightarrow NO_3^-(aq) + NO_2^-(aq) + H_2O(\ell)$
(c) $MnO_4^-(aq) + H_2O(\ell) + NO_2^-(aq) \rightarrow MnO_2(s) + NO_3^-(aq) + OH^-(aq)$
(d) $I^-(aq) + H^+(aq) + NO_2^-(aq) \rightarrow NO(g) + H_2O(\ell) + I_2(s)$
(e) $Hg_2Cl_2(s) + NH_3(aq) \rightarrow Hg(\ell) + HgNH_2Cl(s) + NH_4^+(aq) + Cl^-(aq)$

83. Balance the following ionic equations for reactions in acidic solution. H^+ or H_2O (but not OH^-) may be added as necessary.
(a) $P_4(s) + NO_3^-(aq) \rightarrow H_3PO_4(aq) + NO(g)$
(b) $H_2O_2(aq) + MnO_4^-(aq) \rightarrow Mn^{2+}(aq) + O_2(g)$
(c) $HgS(s) + Cl^-(aq) + NO_3^-(aq) \rightarrow HgCl_4^{2-}(aq) + NO_2(g) + S(s)$
(d) $HBrO(aq) \rightarrow Br^-(aq) + O_2(g)$
(e) $Cl_2(g) \rightarrow ClO_3^-(aq) + Cl^-(aq)$

84. Balance the following ionic equations for reactions in acidic solution. H^+ or H_2O (but not OH^-) may be added as necessary.
(a) $Fe^{2+}(aq) + MnO_4^-(aq) \rightarrow Fe^{3+}(aq) + Mn^{2+}(aq)$
(b) $Br_2(\ell) + SO_2(g) \rightarrow Br^-(aq) + SO_4^{2-}(aq)$
(c) $Cu(s) + NO_3^-(aq) \rightarrow Cu^{2+}(aq) + NO_2(g)$
(d) $PbO_2(s) + Cl^-(aq) \rightarrow PbCl_2(s) + Cl_2(g)$
(e) $Zn(s) + NO_3^-(aq) \rightarrow Zn^{2+}(aq) + N_2(g)$

85. Balance the following ionic equations in basic solution. OH^- or H_2O (but not H^+) may be added as necessary.
(a) $Mn(OH)_2(s) + H_2O_2(aq) \rightarrow MnO_2(s)$
(b) $CN^-(aq) + MnO_4^-(aq) \rightarrow CNO^-(aq) + MnO_2(s)$
(c) $As_2S_3(s) + H_2O_2(aq) \rightarrow AsO_4^{3-}(aq) + SO_4^{2-}(aq)$
(d) $CrI_3(aq) + H_2O_2(aq) \rightarrow CrO_4^{2-}(aq) + IO_4^-(aq)$

86. Balance the following ionic equations in basic solution. OH^- or H_2O (but not H^+) may be added as necessary.
(a) $MnO_4^-(aq) + NO_2^-(aq) \rightarrow MnO_2(s) + NO_3^-(aq)$
(b) $Zn(s) + NO_3^-(aq) \rightarrow NH_3(aq) + Zn(OH)_4^{2-}(aq)$
(c) $N_2H_4(aq) + Cu(OH)_2(s) \rightarrow N_2(g) + Cu(s)$
(d) $Mn^{2+}(aq) + MnO_4^-(aq) \rightarrow MnO_2(s)$
(e) $Cl_2(g) \rightarrow ClO_3^-(aq) + Cl^-(aq)$

Mixed Exercises

The following reactions apply to Exercises 87 through 93.

a. $H_2SO_4(aq) + 2KOH(aq) \rightarrow K_2SO_4(aq) + 2H_2O(\ell)$
b. $2Rb(s) + Br_2(\ell) \xrightarrow{\Delta} 2RbBr(s)$
c. $2KI(aq) + F_2(g) \rightarrow 2KF(aq) + I_2(s)$
d. $CaO(s) + SiO_2(s) \xrightarrow{\Delta} CaSiO_3(s)$
e. $S(s) + O_2(g) \xrightarrow{\Delta} SO_2(g)$
f. $BaCO_3(s) \xrightarrow{\Delta} BaO(s) + CO_2(g)$
g. $HgS(s) + O_2(g) \xrightarrow{\Delta} Hg(\ell) + SO_2(g)$
h. $AgNO_3(aq) + HCl(aq) \rightarrow AgCl(s) + HNO_3(aq)$
i. $Pb(s) + 2HBr(aq) \rightarrow PbBr_2(s) + H_2(g)$
j. $2HI(aq) + H_2O_2(aq) \rightarrow I_2(s) + 2H_2O(\ell)$
k. $RbOH(aq) + HNO_3(aq) \rightarrow RbNO_3(aq) + H_2O(\ell)$
l. $N_2O_5(s) + H_2O(\ell) \rightarrow 2HNO_3(aq)$
m. $H_2O(g) + CO(g) \xrightarrow{\Delta} H_2(g) + CO_2(g)$
n. $MgO(s) + H_2O(\ell) \rightarrow Mg(OH)_2(s)$
o. $PbSO_4(s) + PbS(s) \xrightarrow{\Delta} 2Pb(s) + 2SO_2(g)$

87. Identify the precipitation reactions.

88. Identify the acid–base reactions.

89. Identify the oxidation–reduction reactions.

90. Identify the oxidizing agent and reducing agent for each oxidation–reduction reaction.

91. Identify the oxidation–reduction reactions that are also displacement reactions.

92. Why can some reactions fit into more than one class?

93. Which of these reactions do not fit into any of our classes of reactions?

94. How many moles of oxygen can be obtained by the decomposition of 10.0 grams of reactant in each of the following reactions?
(a) $2KClO_3(s) \rightarrow 2KCl(s) + 3O_2(g)$
(b) $2H_2O_2(aq) \rightarrow 2H_2O(\ell) + O_2(g)$
(c) $2HgO(s) \rightarrow 2Hg(\ell) + O_2(g)$

95. For the formation of 1.00 mol of water, which reaction uses the most nitric acid?
(a) $3Cu(s) + 8HNO_3(aq) \rightarrow 3Cu(NO_3)_2(aq) + 2NO(g) + 4H_2O(\ell)$
(b) $Al_2O_3(s) + 6HNO_3(aq) \rightarrow 2Al(NO_3)_3(aq) + 3H_2O(\ell)$
(c) $4Zn(s) + 10HNO_3(aq) \rightarrow 4Zn(NO_3)_2(aq) + NH_4NO_3(aq) + 3H_2O(\ell)$

96. Balance these equations for reactions in acidic solutions by the ion–electron method.
(a) $MnO_4^- + H_2C_2O_4 \rightarrow Mn^{2+} + CO_2$
(b) $IO_3^- + Cl^- + N_2H_4 \rightarrow ICl_2^- + N_2$
(c) $Zn + NO_3^- \rightarrow Zn^{2+} + NH_4^+$
(d) $I_2 + S_2O_3^{2-} \rightarrow S_4O_6^{2-} + I^-$
(e) $NO_2^- + I^- \rightarrow I_2 + NO$
(f) $Ag^+ + AsH_3 \rightarrow H_3AsO_4 + Ag$

97. Balance these equations for reactions in basic solutions by the ion–electron method.
(a) $MnO_4^- + IO_3^- \rightarrow IO_4^- + MnO_2$
(b) $SO_3^{2-} + MnO_4^- \rightarrow SO_4^{2-} + MnO_4^{2-}$
(c) $Cl_2 \rightarrow Cl^- + ClO_3^-$

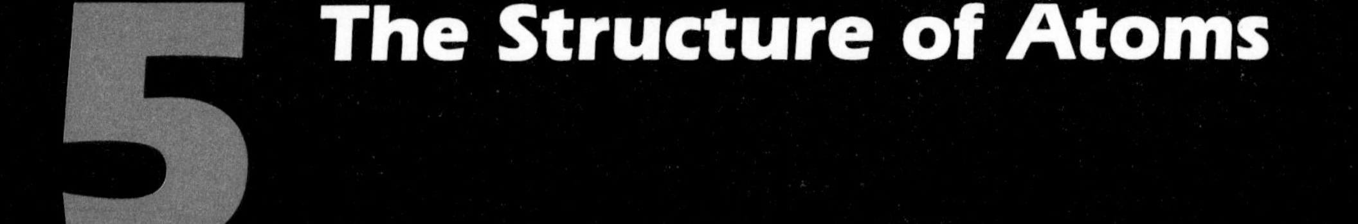

5 The Structure of Atoms

The Rosette Nebula. The red light is given off by ionized atoms, mostly interstellar hydrogen. A study of the wavelengths of light emitted and absorbed gives information about the composition of stars, interstellar gas, and other astronomical objects.

Outline

Objectives

As you study this chapter, you should learn about

- ☐ The evidence for the existence and properties of electrons, protons, and neutrons
- ☐ The arrangements of these particles in atoms
- ☐ Isotopes and their composition
- ☐ The relation between isotopic abundance and observed atomic weights
- ☐ The wave view of light and how wavelength, frequency, and speed are related
- ☐ The particle description of light, and how it is related to the wave description
- ☐ Atomic emission and absorption spectra, and how these were the basis for an important advance in atomic theory
- ☐ The quantum mechanical picture of the atom
- ☐ The four quantum numbers and possible combinations of their values
- ☐ The shapes of orbitals and the usual order of their relative energies
- ☐ Ways to determine electronic configurations of atoms
- ☐ The relation between electronic configurations and the positions of elements in the periodic table

The Dalton theory of the atom and related ideas were the basis for our study of *composition stoichiometry* (Chapter 2) and *reaction stoichiometry* (Chapter 3). But that level of atomic theory leaves many questions unanswered. *Why* do atoms combine to form compounds? *Why* do they combine only in simple numerical ratios? *Why* are particular numerical ratios of atoms observed in compounds? *Why* do different elements have such different properties—gases, liquids, solids, metals, nonmetals, and so on? *Why* do some groups of elements have such similar properties, and form compounds with similar formulas? The answers to these and many other fascinating questions in chemistry are supplied by our modern understanding of the nature of atoms. But how can we study something as small as an atom?

Much of the development of modern atomic theory was based on two broad types of research carried out by dozens of scientists just before and after 1900. The first type dealt with the electrical nature of matter, studies of which led scientists to recognize that atoms are composed of still more fundamental particles, and helped to describe the approximate arrangements of these particles in atoms. The second broad area of research dealt with the interaction of matter with energy in the form of light. Such research included studies of the colors of light that substances give off or absorb. These studies led to a much more detailed understanding of the arrangements of particles in atoms. It became clear that the arrangement of the particles determines the chemical and physical properties of each element. As we learn more about the structures of atoms, we are able to collect chemical facts in ways that help us to understand the behavior of matter.

We shall first study the particles that make up atoms and the basic structure of atoms. Then we shall trace the development of the quantum mechanical theory of atoms and see how this theory describes the arrangement of the electrons in atoms. This will give us the background necessary to describe, in the next few chapters, the forces responsible for chemical bonding. Current atomic theory is considerably less than complete. Even so, it is a powerful tool that helps us to understand the forces holding atoms in chemical combination with each other.

Subatomic Particles

5-1 Fundamental Particles

In our study of atomic theory, we look first at the **fundamental particles**. These are the basic building blocks of all atoms. Atoms, and hence *all* matter, consist principally of three fundamental particles: *electrons, protons,* and *neutrons*. Knowledge of the nature and functions of these particles is essential to understanding chemical interactions. The masses and charges of the three fundamental particles are shown in Table 5-1. The mass of an electron is very small compared with the mass of either a proton or a neutron. The charge on a proton is equal in magnitude, but opposite in sign, to the charge on an electron. Let's examine these particles in more detail.

Many other particles, such as quarks, positrons, neutrinos, pions, and muons, have also been discovered. It is not necessary to study their characteristics to learn the fundamentals of atomic structure that are important in chemical reactions.

Table 5-1
Fundamental Particles of Matter

Particle	Isolated Rest Mass	Charge (relative scale)
electron (e^-)	0.00054858 amu	1−
proton (p or p^+)	1.0073 amu	1+
neutron (n or n^0)	1.0087 amu	none

5-2 The Discovery of Electrons

The process is called chemical electrolysis. *Lysis* means "splitting apart."

Some of the earliest evidence about atomic structure was supplied in the early 1800s by the English chemist Humphrey Davy. He found that when he passed electrical current through some substances, the substances decomposed. This led him to propose that the elements of a chemical compound are held together by electrical forces. In 1832–33, Michael Faraday, Davy's protégé, determined the quantitative relationship between the amount of electricity used in electrolysis and the amount of chemical reaction that occurs. Studies of Faraday's work by George Stoney led him to suggest in 1874 that units of electrical charge are associated with atoms. In 1891 he suggested that they be named *electrons*.

Study Figures 5-1 and 5-2 carefully as you read this section.

The most convincing evidence for the existence of electrons came from experiments using *cathode ray tubes* (Figure 5-1). Two electrodes are sealed in a glass tube containing gas at a very low pressure. When a high voltage is applied, current flows and rays are given off by the cathode (negative electrode). These rays travel in straight lines to the anode (positive electrode)

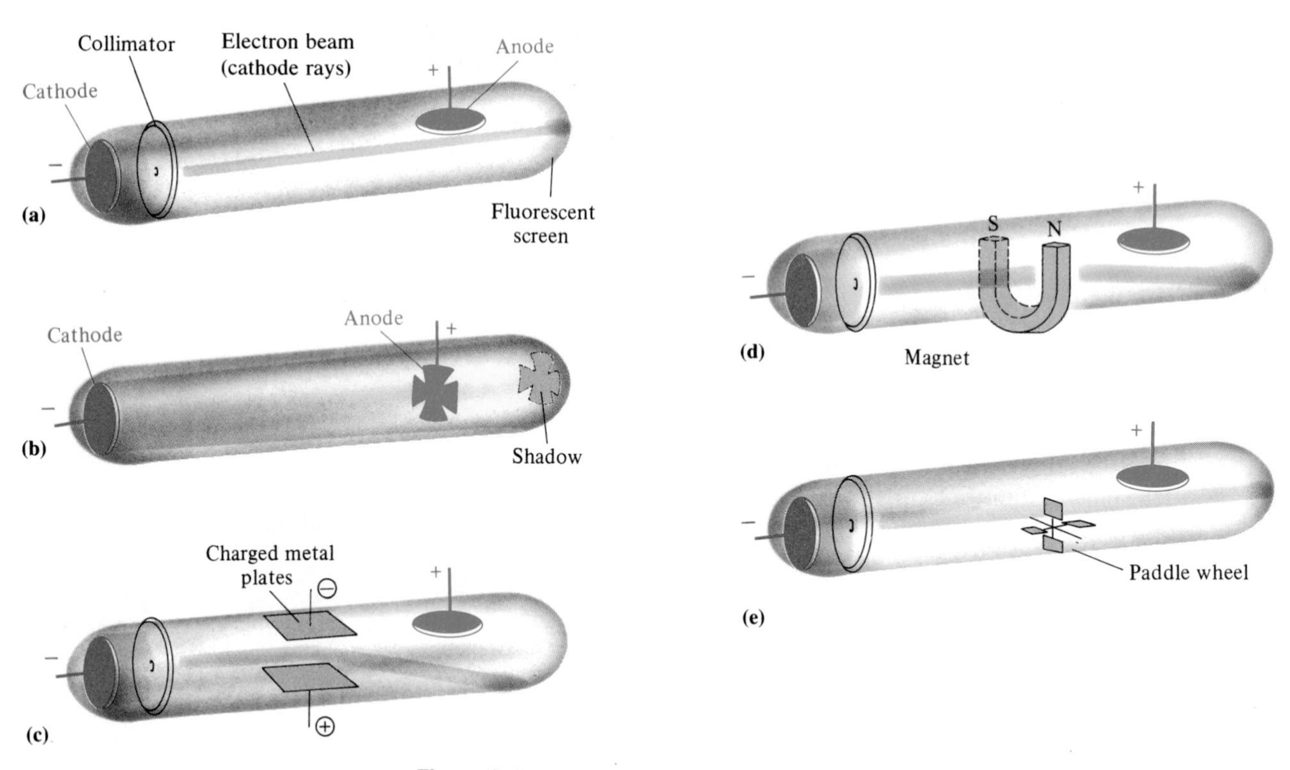

Figure 5-1
Some experiments with cathode ray tubes that show the nature of cathode rays. (a) A cathode ray (discharge) tube, showing the production of a beam of electrons (cathode rays). The beam is detected by observing the glow of a fluorescent screen. (b) A small object placed in a beam of cathode rays casts a shadow. This shows that cathode rays travel in straight lines. (c) Cathode rays have negative electrical charge, as demonstrated by their deflection in an electric field. (The electrically charged plates produce an electric field.) (d) Interaction of cathode rays with a magnetic field is also consistent with negative charge. The magnetic field goes from one pole to the other. (e) Cathode rays have mass, as shown by their ability to turn a small paddle wheel in their path.

and cause the walls opposite the cathode to glow. An object placed in the path of the cathode rays casts a shadow on a zinc sulfide screen placed near the anode. The shadow shows that the rays travel from the cathode toward the anode. Therefore the rays must be negatively charged. Additionally, they are deflected by both magnetic and electrical fields in the directions expected for negatively charged particles.

In 1897, J. J. Thomson studied these negatively charged particles more carefully. He called them **electrons**, the name Stoney had suggested in 1891. By studying the degree of deflections of cathode rays in different magnetic and electric fields, Thomson determined the charge (e) to mass (m) ratio for electrons. The modern value for this ratio is

$$e/m = 1.75881 \times 10^8 \text{ coulomb (C) per gram}$$

The coulomb (C) is the standard unit of *quantity* of electrical charge. It is defined as the quantity of electricity transported in one second by a current of one ampere. It corresponds to the amount of electricity that will deposit 0.001118 g of silver in an apparatus set up for plating silver.

This ratio is the same regardless of the type of gas in the tube, the composition of the electrodes, or the nature of the electrical power source. The clear implication of Thomson's work was that electrons are fundamental particles present in all atoms. We now know that this is true and that all atoms contain integral numbers of electrons.

Once the charge-to-mass ratio for the electron had been determined, additional experiments were necessary to determine the value of either its mass or its charge, so that the other could be calculated. In 1909 Robert Millikan very nicely solved this dilemma with his famous "oil-drop experiment," in which he determined the charge on the electron. This experiment is described in Figure 5-2. All of the charges measured by Millikan turned out to be

X-rays are radiations of much shorter wavelength than visible light (Section 5-9). They are sufficiently energetic to knock electrons out of the atoms in the air. In Millikan's experiment these free electrons became attached to some of the oil droplets.

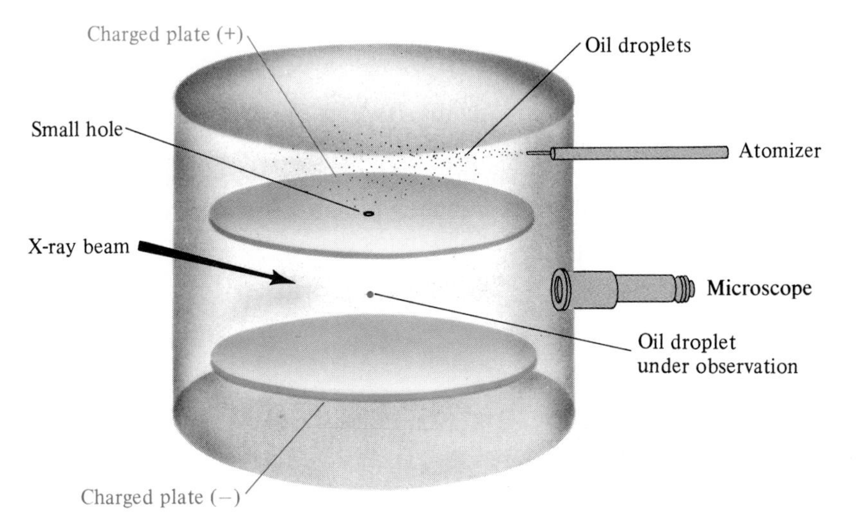

Figure 5-2
The Millikan oil-drop experiment. Tiny oil droplets are produced by an atomizer. A few of them fall through the hole in the upper plate. Irradiation with X-rays gives some of these oil droplets a negative charge. When the voltage between the plates is increased, a negatively charged drop falls more slowly because it is attracted by the positively charged upper plate and repelled by the negatively charged lower plate. At one particular voltage, the electrical force (up) and the gravitational force (down) on the drop are exactly balanced, and the drop remains stationary. If we know this voltage and the mass of the drop, we can calculate the charge on the drop. The mass of the spherical drop can be calculated from its volume (obtained from a measurement of the radius of the drop with a microscope) and the known density of the oil.

Robert A. Millikan (1868–1953) was an American physicist who was a physics professor at the University of Chicago and later director of the physics laboratory at the California Institute of Technology. For his investigations into photoelectric phenomena and the determination of the charge on the electron, he won the 1923 Nobel Prize in physics.

The charge on one mole (Avogadro's number) of electrons is 96,487 coulombs.

integral multiples of the same number. He assumed that this smallest charge was the charge on one electron. This value is 1.60219×10^{-19} coulomb (modern value).

The charge-to-mass ratio, $e/m = 1.75881 \times 10^8$ C/g, can be used in inverse form to calculate the mass of the electron:

$$m = \frac{1\text{ g}}{1.75881 \times 10^8\text{ C}} \times 1.60219 \times 10^{-19}\text{ C}$$
$$= 9.10952 \times 10^{-28}\text{ g per electron}$$

The value of e/m obtained by Thomson and the values of e and m obtained by Millikan differ slightly from the modern values given in this text because early measurements were not as accurate as modern ones.

This is only about 1/1836 the mass of a hydrogen atom, the lightest of all atoms. Millikan's simple oil-drop experiment stands as one of the most clever, yet most fundamental, of all classic scientific experiments. It was the first experiment to suggest that atoms contain integral numbers of electrons, a fact we now know to be true.

5-3 Canal Rays and Protons

In 1886 Eugen Goldstein first observed that a cathode ray tube also generates a stream of positively charged particles that moves toward the cathode. These were called **canal rays** because they were observed occasionally to pass through a channel, or "canal," drilled in the negative electrode (Figure 5-3). These *positive rays,* or *positive ions,* are created when cathode rays knock electrons from the gaseous atoms in the tube, forming positive ions by processes such as

$$\text{atom} \longrightarrow \text{cation}^+ + e^- \qquad \text{or} \qquad X \longrightarrow X^+ + e^-$$

The proton was observed by Rutherford and Chadwick in 1919 as a particle that is emitted by bombardment of certain atoms with alpha particles.

Different elements give positive ions with different e/m ratios. The regularity of the e/m values for different ions led to the idea that there is a unit of positive charge and that it resides in the **proton**. The proton is a fundamental particle with a charge equal in magnitude but opposite in sign to the charge on the electron. Its mass is almost 1836 times that of the electron.

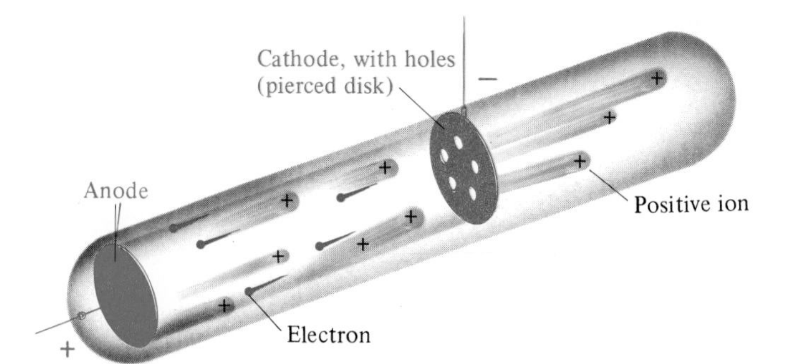

Figure 5-3
A cathode ray tube with a different design and with a perforated cathode. Such a tube was used to produce canal rays and to demonstrate that they travel toward the cathode. Like cathode rays, these *positive* rays are deflected by magnetic or electric fields, but in the opposite direction from cathode rays. Canal ray particles have e/m ratios many times smaller than those of electrons due to their much greater masses. When different elements are in the tube, positive ions with different e/m ratios are observed.

5-4 Rutherford and the Nuclear Atom: Atomic Number

By the first decade of this century, it was clear that each atom contained regions of both positive and negative charge. The question was, how are these charges distributed? The dominant view of that time was summarized in J. J. Thomson's model of the atom, in which the positive charge was assumed to be distributed evenly throughout the atom. The negative charges were pictured as being imbedded in the atom like plums in a pudding (hence the name "plum pudding model").

Soon after Thomson developed his model, tremendous insight into atomic structure was provided by one of Thomson's former students, Ernest Rutherford, certainly the outstanding experimental physicist of his time.

By 1909 Ernest Rutherford had established that alpha (α) particles are positively charged particles. They can be emitted by some radioactive atoms, i.e., atoms that undergo spontaneous disintegration. In 1910 Rutherford's research group carried out a series of experiments that had enormous impact on the scientific world. They bombarded a very thin gold foil with α-particles from a radioactive source. A fluorescent zinc sulfide screen was placed behind the foil to observe the scattering of the α-particles by the gold foil (Figure 5-4). Scintillations (flashes) on the screen, caused by the individual α-particles, were counted to determine the relative numbers of α-particles deflected at various angles. Alpha particles were known to be extremely dense, much denser than gold. Furthermore, they were known to be emitted at high kinetic energies.

Alpha particles are now known to be helium atoms minus their two electrons, or helium nuclei, which have 2+ charges (see Chapter 30).

Radioactivity is contrary to the Daltonian idea of the indivisibility of atoms.

If the Thomson model of the atom were correct, any α-particles passing through the foil would be expected to be deflected by very small angles. Quite unexpectedly, nearly all of the α-particles passed through the foil with little or no deflection. However, a few were deflected through large angles. A very few α-particles even returned from the gold foil in the direction from which they had come! Rutherford was astounded. In his own words.

> It was quite the most incredible event that has ever happened to me in my life. It was almost as if you fired a 15-inch shell into a piece of tissue paper and it came back and hit you.

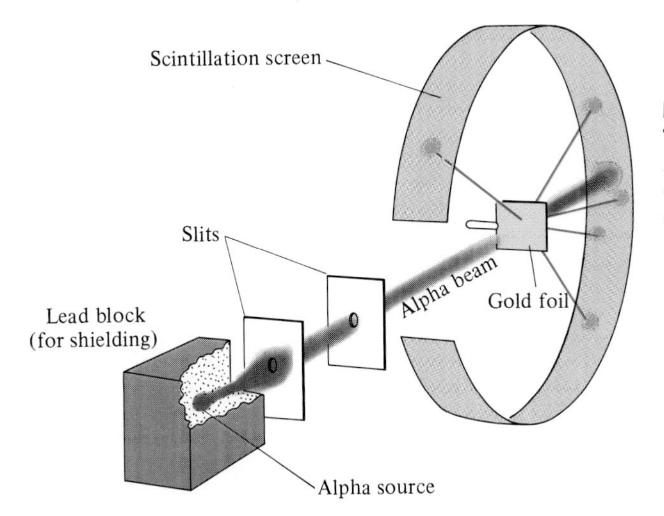

Figure 5-4
The Rutherford scattering experiment. A narrow beam of alpha particles (helium atoms stripped of their electrons) from a radioactive source was directed at a very thin gold foil. Most of the particles passed right through the foil (gold). Many were deflected through moderate angles (shown in red). These deflections were surprises, but the 0.001% of the total that were reflected at acute angles (shown in blue) were totally unexpected. Similar results were observed using foils of other metals.

Ernest Rutherford (1871–1937) was one of the giants in the development of our understanding of atomic structure. A native of New Zealand, Rutherford traveled to England in 1895 where he worked for much of his life. While working with J. J. Thomson at Cambridge University, he discovered α and β radiation. The years 1899–1907 were spent at McGill University in Canada where he proved the nature of these two radiations, for which he received the Nobel Prize in Chemistry in 1908. He returned to England in 1908, and it was there, at Manchester University, that he and his coworkers Geiger and Marsden performed the famous gold foil experiments that revolutionized our view of the atom. Not only did he perform much important research in physics and chemistry, but he also guided the work of ten future recipients of the Nobel Prize.

Rutherford's mathematical analysis of his results showed that the scattering of positively charged α-particles was caused by repulsion from very dense regions of positive charge in the gold foil. He concluded that the mass of one of these regions is nearly equal to that of a gold atom, but that the diameter is no more than 1/10,000 that of an atom. Many experiments with foils of different metals yielded similar results. Realizing that these observations were inconsistent with previous theories about atomic structure, Rutherford discarded the old theory and proposed a better one. He suggested that each atom contains a *tiny, positively charged, massive center* that he called an **atomic nucleus**. Most α-particles pass through metal foils undeflected because atoms are *primarily* empty space populated only by the very light electrons. The few particles that are deflected are the ones that come close to the heavy, highly charged metal nuclei (Figure 5-5).

This representation is *not* to scale. If nuclei were as large as the black dots that represent them, each white region, which represents the size of an atom, would have a diameter of more than 30 feet!

As a somewhat similar experiment, imagine that we shoot pellets from a BB gun at a chain-link fence. Because the "particles" are much smaller than the "empty spaces" in such a fence, most of the pellets would go through the fence undeflected. However, some would glance off the wires of the fence and be deflected through moderate angles; a few would hit a wire directly enough to bounce back to our side of the fence. If we fired a very large number, say a million such "particles," at the fence, and counted how

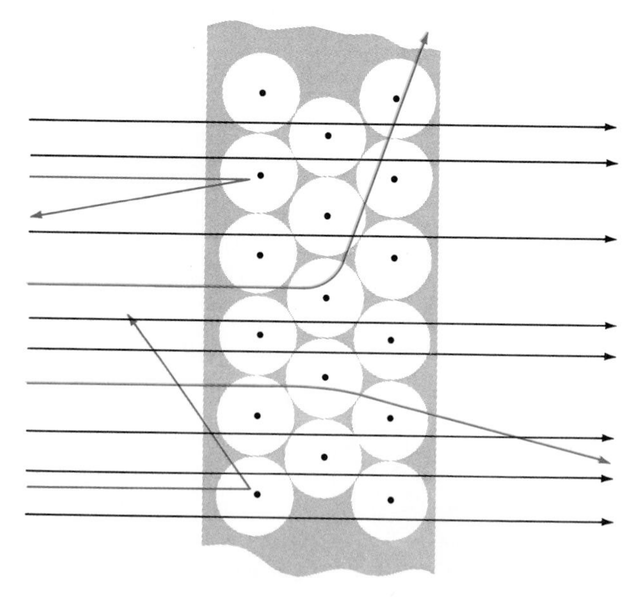

Figure 5-5
An interpretation of the Rutherford scattering experiment. The atom is pictured as consisting mostly of "open" space. At the center is a tiny and extremely dense nucleus that contains all of the atom's positive charge and nearly all of the mass. The electrons are thinly distributed throughout the "open" space. Most of the positively charged alpha particles (shown in black) pass through the open space undeflected, not coming near any gold nuclei. The few that pass fairly close to a nucleus (shown in red) are repelled by electrostatic force and thereby deflected. The very few particles that are on a "collision course" with gold nuclei are repelled backward at acute angles (shown in blue). Calculations based on the results of the experiment indicated that the diameter of the open-space portion of the atom is from 10,000 to 100,000 times greater than the diameter of the nucleus.

many went through undeflected and how many were deflected by various angles, we could perhaps calculate the fraction of the area of the fence that is open space, and even discover something about the size and distribution of the wires. (Gold foil is many atomic layers thick, so we could suppose that there were several such fences, one behind the other.)

Rutherford was able to determine the magnitudes of the positive charges on the atomic nuclei. The picture of atomic structure that he developed is called the Rutherford model of the atom.

> Atoms consist of very small, very dense positively charged nuclei surrounded by clouds of electrons at relatively great distances from the nuclei.

We now know that every nucleus contains an integral number of protons exactly equal to the number of electrons in a neutral atom of element. Every hydrogen atom contains one proton, every helium atom contains two protons, and every lithium atom contains three protons. The number of protons in the nucleus of an atom determines its identity; this number is known as the **atomic number** of that element.

5-5 Neutrons

The third fundamental particle, the neutron, eluded discovery until 1932. James Chadwick correctly interpreted experiments on the bombardment of beryllium with high-energy alpha particles. Later experiments showed that nearly all elements up to potassium, element 19, produce neutrons when they are bombarded with high-energy alpha particles. The **neutron** is an uncharged particle with a mass slightly greater than that of the proton. With its discovery, the picture of the nuclear atom was complete:

This does not mean that elements above number 19 do not have neutrons, only that neutrons are not generally knocked out of atoms of higher atomic number by alpha particle bombardment.

> Atoms consist of very small, very dense nuclei surrounded by clouds of electrons at relatively great distances from the nuclei. All nuclei contain protons; nuclei of all atoms except the common form of hydrogen also contain neutrons.

Nuclear diameters are about 10^{-4} Ångstroms (10^{-5} nanometers); atomic diameters are about 1 Ångstrom (10^{-1} nanometers). To put this difference in perspective, suppose that you wish to build a model of an atom using a basketball (diameter about 9.5 inches) as the nucleus; on this scale, the atomic model would be nearly 6 miles across!

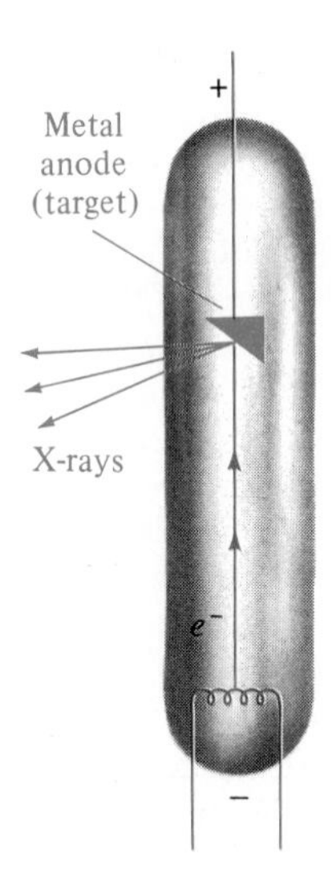

Figure 5-6
A simplified representation of the production of X-rays by bombardment of a solid target with a high-energy beam of electrons.

5-6 Mass Number and Isotopes

Only a few years after Rutherford's scattering experiments, H. G. J. Moseley studied X-rays given off by various elements. Max von Laue had shown that X-rays could be diffracted by crystals into a spectrum in much the same way that visible light can be separated into its component colors. Moseley generated X-rays by aiming a beam of high-energy electrons at a solid target made of a single pure element (Figure 5-6).

Chemistry in Use. . .
Stable Isotope Ratio Analysis

Many elements exist as two or more stable isotopes, although one isotope is usually present in far greater abundance. For example, there are two stable isotopes of carbon, ^{13}C and ^{12}C, of which ^{12}C is the more abundant, constituting 98.89% of all carbon. Similarly, there are two stable isotopes of nitrogen, ^{14}N and ^{15}N, of which ^{14}N makes up 99.63% of all nitrogen.

Differences in chemical and physical properties that arise from differences in atomic mass of an element are known as isotope effects. We know that the extranuclear structure of an element (the number of electrons and their arrangement) essentially determines its chemical behavior, whereas the nucleus has more influence on many of the physical properties of the element. Because all isotopes of a given element contain the same number and arrangement of electrons, it was assumed for a long time that isotopes would behave identically in chemical reactions. In reality, although isotopes behave very similarly in chemical reactions, the correspondence is not perfect. The mass differences between different isotopes of the same element cause them to have slightly different physical and chemical properties. For example, the presence of only one additional neutron in the nucleus of the heavier isotope can cause it to react a little more slowly than its lighter counterpart. Such an effect often results in a ratio of heavy isotope to light isotope in the product of a reaction that is different from the ratio found in the reactant.

Stable isotope ratio analysis (SIRA) is an analytical technique that takes advantage of the different chemical and physical properties of isotopes. In SIRA the isotopic composition of a sample is measured using a mass spectrometer. This composition is then expressed as the relative ratios of two or more of the stable isotopes of a specific element. For instance, the ratio of ^{13}C to ^{12}C in a sample can be determined. This ratio is then compared to the isotope ratio of a defined standard. Because mass differences are most pronounced among the lightest elements, those elements experience the greatest isotope effects. Thus, the isotopes of the elements H, C, N, O, and S are used most frequently for SIRA. These elements have further significance because they are among the most abundant elements in biological systems.

The isotopic composition of a sample is usually expressed as a "del" value (∂), defined as

$$\partial X_{\text{sample}}\ (‰) = \frac{(R_{\text{sample}} - R_{\text{standard}})}{R_{\text{standard}}} \times 1000$$

where, $\partial X_{\text{sample}}$ is the isotope ratio relative to a standard, and R_{sample} and R_{standard} are the absolute isotope ratios of the sample and standard, respectively. Multiplying by 1000 allows the values to be expressed in parts per thousand (‰). If the del value is a positive number, the sample has a greater amount of the heavier isotope than does the standard. In such cases the sample is said to be "heavier" than the standard, or to have been "enriched" in the heavy isotope. Similarly, if the del value is negative, the sample has a higher proportion of the lighter isotope and thus is described as "lighter" than the standard.

The most frequently used element for SIRA is carbon. The first limited data on $^{13}C/^{12}C$ isotope ratios in natural materials were published in 1939. At that time it was established that limestones, atmospheric CO_2, marine plants, and terrestrial plants each possessed characteristic carbon isotope ratios. In the succeeding years, $^{13}C/^{12}C$ ratios were determined for a wide variety of things, including petroleum, coal, diamonds, marine organisms, and terrestrial organisms. Such data led to the important conclusion that a biological organism has an isotope

ratio that depends on the main source of carbon to that organism—that is, its food source. For example, if an herbivore (an animal that feeds on plants) feeds exclusively on one type of plant, that animal's carbon isotope ratio will be almost identical to that of the plant. If another animal were to feed exclusively on that herbivore, it would also have a similar carbon isotope ratio. Suppose now that an animal, say a rabbit, has a diet comprising two different plants, A and B. Plant A has a ∂^{13}C value of -24‰, and plant B has a del value of -10‰. If the rabbit eats equal amounts of the two plants, then the ∂^{13}C value of the rabbit will be the average of the two values, or -17‰. Values more positive than -17‰ would indicate a higher consumption of plant B than of plant A, whereas more negative values would reflect a preference for plant A.

Similar studies have been conducted with the stable isotopes of nitrogen. A major way in which nitrogen differs from carbon in isotopic studies relates to how ∂^{13}C and ∂^{15}N values change as organic matter moves along the food chain—from inorganic nutrient to plant, then to herbivore, to carnivore, and on to higher carnivores. It has been pointed out that ∂^{13}C remains nearly constant throughout successive levels of the food chain. In contrast, on average there is a $+3$ to $+5$‰ shift in the value of ∂^{15}N at each successive level of the food chain. For instance, suppose a plant has a ∂^{15}N value of 1‰. If an herbivore, such as a rabbit, feeds exclusively on that one type of plant, it will have a ∂^{15}N value of 4‰. If another animal, such as a fox, feeds exclusively on that particular type of rabbit, it in turn will have a ∂^{15}N value of 7‰. An important implication of this phenomenon is that an organism's nitrogen isotope ratio can be used as an indicator of the level in the food chain at which that species of animal feeds.

An interesting application of SIRA is the determination of the adulteration of food. As already mentioned, the isotope ratios of different plants and animals have been determined. For instance, corn has a ∂^{13}C value of about -12‰ and most flowering plants have ∂^{13}C values of about -26‰. The difference in these ∂^{13}C values arises because these plants carry out photosynthesis by slightly different chemical reactions. In the first reaction of photosynthesis, corn produces a molecule that contains four carbons, whereas flowering plants produce a molecule that has only three carbons. High-fructose corn syrup (HFCS) is thus derived from a "C_4" plant, whereas the nectar that bees gather comes from "C_3" plants. The slight differences in the photosynthetic pathways of C_3 and C_4 plants create the large differences in their ∂^{13}C values. Brokers who buy and sell huge quantities of "sweet" products are able to monitor HFCS adulteration of honey, maple syrup, apple juice, and so on by taking advantage of the SIRA technique. If the ∂^{13}C value of one of these products is not appropriate, then the product obviously has had other substances added to it, i.e., has been adulterated. The U.S. Department of Agriculture conducts routine isotope analyses to ensure the purity of those products submitted for subsidy programs. Similarly, the honey industry monitors itself with the SIRA technique.

Another interesting use of SIRA is in the determination of the diets of prehistoric human populations. It is known that marine plants have higher ∂^{15}N values than terrestrial plants. This difference in ∂^{15}N is carried up food chains, causing marine animals to have higher ∂^{15}N values than terrestrial animals. The ∂^{15}N values of humans feeding on marine food sources are therefore higher than those of people feeding on terrestrial food. This phenomenon has been used to estimate the marine and terrestrial components of the diets of historic and prehistoric human groups through the simple determination of the ∂^{15}N value of bone collagen collected from excavated skeletons.

Stable isotope ratio analysis is a powerful tool; many of its potential uses are only slowly being recognized by researchers. In the meantime, the use of stable isotope methods in research is becoming increasingly common, and through these methods scientists are attaining new levels of understanding of chemical, biological, and geological processes.

Beth A. Trust
Graduate student in chemistry
University of Texas Marine Sciences Institute

H. G. J. Moseley was one of the many remarkable scientists who worked with Ernest Rutherford. In 1913 Moseley found that the wavelengths of X-rays emitted by an element are related in a precise way to the atomic number of the element. This discovery led to the realization that atomic number, related to electrical properties of the atom, was more fundamental to determining the properties of the elements than atomic weight. This put the ideas of the periodic table on a more fundamental footing. Moseley's scientific career was very short. He was enlisted in the British army during World War I, and died in battle in the Gallipoli campaign in 1915. In subsequent wars, most countries have not allowed their promising scientists to take part in front-line service.

The spectra of X-rays produced by targets of different elements were recorded photographically. Each photograph consisted of a series of lines representing X-rays at various wavelengths. Comparison of results from different elements revealed that corresponding lines were displaced toward shorter wavelengths as atomic weights of the target materials increased, with three exceptions. Moseley showed that the X-ray wavelengths could be better correlated with the atomic number (Section 5-4). On the basis of his mathematical analysis of these X-ray data, he concluded that

> each element differs from the preceding element by having one more positive charge in its nucleus.

For the first time it was possible to arrange all known elements in order of increasing nuclear charge. A plot summarizing this interpretation of Moseley's data appears in Figure 5-7.

Most elements consist of atoms of different masses, called **isotopes**. The isotopes of a given element contain the same number of protons (and also the same number of electrons) because they are atoms of the same element. They differ in mass because they contain different numbers of neutrons in their nuclei.

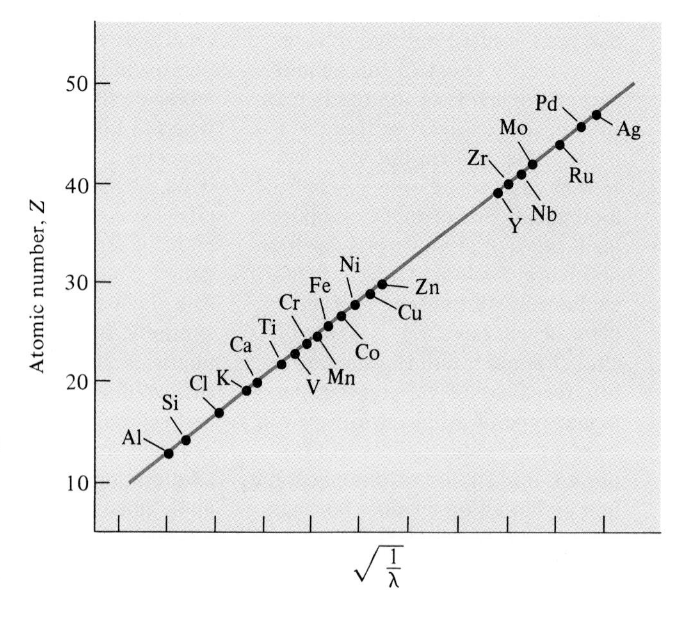

Figure 5-7
A plot of some of Moseley's X-ray data. The atomic number of an element is found to be directly proportional to the square root of the reciprocal of the wavelength of a particular X-ray spectral line. Wavelength (Section 5-9) is represented by λ.

Table 5-2
Make-Up of the Three Isotopes of Hydrogen (neutral atoms)

Name	Symbol	Nuclide Symbol	Atomic Abundance in Nature	No. of Protons	No. of Neutrons	No. of Electrons (in neutral atoms)
hydrogen	H	${}^{1}_{1}H$	99.985%	1	0	1
deuterium	D	${}^{2}_{1}H$	0.015%	1	1	1
tritium*	T	${}^{3}_{1}H$	0.000%	1	2	1

*No known natural sources; produced by decomposition of artificial isotopes.

For example, there are three distinct kinds of hydrogen atoms, commonly called hydrogen, deuterium, and tritium. (This is the only element for which we give each isotope a different name.) Each of these three contains one proton in the atomic nucleus. The predominant form of hydrogen contains no neutrons, but each deuterium atom contains one neutron and each tritium atom contains two neutrons in its nucleus (Table 5-2). All three forms of hydrogen display very similar chemical properties.

The **mass number** of an atom is the sum of the number of protons and the number of neutrons in its nucleus; i.e.,

A mass number is a count of the number of things present, so it must be a whole number. Because the masses of the proton and the neutron are both about 1 amu, the mass number is *approximately* equal to the actual mass of the isotope (which is not a whole number).

$$\begin{aligned} \text{mass number} &= \text{number of protons} + \text{number of neutrons} \\ &= \text{atomic number} \quad\; + \text{neutron number} \end{aligned}$$

The mass number for normal hydrogen atoms is 1, for deuterium 2, and for tritium 3. The composition of a nucleus is indicated by its **nuclide symbol**. This consists of the symbol for the element (E), with the atomic number (Z) written as a subscript at the lower left and the mass number (A) as a superscript at the upper left, ${}^{A}_{Z}E$. By this system, the three isotopes of hydrogen are designated as ${}^{1}_{1}H$, ${}^{2}_{1}H$, and ${}^{3}_{1}H$.

5-7 Mass Spectrometry and Isotopic Abundance

Mass spectrometers are instruments that measure the charge-to-mass ratio of charged particles (Figure 5-8). A gas sample at very low pressure is bombarded with high-energy electrons. This causes electrons to be ejected from some of the gas molecules, creating positive ions. The positive ions are then focused into a very narrow beam and accelerated by an electric field toward a magnetic field. The magnetic field deflects the ions from their straight-line path. The extent to which the beam of ions is deflected depends upon four factors:

1. *Magnitude of the accelerating voltage* (*electric field strength*). The range varies from about 500 to about 2000 volts. Higher voltages result in beams of more rapidly moving particles that are deflected less than the beams of the more slowly moving particles produced by lower voltages.
2. *Magnetic field strength*. Stronger fields deflect a given beam more than weaker fields.
3. *Masses of the particles*. Because of their inertia, heavier particles are deflected less than lighter particles that carry the same charge.

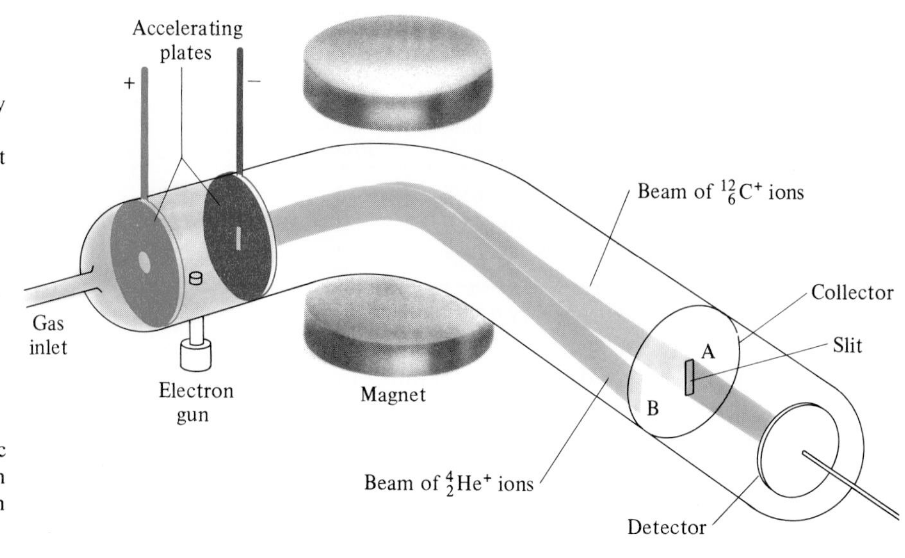

Figure 5-8
The mass spectrometer. In the mass spectrometer, gas molecules at low pressure are ionized and accelerated by an electric field. The ion beam is then passed through a magnetic field. In that field the beam is resolved into components, each containing particles of equal charge-to-mass ratio. Lighter particles are deflected more strongly than heavy ones with the same charge. In a beam containing $^{12}_{6}C^{+}$ and $^{4}_{2}He^{+}$ ions, the lighter $^{4}_{2}He^{+}$ ions would be deflected more than the heavier $^{12}_{6}C^{+}$ ions. The spectrometer shown is adjusted to detect the $^{12}_{6}C^{+}$ ions. By changing the magnitude of the magnetic or electric field, we can move the beam of $^{4}_{2}He^{+}$ ions striking the collector from B to A, where it would be detected.

4. *Charges on the particles*. Highly charged particles interact more strongly with magnetic and electric fields and are thus deflected more than particles of equal mass with smaller charges.

The mass spectrometer is used to measure masses of isotopes as well as isotopic abundances. Helium occurs in nature almost exclusively as $^{4}_{2}He$. Let's see how its atomic mass is measured. To simplify the picture, we'll assume that only ions with 1+ charge are formed in the experiment illustrated in Figure 5-8.

As we saw in Section 2-4, $^{12}_{6}C$ is the (arbitrary) reference point on the atomic weight scale. We first use carbon-12 to calibrate the instrument. A carefully measured, fixed accelerating voltage is applied and a sample of vaporized $^{12}_{6}C$ is fed into the mass spectrometer. The magnetic field strength that causes $^{12}_{6}C^{+}$ ions to arrive at point A is measured. A sample of helium is then fed into the mass spectrometer with exactly the same accelerating voltage. The lighter $^{4}_{2}He^{+}$ ions are deflected more than the heavier $^{12}_{6}C^{+}$ ions to some point B. The magnetic field strength is then decreased slowly until the beam of lighter ions falls at point A. Now that the field strengths required to focus the two kinds of ions at the same point are known, the mass of the lighter ion can be calculated.

The relationship between masses of particles and magnetic field strength, $\mathcal{H}$, is

The $\mathcal{H}$'s are the magnetic field strengths required to focus the beams of ions at point A.

$$\frac{\text{mass } ^{4}_{2}He^{+}}{\text{mass } ^{12}_{6}C^{+}} = \left(\frac{\mathcal{H} \text{ for } ^{4}_{2}He^{+}}{\mathcal{H} \text{ for } ^{12}_{6}C^{+}}\right)^2$$

$(0.5776)^2 = 0.3336$

The ratio of the two magnetic field strengths in the experiment is measured as 0.5776, so the mass of $^{4}_{2}He^{+}$ is found to be 0.3336 times the mass of the $^{12}_{6}C^{+}$ ion. The mass of the $^{12}_{6}C^{+}$ ion is exactly 12 amu, so the mass of the $^{4}_{2}He^{+}$ ion is 0.3336×12 amu $= 4.003$ amu.

A beam of Ne^{+} ions in the mass spectrometer is split into three segments. The mass spectrum of these ions (a graph of the relative numbers of ions of each mass) is shown in Figure 5-9. This indicates that neon occurs in nature

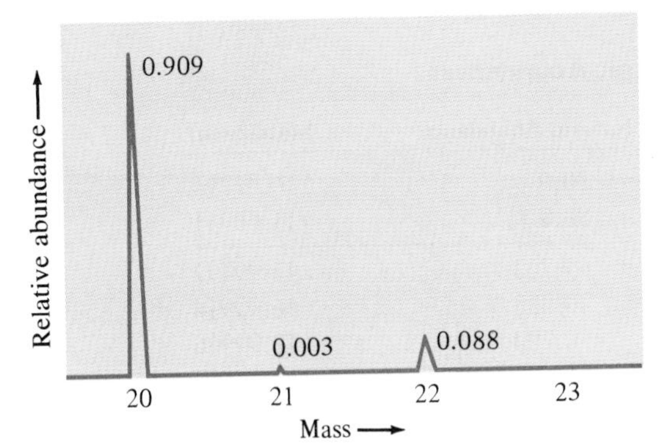

Figure 5-9
Mass spectrum of neon(1+ ions only). Neon consists of three isotopes, of which neon-20 is by far the most abundant (90.9%). The mass of that isotope, to five decimal places, is 19.99244 amu on the carbon-12 scale. The number by each peak represents the number of Ne^+ ions corresponding to that isotope, expressed as a fraction of all Ne^+ ions.

as three isotopes: $^{20}_{10}$Ne, $^{21}_{10}$Ne, and $^{22}_{10}$Ne. In Figure 5-9 we see that the isotope $^{20}_{10}$Ne, mass 19.99244 amu, is the most abundant isotope (has the tallest peak). It accounts for 90.5% of the atoms. $^{22}_{10}$Ne accounts for 9.2% and $^{21}_{10}$Ne for only 0.3% of the atoms.

Figure 5-10 shows a modern mass spectrometer and a typical spectrum of an element. In nature, some elements, such as fluorine and phosphorus, exist in only one form, but most elements occur as isotopic mixtures. Some examples of natural isotopic abundances are given in Table 5-3. The percentages are based on the numbers of naturally occurring atoms of each isotope, *not* on their masses.

The distribution of isotopic masses, while nearly constant, does vary somewhat depending on the source of the element. For example, the abundance of $^{13}_{6}$C in atmospheric CO_2 is slightly different from that in seashells. The chemical history of a compound can be inferred from the small differences in isotope ratios.

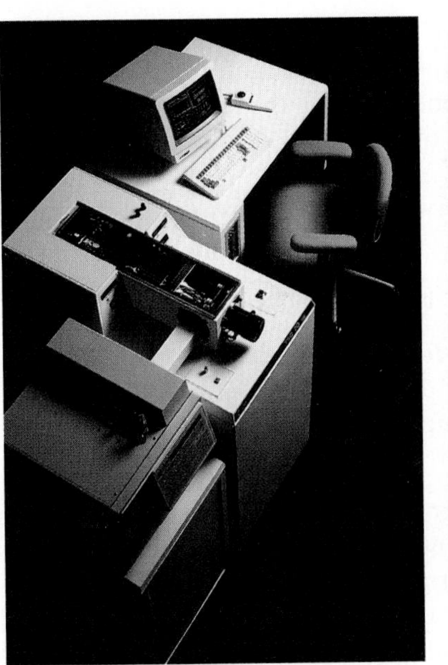

(a)

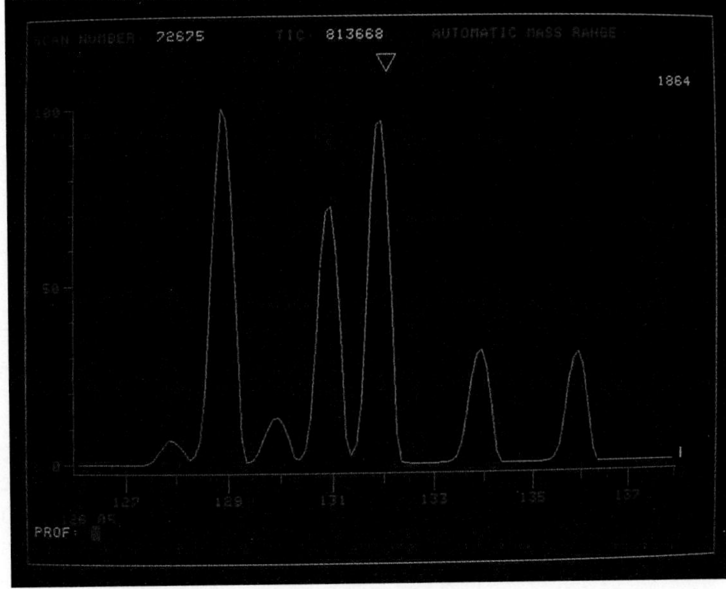

(b)

Figure 5-10
(a) A modern mass spectrometer. (b) The mass spectrum of Xe^{1+} ions, measured on the instrument shown in (a). The isotope ^{126}Xe is at too low an abundance (0.090%) to appear in this experiment.

Table 5-3
Some Naturally Occurring Isotopic Abundances

Element	Isotope	% Natural Abundance	Mass (amu)
boron	${}^{10}_{5}B$	20.0	10.01294
	${}^{11}_{5}B$	80.0	11.00931
oxygen	${}^{16}_{8}O$	99.762	15.99491
	${}^{17}_{8}O$	0.038	16.99914
	${}^{18}_{8}O$	0.200	17.99916
chlorine	${}^{35}_{17}Cl$	75.77	34.96885
	${}^{37}_{17}Cl$	24.23	36.9658
uranium	${}^{234}_{92}U$	0.0057	234.0409
	${}^{235}_{92}U$	0.72	235.0439
	${}^{238}_{92}U$	99.27	238.0508

The 20 elements that have only one naturally occurring isotope are ${}^{9}_{4}Be$, ${}^{19}_{9}F$, ${}^{23}_{11}Na$, ${}^{27}_{13}Al$, ${}^{31}_{15}P$, ${}^{45}_{21}Sc$, ${}^{55}_{25}Mn$, ${}^{59}_{27}Co$, ${}^{75}_{33}As$, ${}^{89}_{39}Y$, ${}^{93}_{41}Nb$, ${}^{103}_{45}Rh$, ${}^{127}_{53}I$, ${}^{133}_{55}Cs$, ${}^{141}_{59}Pr$, ${}^{159}_{65}Tb$, ${}^{165}_{67}Ho$, ${}^{169}_{69}Tm$, ${}^{197}_{79}Au$, and ${}^{209}_{83}Bi$. However, there are other, artificially produced isotopes of these elements.

5-8 The Atomic Weight Scale and Atomic Weights

We saw in Section 2-4 that the **atomic weight scale** is based on the mass of the carbon-12 isotope. As a result of action taken by the International Union of Pure and Applied Chemistry in 1962,

Described another way, the mass of one atom of ${}^{12}_{6}C$ is *exactly* 12 amu.

one **amu** is exactly 1/12 of the mass of a carbon-12 atom.

This is approximately the mass of one atom of ${}^{1}H$, the lightest isotope of the element with lowest mass.

In Section 2-5 we said that one mole of atoms contains 6.022×10^{23} atoms. The mass of one mole of atoms of any element, in grams, is numerically equal to the atomic weight of the element. Because the mass of one carbon-12 atom is exactly 12 amu, the mass of one mole of carbon-12 atoms is exactly 12 grams.

Let us now show the relationship between atomic mass units and grams.

$$\underline{?}\ g = 1\ amu \times \frac{1\ {}^{12}_{6}C\ atom}{12\ amu} \times \frac{1\ mol\ {}^{12}_{6}C\ atoms}{6.022 \times 10^{23}\ {}^{12}_{6}C\ atoms} \times \frac{12\ g\ {}^{12}_{6}C}{1\ mol\ {}^{12}_{6}C\ atoms} = 1.660 \times 10^{-24}\ g$$

Thus we see that *1 amu* $= 1.660 \times 10^{-24}$ *g*. Multiplying both sides by Avogadro's number, we see that *1 g* $= 6.022 \times 10^{23}$ *amu*.

You may wish to verify that the same result is obtained regardless of the element or isotope chosen.

At this point, we should clearly emphasize the differences among the following quantities:

1. The *atomic number, Z*, is an integer equal to the number of protons in the nucleus of an atom of the element. It is also the number of electrons in a neutral atom. It is the same for all atoms of an element.
2. The *mass number, A*, is an integer equal to the *sum* of the number of protons and the number of neutrons in the nucleus of an atom of a

particular isotope of an element. It is different for different isotopes of the same element.

3. The *atomic weight* of an element is the weighted average of the masses of its constituent isotopes. Atomic weights are fractional numbers, not integers.

The atomic weight that we determine experimentally (for an element that consists of more than one isotope) is such a weighted average. The following example shows how such an atomic weight can be calculated from measured isotopic abundances.

Example 5-1

Three isotopes of magnesium occur in nature. Their abundances and masses, determined by mass spectrometry, are listed below. Use this information to calculate the atomic weight of magnesium.

Isotope	% Abundance	Mass (amu)
$^{24}_{12}Mg$	78.99	23.98504
$^{25}_{12}Mg$	10.00	24.98584
$^{26}_{12}Mg$	11.01	25.98259

Plan

We multiply the fraction of each isotope by its mass and add these numbers to obtain the atomic weight of magnesium.

Solution

$$\begin{aligned}\text{atomic weight} &= 0.7899(23.98504\text{ amu}) + 0.1000(24.98584\text{ amu}) + 0.1101(25.98259\text{ amu})\\ &= 18.94\text{ amu} + 2.498\text{ amu} + 2.861\text{ amu}\\ &= 24.30\text{ amu} \quad \text{(to four significant figures)}\end{aligned}$$

The two heavier isotopes make small contributions to the atomic weight of magnesium, because most magnesium atoms are the lightest isotope.

EOC 33, 34

Example 5-2 shows how the process can be reversed. Percent abundances can be calculated from isotopic masses and from the atomic weight of an element that occurs in nature as a mixture of only two isotopes.

Example 5-2

The atomic weight of gallium is 69.72 amu. The masses of the naturally occurring isotopes are 68.9257 amu for $^{69}_{31}Ga$ and 70.9249 amu for $^{71}_{31}Ga$. Calculate the percent abundance of each isotope.

Plan

We can represent the fraction of each isotope algebraically. Atomic weight is the weighted average of the masses of the constituent isotopes. Therefore, the fraction of each isotope is multiplied by its mass and the sum of the results is equal to the atomic weight.

When a quantity is represented by fractions, the sum of the fractions must always be unity. In this case, $x + (1 - x) = 1$.

Solution

Let x = fraction of $^{69}_{31}Ga$. Then $(1 - x)$ = fraction of $^{71}_{31}Ga$.

$$x(68.9257 \text{ amu}) + (1 - x)(70.9249 \text{ amu}) = 69.72 \text{ amu}$$

$$68.9257x + 70.9249 - 70.9249x = 69.72$$

$$-1.9992x = -1.20$$

$$x = 0.600$$

$$x = 0.600 = \text{fraction of } ^{69}_{31}\text{Ga} \quad \therefore \quad 60.0\% \ ^{69}_{31}\text{Ga}$$

$$(1 - x) = 0.400 = \text{fraction of } ^{71}_{31}\text{Ga} \quad \therefore \quad 40.0\% \ ^{71}_{31}\text{Ga}$$

EOC 28, 29

The Electronic Structures of Atoms

The Rutherford model of the atom is consistent with the evidence presented so far, but it has some serious limitations. It does not answer important questions such as the following. *Why* do different elements have such different chemical and physical properties? *Why* does chemical bonding occur at all? *Why* does each element form compounds with characteristic formulas? *How* can atoms of different elements give off or absorb light only of characteristic colors (as was known long before 1900)?

To go further in our understanding, we must first learn more about the arrangements of electrons in atoms. The theory of these arrangements is based largely on the study of the light given off and absorbed by atoms. Then we shall develop a detailed picture of the *electron configurations* of different elements.

5-9 Electromagnetic Radiation

Our ideas about the arrangements of electrons in atoms have evolved slowly. Much of the information has been derived from **atomic emission spectra**. These are the lines, or bands, produced on photographic film by radiation that has passed through a refracting glass prism after being emitted from electrically or thermally excited atoms. To help us understand the nature of atomic spectra, let us first describe electromagnetic radiation in general.

All types of electromagnetic radiation, or radiant energy, can be described in the terminology of waves. To help characterize any wave, we specify its *wavelength* (or its *frequency*). Let us use a familiar kind of wave, that on the surface of water (Figure 5-11), to illustrate these terms. The significant feature of wave motion is its repetitive nature. The **wavelength**, λ, is the distance between any two adjacent identical points of the wave, for instance, two adjacent crests. The **frequency** is the number of wave crests passing a given point per unit time; it is represented by the symbol ν (Greek letter

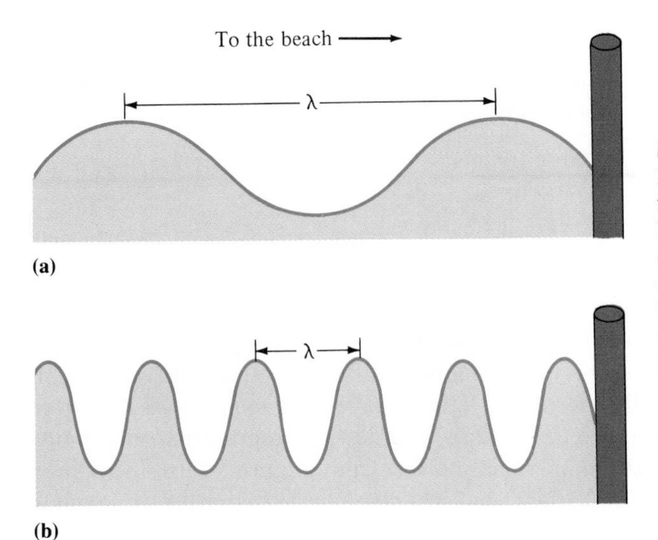

Figure 5-11
Illustrations of the wavelength and frequency of water waves. The distance between any two identical points, e.g., crests, is the wavelength, λ. We could measure the frequency, ν, of the wave by observing the frequency at which the level rises and falls at a fixed point in its path—for instance, at the post. (a) and (b) represent two waves that are traveling at the same speed. In (a) the wave has long wavelength and low frequency; in (b) the wave has shorter wavelength and higher frequency.

''nu'') and is usually expressed in cycles/second or, more commonly, simply as 1/s or s^{-1}. For a wave that is ''traveling'' at some speed, the wavelength and the frequency are related to each other by

One cycle per second is also called one *hertz* (Hz), after Heinrich Hertz. In 1887 Hertz discovered electromagnetic radiation outside the visible range and measured its speed and wavelengths.

$$\lambda\nu = \text{speed of propagation of the wave} \quad \text{or} \quad \lambda\nu = c$$

Thus, wavelength and frequency are inversely proportional to each other; for the same wave speed, the shorter the wavelength, the higher the frequency.

For water waves, it is the surface of the water that changes repetitively; for a vibrating violin string, it is the displacement of any point on the string. Electromagnetic radiation consists of a regular, repetitive variation in electrical and magnetic fields. The electromagnetic radiation most obvious to us is visible light. It has wavelengths ranging from about 4×10^{-7} m (violet) to about 7×10^{-7} m (red). Expressed in frequencies, this range is about 7.5×10^{14} Hz (violet) to about 4.3×10^{14} Hz (red).

In a vacuum, the speed of electromagnetic radiation, c, is the same for all wavelengths, 2.9979249×10^8 m/s. The relationship between the wavelength and frequency of electromagnetic radiation, with c rounded to three significant figures, is

$$\lambda\nu = c = 3.00 \times 10^8 \text{ m/s}$$

Example 5-3
The frequency of violet light is $7.31 \times 10^{14}\ s^{-1}$, and that of red light is $4.57 \times 10^{14}\ s^{-1}$. Calculate the wavelength of each color.

Plan
Frequency and wavelength are inversely proportional to each other, $\lambda = c/\nu$. We can substitute the frequencies into this relationship and calculate wavelengths.

Sir Isaac Newton (1642–1727), one of the giants of science. You probably know of him from his theory of gravitation. In addition, he made enormous contributions to the understanding of many other aspects of physics, including the nature and behavior of light, optics, and the laws of motion. He is credited with the discoveries of differential calculus and of expansions into infinite series.

Solution

$$\text{(violet light) } \lambda = \frac{c}{\nu} = \frac{3.00 \times 10^8 \text{ m/s}}{7.31 \times 10^{14} \text{ s}^{-1}} = 4.10 \times 10^{-7} \text{ m } (4.10 \times 10^3 \text{ Å})$$

$$\text{(red light) } \lambda = \frac{c}{\nu} = \frac{3.00 \times 10^8 \text{ m/s}}{4.57 \times 10^{14} \text{ s}^{-1}} = 6.56 \times 10^{-7} \text{ m } (6.56 \times 10^3 \text{ Å})$$

EOC 39, 40

Isaac Newton first recorded the separation of sunlight into its component colors by allowing it to pass through a prism. Because sunlight (white light) contains all wavelengths of visible light, it gives the *continuous spectrum* observed in a rainbow (Figure 5-12a). Visible light represents only a tiny segment of the electromagnetic radiation spectrum (Figure 5-12b). In addition to all wavelengths of visible light, sunlight also contains shorter wavelength (ultraviolet) radiation as well as longer wavelength (infrared) radiation. Neither of these can be detected by the human eye. Both may be detected and recorded photographically or by detectors designed for that purpose. Many other familiar kinds of radiation are simply electromagnetic radiation of longer or shorter wavelengths.

Thus, we see that light is usually described in terms of wave behavior. Under certain conditions, it is also possible to describe light as composed of *particles*, or **photons**. According to the ideas presented by Max Planck in 1900, each photon of light has a particular amount (a **quantum**) of energy. Furthermore, the amount of energy possessed by a photon depends on the color of the light. The energy of a photon of light is given by Planck's equation

$$E = h\nu \quad \text{or} \quad E = \frac{hc}{\lambda}$$

Violet light has a shorter wavelength and a higher frequency than red light.

where h is Planck's constant, 6.6262×10^{-34} J · s, and ν is the frequency of the light. Thus, energy is directly proportional to frequency. Planck's equation is used in Example 5-4 to show that a photon of violet light has more energy than a photon of red light.

Example 5-4

In Example 5-3 we calculated the wavelengths of violet light of frequency 7.31×10^{14} s^{-1} and of red light of frequency 4.57×10^{14} s^{-1}. Calculate the energy, in joules, of an individual photon in each of these two colors of light.

Plan

We use the frequencies to calculate the energy of a photon from the relationship $E = h\nu$.

Solution

$$\text{(violet light) } E = h\nu = (6.63 \times 10^{-34} \text{ J} \cdot \text{s})(7.31 \times 10^{14} \text{ s}^{-1}) = 4.85 \times 10^{-19} \text{ J}$$

$$\text{(red light) } E = h\nu = (6.63 \times 10^{-34} \text{ J} \cdot \text{s})(4.57 \times 10^{14} \text{ s}^{-1}) = 3.03 \times 10^{-19} \text{ J}$$

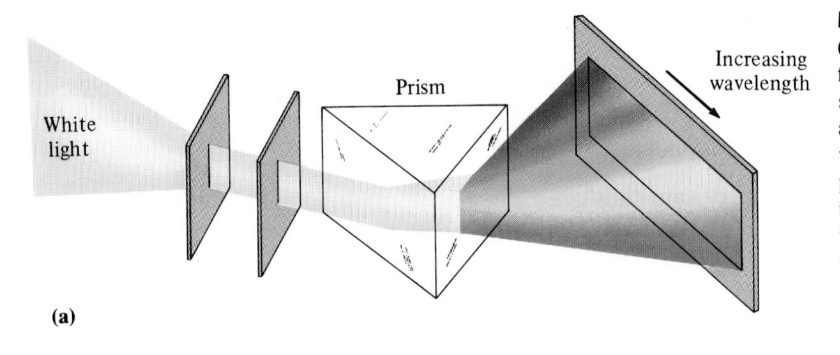

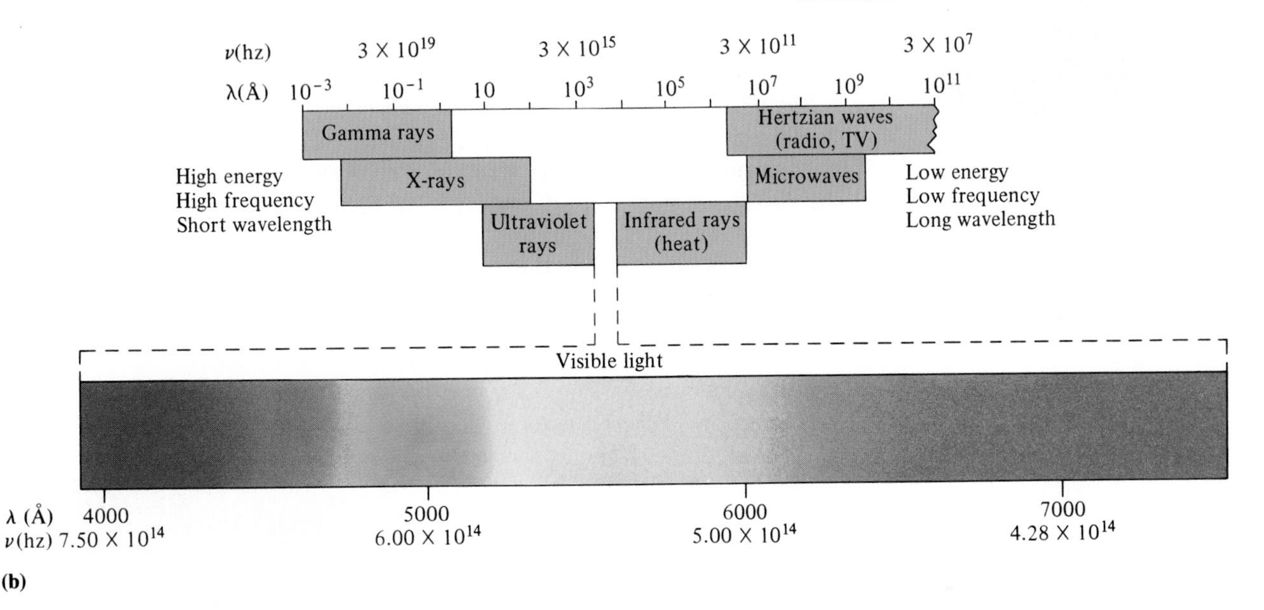

Figure 5-12
(a) Dispersion of visible light by a prism. Light from a source of white light is passed through a slit and then through a prism. It is separated into a continuous spectrum of all wavelengths of visible light. (b) Visible light is only a very small portion of the electromagnetic spectrum. Some radiant energy has longer or shorter wavelengths than our eyes can detect. The upper part shows the approximate ranges of the electromagnetic spectrum on a logarithmic scale. The lower part shows the visible region on an expanded scale. Note that wavelength increases as frequency decreases.

You can check these answers by calculating the energies directly from the wavelengths.

EOC 41

5-10 The Photoelectric Effect

One experiment that had not been satisfactorily explained with the wave model of light was the **photoelectric effect**. The apparatus for the photoelectric effect is shown in Figure 5-13. The negative electrode in the evacuated tube is made of a pure metal such as cesium. When light of a sufficiently high energy strikes the metal, electrons are knocked off its surface. They then travel to the positive electrode and form a current flowing through the circuit. The important observations follow.

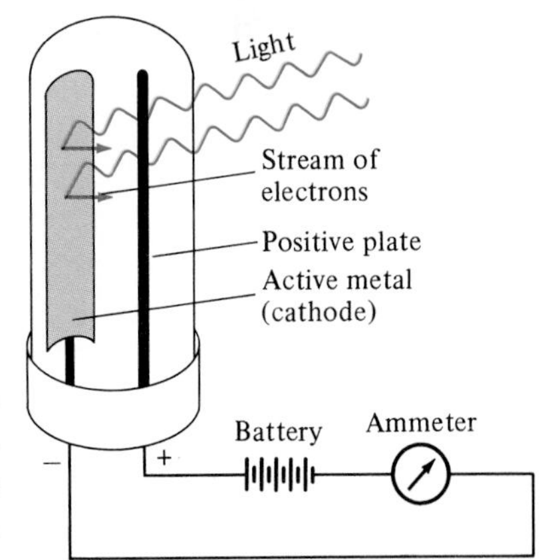

Figure 5-13
The photoelectric effect. When electromagnetic radiation of sufficient minimum energy strikes the surface of a metal (negative electrode) inside an evacuated tube, electrons are stripped off the metal to create an electric current. The current increases with increasing radiation intensity.

1. Electrons can be ejected only if the light is of sufficiently short wavelength (has sufficiently high energy), no matter how long or how brightly the light shines. This wavelength limit is different for different metals.
2. The current (the number of electrons emitted per second) increases with increasing *brightness* (intensity) of the light. However, it does not depend

The intensity of light is the brightness of the light. In wave terms, it is related to the amplitude of the light waves.

on the color of the light as long as the wavelength is short enough (has high enough energy).

Classical theory said that even "low" energy light should cause current to flow if the metal is irradiated long enough. Electrons should accumulate energy and be released when they have enough energy to escape from the metal atoms. According to the old theory, if the light is made more energetic, then the current should increase even though the light intensity remains the same. Such is *not* the case.

The answer to the puzzle was provided by Albert Einstein. In 1905 he extended Planck's idea that light behaves as though it were composed of *photons,* each with a particular amount (a quantum) of energy. According to Einstein, each photon can transfer its energy to a single electron during a collision. When we say that the intensity of light is increased, we mean that the number of photons striking a given area per second is increased. The picture is now one of a particle of light striking an electron near the surface of the metal and giving up its energy to the electron. If that energy is equal to or greater than the amount needed to liberate the electron, it can escape to join the photoelectric current. For this explanation, Einstein received the 1921 Nobel prize in physics.

The photoelectric effect is used in the photoelectric sensors that open some supermarket and elevator doors when the shadow of a person interrupts the light beam. Automatic cameras also use photocells.

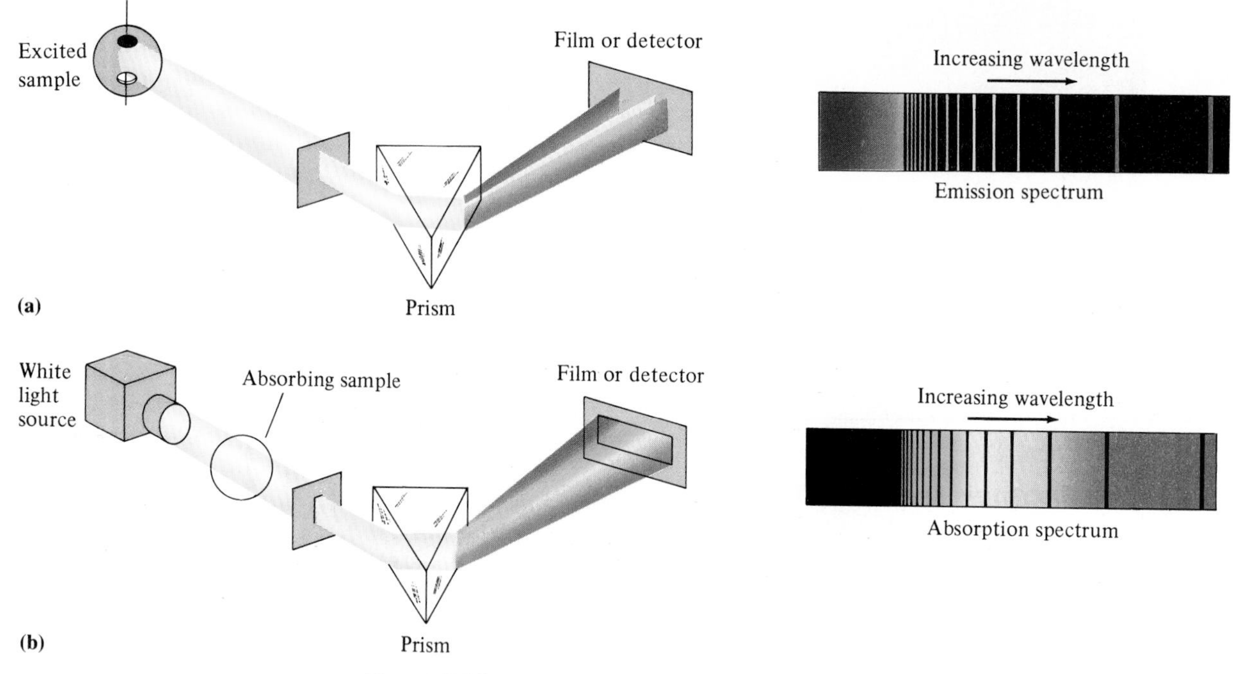

Figure 5-14
(a) *Atomic emission.* The light emitted by a sample of excited hydrogen atoms (or any other element) can be passed through a prism and separated into certain discrete wavelengths. Thus an emission spectrum, which is a photographic recording of the separated wavelengths, is called a line spectrum. Any sample of reasonable size contains an enormous number of atoms. Although a single atom can be in only one excited state at a time, the collection of atoms contains all possible excited states. The light emitted as these atoms fall to lower energy states is responsible for the spectrum. (b) *Atomic absorption.* When white light is passed through unexcited hydrogen and then through a slit and a prism, the transmitted light is lacking in intensity at the same wavelengths as are emitted in (a). The recorded absorption spectrum is also a line spectrum and the photographic negative of the emission spectrum.

5-11 Atomic Spectra and the Bohr Atom

Incandescent (''red hot'' or ''white hot'') solids, liquids, and high-pressure gases give continuous spectra. However, when an electric current is passed through a gas in a vacuum tube at very low pressures, the light that the gas emits is dispersed by a prism into distinct lines (Figure 5-14a). Such **emission spectra** are described as *bright line spectra*. The lines can be recorded photographically, and the wavelength of light that produced each line can be calculated from the position of that line on the photograph.

Similarly, we can shine a beam of white light (containing a continuous distribution of wavelengths) through a gas and analyze the beam that emerges. We find that only certain wavelengths have been absorbed (Figure 5-14b). The wavelengths that are absorbed in this **absorption spectrum** are the same as those given off in the emission experiment. Each element displays its own characteristic set of lines in its emission or absorption spectrum (Figure 5-15). These spectra can serve as ''fingerprints'' to allow us to identify different elements present in a sample, even in trace amounts.

Example 5-5

A green line of wavelength 4.86×10^{-7} m is observed in the emission spectrum of hydrogen. Calculate the energy of one photon of this green light.

Plan

We know the wavelength of the light, and we calculate its frequency so that we can then calculate the energy of each photon.

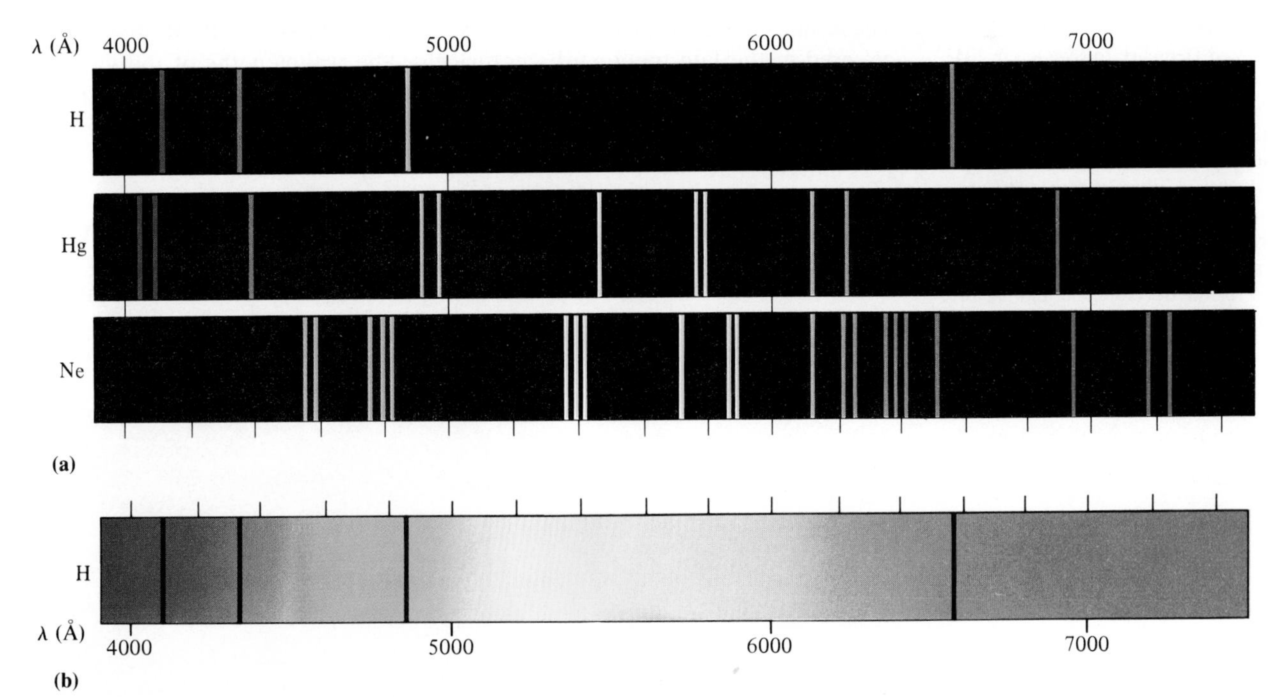

Figure 5-15
Atomic spectra in the visible region for some elements. Figure 5-14(a) shows how such spectra are produced. (a) Emission spectra for some elements. (b) Absorption spectrum for hydrogen. Compare the positions of these lines with those in the emission spectrum for H in (a).

The Danish physicist Niels Bohr (1885–1962) was one of the most influential scientists of the twentieth century. Like many other now-famous physicists of his time, he worked for a time in England with J. J. Thomson and later with Ernest Rutherford. During this period, he began to develop the ideas that led to the publication of his explanation of atomic spectra and his theory of atomic structure, for which he received the Nobel Prize in 1922. After escaping from German-occupied Denmark to Sweden in 1943, he helped to arrange the escape of hundreds of Danish Jews from the Hitler regime. He later went to the United States, where, until 1945, he worked with other scientists at Los Alamos, New Mexico on the development of the atomic bomb. From then until his death in 1962, he worked hard for the development and use of atomic energy for peaceful purposes.

Solution

$$\lambda\nu = c$$

$$\nu = \frac{c}{\lambda} = \frac{3.00 \times 10^8 \text{ m/s}}{4.86 \times 10^{-7} \text{ m}} = 6.17 \times 10^{14} \text{ s}^{-1}$$

$$E = h\nu = (6.63 \times 10^{-34} \text{ J} \cdot \text{s})(6.17 \times 10^{14} \text{ s}^{-1}) = \boxed{4.09 \times 10^{-19} \text{ J/photon}}$$

To gain a better appreciation of the amount of energy involved, let's calculate the total energy, in kilojoules, emitted by one mole of atoms. (Each atom emits one photon.)

$$\frac{?\text{ kJ}}{\text{mol}} = 4.09 \times 10^{-19} \frac{\text{J}}{\text{atom}} \times \frac{1 \text{ kJ}}{1 \times 10^3 \text{ J}} \times \frac{6.02 \times 10^{23} \text{ atoms}}{\text{mol}}$$

$$= \boxed{2.46 \times 10^2 \text{ kJ/mol}}$$

This calculation shows that when each atom in one mole of hydrogen atoms emits light of wavelength 4.86×10^{-7} m, the mole of atoms loses **246 kJ** of energy as green light. (This would be enough energy to operate a 100-watt light bulb for more than 40 minutes.)

EOC 44, 45, 46

When an electric current is passed through hydrogen gas at very low pressures, several series of lines in the spectrum of hydrogen are produced. These lines were studied intensely by many scientists. J. R. Rydberg discovered in the late nineteenth century that the wavelengths of the various lines in the hydrogen spectrum can be related by a mathematical equation:

$$\frac{1}{\lambda} = R\left(\frac{1}{n_1^2} - \frac{1}{n_2^2}\right)$$

Here R is $1.097 \times 10^7 \text{ m}^{-1}$ and is known as the Rydberg constant. The n's are positive integers, and n_1 is smaller than n_2. The Rydberg equation was derived from numerous observations, not theory. It is thus an empirical equation.

The lightning flashes produced in electrical storms and the light produced by neon gas in neon signs are two familiar examples of visible light produced by electronic transitions.

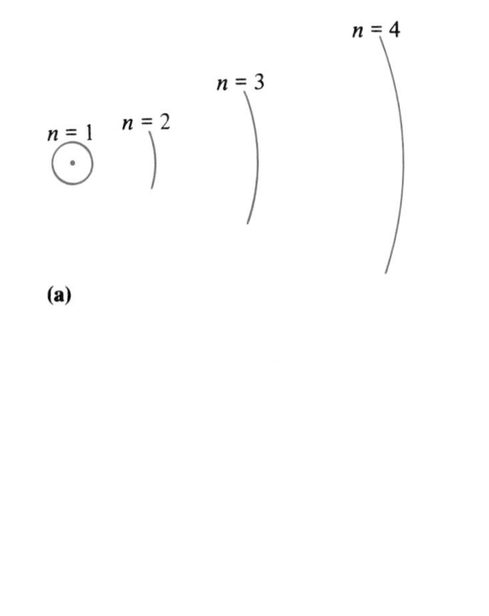

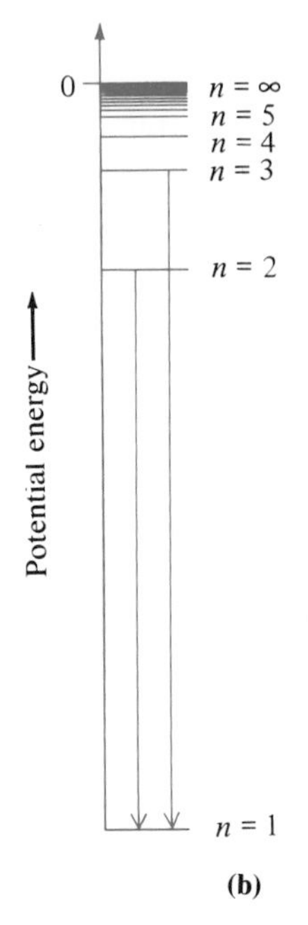

Figure 5-16
(a) The radii of the first four Bohr orbits for a hydrogen atom. The dot at the center represents the nuclear position. The radius of each orbit is proportional to n^2, so these four are in the ratio 1:4:9:16. (b) Relative values for the energies associated with the various energy levels in a hydrogen atom. The energies become closer together as n increases. They are so close together for large values of n that they form a continuum. By convention, potential energy is defined as zero when the electron is at an infinite distance from the atom. Any more stable arrangement would have a lower energy. Therefore, potential energies of electrons in atoms are always negative. Some possible electronic transitions corresponding to lines in the hydrogen emission spectrum are indicated by arrows. Transitions in the opposite directions account for lines in the absorption spectrum.

In 1913 Niels Bohr, a Danish physicist, provided an explanation for Rydberg's observations. He wrote equations that described the electron of a hydrogen atom as revolving around the nucleus of an atom in circular orbits. He included the assumption that the electronic energy is *quantized;* that is, only certain values of electron energy are possible. This led him to the suggestion that electrons can only be in certain discrete orbits, and that they absorb or emit energy in discrete amounts as they move from one orbit to another. Each orbit thus corresponds to a definite *energy level* for the electron. When an electron is promoted from a lower energy level to a higher one, it absorbs a definite (or quantized) amount of energy. When the electron falls back to the original energy level, it emits exactly the same amount of energy it absorbed in moving from the lower to the higher energy level. Figure 5-16 illustrates these transitions schematically. The values of n_1 and n_2 in the Rydberg equation identify the lower and higher levels, respectively, of these electronic transitions.

The Bohr Theory and the Rydberg Equation

Enrichment

From mathematical equations describing the orbits for the hydrogen atoms, together with the assumption of quantization of energy, Bohr was able to determine two significant aspects of each allowed orbit:

1. *Where* (with respect to the nucleus) the electron can be—that is, the radius, r, of the circular orbit. This is given by

$$r = \frac{n^2h^2}{4\pi^2me^2}$$

Note: r is proportional to n^2.

where h = Planck's constant, e = the charge of the electron, m = the mass of the electron, and n is a positive integer (1, 2, 3, . . .) that tells us which orbit is being described.

2. *How stable* the electron would be in that orbit—that is, its potential energy, E. This is given by

Note: E is proportional to $-\frac{1}{n^2}$.

$$E = -\frac{2\pi me^4}{n^2h^2}$$

where the symbols have the same meaning as before. Note that E is always negative.

Results of evaluating these equations for some of the possible values of n (1, 2, 3, . . .) are shown in Figure 5-17. The larger the value of n, the farther from the nucleus is the orbit being described, and the radius of this orbit increases as the *square of n* increases. As n increases, n^2 increases, $1/n^2$ decreases, and thus the electronic energy increases (becomes less negative and smaller in magnitude). For orbits farther from the nucleus, the electronic potential energy is higher (less negative—the electron is in a *higher* energy level or in a less stable state). Going away from the nucleus, the allowable orbits are farther apart in distance, but closer together in energy. Consider the two possible limits of these equations. One limit is when $n = 1$; this describes the electron at the smallest possible distance from the nucleus and at its lowest (most negative) energy. The other limit is for very large values of n, i.e., as n approaches infinity. As this limit is approached, the electron is very far from the nucleus, or effectively removed from the atom; the energy is as high as possible, approaching zero.

With these equations and the relationship that the Planck equation provides between energy and the frequency or wavelength, Bohr was able to predict the wavelengths observed in the hydrogen emission spectrum. Figure 5-17 illustrates the relationship between lines in the emission spectrum and the electronic transitions (changes of energy level) that occur in hydrogen atoms.

Each line in the emission spectrum represents the *difference in energies* between two allowed energy levels for the electron. When the electron goes from energy level n_2 to energy level n_1, the difference in energy is given off as a single photon. The energy of this photon can be calculated from Bohr's equation for the energy, as follows.

$$E \text{ of photon} = E_2 - E_1 = -\frac{2\pi me^4}{n_2{}^2h^2} - \left(-\frac{2\pi me^4}{n_1{}^2h^2}\right)$$

Factoring out the quantity $2\pi me^4/h^2$ and rearranging, we get

$$E \text{ of photon} = \frac{2\pi me^4}{h^2}\left(\frac{1}{n_1{}^2} - \frac{1}{n_2{}^2}\right)$$

The Planck equation, $E = hc/\lambda$, relates the energy of the photon to the wavelength of the light, so

$$\frac{hc}{\lambda} = \frac{2\pi me^4}{h^2}\left(\frac{1}{n_1{}^2} - \frac{1}{n_2{}^2}\right)$$

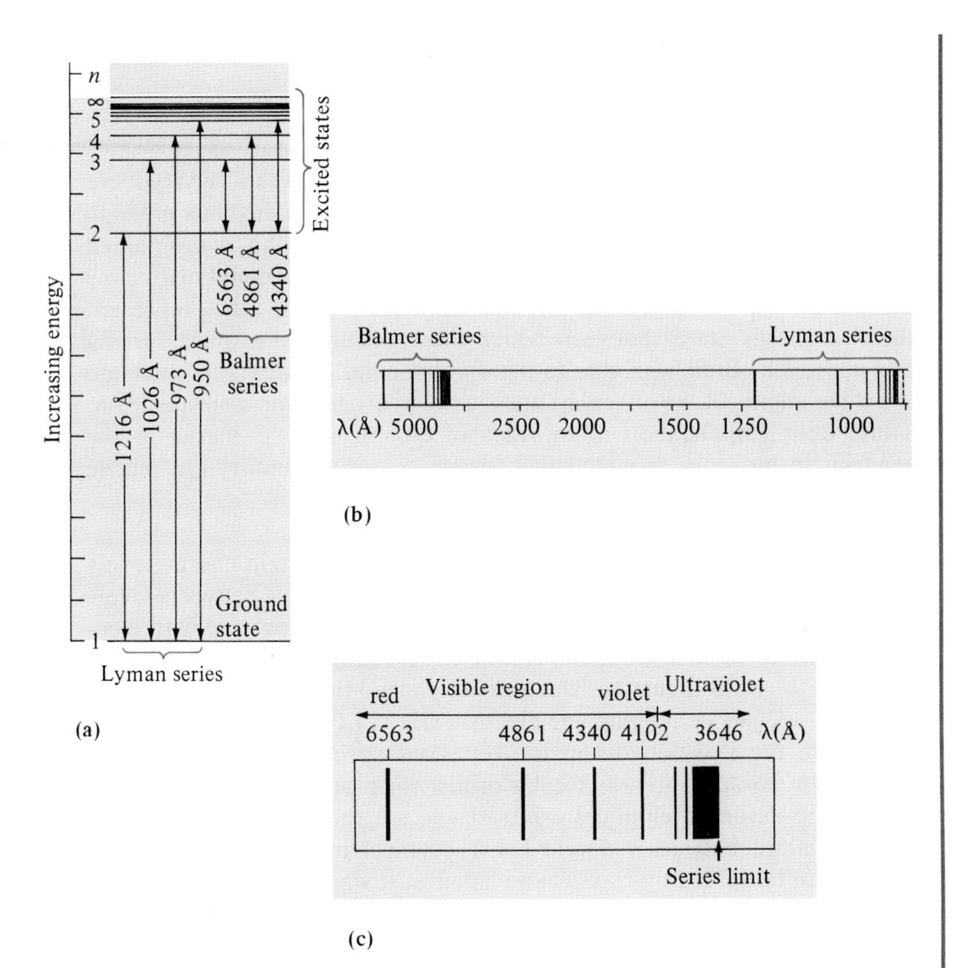

Figure 5-17

(a) The energy levels that the electron can occupy in a hydrogen atom and a few of the transitions that cause the emission spectrum of hydrogen. The numbers on the vertical lines show the wavelengths of light emitted when the electron falls to a lower energy level. (Light of the same wavelength is absorbed when the electron is promoted to the higher energy level.) The difference in energy between two given levels is exactly the same for all hydrogen atoms, so it corresponds to a specific wavelength and to a specific line in the emission spectrum of hydrogen. In a given sample, some hydrogen atoms could have their electrons excited to the $n = 2$ level. Some of these electrons could then fall to the $n = 1$ energy level, giving off the *difference* in energy in the form of light (the 1216-Å transition). Other hydrogen atoms might have their electrons excited to the $n = 3$ level; subsequently some could fall to the $n = 1$ level (the 1026-Å transition). Because higher energy levels become closer and closer in energy, *differences* in energy between successive transitions become smaller and smaller. The corresponding lines in the emission spectrum become closer together and eventually result in a continuum, a series of lines so close together that they are indistinguishable. (b) The emission spectrum of hydrogen. The series of lines produced by the electron falling to the $n = 1$ level is known as the *Lyman series;* it is in the ultraviolet region. A transition in which the electron falls to the $n = 2$ level gives rise to a similar set of lines in the visible region of the spectrum, known as the *Balmer series*. Not shown are series involving transitions to energy levels with higher values of n. (c) The Balmer series shown on an expanded scale. The line of 6563 Å (the $n = 3 \rightarrow n = 2$ transition) is much more intense than the line at 4861 Å (the $n = 4 \rightarrow n = 2$ transition) because the first transition occurs much more frequently than the second. Successive lines in the spectrum become less intense as the series limit is approached because the transitions that correspond to these lines are less probable.

Rearranging for $1/\lambda$, we obtain

$$\frac{1}{\lambda} = \frac{2\pi m e^4}{h^3 c}\left(\frac{1}{n_1^2} - \frac{1}{n_2^2}\right)$$

Comparing this to the Rydberg equation, Bohr showed that the Rydberg constant is equivalent to $2\pi me^4/h^3c$. We could use the constants given in Appendix D to obtain the same value, $1.097 \times 10^7\ m^{-1}$, that was obtained by Rydberg on a solely empirical basis. Further, Bohr could also show the physical meaning of the two whole numbers n_1 and n_2; they represent the two energy states between which the transition takes place. Using this approach, Bohr was able to use fundamental constants to calculate the wavelengths of the observed lines in the hydrogen emission spectrum. Thus, Bohr's application of the idea of quantization of energy to the electron in an atom provided the answer to a half-century-old puzzle concerning the discrete colors given off in the spectrum.

We now accept the fact that electrons occupy only certain energy levels in atoms. In most atoms, some of the energy differences between levels correspond to the energy of visible light. Thus, colors associated with electronic transitions in such elements can be observed by the human eye.

Although the Bohr theory satisfactorily explained the spectra of hydrogen and of other species containing one electron (He^+, Li^{2+}, and so on) it could not calculate the wavelengths in the observed spectra of more complex species. Bohr's assumption of circular orbits was modified in 1916 by Sommerfeld, who assumed elliptical orbits. Even so, the Bohr approach was doomed to failure, because it modified classical mechanics to solve a problem that could not be solved by classical mechanics. It was a contrived solution. There was a need literally "to invent" a new physics, quantum mechanics, to deal with small particles, and Bohr's failure set the stage for the development of this new physics. However, the Bohr theory did introduce the ideas that only certain energy levels are possible, that these energy levels are described by quantum numbers that can have only certain allowed values, and that the quantum numbers indicate something about where and how stable the electrons are in these energy levels. The ideas of modern atomic theory have superseded Bohr's original theory. But his achievement in showing a link between electronic arrangements and Rydberg's empirical description of light absorption, and in establishing the quantization of electronic energy, was a very important step toward an understanding of atomic structure.

Two big questions remained about electrons in atoms: (1) How are electrons arranged in atoms? (2) How do these electrons behave? We now have the background to consider how modern atomic theory answers these questions.

5-12 The Wave Nature of the Electron

The idea that light can exhibit both wave properties and particle properties suggested to Louis de Broglie that very small particles, such as electrons, might also display wave properties under the proper circumstances. In his doctoral thesis in 1925, de Broglie predicted that a particle with a mass m and velocity v should have a wavelength associated with it. The numerical value of this de Broglie wavelength is given by

$$\lambda = h/mv \quad \text{(where } h = \text{Planck's constant)}$$

Be sure to distinguish between velocity, represented by the letter v, and frequency (Section 5-9), represented by ν (Greek letter *nu*).

Two years after de Broglie's prediction, C. Davisson and L. H. Germer at the Bell Telephone Laboratory demonstrated diffraction of electrons by a crystal of nickel. This behavior is an important characteristic of waves. It shows conclusively that electrons do have wave properties. Davisson and Germer found that the wavelength associated with electrons of known energy is exactly that predicted by de Broglie. Similar diffraction experiments have been successfully performed with other particles, such as neutrons.

Example 5-6

(a) Calculate the wavelength of an electron traveling at 1.24×10^7 m/s. The mass of an electron is 9.11×10^{-28} g. (b) Calculate the wavelength of a baseball of mass 5.25 oz traveling at 92.5 mi/h. Recall that $1\ \text{J} = 1\ \text{kg} \cdot \text{m}^2/\text{s}^2$.

Plan

For each calculation, we use the de Broglie equation

$$\lambda = \frac{h}{mv}$$

where

$$h\ (\text{Planck's constant}) = 6.63 \times 10^{-34}\ \text{J} \cdot \text{s} \times \frac{1\ \dfrac{\text{kg} \cdot \text{m}^2}{\text{s}^2}}{1\ \text{J}} = 6.63 \times 10^{-34}\ \frac{\text{kg} \cdot \text{m}^2}{\text{s}}$$

For consistency of units, mass must be expressed in kilograms. In part (b), we must also convert the speed to meters per second.

Solution

(a) $$m = 9.11 \times 10^{-28}\ \text{g} \times \frac{1\ \text{kg}}{1000\ \text{g}} = 9.11 \times 10^{-31}\ \text{kg}$$

Substituting into the de Broglie equation,

$$\lambda = \frac{h}{mv} = \frac{6.63 \times 10^{-34}\ \dfrac{\text{kg} \cdot \text{m}^2}{\text{s}}}{(9.11 \times 10^{-31}\ \text{kg})\left(1.24 \times 10^7\ \dfrac{\text{m}}{\text{s}}\right)} = \boxed{5.87 \times 10^{-11}\ \text{m}}$$

While this seems like a very short wavelength, it is of the same order as the spacing between atoms in many crystals. A stream of such electrons hitting a crystal gives easily measurable diffraction effects.

(b) $$m = 5.25\ \text{oz} \times \frac{1\ \text{lb}}{16\ \text{oz}} \times \frac{1\ \text{kg}}{2.205\ \text{lb}} = 0.149\ \text{kg}$$

$$v = \frac{92.5\ \text{mi}}{\text{h}} \times \frac{1\ \text{h}}{3600\ \text{s}} \times \frac{1.609\ \text{km}}{1\ \text{mi}} \times \frac{1000\ \text{m}}{1\ \text{km}} = 41.3\ \frac{\text{m}}{\text{s}}$$

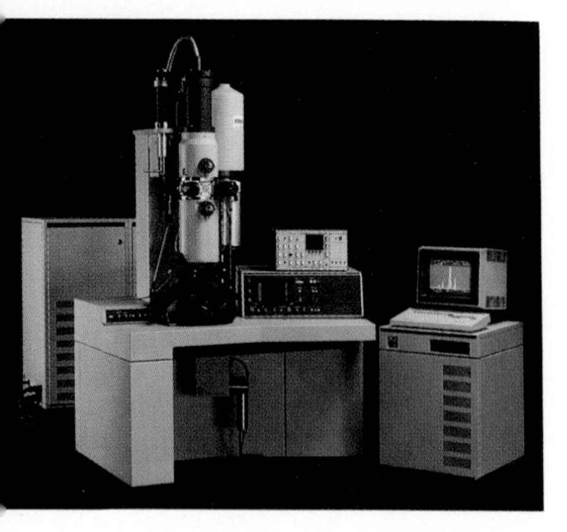

A modern electron microscope.

Now we substitute into the de Broglie equation.

$$\lambda = \frac{h}{mv} = \frac{6.63 \times 10^{-34}\ \frac{\text{kg} \cdot \text{m}^2}{\text{s}}}{(0.149\ \text{kg})\left(41.3\ \frac{\text{m}}{\text{s}}\right)} = 1.08 \times 10^{-34}\ \text{m}$$

This wavelength is far too short to give any measurable effects. Recall that atomic diameters are of the order of 10^{-10} m, which is 24 powers of 10 greater than the baseball "wavelength."

EOC 68

As you can see from the results of Example 5-6, the particles of the subatomic world behave very differently from the macroscopic objects with which we are familiar. To talk about behavior of atoms and their particles, we must give up many of our long-held prejudices about the behavior of matter. We must be willing to visualize a world of new and unfamiliar properties, such as the ability to act in some ways like a particle and in other ways like a wave.

The wave behavior of electrons is exploited in the electron microscope. This instrument allows magnification of objects far too small to be seen with an ordinary light microscope.

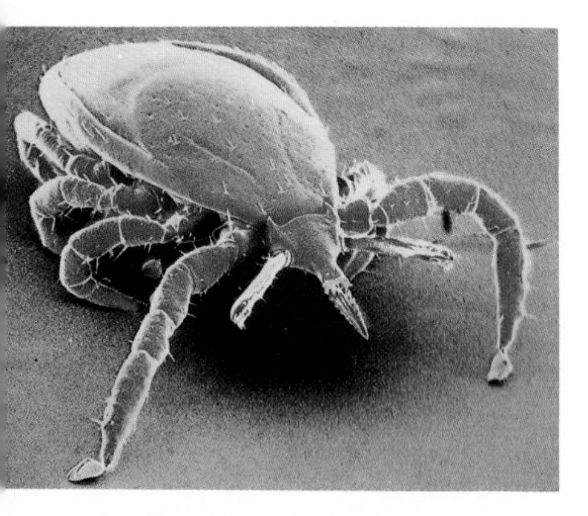

A color-enhanced scanning electron micrograph of a deer tick, the carrier of Lyme disease.

5-13 The Quantum Mechanical Picture of the Atom

Through the work of de Broglie, Davisson and Germer, and others, we now know that electrons in atoms can be treated as waves more effectively than as small compact particles traveling in circular or elliptical orbits. Large objects such as golf balls and moving automobiles obey the laws of classical mechanics (Isaac Newton's laws), but very small particles such as electrons, atoms, and molecules do not. A different kind of mechanics, called **quantum mechanics**, describes the behavior of very small particles much better because it is based on the *wave* properties of matter. Quantization of energy is a consequence of these properties.

One of the underlying principles of quantum mechanics is that we cannot determine precisely the paths that electrons follow as they move about atomic nuclei. The **Heisenberg Uncertainty Principle**, stated in 1927 by Werner Heisenberg, is a theoretical assertion that is consistent with all experimental observations.

This is like trying to locate the position of a moving automobile by driving another automobile into it.

It is impossible to determine accurately both the momentum and the position of an electron (or any other very small particle) simultaneously.

Momentum is mass times velocity, mv. Because electrons are so small and move so rapidly, their motion is usually detected by electromagnetic radiation. Photons that interact with electrons have about the same energies as the electrons. Consequently, the interaction of a photon with an electron severely disturbs the motion of the electron. It is not possible to determine simultaneously both the position and the velocity of an electron, so we resort to a statistical approach and speak of the probability of finding an electron within specified regions in space.

With these ideas in mind, we can now list some basic ideas of quantum mechanics.

1. Atoms and molecules can exist only in certain energy states. In each energy state, the atom or molecule has a definite energy. When an atom or molecule changes its energy state, it must emit or absorb just enough energy to bring it to the new energy state (the quantum condition).

Atoms and molecules possess various forms of energy. Let us focus our attention on their *electronic energies*.

2. Atoms or molecules emit or absorb radiation (light) as they change their energies. The frequency of the light emitted or absorbed is related to the energy change by a single equation:

$$\Delta E = h\nu \qquad \text{or} \qquad \Delta E = hc/\lambda$$

Recall that $\lambda\nu = c$, so $\nu = c/\lambda$.

This gives a relationship between the energy change, ΔE, and the wavelength, λ, of the radiation emitted or absorbed. *The energy lost (or gained) by an atom as it goes from higher to lower (or lower to higher) energy states is equal to the energy of the photon emitted (or absorbed) during the transition.*

3. The allowed energy states of atoms and molecules can be described by sets of numbers called *quantum numbers*.

The mathematical approach of quantum mechanics involves treating the electron in an atom as a *standing wave*. A standing wave is a wave that does not travel and therefore has at least one point at which it has zero amplitude, called a node. As an example, consider the various ways that a guitar string can vibrate when it is plucked (Figure 5-18). Because both ends are fixed (nodes), the string can vibrate only in ways in which there is a whole number of *half-wavelengths* in the length of the string (Figure 5-18a). Any actual motion of the string can be described as some combination of these allowed vibrations. In a similar way, we can imagine that the electron in the hydrogen atom behaves as a wave (recall the de Broglie relationship in the last section). The electron can be described by the same kind of standing-wave mathematics that is applied to the vibrating guitar string. In a given space around the nucleus, only certain "waves" can exist. Each "allowed wave" corresponds to a stable energy state for the electron, and is described by a particular set of quantum numbers.

The quantum mechanical treatment of atoms and molecules is highly mathematical. The important point is that each solution of the Schrödinger wave equation (see Enrichment section) describes a possible energy state for the electrons in the atom. Each solution is described by a set of three **quantum numbers**. These numbers are in accord with those deduced from experiment and from empirical equations such as the Rydberg equation. Solutions of the Schrödinger equation also tell us about the shapes and orientations of the statistical probability distributions of the electrons. (The Heisenberg Principle implies that this is how we must describe the positions of the electrons.) These *atomic orbitals* (which are described in Section 5-15) are deduced from the solutions of the Schrödinger equation. The orbitals

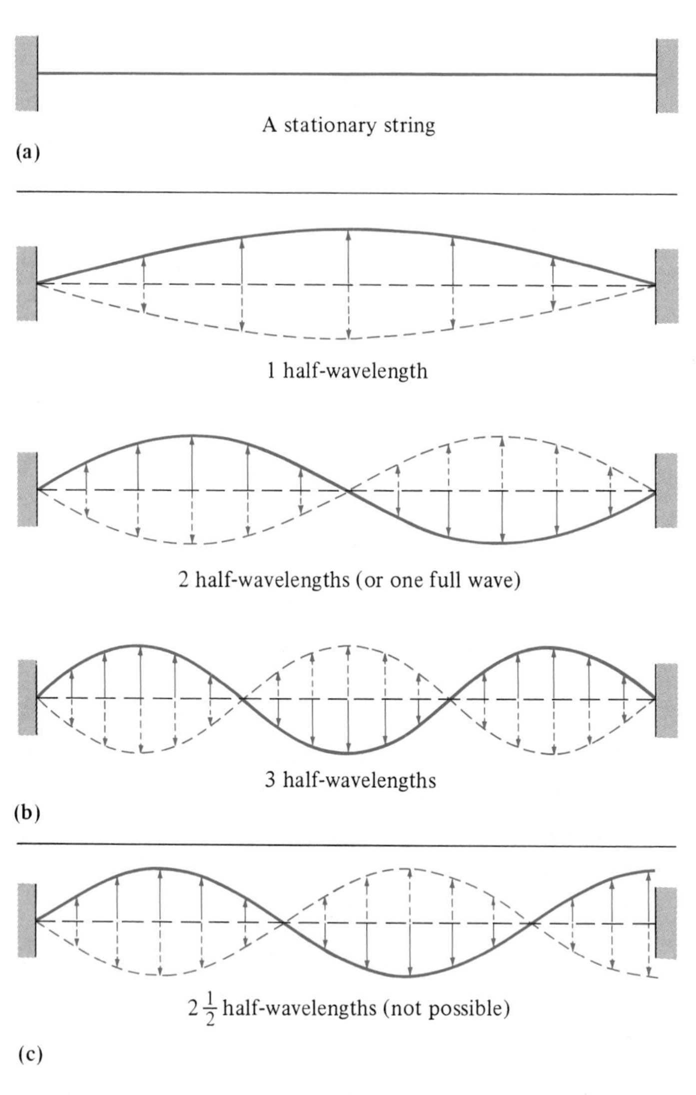

Figure 5-18
When a string that is fixed at both ends—such as a guitar string (a)—is plucked, it has a number of natural patterns of vibration, called normal modes. Because the string is fixed at both ends, the ends must be stationary. Each different possible vibration is a standing wave, and can be described by a wave function. The only waves that are possible are those in which a whole number of half-wavelengths fits into the string length. These allowed waves comprise a harmonic series. Any total motion of the string is some combination of these allowed harmonics. (b) Some of the ways in which a plucked guitar string can vibrate. The position of the string at one extreme of each vibration is shown as a solid line, and at the other extreme as a dashed line. (c) An example of a vibration that is *not* possible for a plucked string. In such a vibration, an end of the string would move; this is not possible because the ends are fixed.

are directly related to the quantum numbers. In 1928 Paul A. M. Dirac reformulated electron quantum mechanics to take into account the effects of relativity. This gave rise to a fourth quantum number.

Enrichment

The Schrödinger Equation

In 1926 Erwin Schrödinger modified an existing equation that described a three-dimensional standing wave by imposing wavelength restrictions suggested by de Broglie's ideas. The modified equation allowed him to calculate the energy levels in the hydrogen atom. It is a second-order differential equation that need not be memorized or even understood to read this book. A knowledge of differential calculus would be necessary.

$$-\frac{h^2}{8\pi^2 m}\left(\frac{\partial^2\psi}{\partial x^2}+\frac{\partial^2\psi}{\partial y^2}+\frac{\partial^2\psi}{\partial z^2}\right)+V\psi=E\psi$$

This equation has been solved exactly only for one-electron species such as the hydrogen atom and the ions He^+ and Li^{2+}. Simplifying assumptions are necessary to solve the equation for more complex atoms and molecules. However, chemists and physicists have used their intuition and ingenuity (and modern computers) to apply this equation to more complex systems, to the general benefit of science.

5-14 Quantum Numbers

The solutions of the Schrödinger and Dirac equations for hydrogen atoms give wave functions that describe the various states available to hydrogen's single electron. Each of these possible states is described by four quantum numbers. We can use these quantum numbers to describe the electronic arrangements in all atoms, their so-called **electronic configurations**. These quantum numbers play important roles in describing the energy levels of electrons and the shapes of the orbitals that describe distributions of electrons in space. This interpretation will become clearer when we discuss atomic orbitals in the following section. For now, let's say that

> an **atomic orbital** is a region of space in which the probability of finding an electron is high.

We define each quantum number and describe the range of values it may take.

1. The **principal quantum number**, n, describes the *main energy level* an electron occupies. It may be any positive integer:

$$n = 1, 2, 3, 4, \ldots$$

2. The **subsidiary** (or **azimuthal**) **quantum number**, ℓ, designates the *shape of the region* in space an electron occupies. Within each energy level (defined by the value of n, the principal quantum number), ℓ may take integral values from 0 up to and including $(n - 1)$:

$$\ell = 0, 1, 2, \ldots, (n - 1)$$

The maximum value of ℓ thus is $(n - 1)$. The subsidiary quantum number designates a **sublevel**, or a specific *kind* of atomic orbital, that an electron may occupy. We give a letter notation to each value of ℓ. Each letter corresponds to a different kind of atomic orbital:

$$\ell = 0, 1, 2, 3, \ldots, (n - 1)$$

$s \quad p \quad d \quad f$ kind of sublevel

The s, p, d, f designations arise from the characteristics of spectral emission lines produced by electrons occupying the orbitals: s (sharp), p (principal), d (diffuse), and f (fundamental).

In the first energy level, the maximum value of ℓ is zero, which tells us that there is only an s sublevel and no p sublevel. In the second energy level, the permissible values of ℓ are 0 and 1, which tells us that there are only s and p sublevels.

Table 5-4
Permissible Values of the Quantum Numbers Through $n = 4$

n	ℓ	m_ℓ	m_s	Electron Capacity of Sublevel = $4\ell + 2$	Electron Capacity of Energy Level = $2n^2$
1 (K)	0 (1*s*)	0	$+\frac{1}{2}, -\frac{1}{2}$	2	2
2 (L)	0 (2*s*)	0	$+\frac{1}{2}, -\frac{1}{2}$	2	8
	1 (2*p*)	−1, 0, +1	$\pm\frac{1}{2}$ for each value of m_ℓ	6	
3 (M)	0 (3*s*)	0	$+\frac{1}{2}, -\frac{1}{2}$	2	18
	1 (3*p*)	−1, 0, +1	$\pm\frac{1}{2}$ for each value of m_ℓ	6	
	2 (3*d*)	−2, −1, 0, +1, +2	$\pm\frac{1}{2}$ for each value of m_ℓ	10	
4 (N)	0 (4*s*)	0	$+\frac{1}{2}, -\frac{1}{2}$	2	32
	1 (4*p*)	−1, 0, +1	$\pm\frac{1}{2}$ for each value of m_ℓ	6	
	2 (4*d*)	−2, −1, 0, +1, +2	$\pm\frac{1}{2}$ for each value of m_ℓ	10	
	3 (4*f*)	−3, −2, −1, 0, +1, +2, +3	$\pm\frac{1}{2}$ for each value of m_ℓ	14	

3. The **magnetic quantum number**, m_ℓ, designates the spatial orientation of an atomic orbital. Within each sublevel, m_ℓ may take any integral values from $-\ell$ through zero up to and including $+\ell$:

$$m_\ell = (-\ell), \ldots, 0, \ldots, (+\ell)$$

The maximum value of m_ℓ depends on the value of ℓ. For example, when $\ell = 1$, which designates the *p* sublevel, there are three permissible values of m_ℓ: -1, 0, and $+1$. There are thus three distinct regions of space, called atomic orbitals, associated with a *p* sublevel. We refer to these orbitals as the p_x, p_y, and p_z orbitals (see the next section).

4. The **spin quantum number**, m_s, refers to the spin of an electron and the orientation of the magnetic field produced by this spin. For every set of n, ℓ, and m_ℓ values, m_s can take the value $+\frac{1}{2}$ or $-\frac{1}{2}$:

$$m_s = \pm\tfrac{1}{2}$$

The values of n, ℓ, and m_ℓ describe a particular atomic orbital. Each atomic orbital can accommodate no more than two electrons, one with $m_s = +\frac{1}{2}$ and another with $m_s = -\frac{1}{2}$.

Table 5-4 summarizes some permissible values for the four quantum numbers. Spectroscopic evidence confirms the quantum mechanical predictions about the number of atomic orbitals in each energy level.

5-15 Atomic Orbitals

Let us now describe the distributions of electrons in atoms. For each neutral atom, we must account for a number of electrons equal to the number of protons in the nucleus, i.e., the atomic number of the atom. Each electron is said to occupy an atomic orbital defined by a set of quantum numbers n, ℓ, and m_ℓ. An orbital is the space in which there is a high probability of

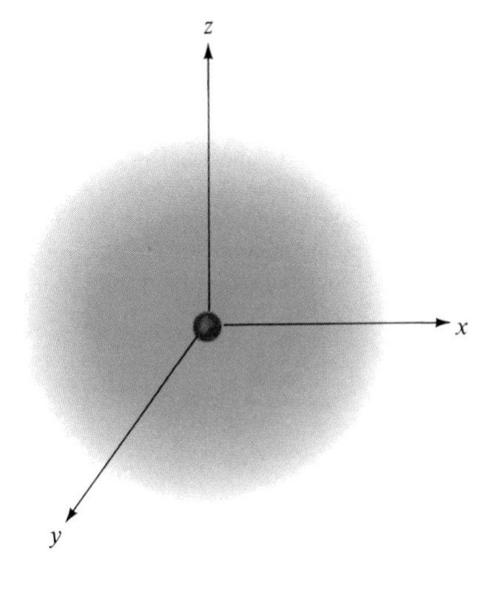

Figure 5-19
An electron cloud surrounding an atomic nucleus. The electron density drops off rapidly but smoothly as distance from the nucleus increases.

finding the electron. In any atom, each orbital can hold a maximum of two electrons. Within each atom, these atomic orbitals, taken together, can be represented as a diffuse cloud of electrons (Figure 5-19).

The energy level of each atomic orbital in an atom is indicated by the principal quantum number n (from the Schrödinger equation). As we have seen, the principal quantum number takes integral values: $n = 1, 2, 3, 4, \ldots$. The value $n = 1$ describes the first, or lowest, energy level. In the past, these energy levels have been referred to as electron shells and designated K, L, M, N, ... shells (Table 5-4). The correspondence between the two notations is

The letter notation comes from early experiments in spectroscopy and with X-rays.

$$\begin{array}{rl} \text{principal quantum number } n = & 1, \quad 2, \quad 3, \quad 4, \ldots \\ \text{shell} & \text{K}, \quad \text{L}, \quad \text{M}, \quad \text{N}, \ldots \\ & \xrightarrow[\text{from nucleus}]{\text{increasing distance}} \end{array}$$

Successive shells are at increasingly greater distances from the nucleus. For example, the $n = 2$ (or L) shell is farther than the $n = 1$ (or K) shell. The electron capacity of each energy level is indicated in the right-hand column of Table 5-4. For a given n, the capacity is $2n^2$.

By the rules in Section 5-14, each energy level has one s sublevel (defined by $\ell = 0$) consisting of one s atomic orbital (defined by $m_\ell = 0$). We distinguish among orbitals in different principal shells (main energy levels) by using the principal quantum number as a coefficient; $1s$ indicates the s orbital in the first energy level, $2s$ is the s orbital in the second energy level, $2p$ is a p orbital in the second energy level, and so on (Table 5-4).

For each solution to the quantum mechanical equation, we can calculate the electron probability density (sometimes just called the electron density) at each point in the atom. This is the probability of finding an electron at that point. It can be shown that this electron density is proportional to $r^2\psi^2$, where r is the distance from the nucleus.

In the graphs in Figure 5-20, the electron probability density at a given distance from the nucleus is plotted against distance from the nucleus, for

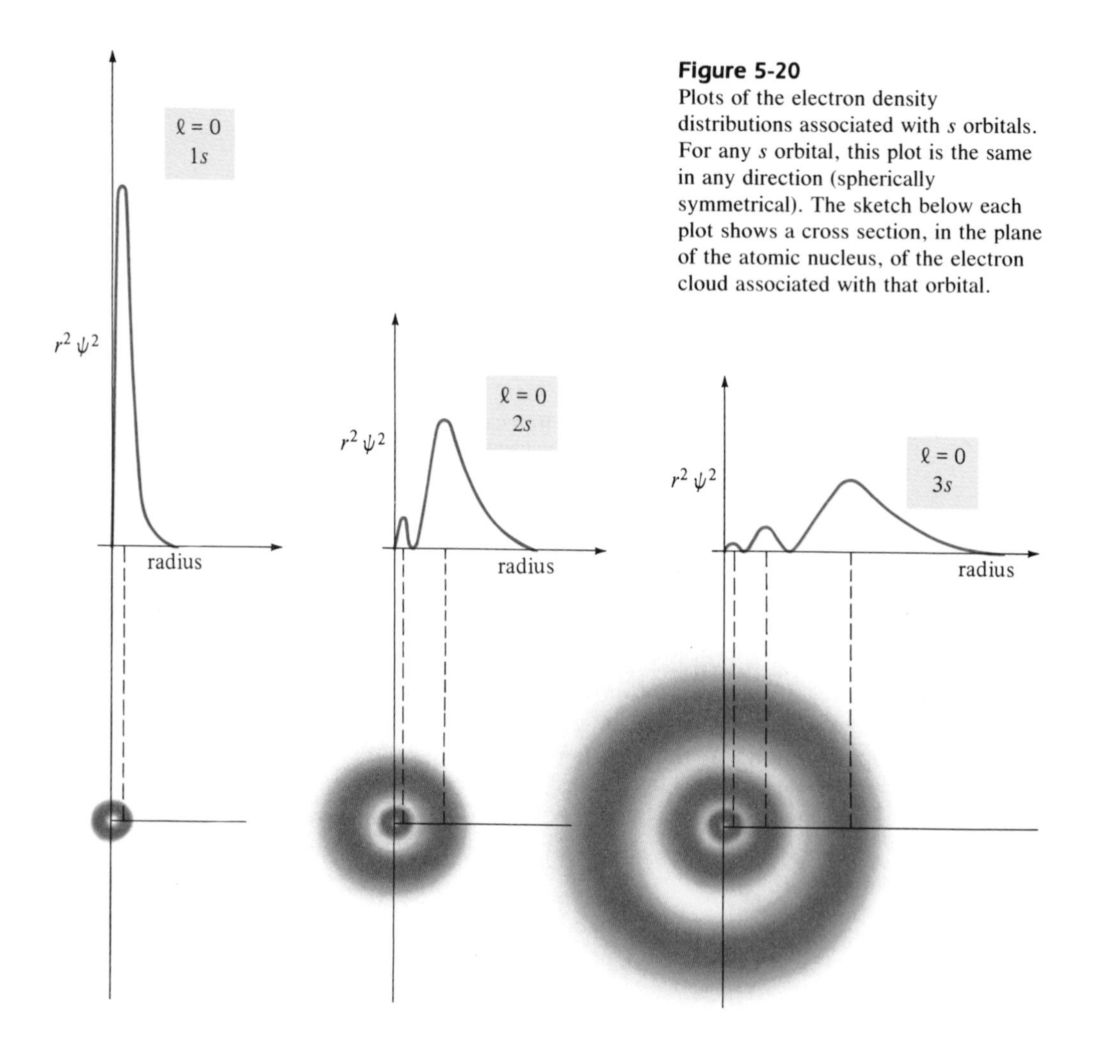

Figure 5-20
Plots of the electron density distributions associated with *s* orbitals. For any *s* orbital, this plot is the same in any direction (spherically symmetrical). The sketch below each plot shows a cross section, in the plane of the atomic nucleus, of the electron cloud associated with that orbital.

s orbitals. It is found that the probability density curve is the same regardless of the direction in the atom. We describe an *s* orbital as *spherically symmetrical;* i.e., it is round like a basketball (Figure 5-21). The electron clouds (electron densities) associated with the 1*s*, 2*s*, and 3*s* atomic orbitals are shown just below the plots. The electron clouds are three-dimensional, and only cross-sections are shown here. The regions shown in some figures (Figures 5-21, 5-22, 5-23, 5-24, and 5-25) appear to have surfaces or skins only because they are arbitrarily ''cut off'' so that there is a 90% probability of finding an electron occupying the orbital somewhere within the surfaces.

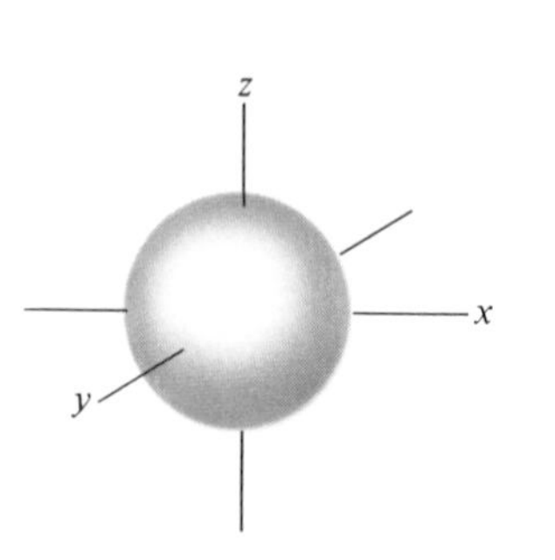

Figure 5-21
The shape of an atomic *s* orbital.

Beginning with the second energy level, each level contains a *p* sublevel, defined by $\ell = 1$. Each of these sublevels consists of a set of *three p* atomic orbitals, corresponding to the three allowed values of m_ℓ (-1, 0, and $+1$) when $\ell = 1$. The sets are referred to as 2*p*, 3*p*, 4*p*, 5*p*, . . . orbitals to indicate the main energy levels in which they occur. Each set of atomic *p* orbitals resembles three mutually perpendicular equal-arm dumbbells (Figure 5-22). The nucleus defines the origin of a set of Cartesian coordinates with the usual *x*, *y*, and *z* axes (Figure 5-23a). The subscript *x*, *y*, or *z* indicates the

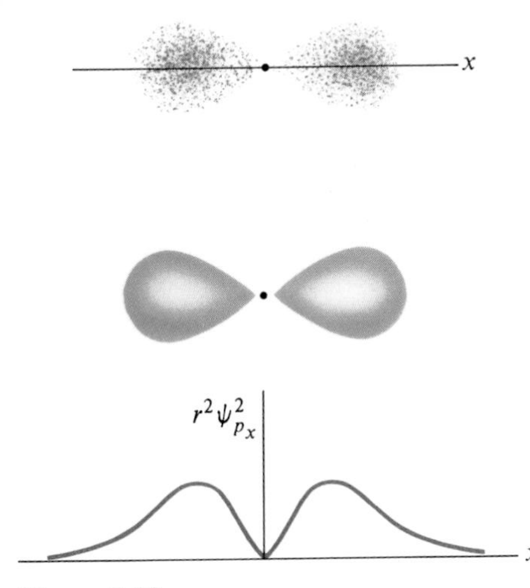

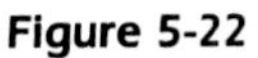

Figure 5-22
Three representations of the shape of a p orbital. The plot at the bottom is along the axis of maximum electron density for this orbital. A plot along any other direction would be different, because a p orbital is *not* spherically symmetrical.

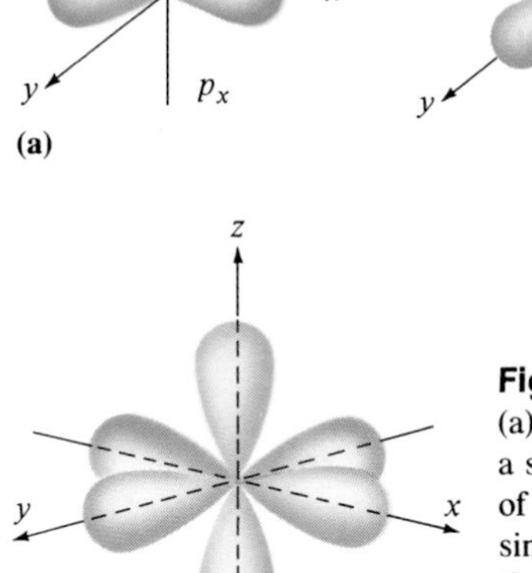

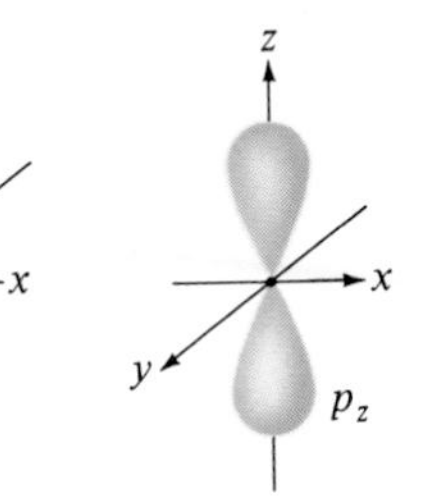

Figure 5-23
(a) The relative directional character of a set of atomic p orbitals. (b) A model of three p orbitals (p_x, p_y, and p_z) of a single set of orbitals. The nucleus is at the center. (The lobes are actually more diffuse ("fatter") than depicted. See Figure 5-26.)

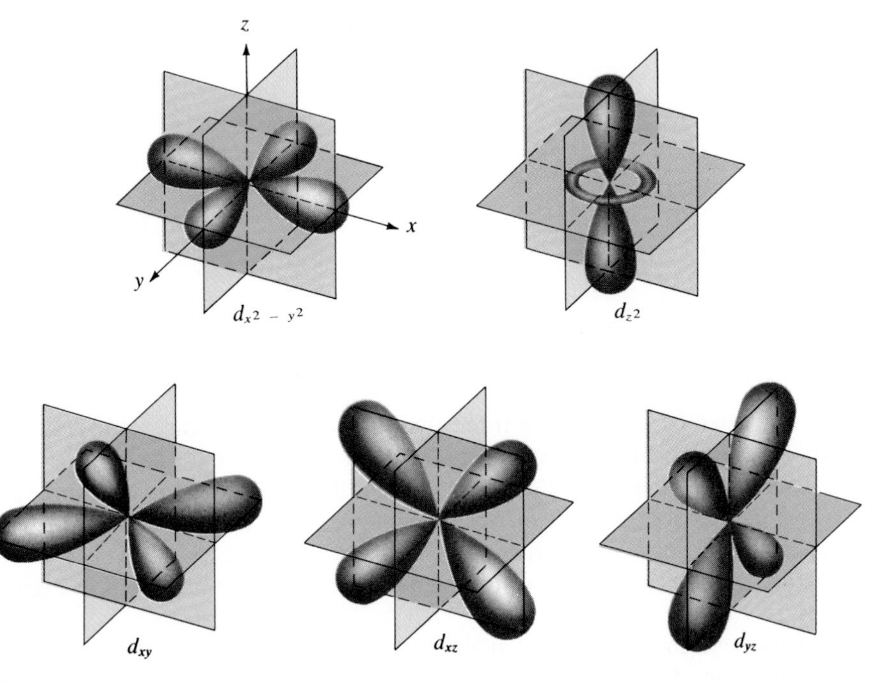

Figure 5-24
Spatial orientation of d orbitals. Note that the lobes of the $d_{x^2-y^2}$ and d_{z^2} orbitals lie along the axes, whereas the lobes of the others lie along diagonals between the axes.

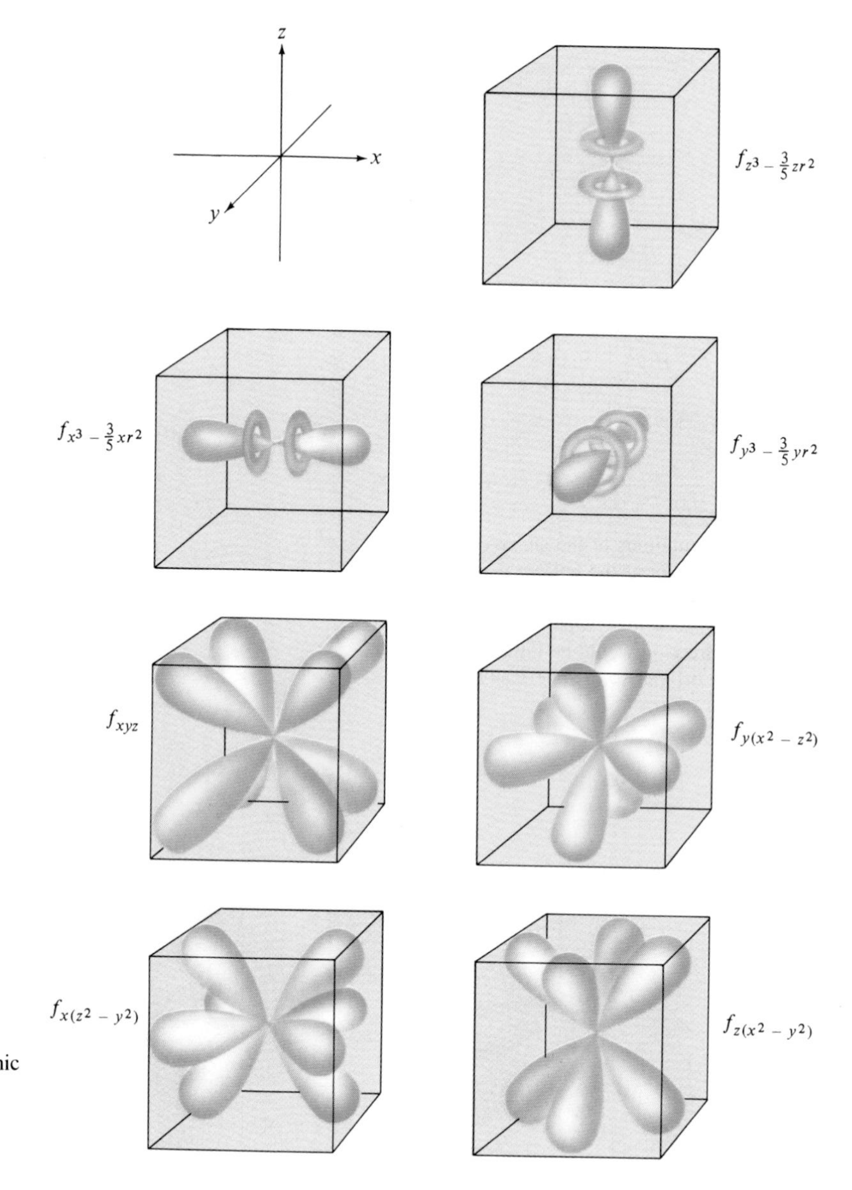

Figure 5-25
Relative directional character of atomic f orbitals. The seven orbitals are shown within cubes as an aid to visualization.

axis along which each of the three two-lobed orbitals is directed. A set of three p atomic orbitals may be represented as in Figure 5-23b.

Beginning at the third energy level, each level contains a third sublevel ($\ell = 2$) composed of a set of *five* d atomic orbitals ($m_\ell = -2, -1, 0, +1, +2$). They are designated $3d$, $4d$, $5d$, . . . to indicate the energy level in which they are found. The shapes of the members of a set are indicated in Figure 5-24.

In each of the fourth and higher energy levels, there is also a fourth sublevel, containing a set of *seven* f atomic orbitals ($\ell = 3$, $m_\ell = -3, -2, -1, 0, +1, +2, +3$). These are shown in Figure 5-25.

Thus we see the first energy level contains only the $1s$ orbital; the second energy level contains the $2s$ and three $2p$ orbitals; the third energy level contains the $3s$, three $3p$, and five $3d$ orbitals; and the fourth energy level consists of a $4s$, three $4p$, five $4d$, and seven $4f$ orbitals. All subsequent energy levels contain s, p, d, and f sublevels as well as others that are not occupied in any presently known elements in their lowest energy states.

The sizes of orbitals increase with increasing n, as shown in Figure 5-26. Slender representations of orbitals, such as those in Figures 5-23 through 5-25, are usually used for convenience. The true shapes are actually more like those in Figure 5-26.

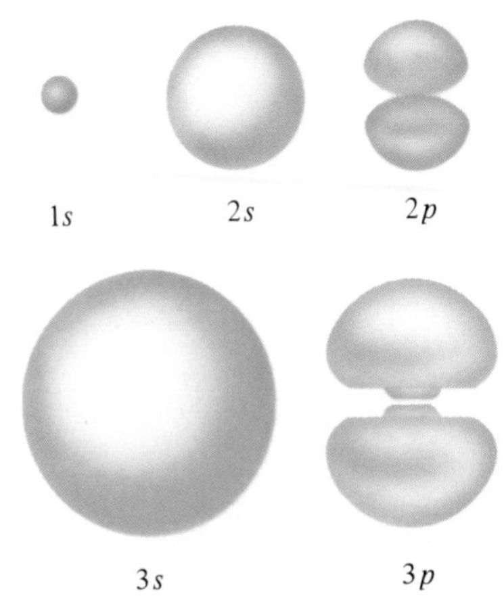

Figure 5-26
Shapes and approximate relative sizes of several orbitals in an atom.

Let us summarize in tabular form some of the information we have developed to this point. The principal quantum number n indicates the energy level. The number of sublevels per energy level is equal to n, the number of atomic orbitals per energy level is n^2, and the maximum number of electrons per energy level is $2n^2$, because each atomic orbital can hold two electrons.

Energy Level n	Number of Sublevels per Energy Level n	Number of Atomic Orbitals n^2	Maximum Number of Electrons $2n^2$
1	1	1 ($1s$)	2
2	2	4 ($2s$, $2p_x$, $2p_y$, $2p_z$)	8
3	3	9 ($3s$, three $3p$'s, five $3d$'s)	18
4	4	16	32
5	5	25	50

In this section, we haven't yet discussed the fourth quantum number, the spin quantum number, m_s. Because m_s has two possible values, $+\frac{1}{2}$ and $-\frac{1}{2}$, each atomic orbital, defined by the values of n, ℓ, and m_ℓ, has a capacity of two electrons. Electrons are negatively charged, and they behave as though they were spinning about axes through their centers, so they act like tiny magnets. The motions of electrons produce magnetic fields, and these can interact with one another. Two electrons in the same orbital having opposite m_s values are said to be **spin-paired**, or simply **paired** (Figure 5-27).

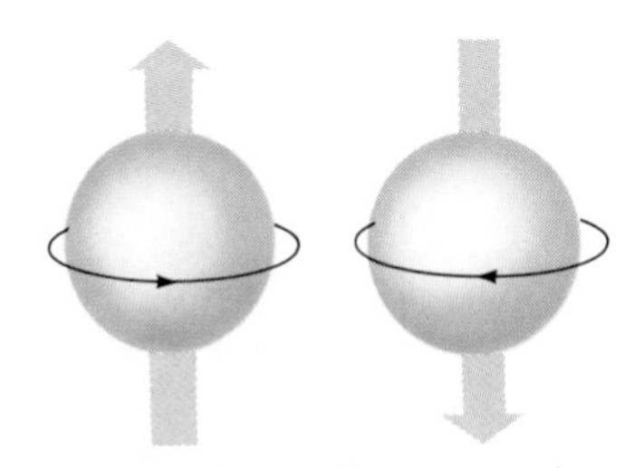

One electron has $m_s = +\frac{1}{2}$;
the other has $m_s = -\frac{1}{2}$.

Figure 5-27
Electron spin. Electrons act as though they spin about an axis through their centers. Because there are two directions in which an electron may spin, the spin quantum number has two possible values, $+\frac{1}{2}$ and $-\frac{1}{2}$. Each electron spin produces a magnetic field. When two electrons have opposite spins, the attraction due to their opposite magnetic fields helps to overcome the repulsion of their like charges. This permits two electrons to occupy the same region (orbital).

5-16 Electron Configurations

The wave function for an atom simultaneously depends on (describes) all of the electrons in the atom. The Schrödinger equation is much more complicated for atoms with more than one electron than for a one-electron species such as hydrogen, and an explicit solution to this equation is not possible even for helium, let alone for more complicated atoms. Therefore we must rely on approximations to solutions of the many-electron Schrödinger equation. We shall use one of the most common and useful, called the **orbital approximation**. In this approximation, the electron cloud of an atom is assumed to be the superposition of charge clouds, or orbitals, arising from the individual electrons; these orbitals resemble the atomic orbitals of hydrogen (for which exact solutions are known), which we described in some detail in the last section. Each electron is described by the same allowed combinations of quantum numbers (n, ℓ, m_ℓ, and m_s) that we used for the hydrogen atom; however, the order of energies of the orbitals is often different from that in hydrogen.

The great power of modern computers has allowed scientists to make numerical approximations to this solution to very high accuracy for simple atoms such as helium. However, as the number of electrons increases, even such numerical approaches become quite difficult to apply and interpret. For many purposes, more qualitative approximations are suitable.

Let us now examine the electronic structures of atoms of different elements. The electronic arrangement that we shall describe for each atom is called the **ground state electron configuration**. This corresponds to the isolated atom in its lowest-energy, or unexcited, state. We shall consider the elements in order of increasing atomic number, using as our guide the periodic table inside the front cover of this text. For simplicity, we shall indicate atomic orbitals as $\underline{\quad}$ and show an unpaired electron as $\underline{\uparrow}$ and spin-paired electrons as $\underline{\uparrow\downarrow}$. By "unpaired electron" we mean an electron that occupies an orbital singly.

In building up ground state electron configurations, the guiding idea is that the *total energy* of the atom is as low as possible. To determine these configurations, we use the **Aufbau Principle** as a guide:

The German verb *aufbauen* means "to build up."

> Each atom is "built up" by (1) adding the appropriate numbers of protons and neutrons as specified by the atomic number and the mass number, and (2) adding the necessary number of electrons into orbitals in the way that gives the lowest *total* energy for the atom.

As we apply this principle, we shall focus on the difference in electronic arrangement between a given element and the element with an atomic number

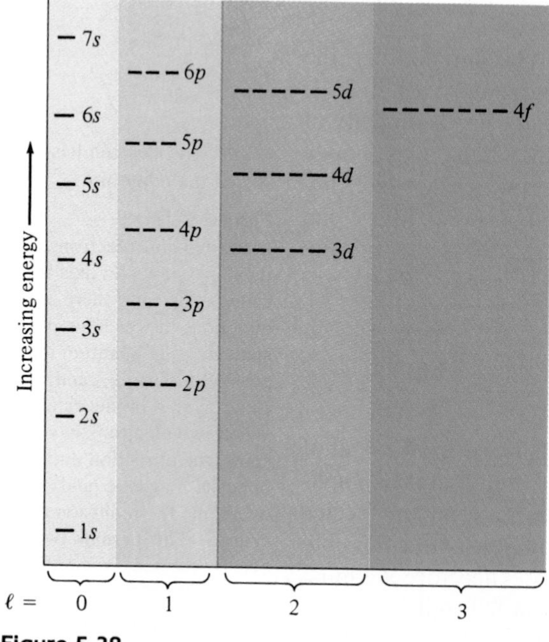

Figure 5-28
The usual order of filling (Aufbau order) of the orbitals of an atom. The energy scale varies for different elements, but the following main features should be noted: (1) The largest energy gap is between the 1*s* and 2*s* orbitals. (2) The energies of orbitals are generally closer together at higher energies. (3) The gap between *np* and $(n + 1)s$ (e.g., between 2*p* and 3*s* or between 3*p* and 4*s*) is usually fairly large. (4) The gap between $(n - 1)d$ and *ns* (e.g., between 3*d* and 4*s*) is quite small. (5) The gap between $(n - 2)f$ and *ns* (e.g., between 4*f* and 6*s*) is even smaller.

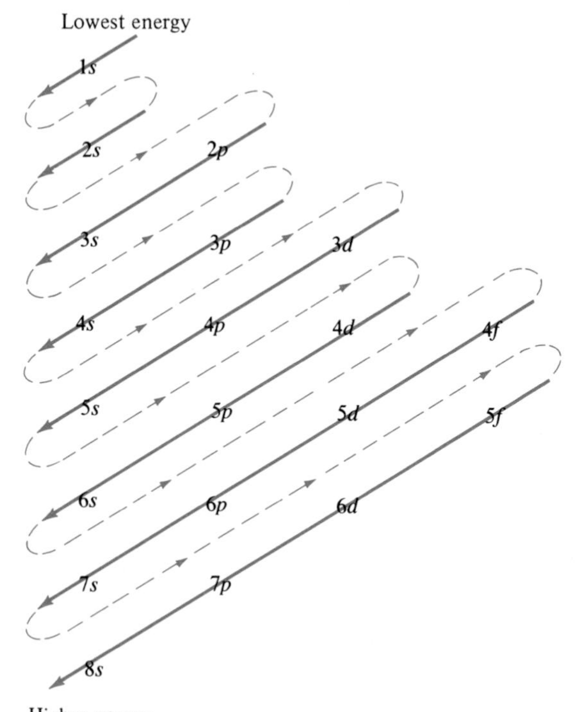

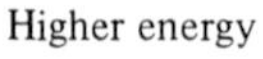

Figure 5-29
An aid to remembering the Aufbau order of atomic orbitals. Write all sublevels in the same major energy level on the same horizontal line. Write all like sublevels in the same vertical column. Draw parallel arrows diagonally from upper right to lower left. The arrows are read from top to bottom, and tail to head. The order is 1*s*, 2*s*, 2*p*, 3*s*, 3*p*, 4*s*, 3*d*, 4*p*, 5*s*, 4*d*, . . . and so on.

that is one lower. Though we do not always point it out, we *must* keep in mind that the atomic number (the charge on the nucleus) is different. We also emphasize the particular electron that distinguishes each element from the previous one; however, we should remember that this distinction is artificial, because electrons are not really distinguishable.

The orbitals increase in energy with increasing value of the quantum number n. For a given value of n, energy increases with increasing value of ℓ. In other words, within a particular major energy level, the s sublevel is lowest in energy, the p sublevel is the next lowest, then the d, then the f, and so on. As a result of changes in the nuclear charge and interactions among the electrons in the atom, the order of energies of the orbitals can vary somewhat from atom to atom. The *usual* order of energies of the orbitals of an atom and a helpful device for remembering this order are shown in Figures 5-28 and 5-29.

The electronic structures of atoms are governed by the **Pauli Exclusion Principle**:

> No two electrons in an atom may have identical sets of four quantum numbers.

An orbital is described by a particular allowed set of values for n, ℓ, and m_ℓ. Thus, two electrons can occupy the same orbital only if they have opposite spins, m_s. Two such electrons in the same orbital are *paired*.

Row 1 The first energy level consists of only one atomic orbital, $1s$. This can hold a maximum of two electrons. Hydrogen, as we have already noted, contains just one electron. Helium, a noble gas, has a filled first energy level (two electrons). The atom is so stable that no chemical reactions of helium are known.

Helium's electrons can be displaced only by electrical forces, as in excitation by high-voltage discharge.

	Orbital Notation	
	$1s$	Simplified Notation
$_1H$	↑	$1s^1$
$_2He$	↑↓	$1s^2$

In the simplified notation, we indicate with superscripts the number of electrons in each sublevel.

Row 2 Elements of atomic numbers 3 through 10 occupy the second period, or horizontal row, in the periodic table. In neon atoms the second energy level is filled completely. Neon, a noble gas, is extremely stable. No reactions of it are known.

	Orbital Notation			Simplified Notation		
	$1s$	$2s$	$2p$			
$_3Li$	↑↓	↑		$1s^22s^1$	or	[He] $2s^1$
$_4Be$	↑↓	↑↓		$1s^22s^2$		[He] $2s^2$
$_5B$	↑↓	↑↓	↑ _ _	$1s^22s^22p^1$		[He] $2s^22p^1$
$_6C$	↑↓	↑↓	↑ ↑ _	$1s^22s^22p^2$		[He] $2s^22p^2$
$_7N$	↑↓	↑↓	↑ ↑ ↑	$1s^22s^22p^3$		[He] $2s^22p^3$
$_8O$	↑↓	↑↓	↑↓ ↑ ↑	$1s^22s^22p^4$		[He] $2s^22p^4$
$_9F$	↑↓	↑↓	↑↓ ↑↓ ↑	$1s^22s^22p^5$		[He] $2s^22p^5$
$_{10}Ne$	↑↓	↑↓	↑↓ ↑↓ ↑↓	$1s^22s^22p^6$		[He] $2s^22p^6$

In writing electronic structures of atoms, we frequently simplify notations. The abbreviation [He] indicates that the $1s$ orbital is completely filled, $1s^2$, as in helium.

As with helium, neon's electrons can be displaced by high-voltage electrical discharge, as is observed in neon signs.

We see that some atoms have unpaired electrons in the same set of energetically equivalent, or **degenerate**, orbitals. We have already seen that two electrons can occupy a given atomic orbital (with the same values of n, ℓ, and m_ℓ) *only* if their spins are paired (have opposite values of m_s). Even with pairing of spins, however, two electrons that are in the same orbital repel each other more strongly than do two electrons in different (but equal-energy) orbitals. Thus, both theory and experimental observations (see Enrichment section) lead to **Hund's Rule**:

> Electrons must occupy all the orbitals of a given sublevel singly before pairing begins. These unpaired electrons have parallel spins.

Thus, carbon has two unpaired electrons in its $2p$ orbitals, and nitrogen has three.

Enrichment

Paramagnetism and Diamagnetism

Both paramagnetism and diamagnetism are hundreds to thousands of times weaker than *ferromagnetism*, the effect seen in iron bar magnets.

Substances that contain unpaired electrons are weakly *attracted* into magnetic fields and are said to be **paramagnetic**. By contrast, those in which all electrons are paired are very weakly repelled by magnetic fields and are called **diamagnetic**. The magnetic effect can be measured by hanging a test tube full of a substance on a balance by a long thread and suspending it above the gap of an electromagnet (Figure 5-30). When the current is switched on, a paramagnetic substance such as copper(II) sulfate is pulled into the strong field. The paramagnetic attraction per mole of substance can be measured by weighing the sample before and after energizing the magnet. The paramagnetism per mole increases with increasing number of unpaired electrons per formula unit.

Row 3 The next element beyond neon is sodium. Here we begin to add electrons to the third energy level. Elements 11 through 18 occupy the third period in the periodic table.

	Orbital Notation		
	$3s$	$3p$	Simplified Notation
$_{11}$Na	[Ne] ↑		[Ne] $3s^1$
$_{12}$Mg	[Ne] ⇅		[Ne] $3s^2$
$_{13}$Al	[Ne] ⇅	↑ _ _	[Ne] $3s^23p^1$
$_{14}$Si	[Ne] ⇅	↑ ↑ _	[Ne] $3s^23p^2$
$_{15}$P	[Ne] ⇅	↑ ↑ ↑	[Ne] $3s^23p^3$
$_{16}$S	[Ne] ⇅	⇅ ↑ ↑	[Ne] $3s^23p^4$
$_{17}$Cl	[Ne] ⇅	⇅ ⇅ ↑	[Ne] $3s^23p^5$
$_{18}$Ar	[Ne] ⇅	⇅ ⇅ ⇅	[Ne] $3s^23p^6$

Although the third energy level is not yet filled (the d orbitals are still empty), argon is a noble gas. All noble gases except helium have ns^2np^6 electronic configurations (where n indicates the highest occupied energy level). The noble gases are quite unreactive elements.

Rows 4 and 5 It is an experimentally observed fact that *an electron occupies the available orbital that gives the atom the lowest total energy*. It is observed that filling the $4s$ orbitals before electrons enter the $3d$ orbitals *usually* leads to a lower total energy for the atom than some other arrangement.

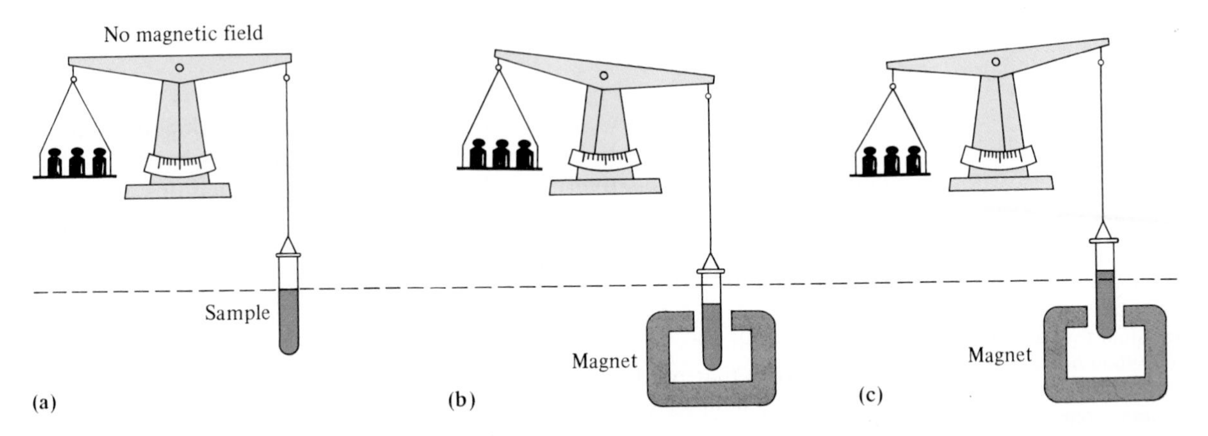

Figure 5-30
Diagram of an apparatus for measuring the paramagnetism of a substance. The tube contains a measured amount of the substance, often in solution. (a) Before the magnetic field is turned on, the position and mass of the sample are determined. (b) When the field is on, a paramagnetic substance is attracted *into* the field. (c) A diamagnetic substance would be repelled *very weakly* by the field.

Therefore we fill the orbitals in this order (see Figure 5-28). According to the normal Aufbau order (recall Figures 5-28 and 5-29), 4*s* fills before 3*d*. In general, *the (n+1)s orbital fills before the nd orbital*. This is sometimes referred to as the *n+1 rule*.

After the 3*d* sublevel is filled to its capacity of 10 electrons, the 4*p* orbitals fill next, taking us to the noble gas krypton. Then the 5*s* orbital, the five 4*d* orbitals, and the three 5*p* orbitals fill to take us to xenon, a noble gas.

Let us now examine the electronic structures of the 18 elements in the fourth period in some detail. Some of these have electrons in *d* orbitals.

	Orbital Notation				
		3*d*	4*s*	4*p*	Simplified Notation
$_{19}K$	[Ar]		↑		[Ar] $4s^1$
$_{20}Ca$	[Ar]		↑↓		[Ar] $4s^2$
$_{21}Sc$	[Ar]	↑ _ _ _ _	↑↓		[Ar] $3d^14s^2$
$_{22}Ti$	[Ar]	↑ ↑ _ _ _	↑↓		[Ar] $3d^24s^2$
$_{23}V$	[Ar]	↑ ↑ ↑ _ _	↑↓		[Ar] $3d^34s^2$
$_{24}Cr$	[Ar]	↑ ↑ ↑ ↑ ↑	↑		[Ar] $3d^54s^1$
$_{25}Mn$	[Ar]	↑ ↑ ↑ ↑ ↑	↑↓		[Ar] $3d^54s^2$
$_{26}Fe$	[Ar]	↑↓ ↑ ↑ ↑ ↑	↑↓		[Ar] $3d^64s^2$
$_{27}Co$	[Ar]	↑↓ ↑↓ ↑ ↑ ↑	↑↓		[Ar] $3d^74s^2$
$_{28}Ni$	[Ar]	↑↓ ↑↓ ↑↓ ↑ ↑	↑↓		[Ar] $3d^84s^2$
$_{29}Cu$	[Ar]	↑↓ ↑↓ ↑↓ ↑↓ ↑↓	↑		[Ar] $3d^{10}4s^1$
$_{30}Zn$	[Ar]	↑↓ ↑↓ ↑↓ ↑↓ ↑↓	↑↓		[Ar] $3d^{10}4s^2$
$_{31}Ga$	[Ar]	↑↓ ↑↓ ↑↓ ↑↓ ↑↓	↑↓	↑ _ _	[Ar] $3d^{10}4s^24p^1$
$_{32}Ge$	[Ar]	↑↓ ↑↓ ↑↓ ↑↓ ↑↓	↑↓	↑ ↑ _	[Ar] $3d^{10}4s^24p^2$
$_{33}As$	[Ar]	↑↓ ↑↓ ↑↓ ↑↓ ↑↓	↑↓	↑ ↑ ↑	[Ar] $3d^{10}4s^24p^3$
$_{34}Se$	[Ar]	↑↓ ↑↓ ↑↓ ↑↓ ↑↓	↑↓	↑↓ ↑ ↑	[Ar] $3d^{10}4s^24p^4$
$_{35}Br$	[Ar]	↑↓ ↑↓ ↑↓ ↑↓ ↑↓	↑↓	↑↓ ↑↓ ↑	[Ar] $3d^{10}4s^24p^5$
$_{36}Kr$	[Ar]	↑↓ ↑↓ ↑↓ ↑↓ ↑↓	↑↓	↑↓ ↑↓ ↑↓	[Ar] $3d^{10}4s^24p^6$

As you study these electronic configurations, you should be able to see how most of them are predicted from the Aufbau order. However, as we fill the 3*d* set of orbitals, from $_{21}Sc$ to $_{30}Zn$, we see that these orbitals are not filled quite regularly. Some sets of orbitals are so close in energy (e.g., 4*s* and 3*d*) that minor changes in their relative energies may occasionally change the order of filling.

End-of-chapter Exercises 84–107 provide much valuable practice in writing electron configurations.

These two elements illustrate an exception to the $n + 1$ rule. You should realize that such statements as the Aufbau Principle and the $n + 1$ rule merely represent general guidelines and should not be viewed as hard-and-fast rules. It is the *total energy* of the atom that is as low as possible. The Aufbau order of orbital energies is based on calculations for the hydrogen atom. The orbital energies also depend on additional factors such as the nuclear charge and interactions of different occupied orbitals.

Chemical and spectroscopic evidence indicates that the configurations of Cr and Cu have only one electron in the 4*s* orbital. Their 3*d* sets are half-filled and filled, respectively, in the ground state. Calculations from the quantum mechanical equations also indicate that *half-filled and filled sets of equivalent orbitals have a special stability*. In $_{24}$Cr, for example, this increased stability is apparently sufficient to make the *total* energy of [Ar] 3*d* ↿ ↿ ↿ ↿ ↿ 4*s* ↿ lower than that of [Ar] 3*d* ↿ ↿ ↿ ↿ _ 4*s* ⇅. Similar reasoning helps us understand the apparent exception of the configuration of $_{29}$Cu from that predicted by the Aufbau Principle.

You may wonder why such an exception does not occur in, for example, $_{32}$Ge or $_{14}$Si, where we could have an s^1p^3 configuration that would have half-filled sets of *s* and *p* orbitals. It does not occur because of the very large energy gap between *ns* and *np* orbitals. We shall see evidence in Chapter 6 that does, however, illustrate the enhanced stability of half-filled sets of *p* orbitals.

Let us now write the quantum numbers to describe each electron in an atom of nitrogen. Keep in mind the fact that Hund's Rule must be obeyed. Thus, there is only one (unpaired) electron in each 2*p* orbital in a nitrogen atom.

Example 5-7

Write an acceptable set of four quantum numbers for each of the electrons of a nitrogen atom.

Plan

Nitrogen has seven electrons, which occupy the lowest-energy orbitals available. Two electrons can occupy the first energy level, $n = 1$, in which there is only one *s* orbital; when $n = 1$, then ℓ must be zero, and therefore $m_\ell = 0$. The two electrons differ only in spin quantum number, m_s. The next five electrons can all fit into the second energy level, for which $n = 2$ and ℓ may be either 0 or 1. The $\ell = 0$ (*s*) sublevel fills first, and the $\ell = 1$ (*p*) sublevel is occupied next.

Solution

Electron	n	ℓ	m_ℓ	m_s	e^- Configuration
1, 2	1	0	0	$+\frac{1}{2}$	$1s^2$
	1	0	0	$-\frac{1}{2}$	
3, 4	2	0	0	$+\frac{1}{2}$	$2s^2$
	2	0	0	$-\frac{1}{2}$	
5, 6, 7	2	1	-1	$+\frac{1}{2}$ or $-\frac{1}{2}$	$2p_x^1$
	2	1	0	$+\frac{1}{2}$ or $-\frac{1}{2}$	$2p_y^1$ or $2p^3$
	2	1	$+1$	$+\frac{1}{2}$ or $-\frac{1}{2}$	$2p_z^1$

Electrons are indistinguishable. We have numbered them 1, 2, 3, and so on as an aid to counting them.

In the lowest-energy configurations, the three 2*p* electrons either all have $m_s = +\frac{1}{2}$ or all have $m_s = -\frac{1}{2}$.

Example 5-8

Write an acceptable set of four quantum numbers that decribe each electron in a chlorine atom.

Plan

Chlorine is element number 17. Its first seven electrons have the same quantum numbers as those of nitrogen in Example 5-7. Electrons 8, 9, and 10 complete the filling of the $2p$ sublevel ($n = 2$, $\ell = 1$) and therefore also the second energy level. Electrons 11 through 17 fill the $3s$ sublevel ($n = 3$, $\ell = 0$) and partially fill the $3p$ sublevel ($n = 3$, $\ell = 1$).

Solution

Electron	n	ℓ	m_ℓ	m_s	e^- Configuration
1, 2	1	0	0	$\pm\frac{1}{2}$	$1s^2$
3, 4	2	0	0	$\pm\frac{1}{2}$	$2s^2$
5–10	2	1	−1	$\pm\frac{1}{2}$	$2p^6$
	2	1	0	$\pm\frac{1}{2}$	
	2	1	+1	$\pm\frac{1}{2}$	
11, 12	3	0	0	$\pm\frac{1}{2}$	$3s^2$
13–17	3	1	−1	$\pm\frac{1}{2}$	$3p^5$
	3	1	0	$\pm\frac{1}{2}$	
	3	1	+1	$+\frac{1}{2}$ or $-\frac{1}{2}$*	

*The $3p$ orbital with only a single electron can be any one of the set, not *necessarily* the one with $m_\ell = +1$.

EOC 101, 105

5-17 The Periodic Table and Electron Configurations

In this section, we view the *periodic table* (see inside front cover and Section 4-1) from a modern, much more useful perspective—as a systematic representation of the electronic configurations of the elements. In the long form of the periodic table, used throughout this text, elements are arranged in blocks based on the kinds of atomic orbitals that are being filled (Figure 5-31). The periodic tables in this text are divided into "A" and "B" groups. The A groups contain elements in which s and p orbitals are being filled. Elements within any particular A group have similar electronic configurations and chemical properties, as we shall see in the next chapter. The B groups are those in which there are one or two electrons in the s orbital of the highest occupied energy level, and the d orbitals one energy level lower are being filled.

Lithium, sodium, and potassium, elements of the leftmost column of the periodic table (Group IA), have a single electron in their outermost s orbital (ns^1). Beryllium and magnesium, of Group IIA, have two electrons in their highest energy level, ns^2, while boron and aluminum (Group IIIA) have three electrons in their highest energy level, ns^2np^1. Similar observations can be made for each A group.

The electronic configurations of the A group elements and the noble gases can be predicted reliably from Figures 5-28 and 5-29. However, there are some irregularities in the B groups below the fourth period that require special attention. In the heavier B group elements, the higher-energy sublevels in different principal energy levels have energies that are very nearly equal (Figure 5-29). It is easy for an electron to jump from one orbital to

H is shown in the 1*s* block in Figure 5-31. It is usually shown in Group IA.

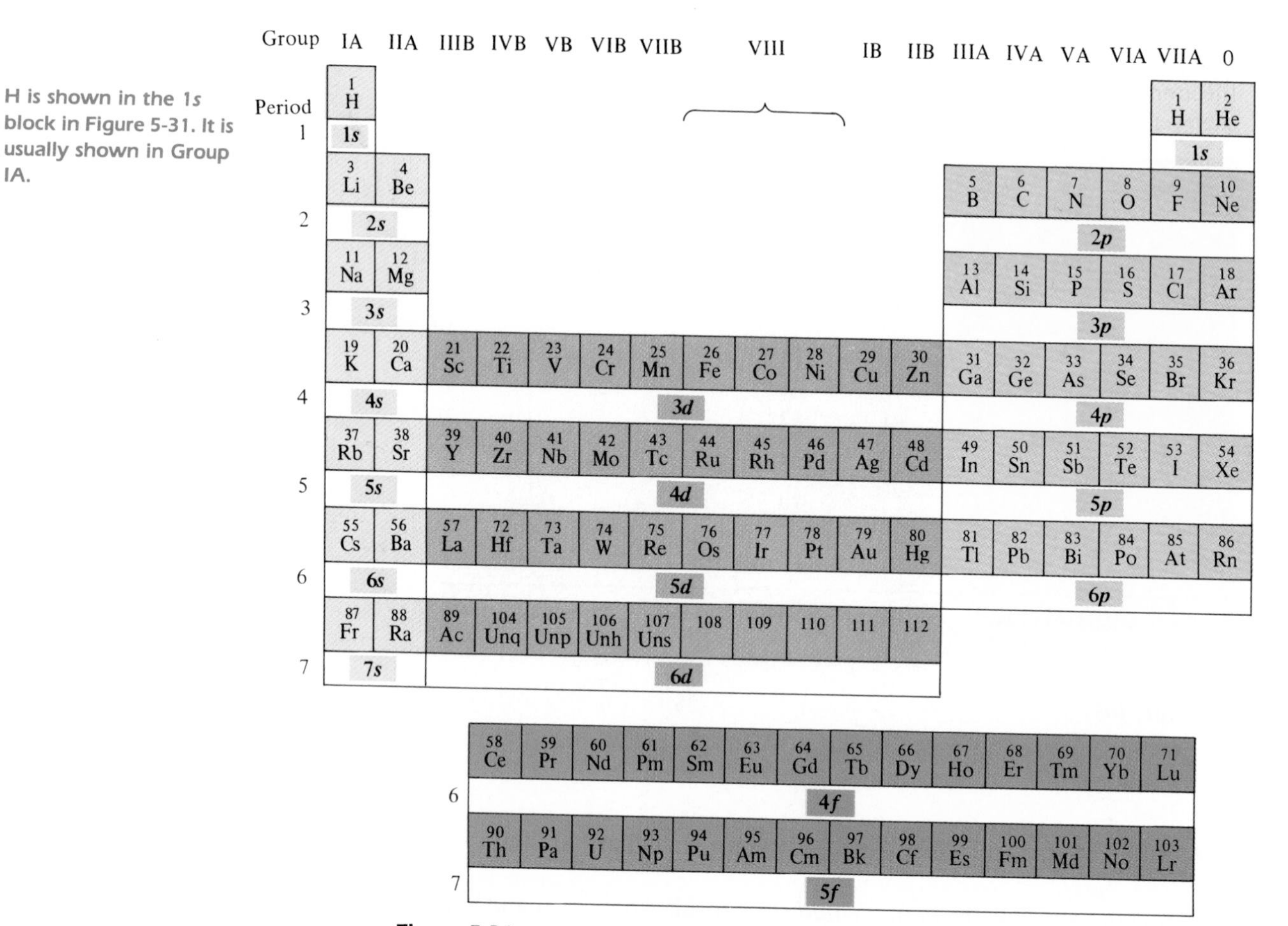

Figure 5-31

A periodic table colored to show the kinds of atomic orbitals (sublevels) being filled, below the symbols of blocks of elements. The electronic structures of the A group and 0 group elements are perfectly regular and can be predicted from their positions in the periodic table, but there are many exceptions in the *d* and *f* blocks. The colors in this figure are the same as those in Figure 5-28.

Hydrogen and helium are shown here in their usual positions in the periodic table. These may seem somewhat unusual based just on their electronic configurations. However, we should remember that the first main energy level ($n = 1$) can hold a maximum of only two electrons. This shell is entirely filled in helium, so He behaves as a noble gas, and we put it in the column with the other noble gases (Group 0). Hydrogen has one electron that is easily lost, like the metals in Group IA, so we put it in Group IA even though it is not a metal. Furthermore, hydrogen is one electron short of a noble gas configuration (He), so we could also place it with the other such elements in Group VIIA.

another of nearly the same energy in a different set. This is because the orbital energies are *perturbed* (they change slightly) as the nuclear charge changes, and an extra electron is added in going from one element to the next. This phenomenon gives rise to other irregularities that are analogous to those of Cr and Cu, described earlier.

We can extend the information in Figure 5-31 to indicate the electron configurations that are represented by each *group* (column) of the periodic table. Table 5-5 shows this interpretation of the periodic table, along with the most important exceptions. (A more complete listing of electron config-

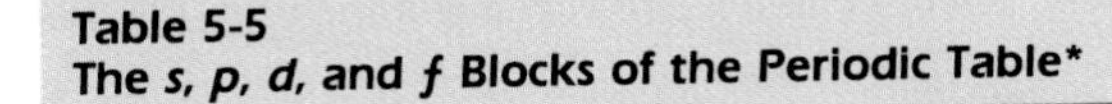

Table 5-5
The *s*, *p*, *d*, and *f* Blocks of the Periodic Table*

s orbital block · GROUPS · *p* orbital block · *d* orbital block · *f* orbital block

	IA s^1	IIA s^2	IIIB d^1s^2	*f* orbital block	IVB d^2s^2	VB d^3s^2	VIB d^5s^1	VIIB d^5s^2	VIII d^6s^2	VIII d^7s^2	VIII d^8s^2	IB $d^{10}s^1$	IIB $d^{10}s^2$	IIIA s^2p^1	IVA s^2p^2	VA s^2p^3	VIA s^2p^4	VIIA s^2p^5	0 s^2 / s^2p^6
$n = 1$	1 H																		2 He
$n = 2$	3 Li	4 Be												5 B	6 C	7 N	8 O	9 F	10 Ne
$n = 3$	11 Na	12 Mg												13 Al	14 Si	15 P	16 S	17 Cl	18 Ar
$n = 4$	19 K	20 Ca	21 Sc		22 Ti	23 V	24 Cr	25 Mn	26 Fe	27 Co	28 Ni	29 Cu	30 Zn	31 Ga	32 Ge	33 As	34 Se	35 Br	36 Kr
$n = 5$	37 Rb	38 Sr	39 Y		40 Zr	41 Nb d^4s^1	42 Mo	43 Tc	44 Ru d^7s^1	45 Rh d^8s^1	46 Pd	47 Ag	48 Cd	49 In	50 Sn	51 Sb	52 Te	53 I	54 Xe
$n = 6$	55 Cs	56 Ba	57 La	58 Ce→71 Lu	72 Hf	73 Ta	74 W d^4s^2	75 Re	76 Os	77 Ir	78 Pt d^9s^1	79 Au	80 Hg	81 Tl	82 Pb	83 Bi	84 Po	85 At	86 Rn
$n = 7$	87 Fr	88 Ra	89 Ac	90 Th→103 Lr	104 Unq	105 Unp	106 Unh	107 Uns	108	109	110	111	112						

$n = 6$	LANTHANIDE SERIES	58 Ce	59 Pr	60 Nd	61 Pm	62 Sm	63 Eu	64 Gd	65 Tb	66 Dy	67 Ho	68 Er	69 Tm	70 Yb	71 Lu	4*f* subshell being filled
$n = 7$	ACTINIDE SERIES	90 Th	91 Pa	92 U	93 Np	94 Pu	95 Am	96 Cm	97 Bk	98 Cf	99 Es	100 Fm	101 Md	102 No	103 Lr	5*f* subshell being filled

*n is the principal quantum number. The d^1s^2, d^2s^2, . . . designations represent *known* configurations. They refer to $(n - 1)d$ and ns orbitals. Several exceptions to the configurations indicated above each group are shown in gray.

urations is given in Appendix B.) We can use this interpretation of the periodic table to write, quickly and reliably, the electron configurations for elements.

Example 5-9

Use Table 5-5 to determine the electronic configurations of (a) germanium, Ge; (b) magnesium, Mg; and (c) molybdenum, Mo.

Plan

We use the electron configurations indicated in Table 5-5 for each group. Each *period* (row) begins filling a new shell (new value of n). Elements to the right of the d orbital block have the d orbitals in the $(n - 1)$ shell already filled. We often find it convenient to collect all sets of orbitals with the same value of n together, to emphasize the number of electrons in the *outermost* shell, i.e., the shell with the highest value of n.

Solution

(a) Germanium, Ge, is in Group IVA, for which Table 5-5 shows the general configuration s^2p^2. It is in Period 4 (the 4th row), so we interpret this as $4s^24p^2$. The last filled noble gas configuration is that of argon, Ar, accounting for 18 electrons. In addition, Ge lies beyond the d orbital block, so we know that the 3d orbitals are completely filled. The electron configuration of Ge is [Ar] $4s^23d^{10}4p^2$ or [Ar] $3d^{10}4s^24p^2$.

(b) Magnesium, Mg, is in Group IIA, which has the general configuration s^2; it is in Period 3 (3rd row). The last filled noble gas configuration is that of neon, or [Ne]. The electron configuration of Mg is [Ne] $3s^2$.

(c) Molybdenum, Mo, is in Group VIB, with general configuration d^5s^1; it is in Period 5, which begins with 5s and is beyond the noble gas krypton. The electron configuration of Mo is [Kr] $5s^14d^5$ or [Kr] $4d^55s^1$. The electron configuration of molybdenum is analogous to that of chromium, Cr, the element just above it. The configuration of Cr was discussed in Section 5-16 as one of the exceptions to the Aufbau order of filling.

EOC 102

Example 5-10

Determine the number of unpaired electrons in an atom of tellurium, Te.

Plan

Te is in Group IVA in the periodic table, which tells us that its configuration is s^2p^4. All other shells are completely filled, so they contain only paired electrons. We need only to find out how many unpaired electrons are represented by s^2p^4.

Solution

The notation s^2p^4 is a short representation for $s\underline{\uparrow\downarrow}\ p\underline{\uparrow\downarrow}\ \underline{\uparrow}\ \underline{\uparrow}$. This shows that an atom of Te contains two unpaired electrons.

The periodic table has been described as "the chemist's best friend." Notice how easy it is to use the periodic table to determine many important aspects of the electron configurations of atoms. You should practice until you can use the periodic table with confidence to answer many questions about electron configurations. As we continue our study of chemistry, we shall learn many other useful ways to interpret the periodic table.

Key Terms

Absorption spectrum The spectrum associated with absorption of electromagnetic radiation by atoms (or other species) resulting from transitions from lower to higher energy states.

Alpha (α) particle A helium ion with 2+ charge; an assembly of two protons and two neutrons.

Anode In a cathode ray tube, the positive electrode.

Atomic mass unit An arbitrary mass unit defined to be exactly one-twelfth the mass of the carbon-12 isotope.

Atomic number The integral number of protons in the nucleus; defines the identity of an element.

Atomic orbital The region or volume in space in which the probability of finding electrons is highest.

Aufbau ("building up") Principle Describes the order in which electrons fill orbitals in atoms.

Canal ray A stream of positively charged particles (cations) that moves toward the negative electrode in a cathode ray tube; observed to pass through canals in the negative electrode.

Cathode In a cathode ray tube, the negative electrode.

Cathode ray The beam of electrons going from the negative electrode toward the positive electrode in a cathode ray tube.

Cathode ray tube A closed glass tube containing a gas under low pressure, with electrodes near the ends and a luminescent screen at the end near the positive electrode; produces cathode rays when high voltage is applied.

Continuous spectrum The spectrum that contains all wavelengths in a specified region of the electromagnetic spectrum.

Degenerate Of the same energy.

Diamagnetism *Weak* repulsion by a magnetic field.

***d* orbitals** Beginning in the third energy level, a set of five degenerate orbitals per energy level, higher in energy than *s* and *p* orbitals of the same energy level.

Electromagnetic radiation Energy that is propagated by means of electric and magnetic fields that oscillate in directions perpendicular to the direction of travel of the energy.

Electron A subatomic particle having a mass of 0.00054858 amu and a charge of 1−.

Electron configuration The specific distribution of electrons in the atomic orbitals of atoms or ions.

Electronic transition The transfer of an electron from one energy level to another.

Emission spectrum The spectrum associated with emission of electromagnetic radiation by atoms (or other species) resulting from electronic transitions from higher to lower energy states.

Excited state Any state other than the ground state of an atom or molecule.

***f* orbitals** Beginning in the fourth energy level, a set of seven degenerate orbitals per energy level, higher in energy than *s*, *p*, and *d* orbitals of the same energy level.

Frequency The number of repeating corresponding points on a wave that pass a given observation point per unit time.

Ground state The lowest energy state or most stable state of an atom, molecule, or ion.

Group A vertical column in the periodic table; also called a family.

Heisenberg Uncertainty Principle It is impossible to determine accurately both the momentum and position of an electron simultaneously.

Hund's Rule All orbitals of a given sublevel must be occupied by single electrons before pairing begins. See *Aufbau Principle*.

Isotopes Two or more forms of atoms of the same element with different masses; atoms containing the same number of protons but different numbers of neutrons.

Line spectrum An atomic emission or absorption spectrum.

Magnetic quantum number (m_ℓ) Quantum mechanical solution to a wave equation that designates the particular orbital within a given set (*s*, *p*, *d*, *f*) in which an electron resides.

Mass number The integral sum of the numbers of protons and neutrons in an atom.

Mass spectrometer An instrument that measures the charge-to-mass ratios of charged particles.

Natural radioactivity Spontaneous decomposition of an atom.

Neutron A neutral subatomic nuclear particle having a mass of 1.0087 amu.

Nucleus The very small, very dense, positively charged center of an atom containing protons and neutrons, as well as other subatomic particles.

Nuclide symbol The symbol for an atom, ${}^{A}_{Z}E$, in which E is the symbol for an element, Z is its atomic number, and A is its mass number.

Pairing of electrons Favorable interaction of two electrons with opposite m_s values in the same orbital.

Paramagnetism Attraction toward a magnetic field, stronger than diamagnetism, but still weak compared with ferromagnetism.

Pauli Exclusion Principle No two electrons in the same atom may have identical sets of four quantum numbers.

Period A horizontal row in the periodic table.

Photoelectric effect Emission of an electron from the surface of a metal, caused by impinging electromagnetic radiation of certain minimum energy; current increases with increasing intensity of radiation.

Photon A "packet" of light or electromagnetic radiation; also called a quantum of light.

***p* orbitals** Beginning with the second energy level, a set of three mutually perpendicular, equal-arm, dumbbell-shaped atomic orbitals per energy level.

Principal quantum number (*n*) The quantum mechanical solution to a wave equation that designates the major energy level, or shell, in which an electron resides.

Proton A subatomic particle having a mass of 1.0073 amu and a charge of +1, found in the nuclei of atoms.

Quantum A "packet" of energy.

Quantum mechanics A mathematical method of treating particles on the basis of quantum theory, which assumes that energy (of small particles) is not infinitely divisible.

Quantum numbers Numbers that describe the energies of electrons in atoms; derived from quantum mechanical treatment.

Radiant energy See *Electromagnetic radiation*.

Rydberg equation An empirical equation that relates wavelengths in the hydrogen emission spectrum to integers.

***s* orbital** A spherically symmetrical atomic orbital; one per energy level.

Spectral line Any of a number of lines corresponding to definite wavelengths in an atomic emission or absorption spectrum; represents the energy difference between two energy levels.

Spectrum Display of component wavelengths (colors) of electromagnetic radiation.

Spin quantum number (m_s) The quantum mechanical solution to a wave equation that indicates the relative spins of electrons.

Subsidiary quantum number (ℓ) The quantum mechanical solution to a wave equation that designates the sublevel, or set of orbitals (*s*, *p*, *d*, *f*), within a given major energy level in which an electron resides.

Wavelength The distance between two corresponding points of a wave.

Exercises

Particles and the Nuclear Atom

1. List the three fundamental particles of matter and indicate the mass and charge associated with each.
2. (a) Describe the cathode ray experiment. How can we detect where the rays strike?
   (b) Describe an experiment in which it can be established that the streams of electrons in an operating cathode ray tube travel toward the anode rather than the cathode.
3. Describe how Thomson determined the charge-to-mass ratio for the electron.
4. In the oil-drop experiment, how did Millikan know that none of the oil droplets he observed were ones that had a deficiency of electrons rather than an excess?
5. How many electrons carry a total charge of 1.00 coulomb?
6. (a) How do we know that canal rays have charges opposite in sign to cathode rays? What are canal rays?
   (b) Why are cathode rays from all samples of gases identical, whereas canal rays are not?

*7. The following data are measurements of the charges on oil droplets using an apparatus similar to that used by Millikan:

11.215×10^{-19} C	14.423×10^{-19} C
12.811×10^{-19} C	24.037×10^{-19} C
14.419×10^{-19} C	9.621×10^{-19} C
12.815×10^{-19} C	16.012×10^{-19} C

Each should be a whole-number ratio of some fundamental charge. Using these data, determine the value of the fundamental charge.

*8. Suppose we discover a new positively charged particle, which we call the "whizatron." We want to determine its charge.
   (a) What modifications would we have to make to the Millikan oil-drop apparatus to carry out the corresponding experiment on whizatrons?
   (b) In such an experiment, we observe the following charges on five different droplets:

3.26×10^{-19} C
4.08×10^{-19} C
1.63×10^{-19} C
5.70×10^{-19} C
4.89×10^{-19} C

What is the charge on the whizatron?

9. What are alpha particles? Characterize them as to mass and charge.
10. (a) What do we mean when we refer to the nuclear atom?
    (b) Outline Rutherford's contribution to understanding the nature of atoms.
11. If the mass and electrical charge were uniformly distributed throughout an atom, what would be the expected results of an α-particle scattering experiment? What was the major conclusion drawn from the results of the α-particle scattering experiments?
12. The approximate radius of a hydrogen atom is 0.0529 nm, and that of a proton is 1.5×10^{-15} m. Assuming both the hydrogen atom and the proton to be spherical, calculate the fraction of the space in an atom of hydrogen that is occupied by the nucleus. $V = (4/3)\pi r^3$ for a sphere.
13. The approximate radius of a neutron is 1.5×10^{-15} m, and the mass is 1.675×10^{-27} kg. Calculate the density of a neutron. $V = (4/3)\pi r^3$ for a sphere.
14. Arrange the following in order of increasing ratio of charge to mass: $^{12}C^{+}$, $^{12}C^{2+}$, $^{13}C^{+}$, $^{13}C^{2+}$.
15. Refer to Exercise 14. Suppose all of these high-energy ions are present in a mass spectrometer. For which one will its path be changed (a) the most and (b) the least by increasing the external magnetic field?

Atom Composition, Isotopes, and Atomic Weights

16. Estimate the percentage of the total mass of a $^{195}_{78}Pt$ atom that is due to (a) electrons, (b) protons, and (c) neutrons by *assuming* that the mass of the atom is simply the sum of the masses of the appropriate numbers of subatomic particles.
17. (a) How are isotopic abundances determined experimentally?
    (b) How do the isotopes of a given element differ?
18. Define and illustrate the following terms clearly and concisely: (a) atomic number, (b) isotope, (c) mass number, (d) nuclear charge.
19. Write the composition of one atom of each of the three isotopes of silicon: ^{28}Si, ^{29}Si, and ^{30}Si.
20. Write the composition of one atom of each of the four isotopes of iron: ^{54}Fe, ^{56}Fe, ^{57}Fe, and ^{58}Fe.
21. Complete Chart A for neutral atoms.
22. Complete Chart B for neutral atoms.
23. The element iodine (I) occurs naturally as a single isotope of mass number 127; its atomic number is 53. How many protons and how many neutrons does it have in its nucleus?
24. On the average, a silver atom weighs 107.02 times as much as an atom of hydrogen. Using 1.0079 amu as the atomic weight of hydrogen, calculate the atomic weight of silver in atomic mass units.
25. Prior to 1962 the atomic weight scale was based on the assignment of an atomic weight of exactly 16 amu to the *naturally occurring* mixture of oxygen. The atomic weight of cadmium is 112.411 amu on the carbon-12 scale. What was it on the older scale?
26. What is the nuclear composition of each of the following isotopes of platinum, $_{78}Pt$—mass numbers 192, 194,

Chart A

Kind of Atom	Atomic Number	Mass Number	Isotope	Number of Protons	Number of Electrons	Number of Neutrons
			$^{44}_{20}Ca$			
potassium		39				
	16	32				
		174		70		

Chart B

Kind of Atom	Atomic Number	Mass Number	Isotope	Number of Protons	Number of Electrons	Number of Neutrons
nickel						32
			$^{191}_{77}Ir$			
					26	28
		195			78	

195, 196, 198? How many electrons are there in each of these atoms?

27. How many neutrons, protons, and electrons are in each of the following atoms or ions? $^{209}_{83}Bi^{3+}$, $^{193}_{77}Ir$, $^{51}_{23}V^{5+}$, $^{81}_{35}Br^{-}$, $^{98}_{42}Mo^{4+}$, $^{32}_{16}S^{2-}$.

28. The atomic weight of copper is 63.546 amu. The two naturally occurring isotopes of copper have the following masses: ^{63}Cu, 62.9298 amu; ^{65}Cu, 64.9278 amu. Calculate the percent of ^{63}Cu in naturally occurring copper.

29. The atomic weight of chlorine is 35.453 amu. There are only two isotopes in naturally occurring chlorine: ^{35}Cl (34.96885 amu) and ^{37}Cl (36.9658 amu). Calculate the percent composition of naturally occurring chlorine.

30. Determine the charge on each of the following species: (a) Ca with 18 electrons, (b) Cu with 28 electrons, (c) Cu with 26 electrons, (d) Pt with 74 electrons, (e) F with 10 electrons, (f) C with 6 electrons, (g) C with 7 electrons, (h) C with 5 electrons.

31. The following is a mass spectrum of the 1+ charged ions of an element. Calculate the atomic weight of the element. What is the element?

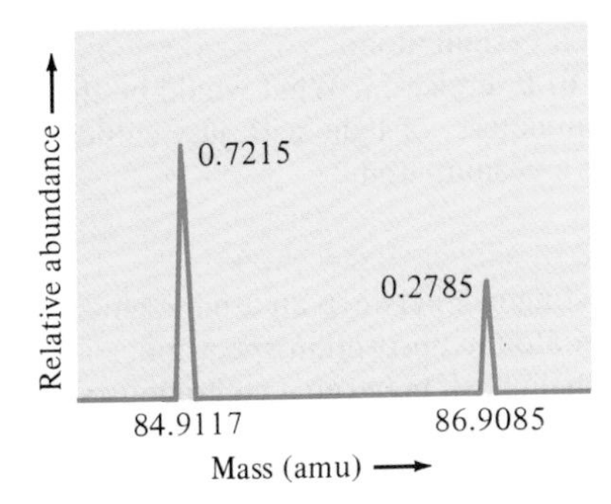

32. Suppose you measure the mass spectrum of the 1+ charged ions of germanium, atomic weight 72.61 amu. Unfortunately, the recorder on the mass spectrometer jams at the beginning and again at the end of your experiment. You obtain only the partial spectrum shown below, which *may or may not be complete*. From the information given here, can you tell whether one of the germanium isotopes is missing? If one is missing, at which end of the plot should it appear?

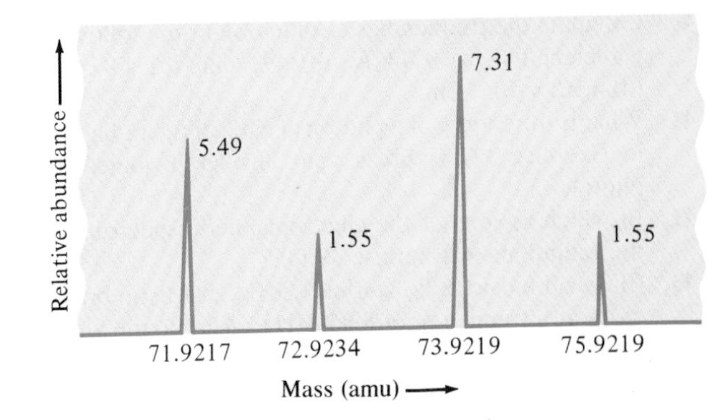

33. Calculate the atomic weight of zinc using the following percent of natural abundance and mass of each isotope: 48.6% $^{64}_{30}Zn$ (63.929 amu), 27.9% $^{66}_{30}Zn$ (65.9260 amu), 4.1% $^{67}_{30}Zn$ (66.9271 amu), and 18.8% $^{68}_{30}Zn$ (67.9298 amu).

34. Calculate the atomic weight of strontium using the following data for the percent of natural abundance and mass of each isotope: 0.5% of ^{84}Sr (83.9134 amu), 9.9% of ^{86}Sr (85.9094 amu), 7.0% of ^{87}Sr (86.9089 amu), and 82.6% of ^{88}Sr (87.9056 amu).

***35.** There are two naturally occurring isotopes of hydrogen (^{1}H, >99%, and ^{2}H, <1%) and two of chlorine (^{35}Cl, 76%, and ^{37}Cl, 24%).

(a) How many different HCl molecules can be formed from these isotopes?

(b) What is the approximate mass of each of the molecules, expressed in atomic mass units?

(c) List the molecules in order of decreasing relative abundance—as would be observed on a mass spectrometer—assuming only the formation of $(HCl)^+$ ions.

*36. In a suitable reference such as the Table of Isotopes in the *Handbook of Chemistry and Physics* (The Chemical Rubber Co.), look up the following information for selenium: (a) the total number of known isotopes, (b) the atomic mass, and (c) the percentage of natural abundance and mass of each of the stable isotopes. (d) Calculate the atomic weight of selenium.

37. Consider the ions ${}^{16}_{8}O^+$, ${}^{17}_{8}O^+$, ${}^{16}_{8}O^{2+}$, ${}^{17}_{8}O^{2+}$ produced in a mass spectrometer. Which ion's path would be deflected (a) most and (b) least by a magnetic field?

38. Consider the ions in Exercise 36. Which one would travel (a) most rapidly and (b) least rapidly under the influence of a particular accelerating voltage? Justify your answers.

Electromagnetic Radiation

39. Calculate the wavelengths, in meters, of radiation of the following frequencies:
(a) 4.80×10^{15} s^{-1}
(b) 1.18×10^{14} s^{-1}
(c) 5.44×10^{12} s^{-1}

40. Calculate the frequencies of radiation of the following wavelengths: (a) 9774 Å, (b) 492 nm, (c) 4.92 cm, (d) 4.92×10^{-9} cm.

41. What is the energy of a photon of each of the radiations in Exercise 39? Express your answer in joules per photon.

42. In which regions of the electromagnetic spectrum do the radiations in Exercise 39 fall?

43. Classical music radio station KMFA in Austin broadcasts at a frequency of 89.5 MHz. What is the wavelength of its signal in meters?

44. Excited lithium ions emit radiation at a wavelength of 670.8 nm in the visible range of the spectrum. (This characteristic color is often used as a qualitative analysis test for the presence of Li^+.) Calculate (a) the frequency and (b) the energy of a photon of this radiation. (c) What color is this light?

45. Find the energy of the photons corresponding to the red line, 6573 Å, in the spectrum of the Ca atom.

46. Ozone in the upper atmosphere absorbs ultraviolet radiation, which induces the following chemical reaction:

$$O_3(g) \longrightarrow O_2(g) + O(g)$$

What is the energy of a 3400-Å photon that is absorbed? What is the energy of a mole of these photons?

*47. During photosynthesis, chlorophyll-*a* absorbs light of wavelength 440 nm and emits light of wavelength 670 nm. What is the energy available for photosynthesis from the absorption–emission of a mole of photons?

*48. Assume that 10^{-17} J of light energy is needed by the interior of the human eye to "see" an object. How many photons of green light (wavelength = 550 nm) are needed to generate this minimum energy?

*49. The human eye receives a 2.500×10^{-14} J-signal consisting of photons of blue light, λ = 4700 Å. How many photons reach the eye?

*50. Water absorbs microwave radiation of wavelength 3 mm. How many photons are needed to raise the temperature of a cup of water (250 g) from 25°C to 85°C in a microwave oven, using this radiation? The specific heat of water is 4.184 J/g · °C.

The Photoelectric Effect

51. What evidence supports the idea that electromagnetic radiation is (a) wave-like; (b) particle-like?

52. Describe the influence of frequency and intensity of electromagnetic radiation on the current in the photoelectric effect.

*53. Cesium is often used in "electric eyes" for self-opening doors in an application of the photoelectric effect. The amount of energy required to ionize (remove an electron from) a cesium atom is 3.89 electron volts (1 eV = 1.60×10^{-19} J). Show by calculation whether a beam of yellow light with wavelength 5230 Å would ionize a cesium atom.

*54. Refer to Exercise 53. What would be the wavelength, in nanometers, of light with just sufficient energy to ionize a cesium atom?

Atomic Spectra and the Bohr Theory

55. (a) Distinguish between an atomic emission spectrum and an atomic absorption spectrum.
(b) Distinguish between a continuous spectrum and a line spectrum.

56. Prepare a sketch similar to Figure 5-16b that shows a ground energy state and three excited energy states. Using vertical arrows, indicate the transitions that would correspond to the absorption spectrum for this system.

57. What is the Rydberg equation? Why is it called an empirical equation?

58. Hydrogen atoms absorb energy so that the electrons are excited to the energy level $n = 7$. Electrons then undergo these transitions: (1) $n = 7 \rightarrow n = 1$, (2) $n = 7 \rightarrow n = 6$, (3) $n = 2 \rightarrow n = 1$. Which transition will produce the photon with (a) the smallest energy; (b) the highest frequency; (c) the shortest wavelength? (d) What is the frequency of a photon resulting from the transition $n = 6 \rightarrow n = 1$?

*59. Five energy levels of the He atom are given in J/atom above an *arbitrary* reference energy: (1) 6.000×10^{-19}, (2) 8.812×10^{-19}, (3) 9.381×10^{-19}, (4) 10.443×10^{-19}, (5) 10.934×10^{-19}. Construct an energy-level diagram for He and find the energy of the photon (a) absorbed for the electron transition from level 1 to level 4 and (b) emitted for the electron transition from level 5 to level 2.

*60. The *Lyman series* is the name given to the series of lines in the ultraviolet portion of the emission spectrum of hydrogen. These lines correspond to transitions of electrons from higher energy states to the lowest ($n_1 = 1$) energy state. Use the Rydberg equation to calculate the wavelengths, in nm, of the three lowest energy lines in the Lyman series.

*61. The Balmer series of lines in the hydrogen emission spectrum is in the visible to ultraviolet range. These lines correspond to transitions of electrons from a higher energy state to the second-lowest ($n_1 = 2$) energy state. One line of the Balmer series has a wavelength of 410.2 nm. What is the quantum number of the upper energy state?

62. The following are prominent lines in the visible region of the emission spectra of the elements listed. The lines can be used to identify the elements. What color is the light responsible for each line? (a) lithium, 6708 Å; (b) neon, 616.0 nm; (c) mercury, 4540 Å; (d) cesium, $\nu = 3.45 \times 10^{14}$ Hz; (e) potassium, $\nu = 3.90 \times 10^{14}$ Hz.

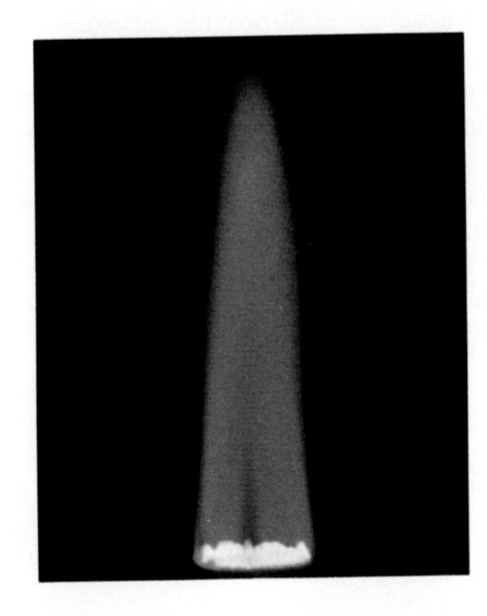

63. Hydrogen atoms have an absorption line at 973 Å. What is the frequency of the photons absorbed, and what is the energy difference, in joules, between the ground state and this excited state of the atom?

*64. If each atom in one mole of atoms emits a photon of wavelength 6.24×10^3 Å, how much energy is lost? Express the answer in kJ/mol. As a reference point, burning one mole (16 g) of CH_4 produces 819 kJ of heat.

*65. Suppose we could excite all of the electrons in a sample of hydrogen atoms to the $n = 5$ level. They would then emit light as they relaxed to lower energy states. Some atoms might undergo the transition $n = 5$ to $n = 1$, while others might go from $n = 5$ to $n = 4$, then from $n = 4$ to $n = 3$, and so on. How many lines would we expect to observe in the resulting emission spectrum?

*66. An argon laser emits blue light with a wavelength of 488.0 nm. How many photons are emitted by this laser in 2.00 seconds, operating at a power of 515 milliwatts? One watt (a unit of power) is equal to 1 joule/second.

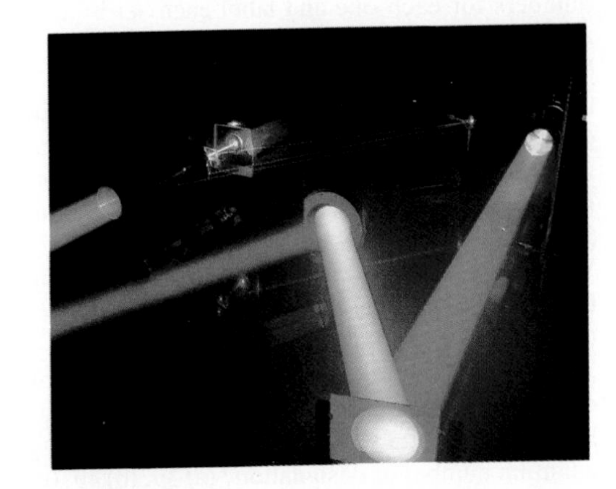

The Wave–Particle View of Matter

67. (a) What evidence supports the idea that electrons are particle-like?
(b) What evidence supports the idea that electrons are wave-like?

68. (a) What is the de Broglie wavelength of a proton moving at a speed of 3.00×10^7 m/s? The proton mass is 1.67×10^{-24} g.
(b) What is the de Broglie wavelength of a stone with a mass of 30.0 g moving at 2.00×10^3 m/h ($\approx$ 100 mi/h)?
(c) How do the wavelengths in (a) and (b) compare with the typical radii of atoms? (See the atomic radii in Figure 6-2.)

69. What is the wavelength corresponding to a neutron of mass 1.67×10^{-27} kg moving at 2200 m/s?

*70. The energy of a photon in the X-ray region of the spectrum is 7×10^{-16} J. According to de Broglie's equation, what is the mass of this photon?

Quantum Numbers and Atomic Orbitals

71. (a) What is a quantum number? What is an atomic orbital?
(b) How many quantum numbers are required to specify a single atomic orbital? What are they?

72. How are the possible values for the subsidiary quantum number for a given electron restricted by the value of n?

73. Without giving the ranges of possible values of the four quantum numbers, n, ℓ, m_ℓ and m_s, describe briefly what information each one gives.

74. (a) How are the values of m_ℓ for a particular electron restricted by the value of ℓ?
(b) What are the letter designations for the values $n = 1, 2, 3, 4$?
(c) What are the letter designations for the values $\ell = 0, 1, 2, 3$?

75. What are the values of n and ℓ for the following sublevels? (a) $2s$, (b) $3d$, (c) $4p$, (d) $5s$, (e) $4f$.

76. How many individual orbitals are there in the third major energy level? Write out n, ℓ and m_ℓ quantum numbers for each one and label each set by the s, p, d, f designations.

77. (a) Write the possible values of ℓ when $n = 5$.
(b) Write the allowed number of orbitals (1) with the quantum numbers $n = 4$, $\ell = 3$; (2) with the quantum number $n = 4$; (3) with the quantum numbers $n = 7$, $\ell = 6$, $m_\ell = 6$; (4) with quantum numbers $n = 6$, $\ell = 5$.

78. Write the subshell notations that correspond to (a) $n = 2$, $\ell = 0$; (b) $n = 4$, $\ell = 2$; (c) $n = 7$, $\ell = 0$; (d) $n = 5$, $\ell = 3$.

79. What values can m_ℓ take for (a) a $4d$ orbital, (b) a $1s$ orbital, and (c) a $3p$ orbital?

80. How many orbitals in any atom can have the given quantum number or designation? (a) $3p$, (b) $4p$, (c) $4p_x$, (d) $n = 5$, (e) $6d$, (f) $5d$, (g) $5f$, (h) $7s$.

81. The following incorrect sets of quantum numbers in the order n, ℓ, m_ℓ, m_s are written for paired electrons or for one electron in an orbital. Correct them, assuming n values are correct. (a) 1, 0, 0, $+\frac{1}{2}$, $+\frac{1}{2}$; (b) 2, 2, 1, $\pm\frac{1}{2}$; (c) 3, 2, 3, $\pm\frac{1}{2}$; (d) 3, 1, 2, $+\frac{1}{2}$; (e) 2, 1, -1, 0; (f) 3, 0, -1, $-\frac{1}{2}$.

82. (a) What do we mean when we refer to the "spin" of an electron?
(b) What are spin-paired electrons?

83. In an atom, how many electrons could have principal quantum number $n = 5$?

84. (a) How are a $1s$ orbital and a $2s$ orbital in an atom similar? How do they differ?
(b) How are a $2p_x$ orbital and a $2p_y$ orbital in an atom similar? How do they differ?

Electron Configurations and the Periodic Table

You should be able to use the positions of elements in the periodic table to answer the remaining exercises.

85. Draw representations of ground state electron configurations using the orbital notation ($\underline{\uparrow\downarrow}$) for the following elements. (a) C, (b) Fe, (c) P, (d) Rh.

86. Draw representations of ground state electron configurations using the orbital notation ($\underline{\uparrow\downarrow}$) for the following elements. (a) S, (b) Ni, (c) Mg, (d) Zr.

87. Give the ground state electron configurations for the elements of Exercise 85 using shorthand notation—that is, $1s^22s^22p^6$, and so on.

88. Give the ground state electron configurations for the elements of Exercise 86 using shorthand notation—that is, $1s^22s^22p^6$, and so on.

89. State the Pauli Exclusion Principle. Would any of the following electron configurations violate this rule: (a) $1s^2$, (b) $1s^22p^1$, (c) $1s^3$? Explain.

90. State Hund's Rule. Would any of the following electron configurations violate this rule: (a) $1s^2$, (b) $1s^22s^22p_x^2$, (c) $1s^22s^22p_x^12p_y^1$, (d) $1s^22s^22p_x^12p_z^1$, (e) $1s^22s^22p_x^22p_y^12p_z^1$? Explain.

***91.** Classify each of the following atomic electron configurations as (i) a ground state, (ii) an excited state, or (iii) a forbidden state: (a) $1s^22s^22p^53s^1$, (b) [Kr] $4d^{10}5s^3$, (c) $1s^22s^22p^63s^23p^63d^84s^2$, (d) $1s^22s^22p^63s^23p^63d^1$, (e) $1s^22s^22p^{10}3s^23p^5$.

92. Which elements are represented by the following electron configurations?
(a) $1s^22s^22p^63s^23p^63d^{10}4s^24p^3$
(b) [Kr] $4d^{10}4f^{14}5s^25p^65d^{10}5f^{14}6s^26p^66d^27s^2$
(c) [Kr] $4d^{10}4f^{14}5s^25p^65d^{10}6s^26p^4$
(d) [Kr] $4d^55s^2$
(e) $1s^22s^22p^63s^23p^63d^34s^2$

93. Repeat Exercise 92 for
(a) $1s^22s^22p^63s^23p^63d^54s^1$
(b) [Kr] $4d^{10}4f^{14}5s^25p^65d^{10}6s^26p^1$
(c) $1s^22s^22p^63s^23p^6$
(d) [Kr] $4d^{10}4f^{14}5s^25p^65d^{10}6s^26p^67s^2$

94. Find the total number of s, p, and d electrons in each of the following: (a) Si, (b) Ar, (c) Ni, (d) Zn, (e) Rb.

95. (a) Distinguish between the terms "diamagnetic" and "paramagnetic," and provide an example that illustrates the meaning of each.
(b) How is paramagnetism measured experimentally?

96. How many unpaired electrons are in atoms of Na, Ne, B, Be, Se, and Ti?

97. Which of the following ions or atoms possess paramagnetic properties? (a) F^-, (b) Na^+, (c) Co, (d) Ar^-, (e) S.

98. Which of the following ions or atoms possess paramagnetic properties? (a) F, (b) Ar, (c) Ar^+, (d) Zn, (e) S^{2-}.

99. Write the electron configurations of the Group IA elements Li, Na, and K (see inside front cover). What similarity do you observe?

100. Construct a table in which you list a possible set of values for the four quantum numbers for each electron in the following atoms in their ground states. (a) N, (b) P, (c) Cu.

101. Construct a table in which you list a possible set of values for the four quantum numbers for each electron in the following atoms in their ground states. (a) B, (b) Cl, (c) Ni.

102. Draw general electronic structures for the A group elements using the ⇅ notation, where n is the principal quantum number for the highest occupied energy level.

	ns	*np*
IA	_	_ _ _
IIA	_	_ _ _
and so on		

103. Repeat Exercise 102 using $ns^x np^y$ notation.

104. List n, ℓ, and m_ℓ quantum numbers for the highest-energy electron (or one of the highest-energy electrons if there are more than one) in the following atoms in their ground states (a) Si, (b) Nb, (c) Br, (d) Pr.

105. List n, ℓ, and m_ℓ quantum numbers for the highest-energy electron (or one of the highest-energy electrons if there are more than one) in the following atoms in their ground states. (a) As, (b) Ag, (c) Mg, (d) Pu.

106. Write the ground state electron configurations for elements A–E.

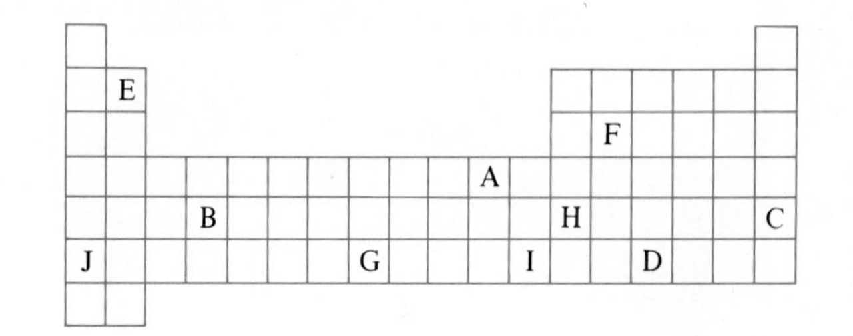

107. Repeat Exercise 105 for elements F–J.

Chemical Periodicity

Outline

Objectives

As you study this chapter, you should learn

- ☐ More about the periodic table (it is so useful)
- ☐ About chemical periodicity in physical properties:
  - Atomic radii
  - Ionization energy
  - Electron affinity
  - Ionic radii
  - Electronegativity
- ☐ About chemical periodicity in the reactions of
  - Hydrogen
  - Oxygen
- ☐ About chemical periodicity in the compounds of
  - Hydrogen
  - Oxygen

The properties of elements are correlated with their positions in the periodic table. Chemists use the periodic table as an invaluable guide in their search for new, useful materials. A barium sodium niobate crystal can convert infrared laser light into visible light. This harmonic generation or "frequency doubling" is very important in chemical research using lasers and in the telecommunications industry.

6-1 More about the Periodic Table

In Chapter 4 we described the development of the periodic table, some terminology for it, and its guiding principle, the *periodic law*.

> The properties of the elements are periodic functions of their atomic numbers.

In Chapter 5 we described electronic configurations of the elements. In the long form of the periodic table, elements are arranged in blocks based on the kinds of atomic orbitals being filled. (You may wish to review Table 5-5 and Figure 5-31 carefully.) We saw that electronic configurations of elements in the A groups and in Group 0 are entirely predictable from their positions in the periodic table. There are some irregularities among the transition elements.

Now we classify the elements according to their electronic configurations, which is a very useful system.

Noble Gases For many years the Group 0 elements—the noble gases—were called inert gases because no chemical reactions were known for them. We

[He] = $1s^2$

now know that the heavier members do form compounds, mostly with fluorine and oxygen. Except for helium, each of these elements has eight electrons in its highest occupied energy level. Their structures may be represented as . . . ns^2np^6.

Representative Elements The A group elements in the periodic table are called representative elements. They have partially occupied highest energy levels. Their "last" electron was added to an *s* or *p* orbital. These elements show distinct and fairly regular variations in their properties with changes in atomic number.

***d*-Transition Elements** Elements in the B groups (except IIB) in the periodic table are known as the *d*-transition elements or, more simply, as transition elements or transition metals. They were considered to be transitions between the alkaline elements (base-formers) on the left and the acid-formers on the right. All are metals and are characterized by electrons being added to *d* orbitals. Stated differently, the *d*-transition elements are building an inner (next to highest occupied) energy level from 8 to 18 electrons. They are referred to as

First Transition Series:	$_{21}Sc$ through $_{29}Cu$
Second Transition Series:	$_{39}Y$ through $_{47}Ag$
Third Transition Series:	$_{57}La$ and $_{72}Hf$ through $_{79}Au$
Fourth Transition Series:	(not complete) $_{89}Ac$ and elements 104 through 111

Strictly speaking, the Group IIB elements—zinc, cadmium, and mercury—are not *d*-transition metals because their "last" electrons go into *s* orbitals. They are usually discussed with the *d*-transition metals because their chemical properties are similar.

Inner Transition Elements Sometimes known as *f-transition elements,* these are elements in which electrons are being added to *f* orbitals. In these

Some transition metals (left to right): Ti, V, Cr, Mn, Fe, Co, Ni, Cu.

The elements of Period 3. Properties progress (left to right) from solids (Na, Mg, Al, Si, P, S) to gases (Cl, Ar) and from the most metallic (Na) to the most nonmetallic (Ar).

elements, the second from the highest occupied energy level is building from 18 to 32 electrons. All are metals. The inner transition elements are located between Groups IIIB and IVB in the periodic table. They are

First Inner Transition Series (lanthanides): $_{58}Ce$ through $_{71}Lu$
Second Inner Transition Series (actinides): $_{90}Th$ through $_{103}Lr$

The A and B designations for groups of elements in the periodic table are arbitrary, and they are reversed in some periodic tables. In another designation, the groups are numbered 1 through 18. The system used in this text is the one commonly used in the United States. Elements with the same group numbers, but with different letters, have relatively few similar properties. The origin of the A and B designations is the fact that some compounds of elements with the same group numbers have similar formulas but quite different properties, e.g., NaCl (IA) and AgCl (IB), $MgCl_2$ (IIA) and $ZnCl_2$ (IIB). As we shall see, variations in the properties of the B groups across a row are not nearly as dramatic as the variations observed across a row of A group elements. In the B groups, electrons are being added to $(n - 1)d$ orbitals, where n represents the highest energy level that contains electrons.

The *outermost* electrons have the greatest influence on the properties of elements. Adding an electron to an *inner d* orbital results in less striking changes in properties than adding an electron to an *outer s* or *p* orbital.

Periodic Properties of the Elements

We shall now investigate the nature of periodicity in some detail. Some knowledge of periodicity is valuable in understanding bonding in simple compounds. Many physical properties, such as melting points, boiling points, and atomic volumes, show periodic variations. For now, we describe the variations that are most useful in predicting chemical properties. The variations in these properties depend on electronic configurations, especially the configurations in the outermost occupied shell, and on how far away that shell is from the nucleus.

6-2 Atomic Radii

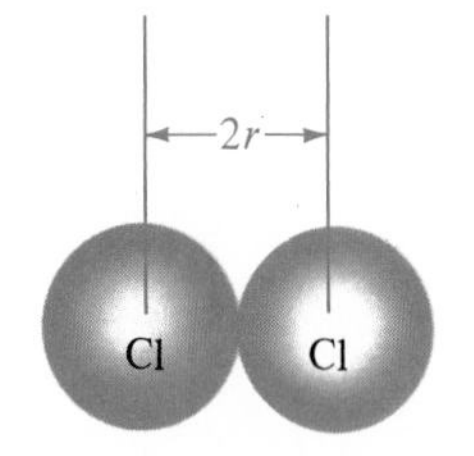

The radius of an atom, *r*, is taken as half of the distance between nuclei in *homonuclear* molecules such as Cl_2.

In Section 5-15 we described atomic orbitals in terms of probabilities of distributions of electrons over certain regions in space. We can visualize the electron cloud that surrounds an atomic nucleus in a similar way, i.e., as somewhat indefinite. Further, we cannot isolate a single atom and measure its diameter the way we can measure the diameter of a golf ball. For all practical purposes, the size of an individual atom cannot be uniquely defined. An indirect approach is required. The size of an atom is determined by its immediate environment, especially its interaction with surrounding atoms. By analogy, suppose we arrange some golf balls in an orderly array in a box. If we know how the balls are positioned, the number of balls, and the dimensions of the box, we can calculate the diameter of an individual ball. Application of this reasoning to solids and their densities leads us to values for the atomic sizes of many elements. In other cases, we derive atomic radii from the observed distances between atoms that are combined with each other. For example, the distance between atomic centers (nuclei) in the Cl_2 molecule is measured to be 1.98 Å. We take the radius of *each* Cl atom to be half the interatomic distance, or 0.99 Å. We collect the data obtained from many such measurements to indicate the *relative* sizes of individual atoms.

The top of Figure 6-1 displays the relative sizes of atoms of the representative elements and the noble gases. It shows the periodicity in atomic radii. (The ionic radii at the bottom of Figure 6-1 are discussed in Section 6-5.)

The **effective nuclear charge**, Z_{eff}, experienced by an electron in an outer energy level is less than the actual nuclear charge, Z. This is because the *attraction* of outer shell electrons by the nucleus is partly counteracted by the *repulsion* of these outer shell electrons by electrons in filled inner shells. We might say that the electrons in filled sets of energy levels *screen* or *shield* electrons in outer energy levels from the full effect of the nuclear charge. This **screening**, or **shielding**, effect helps us to understand many periodic trends in atomic properties.

Consider the Group IA metals, Li–Cs. Lithium, element number 3, has two electrons in a filled energy level, $1s^2$, and one electron in the $2s$ orbital, $2s^1$. The electron in the $2s$ orbital is fairly effectively screened from the nucleus by the two electrons in the filled $1s$ orbital, the He configuration. The electron in the $2s$ orbital "feels" an effective nuclear charge of about 1+, rather than the full nuclear charge of 3+. Sodium, element number 11, has ten electrons in filled sets of orbitals, $1s^22s^22p^6$, the Ne configuration. These ten electrons in a noble gas configuration fairly effectively screen (shield) its outer electron ($3s^1$) from the nucleus. Thus, the $3s$ electron in sodium also "feels" an effective nuclear charge of about 1+ rather than 11+. A sodium atom is larger than a lithium atom because (1) the effective nuclear charge is *approximately* the same for both atoms, and (2) the "outer" electron in a sodium atom is in the third energy level. A similar argument explains why potassium atoms are larger than sodium atoms.

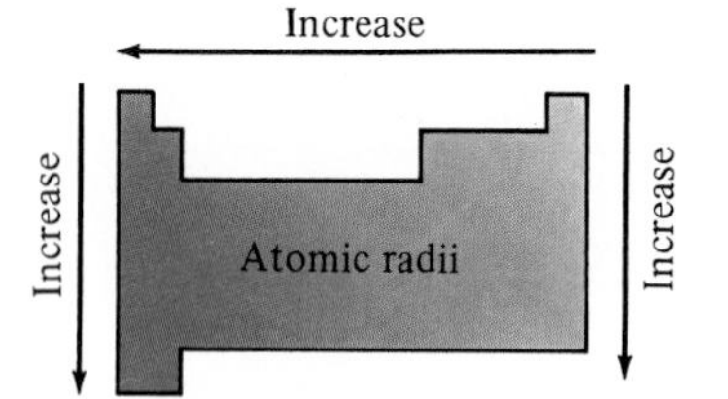

General trends in atomic radii of A group elements with position in the periodic table

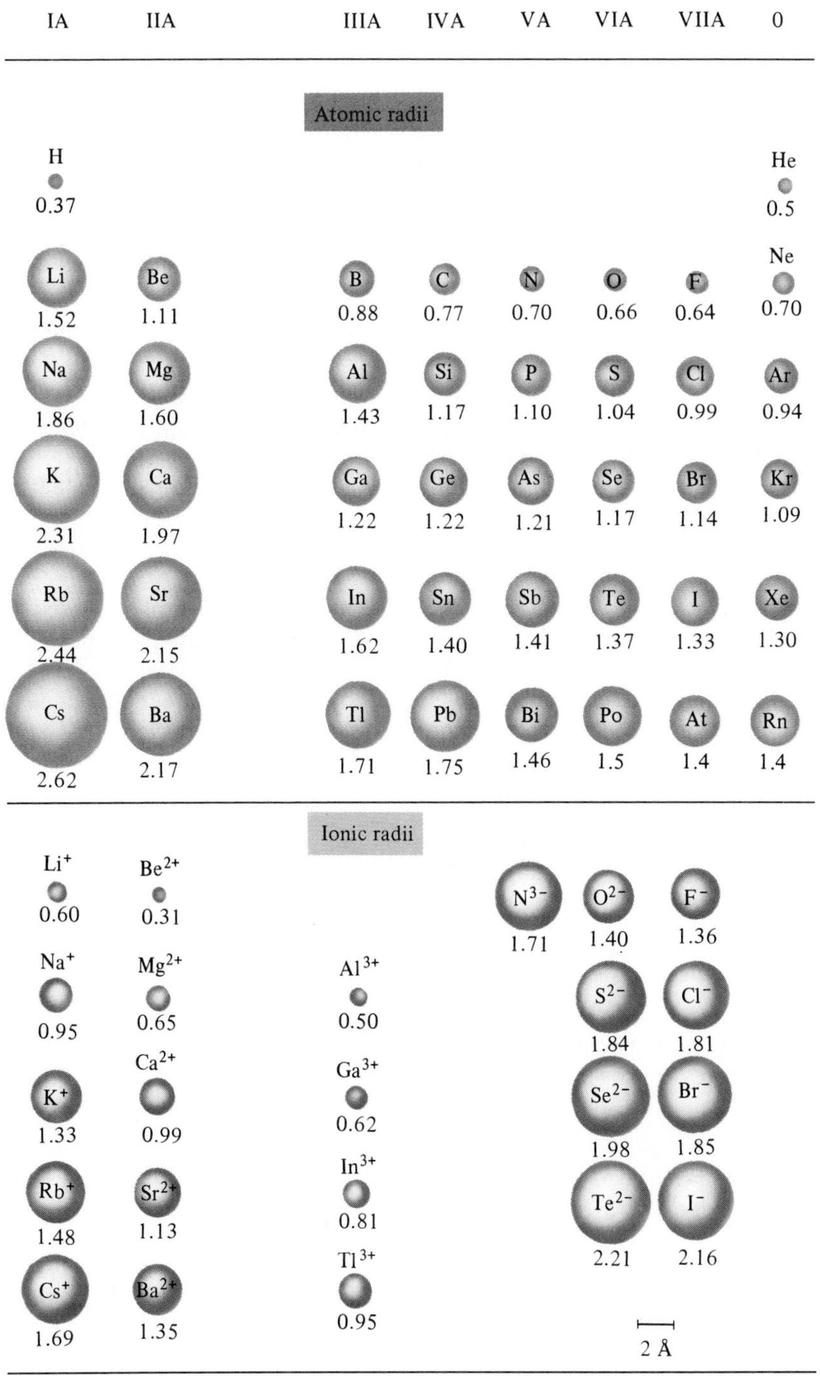

IA	IIA	IIIA	IVA	VA	VIA	VIIA	0
Atomic radii							
H 0.37							He 0.5
Li 1.52	Be 1.11	B 0.88	C 0.77	N 0.70	O 0.66	F 0.64	Ne 0.70
Na 1.86	Mg 1.60	Al 1.43	Si 1.17	P 1.10	S 1.04	Cl 0.99	Ar 0.94
K 2.31	Ca 1.97	Ga 1.22	Ge 1.22	As 1.21	Se 1.17	Br 1.14	Kr 1.09
Rb 2.44	Sr 2.15	In 1.62	Sn 1.40	Sb 1.41	Te 1.37	I 1.33	Xe 1.30
Cs 2.62	Ba 2.17	Tl 1.71	Pb 1.75	Bi 1.46	Po 1.5	At 1.4	Rn 1.4
Ionic radii							
Li^+ 0.60	Be^{2+} 0.31			N^{3-} 1.71	O^{2-} 1.40	F^- 1.36	
Na^+ 0.95	Mg^{2+} 0.65	Al^{3+} 0.50			S^{2-} 1.84	Cl^- 1.81	
K^+ 1.33	Ca^{2+} 0.99	Ga^{3+} 0.62			Se^{2-} 1.98	Br^- 1.85	
Rb^+ 1.48	Sr^{2+} 1.13	In^{3+} 0.81			Te^{2-} 2.21	I^- 2.16	
Cs^+ 1.69	Ba^{2+} 1.35	Tl^{3+} 0.95					

2 Å

Figure 6-1
(Top) Atomic radii of the A group (representative) elements and the noble gases, in angstroms. Atomic radii *increase as a group is descended* because electrons are being added to shells farther from the nucleus. Atomic radii *decrease from left to right within a given row* owing to increasing effective nuclear charge. Hydrogen atoms are the smallest and cesium atoms are the largest naturally occurring atoms. 1 Å = 1×10^{-10} m.

(Bottom) Sizes of ions of the A group elements, in angstroms. Positive ions (cations) are always *smaller* than the neutral atoms from which they are formed. Negative ions (anions) are always *larger* than the neutral atoms from which they are formed.

Atomic radii are often stated in **angstroms** (1 Å = 10^{-10} m) or in the SI units **nanometers** (1 nm = 10^{-9} m) or **picometers** (1 pm = 10^{-12} m). To convert from Å to nm, move the decimal point to the left one place (1 Å = 0.1 nm). For example, the atomic radius of Li is 1.52 Å, or 0.152 nm.

Within a group of representative elements, atomic radii *increase* from top to bottom as electrons are added to higher energy levels.

As we move *across* the periodic table, atoms become smaller due to increasing effective nuclear charges. Consider the elements B ($Z = 5$, $1s^22s^22p^1$) to F ($Z = 9$, $1s^22s^22p^5$). In B there are two electrons in a noble gas configuration, $1s^2$, and three electrons in the second energy level, $2s^22p^1$. The two electrons in the noble gas configuration fairly effectively screen out the effect of two protons in the nucleus. So the three electrons in the second energy level of B "feel" an effective nuclear charge of approximately 3+. By similar arguments, we see that in carbon ($Z = 6$, $1s^22s^22p^2$) the four electrons in the second energy level "feel" an effective nuclear charge of approximately 4+. So we expect C atoms to be smaller than B atoms, and they are. In nitrogen ($Z = 7$, $1s^22s^22p^3$) the five electrons in the second energy level "feel" an effective nuclear charge of approximately 5+, and so N atoms are smaller than C atoms.

As we move from left to right *across a period* in the periodic table, atomic radii of representative elements *decrease* as a proton is added to the nucleus and an electron is added to a particular energy level.

For the transition elements, the variations are not so regular because electrons are being added to an inner shell. All transition elements have smaller radii than the preceding Group IA and IIA elements in the same period.

Example 6-1

Arrange the following elements in order of increasing atomic radii. Justify your order.

Cs, F, Li, Cl

Plan

Both Li and Cs are Group IA metals, while F and Cl are halogens (VIIA nonmetals). Figure 6-1 shows that atomic radii increase as a group is descended, so Li < Cs and F < Cl. Atomic radii decrease from left to right.

Solution

The order of increasing atomic radii is

F < Cl < Li < Cs

EOC 19, 20

6-3 Ionization Energy

The **first ionization energy (IE_1)**, also called **first ionization potential**, is

the minimum amount of energy required to remove the most loosely bound electron from an isolated gaseous atom to form an ion with a 1+ charge.

Table 6-1
First Ionization Energies (kJ/mol of atoms) of Some Elements

H 1312																	He 2372
Li 520	Be 899											B 801	C 1086	N 1402	O 1314	F 1681	Ne 2081
Na 497	Mg 738											Al 578	Si 786	P 1012	S 1000	Cl 1251	Ar 1521
K 419	Ca 590	Sc 631	Ti 658	V 650	Cr 653	Mn 717	Fe 759	Co 758	Ni 737	Cu 745	Zn 906	Ga 579	Ge 762	As 947	Se 941	Br 1140	Kr 1351
Rb 403	Sr 549	Y 616	Zr 660	Nb 664	Mo 685	Tc 702	Ru 711	Rh 720	Pd 805	Ag 731	Cd 868	In 558	Sn 709	Sb 834	Te 869	I 1008	Xe 1170
Cs 376	Ba 503	La 538	Hf 675	Ta 761	W 770	Re 760	Os 840	Ir 878	Pt 870	Au 890	Hg 1007	Tl 589	Pb 716	Bi 703	Po 812	At 920	Rn 1037

For calcium, for example, the first ionization energy, IE_1, is 590 kJ/mol:

$$Ca(g) + 590\ kJ \longrightarrow Ca^+(g) + e^-$$

The **second ionization energy (IE_2)** is the amount of energy required to remove the second electron. For calcium, it may be represented as

$$Ca^+(g) + 1145\ kJ \longrightarrow Ca^{2+}(g) + e^-$$

For a given element, *IE_2 is always greater than IE_1* because it is always more difficult to remove an electron from a positively charged ion than from the corresponding neutral atom. Table 6-1 gives first ionization energies.

Ionization energies measure how tightly electrons are bound to atoms. Ionization always requires energy to remove an electron from the attractive force of the nucleus. Low ionization energies indicate ease of removal of electrons, and hence ease of positive ion (cation) formation. Figure 6-2 shows a plot of first ionization energy versus atomic number for several elements.

We see that in each period of Figure 6-2, the noble gases have the highest first ionization energies. This should not be surprising, because the noble

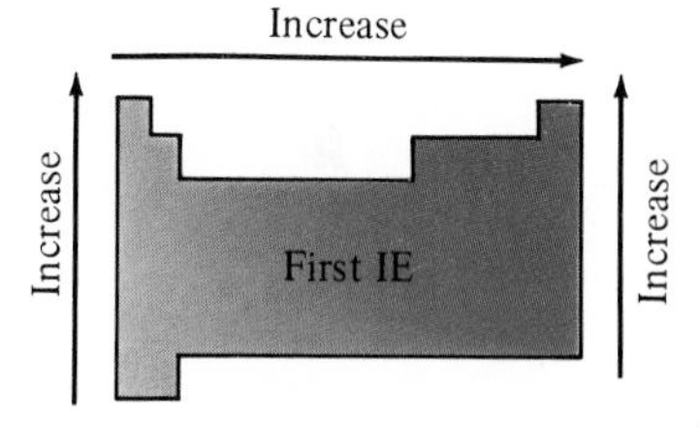

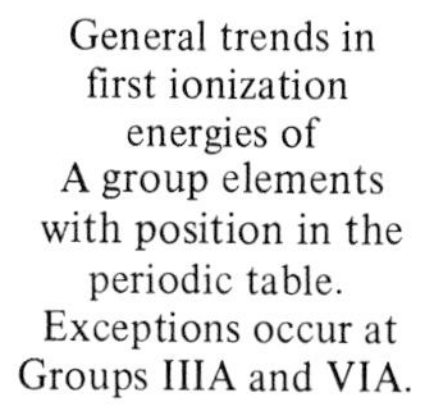

General trends in first ionization energies of A group elements with position in the periodic table. Exceptions occur at Groups IIIA and VIA.

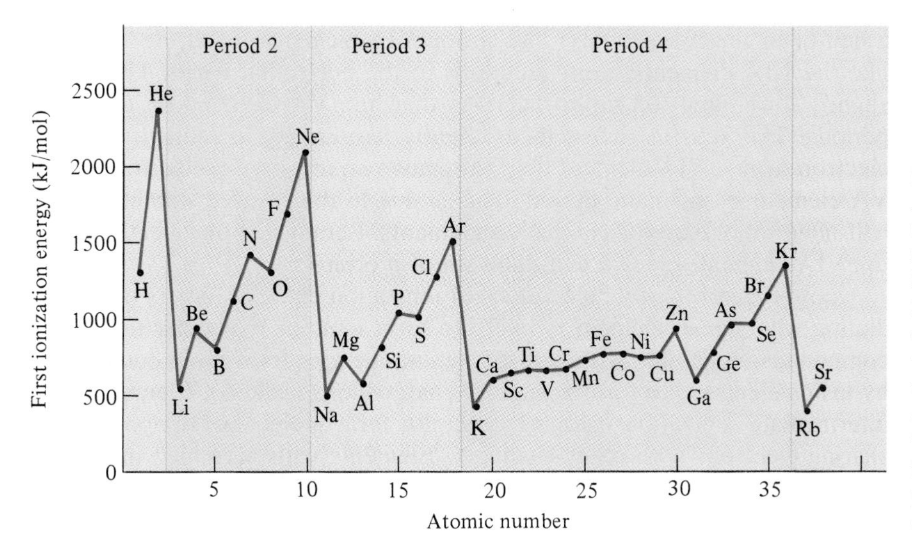

Figure 6-2
A plot of first ionization energies for the first 38 elements versus atomic number. The noble gases have very high first ionization energies, and the IA metals have low first ionization energies. Note the similarities in the variations for the Period 2 elements, 3 through 10, to those for the Period 3 elements, 11 through 18, as well as for the later A group elements. Variations for B group elements are not nearly so pronounced as those for A group elements.

gases are known to be very unreactive elements. It requires more energy to remove an electron from a helium atom (slightly less than 4.0×10^{-18} J/atom or 2372 kJ/mol) than to remove one from a neutral atom of any other element:

$$\text{He(g)} + 2372 \text{ kJ} \longrightarrow \text{He}^+\text{(g)} + e^-$$

The Group IA metals (Li, Na, K, Rb, Cs) have very low first ionization energies. Each of these elements has only one electron in its highest energy level (... ns^1), and they are the largest atoms in their periods. The first electron added to a principal energy level is easily removed to form a noble gas configuration. As we move down the group, the first ionization energies become smaller. The force of attraction of the positively charged nucleus for electrons decreases as the square of the separation between them increases. So as atomic radii increase in a given group, first ionization energies decrease because the valence electrons are farther from the nucleus. The shielding effect of the electrons in filled inner shells, and the decreased Z_{eff} as the group is descended, further weaken the attraction for the outer shell electrons.

By Coulomb's Law, $F \propto \frac{(q^+)(q^-)}{d^2}$, the attraction for the outer shell electrons is directly proportional to the *effective* charges and inversely proportional to the square of the distance between the charges.

The first ionization energies of the Group IIA elements (Be, Mg, Ca, Sr, Ba) are significantly higher than those of the Group IA elements in the same periods. This is because the Group IIA elements have smaller atomic radii and higher Z_{eff} values. Thus, their valence electrons are held more tightly than those of the neighboring IA metals. It is harder to remove an electron from a pair in the filled outermost *s* orbitals of the Group IIA elements than to remove the single electron from the half-filled outermost *s* orbitals of the Group IA elements.

The first ionization energies for the Group IIIA elements (B, Al, Ga, In, Tl) are exceptions to the general horizontal trends. They are *lower* than those of the IIA elements in the same periods because the IIIA elements have only a single electron in their outermost *p* orbitals. It requires less energy to remove the first *p* than to remove the second *s* electron from the same principal energy level because an *ns* orbital is lower in energy (more stable) than an *np* orbital.

The second peak for each period in the ionization energy curve occurs at the Group VA elements (N, P, As, Sb, Bi). These elements have three unpaired electrons in the three outermost *p* orbitals, that is, $ns \, \underline{\uparrow\downarrow} \; np \, \underline{\uparrow} \, \underline{\uparrow} \, \underline{\uparrow}$, a half-filled set of *p* orbitals. The Group VIA elements (O, S, Se, Te, Po), like the IIIA elements, are exceptions to the horizontal trend. They have slightly *lower* first ionization energies than the VA elements in the same periods. This tells us that it takes slightly less energy to remove a paired electron from a VIA element than to remove an unpaired *p* electron from a VA element in the same period. This is due to the relative stability of the half-filled set of *p* orbitals in the VA elements. Removal of one electron from the VIA elements gives a half-filled set of *p* orbitals.

We have seen other consequences of the special stability of half-filled sets of equivalent orbitals, i.e., exceptions to the Aufbau Principle (Section 5-16).

Here is one reason why trends in ionization energies are important.

Knowledge of the relative values of ionization energies assists us in predicting whether an element is likely to form ionic or molecular (covalent) compounds. Elements with low ionization energies form ionic compounds by losing electrons to form positively charged ions (**cations**). Elements with intermediate ionization energies generally form molecular compounds by sharing electrons with other elements. Elements with very high ionization energies, e.g., Groups VIA and VIIA, often gain electrons to form negatively charged ions (**anions**).

One factor that favors an atom of a *representative* element forming a monatomic ion in a compound is the formation of a stable noble gas configuration. Energy considerations are consistent with this observation. For example, as one mole of Li from Group IA forms one mole of Li^+ ions, it absorbs 520 kJ per mole of Li atoms. The IE_2 value is 14 times greater, 7298 kJ/mol, and is prohibitively large for the formation of Li^{2+} ions under ordinary conditions. For Li^{2+} ions to form, an electron would have to be removed from the filled first energy level. We recognize that this is unlikely. The other alkali metals behave in the same way, for the same reason.

Noble gas configurations are stable only for ions in *compounds*. In fact, $Li^+(g)$ is less stable than $Li(g)$ by 520 kJ/mol.

Likewise, the first two ionization energies of Be are 899 and 1757 kJ/mol, but IE_3 is more than eight times larger, 14,849 kJ/mol. So Be forms Be^{2+} ions, but not Be^{3+} ions. The other alkaline earth metals—Mg, Ca, Sr, Ba, and Ra—behave in a similar way. Owing to the high energy required, *simple monatomic cations with charges greater than 3+ do not form under ordinary circumstances*. Only the lower members of Group IIIA, beginning with Al, form 3+ ions. Bi and some *d*- and *f*-transition metals do so, too. We see that the magnitudes of successive ionization energies support the ideas of electronic configurations discussed in Chapter 5.

Example 6-2

Arrange the following in order of increasing first ionization energy. Justify your order.

Na, Mg, Al, Si

Plan

Table 6-1 shows that first ionization energies generally increase from left to right in the periodic table, but there are exceptions at Groups IIIA and VIA. Al is a IIIA element with only one electron in its outer *p* orbitals, $1s^2 2s^2 2p^1$.

Solution

There is a slight dip at Group IIIA in the plot of first IE versus atomic number. The order of increasing first ionization energy is

$$Na < Al < Mg < Si$$

EOC 28–30

6-4 Electron Affinity

The **electron affinity (EA)** of an element is defined as

the amount of energy *absorbed* when an electron is added to an isolated gaseous atom to form an ion with a 1− charge.

The convention is to assign a positive value when energy is absorbed and a negative value when energy is released. For most elements, energy is absorbed. We can represent the electron affinities of beryllium and chlorine as follows:

This is consistent with thermodynamic convention.

$$Be(g) + e^- + 241\text{ kJ} \longrightarrow Be^-(g) \qquad EA = 241\text{ kJ/mol}$$

$$Cl(g) + e^- \longrightarrow Cl^-(g) + 348\text{ kJ} \qquad EA = -348\text{ kJ/mol}$$

The value of *EA* for Cl can also be represented as -5.78×10^{-19} J/atom or -3.61 eV/atom. The electron volt (eV) is a unit of energy (1 eV = 1.60222×10^{-19} J).

The first equation tells us that when one mole of gaseous beryllium atoms gain one electron each to form gaseous Be^- ions, 241 kJ/mol of ions is *absorbed* (*endothermic*). The second equation tells us that when one mole of gaseous chlorine atoms gain one electron each to form gaseous chloride ions, 348 kJ of energy is *released* (*exothermic*). Figure 6-3 shows a plot of electron affinity versus atomic number for several elements.

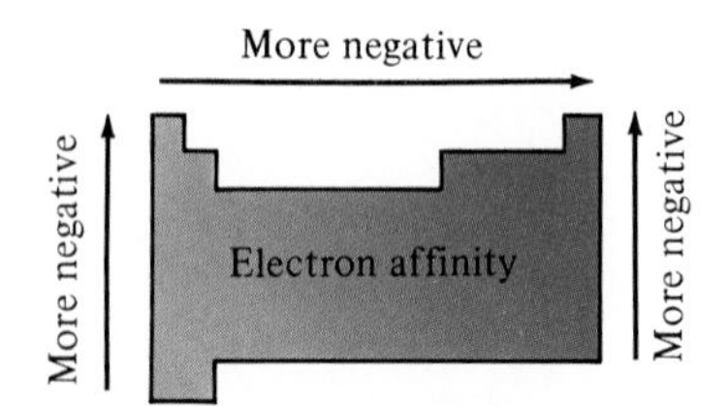

General trends in electron affinities of A group elements with position in the periodic table. There are many exceptions.

Electron affinity involves the *addition* of an electron to a neutral gaseous atom. The process by which a neutral atom X gains an electron (EA),

$$X(g) + e^- \longrightarrow X^-(g) \qquad (EA)$$

is *not* the reverse of the ionization process,

$$X^+(g) + e^- \longrightarrow X(g) \qquad (\text{reverse of } IE_1)$$

The first process begins with a neutral atom, whereas the second begins with a positive ion. Thus, IE_1 and EA are *not* simply equal in value with the signs reversed.

Elements with very negative electron affinities gain electrons easily to form negative ions (anions). We see from Figure 6-3 that electron affinities generally become more negative from left to right across a row in the periodic table (excluding the noble gases). This means that most representative elements in Groups IA to VIIA show a greater attraction for an extra electron from left to right. The halogens, which have the outer electronic configuration ns^2np^5, have the most negative electron affinities. They form stable anions with noble gas configurations, . . . ns^2np^6, by gaining one electron.

"Electron affinity" is a precise and quantitative term, like "ionization energy," but it is difficult to measure. Table 6-2 shows electron affinities for the representative elements.

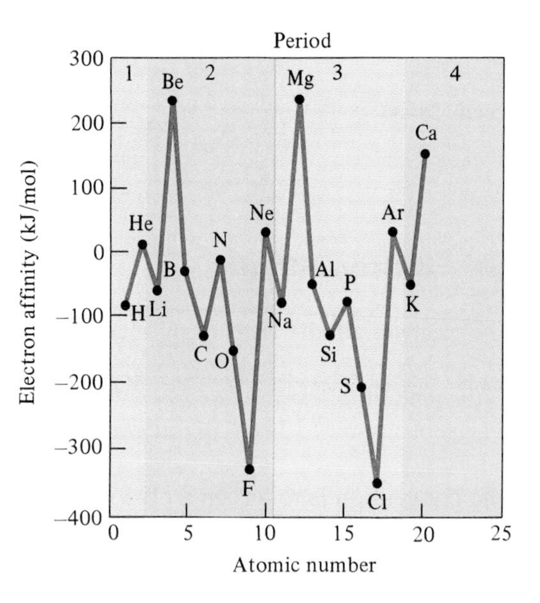

Figure 6-3
A plot of electron affinity versus atomic number for the first 20 elements. The general horizontal trend is that electron affinities become more negative (more energy is released as an extra electron is added) from Group IA through Group VIIA for a given period. Exceptions occur at the IIA and VA elements.

Table 6-2
Electron Affinity Values (kJ/mol) of Some Elements*

1	H −72								He (21)
2	Li −60	Be (241)		B −23	C −122	N 0	O −142	F −322	Ne (29)
3	Na −53	Mg (231)	Cu −123	Al −44	Si −119	P −74	S −200	Cl −348	Ar (35)
4	K −48	Ca (156)	Ag −125	Ga (−36)	Ge −116	As −77	Se −194	Br −323	Kr (39)
5	Rb −47	Sr (119)	Au −222	In (−34)	Sn −120	Sb −101	Te −190	I −295	Xe (40)
6	Cs −45	Ba (52)		Tl (−48)	Pb −101	Bi −101	Po (−173)	At (−270)	Rn (40)
7	Fr (−44)								

* Estimated values are in parentheses.

For many reasons, the variations in electron affinities are not regular across a period. The general trend is for the electron affinities of the elements to become more negative from left to right in each period. Noteworthy exceptions are the elements of Groups IIA and VA, which have less negative (more positive) values than the trends suggest (Figure 6-3). It is very difficult to add an electron to a IIA metal atom because its outer *s* subshell is filled. The values for the VA elements are slightly less negative than expected because they apply to the addition of an electron to a relatively stable half-filled set of *np* orbitals ($ns^2np^3 \rightarrow ns^2np^4$).

The addition of a second electron to form an ion with a 2− charge is always endothermic. So electron affinities of anions are always positive.

Example 6-3

Arrange the following elements in order of increasing values of electron affinity, i.e., from most negative to most positive.

Be, N, Na, Cl

Plan

Table 6-2 shows that electron affinity values generally become more negative from left to right across a period with major exceptions at Groups IIA (Be) and VA (N). They generally become more negative from bottom to top.

Solution

The order of increasing values of electron affinity is

(most negative EA) Cl < Na < N < Be (most positive EA)

EOC 39

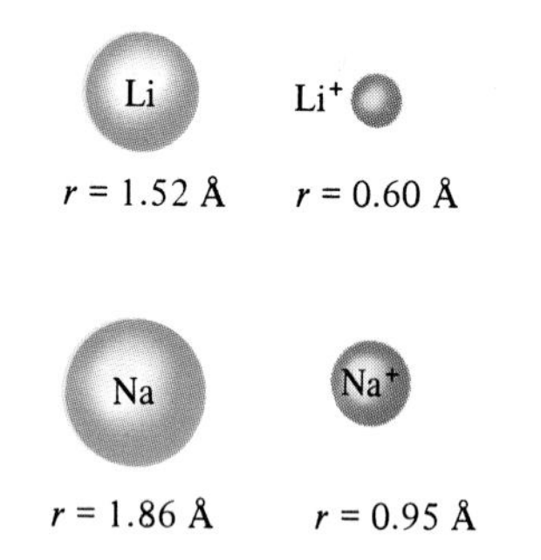

The nuclear charge remains constant when the ion is formed.

6-5 Ionic Radii

Many elements on the left side of the periodic table react with other elements by *losing* electrons to form positively charged ions. Each of the Group IA elements (Li, Na, K, Rb, Cs) has only one electron in its highest energy level (electronic configuration ... ns^1). These elements react with other elements by losing one electron to attain noble gas configurations. They form the ions Li^+, Na^+, K^+, Rb^+, and Cs^+. A neutral lithium atom, Li, contains three protons in its nucleus and three electrons, with its outermost electron in the 2*s* orbital. However, a lithium ion, Li^+, contains three protons in its nucleus but only two electrons, both in the 1*s* orbital. So a Li^+ ion is much smaller than a neutral Li atom (see margin). Likewise, a sodium ion, Na^+, is smaller than a sodium atom, Na. The relative sizes of atoms and common ions of some representative elements are shown in Figure 6-1.

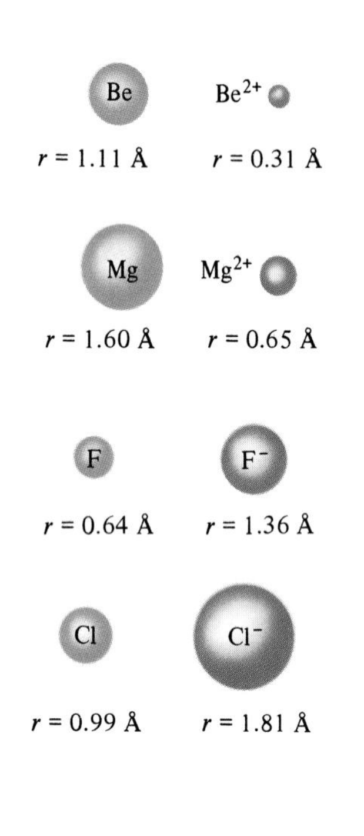

Isoelectronic species have the same number of electrons. We see that the ions formed by the Group IIA elements (Be^{2+}, Mg^{2+}, Ca^{2+}, Sr^{2+}, Ba^{2+}) are significantly smaller than the *isoelectronic* ions formed by the Group IA elements in the same period. The radius of the Li^+ ion is 0.60 Å, while the radius of the Be^{2+} ion is only 0.31 Å. This is just what we might expect. A beryllium ion, Be^{2+}, is formed when a beryllium atom, Be, loses both of its 2*s* electrons while the 4+ nuclear charge remains constant. We expect the 4+ nuclear charge in Be^{2+} to attract the remaining two electrons quite strongly. Comparison of the ionic radii of the IIA elements with their atomic radii indicates the validity of our reasoning. Similar reasoning indicates that the ions of the Group IIIA metals (Al^{3+}, Ga^{3+}, In^{3+}, Tl^{3+}) should be even smaller than the ions of Group IA and Group IIA elements in the same periods.

Now consider the Group VIIA elements (F, Cl, Br, I). These have the outermost electronic configuration ... ns^2np^5. These elements can completely fill their outermost *p* orbitals by *gaining* one electron each to attain noble gas configurations. Thus, when a fluorine atom (with seven electrons in its highest energy level) gains one electron, it becomes a fluoride ion, F^-, with eight electrons in its highest energy level. These eight electrons repel one another more strongly than the original seven, so the electron cloud expands. The F^- ion is much larger than the neutral F atom (see margin). Similar reasoning indicates that a chloride ion, Cl^-, should be larger than a neutral chlorine atom, Cl. Observed ionic radii (see Figure 6-1) verify this prediction.

Comparing the sizes of an oxygen atom (Group VIA) and an oxide ion, O^{2-}, we find that the negatively charged ion is larger than the neutral atom. The oxide ion is also larger than the fluoride ion because the oxide ion contains ten electrons held by a nuclear charge of only 8+, whereas the fluoride ion has ten electrons held by a nuclear charge of 9+.

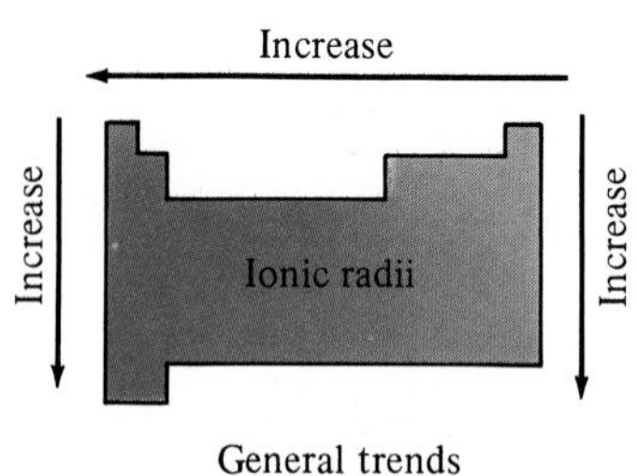

General trends in ionic radii of A group elements with position in the periodic table

1. Simple positively charged ions (cations) are always smaller than the neutral atoms from which they are formed.
2. Simple negatively charged ions (anions) are always larger than the neutral atoms from which they are formed.
3. Within an isoelectronic series of ions, ionic radii decrease with increasing atomic number.

An isoelectronic series of ions

	N^{3-}	O^{2-}	F^-	Na^+	Mg^{2+}	Al^{3+}
Ionic radius (Å)	1.71	1.40	1.36	0.95	0.65	0.50
No. of electrons	10	10	10	10	10	10
Nuclear charge	+7	+8	+9	+11	+12	+13

Example 6-4
Arrange the following ions in order of increasing ionic radii: (a) Ca^{2+}, K^+, Al^{3+}; (b) S^{2-}, Cl^-, Te^{2-}.

Plan
Figure 6-1 shows that ionic radii increase from right to left across a period and from top to bottom within a group.

Solution
(a) Cations are always smaller than the neutral atoms from which they are formed.

$$Al^{3+} < Ca^{2+} < K^+$$

(b) Anions are always larger than the neutral atoms from which they are formed.

$$Cl^- < S^{2-} < Te^{2-}$$

EOC 43–45

6-6 Electronegativity

The **electronegativity** of an element is a measure of the relative tendency of an atom to attract electrons to itself when it is chemically combined with another atom. Electronegativities of the elements are expressed on a somewhat arbitrary scale, called the Pauling scale (Table 6-3). The electronegativity of fluorine (4.0) is higher than that of any other element. This tells us that when fluorine is chemically bonded to other elements, it has a greater tendency to attract electron density to itself than does any other element. Oxygen is the second most electronegative element.

Because the noble gases form few compounds, they are not included in this discussion.

For the representative elements, electronegativities usually increase from left to right across periods and from bottom to top within groups.

Variations among the transition elements are not as regular. In general, both ionization energies and electronegativities are low for elements at the lower left of the periodic table and high for those at the upper right.

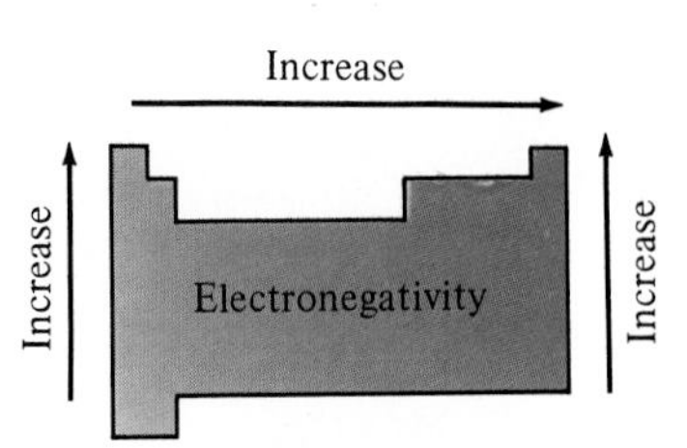

General trends in electronegativities of A group elements with position in the periodic table

Table 6-3
Electronegativity Values of the Elements*

Metals / Nonmetals / Metalloids

	IA	IIA	IIIB	IVB	VB	VIB	VIIB	VIIIB			IB	IIB	IIIA	IVA	VA	VIA	VIIA	0
1	1 H 2.1																	2 He
2	3 Li 1.0	4 Be 1.5											5 B 2.0	6 C 2.5	7 N 3.0	8 O 3.5	9 F 4.0	10 Ne
3	11 Na 1.0	12 Mg 1.2											13 Al 1.5	14 Si 1.8	15 P 2.1	16 S 2.5	17 Cl 3.0	18 Ar
4	19 K 0.9	20 Ca 1.0	21 Sc 1.3	22 Ti 1.4	23 V 1.5	24 Cr 1.6	25 Mn 1.6	26 Fe 1.7	27 Co 1.7	28 Ni 1.8	29 Cu 1.8	30 Zn 1.6	31 Ga 1.7	32 Ge 1.9	33 As 2.1	34 Se 2.4	35 Br 2.8	36 Kr
5	37 Rb 0.9	38 Sr 1.0	39 Y 1.2	40 Zr 1.3	41 Nb 1.5	42 Mo 1.6	43 Tc 1.7	44 Ru 1.8	45 Rh 1.8	46 Pd 1.8	47 Ag 1.6	48 Cd 1.6	49 In 1.6	50 Sn 1.8	51 Sb 1.9	52 Te 2.1	53 I 2.5	54 Xe
6	55 Cs 0.8	56 Ba 1.0	57 La 1.1 *	72 Hf 1.3	73 Ta 1.4	74 W 1.5	75 Re 1.7	76 Os 1.9	77 Ir 1.9	78 Pt 1.8	79 Au 1.9	80 Hg 1.7	81 Tl 1.6	82 Pb 1.7	83 Bi 1.8	84 Po 1.9	85 At 2.1	86 Rn
7	87 Fr 0.8	88 Ra 1.0	89 Ac 1.1 †															

*	58 Ce 1.1	59 Pr 1.1	60 Nd 1.1	61 Pm 1.1	62 Sm 1.1	63 Eu 1.1	64 Gd 1.1	65 Tb 1.1	66 Dy 1.1	67 Ho 1.1	68 Er 1.1	69 Tm 1.1	70 Yb 1.0	71 Lu 1.2
†	90 Th 1.2	91 Pa 1.3	92 U 1.5	93 Np 1.3	94 Pu 1.3	95 Am 1.3	96 Cm 1.3	97 Bk 1.3	98 Cf 1.3	99 Es 1.3	100 Fm 1.3	101 Md 1.3	102 No 1.3	103 Lr 1.5

* Electronegativity values are given at the bottoms of the boxes. The noble gases are not included in this discussion. The heavy "stair step" line approximately separates the metallic elements, to the left, from the nonmetallic elements, to the right.

Example 6-5

Arrange the following elements in order of increasing electronegativity.

B, Na, F, S

Plan

Table 6-3 shows that electronegativities increase from left to right across a period and from bottom to top within a group.

Solution

The order of increasing electronegativity is

Na < B < S < F

EOC 50

Although the electronegativity scale is somewhat arbitrary, we can use it with reasonable confidence to make predictions about bonding. Two elements with quite different electronegativities tend to react with each other to form ionic compounds. The less electronegative element gives up its electron(s) to the more electronegative element. Two elements with similar electronegativities tend to form covalent bonds with each other. That is, they share their electrons. In this sharing, the more electronegative element attains a greater share. This is discussed in detail in Chapters 7 and 8.

Chemical Reactions and Periodicity

Now we shall illustrate the periodicity of chemical properties by considering some reactions of hydrogen, oxygen, and their compounds. We choose to discuss hydrogen and oxygen because, of all the elements, they form the most kinds of compounds with other elements. Additionally, compounds of hydrogen and oxygen are very important in such diverse phenomena as all life processes and most corrosion processes.

6-7 Hydrogen and the Hydrides

1 Hydrogen

Elemental hydrogen is a colorless, odorless, tasteless diatomic gas with the lowest atomic weight and density of any known substance. Discovery of the element is attributed to the Englishman Henry Cavendish, who prepared it in 1766 by passing steam through a red-hot gun barrel (mostly iron) and by the reaction of acids with active metals. The latter is still the method commonly used for the preparation of small amounts of H_2 in the laboratory. In each case, H_2 is liberated by a displacement (and redox) reaction, of the kind described in Section 4-6. (See also the activity series, Table 4-11.)

The name "hydrogen" means "water former."

$$3Fe(s) + 4H_2O(g) \xrightarrow{\Delta} Fe_3O_4(s) + 4H_2(g)$$

$$Zn(s) + 2HCl(aq) \longrightarrow ZnCl_2(aq) + H_2(g)$$

Can you write the net ionic equation for the reaction of Zn with HCl(aq)?

Hydrogen also can be prepared by electrolysis of water.

$$2H_2O(\ell) \xrightarrow{\text{electricity}} 2H_2(g) + O_2(g)$$

In the future, if it becomes economical to convert solar energy into electrical energy that can be used to electrolyze water, H_2 could become an important fuel (although the dangers of storage and transportation would have to be overcome). The *combustion* of H_2 liberates a great deal of heat. **Combustion** is the highly exothermic combination of a substance with oxygen, usually with a flame. (See Section 6-8, part 3.)

$$2H_2(g) + O_2(g) \xrightarrow[\text{or } \Delta]{\text{spark}} 2H_2O(\ell) + \text{energy}$$

This is the reverse of the decomposition of H_2O.

Hydrogen is very flammable; it was responsible for the Hindenburg airship disaster in 1937. A spark is all it takes to initiate the **combustion reaction**, which is exothermic enough to provide the heat necessary to sustain the reaction.

Hydrogen is no longer used in blimps and dirigibles. It has been replaced by helium, which is slightly denser, nonflammable, and much safer.

Hydrogen is prepared by the "water gas reaction," which results from the passage of steam over white-hot coke (impure carbon, a nonmetal) at 1500°C. "Water gas" is used industrially as a fuel. Both components, CO and H_2, undergo combustion.

$$\underset{\text{in coke}}{C(s)} + \underset{\text{steam}}{H_2O(g)} \longrightarrow \underbrace{CO(g) + H_2(g)}_{\text{"water gas"}}$$

Vast quantities of hydrogen are produced commercially each year by the reaction of methane with steam at 830°C in the presence of a nickel catalyst:

$$CH_4(g) + H_2O(g) \xrightarrow[Ni]{\Delta} CO(g) + 3H_2(g)$$

The process is called *steam cracking.*

2 Reactions of Hydrogen and Hydrides

The use of the term "hydride" does not necessarily imply the presence of the hydride ion, H^-.

Now let us turn to some reactions of hydrogen with metals and other nonmetals to form binary compounds called **hydrides**. We shall also discuss characteristic reactions of the hydrides. Atomic hydrogen has the $1s^1$ electronic configuration. It can form (1) **ionic hydrides** containing hydride ions, H^-, by gaining one electron per atom from an active metal; or (2) **molecular hydrides** by sharing its electrons with an atom of another nonmetal.

The ionic or molecular character of the binary compounds of hydrogen depends upon the position of the other element in the periodic table (Figure 6-4.) The reactions of H_2 with the *alkali* (IA) and the heavier (more active) *alkaline earth* (IIA) *metals* result in solid *ionic hydrides,* often called *saline* or *salt-like hydrides*. The reaction with the molten (liquid) IA metals may be represented in general terms as

$$2M(\ell) + H_2(g) \xrightarrow[\text{high pressures of } H_2]{\text{high temperatures}} 2(M^+, H^-)(s) \quad M = Li, Na, K, Rb, Cs$$

The ionic hydrides are named by naming the metal first, followed by "hydride."

Thus, hydrogen combines with lithium to form lithium hydride and with sodium to form sodium hydride:

$$2Li(\ell) + H_2(g) \longrightarrow 2LiH(s) \qquad \text{lithium hydride (mp 680°C)}$$

$$2Na(\ell) + H_2(g) \longrightarrow 2NaH(s) \qquad \text{sodium hydride (mp 800°C)}$$

In general terms, the reactions of the heavier (most active) IIA metals may be represented as

$$M(\ell) + H_2(g) \longrightarrow (M^{2+}, 2H^-)(s) \qquad M = Ca, Sr, Ba$$

Thus, calcium combines with hydrogen to form calcium hydride:

$$Ca(\ell) + H_2(g) \longrightarrow CaH_2(s) \qquad \text{calcium hydride (mp 816°C)}$$

Ionic hydrides can serve as sources of hydrogen. However, they must be stored in environments free of moisture and O_2.

These *ionic hydrides are all basic* because hydride ions reduce water to form hydroxide ions and hydrogen. When water is added by drops to lithium hydride, for example, lithium hydroxide and hydrogen are produced. The reaction of calcium hydride is similar:

We show LiOH and $Ca(OH)_2$ as solids here because not enough water is available to act as a solvent.

$$LiH(s) + H_2O(\ell) \longrightarrow LiOH(s) + H_2(g)$$

$$CaH_2(s) + 2H_2O(\ell) \longrightarrow Ca(OH)_2(s) + 2H_2(g)$$

IA	IIA	IIIA	IVA	VA	VIA	VIIA
LiH	BeH_2	B_2H_6	CH_4	NH_3	H_2O	HF
NaH	MgH_2	$(AlH_3)_x$	SnH_4	PH_3	H_2S	HCl
KH	CaH_2	Ga_2H_6	GeH_4	AsH_3	H_2Se	HBr
RbH	SrH_2	InH_3	SnH_4	SbH_3	H_2Te	HI
CsH	BaH_2	TlH	PbH_4	BiH_3	H_2Po	HAt

Figure 6-4
Common hydrides of the representative elements. The ionic hydrides are shaded blue, molecular hydrides are shaded red, and those of intermediate character are shaded purple.

Hydrogen reacts with *nonmetals* to form binary *molecular hydrides*. For example, H_2 combines with the halogens to form colorless, gaseous hydrogen halides (Figure 6-5):

$$H_2(g) + X_2 \longrightarrow \underset{\text{hydrogen halides}}{2HX(g)} \qquad X = F, Cl, Br, I$$

Specifically, hydrogen reacts with fluorine to form hydrogen fluoride and with chlorine to form hydrogen chloride:

The hydrogen halides are named by the word "hydrogen" followed by the stem for the halogen with an "-ide" ending.

$$H_2(g) + F_2(g) \longrightarrow 2HF(g) \qquad \text{hydrogen fluoride}$$

$$H_2(g) + Cl_2(g) \longrightarrow 2HCl(g) \qquad \text{hydrogen chloride}$$

Hydrogen also combines with the Group VIA elements to form molecular compounds:

These compounds are named:
H_2O, hydrogen oxide (water)
H_2S, hydrogen sulfide
H_2Se, hydrogen selenide
H_2Te, hydrogen telluride
All except H_2O are very toxic.

$$2H_2(g) + O_2(g) \xrightarrow{\Delta} 2H_2O(g)$$

The heavier members of this family also combine with hydrogen to form binary compounds that are gases at room temperature. Their formulas resemble that of water.

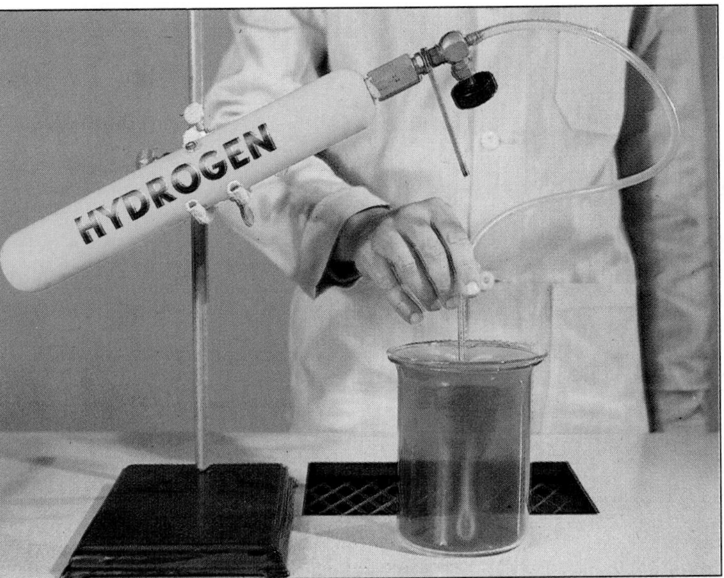

Figure 6-5
Hydrogen, H_2, burns in an atmosphere of bromine vapor, Br_2. Hydrogen bromide, HBr, is formed.

Figure 6-6
Ammonia may be applied directly to the soil as a fertilizer.

The primary industrial use of H_2 is in the synthesis of ammonia, a molecular hydride, by the Haber process (Section 17-6). Most of the NH_3 is used as liquid ammonia as a fertilizer (Figure 6-6) or to make other fertilizers, such as ammonium nitrate, NH_4NO_3, and ammonium sulfate, $(NH_4)_2SO_4$:

$$N_2(g) + 3H_2(g) \xrightarrow[\Delta,\ \text{high pressure}]{\text{catalysts}} 2NH_3(g)$$

As we shall see in Chapter 10, even H_2O is weakly acidic.

Many of the molecular (nonmetal) hydrides are acidic; their aqueous solutions produce hydrogen ions. These include HF, HCl, HBr, HI, H_2S, H_2Se, and H_2Te.

Example 6-6

Predict the products of the reactions involving the reactants shown. Write a balanced formula unit equation for each one.

(a) $H_2(g) + I_2(g) \xrightarrow{\Delta}$
(b) $K(\ell) + H_2(g) \xrightarrow{\Delta}$
(c) $NaH(s) + H_2O(\ell)$ (excess) $\longrightarrow$

Plan

(a) Hydrogen reacts with the halogens (Group VIIA) to form hydrogen halides—in this example, HI.
(b) Hydrogen reacts with active metals to produce hydrides—in this case, KH.
(c) Active metal hydrides reduce water to produce H_2 and a metal hydroxide.

Remember that hydride ions, H^-, react with (reduce) water to produce OH^- ions and $H_2(g)$.

Solution

(a) $H_2(g) + I_2(g) \xrightarrow{\Delta} 2HI(g)$
(b) $2K(\ell) + H_2(g) \xrightarrow{\Delta} 2KH(s)$
(c) $NaH(s) + H_2O(\ell) \longrightarrow NaOH(aq) + H_2(g)$

Example 6-7

Predict the ionic or molecular character of the products in Example 6-6.

Plan

We refer to Figure 6-4, which displays the nature of hydrides.

Solution

Reaction (a) is a reaction between hydrogen and another nonmetal. The product, HI, must be molecular. Reaction (b) is the reaction of hydrogen with an active Group IA metal. Thus, KH must be ionic. The products of reaction (c) are molecular $H_2(g)$ and the strong soluble base, NaOH, which is ionic.

EOC 65–67

6-8 Oxygen and the Oxides

1 Oxygen and Ozone

Oxygen was discovered in 1774 by an English minister and scientist, Joseph Priestley. He observed the thermal decomposition of mercury(II) oxide, a red powder:

The name "oxygen" means "acid former."

$$2HgO(s) \xrightarrow{\Delta} 2Hg(\ell) + O_2(g)$$

That part of the earth we see—land, water, and air—is approximately 50% oxygen by mass. About two thirds of the mass of the human body is due to oxygen in H_2O. Elemental oxygen, O_2, is an odorless and nearly colorless gas that makes up about 21% by volume of dry air. In the liquid and solid states it is pale blue. Oxygen is only very slightly soluble in water; only about 0.04 gram dissolves in 1 liter of water at 25°C. This is sufficient to sustain fish and other marine organisms. The greatest single industrial use of O_2 is for oxygen-enrichment in blast furnaces for the conversion of pig iron to steel. Oxygen is obtained commercially by the fractional distillation of liquid air.

Liquid O_2 is used as an oxidizer for rocket fuels. O_2 also is used in the health fields for oxygen-enriched air.

Oxygen also exists in a second allotropic form, ozone, O_3. Ozone is an unstable, pale blue gas at room temperature. It is formed by passing an electrical discharge through gaseous oxygen. Its unique, pungent odor is often noticed during electrical storms and in the vicinity of electrical equipment. Not surprisingly, its density is about $1\frac{1}{2}$ times that of O_2. At −112°C it condenses to a deep blue liquid. It is a very strong oxidizing agent. As a concentrated gas or a liquid, ozone can easily decompose explosively:

Allotropes are different forms of the same element in the same physical state (Section 2-2).

$$2O_3(g) \longrightarrow 3O_2(g)$$

Oxygen atoms, or **radicals**, are intermediates in this exothermic decomposition of O_3 to O_2. They act as strong oxidizing agents in such applications as destroying bacteria in water purification.

A *radical* is a species containing one or more unpaired electrons; many radicals are very reactive.

The ozone molecule is angular and diamagnetic. Both oxygen—oxygen bond lengths (1.28 Å) are identical and are intermediate between typical single and double bond lengths.

Ozone is formed in the upper atmosphere as O_2 molecules absorb high-energy electromagnetic radiation from the sun. Its concentration in the stratosphere is about 10 ppm, whereas it is only about 0.04 ppm near the earth's surface. The ozone layer is responsible for absorbing some of the ultraviolet light from the sun that, if it reached the surface of the earth in higher intensity, could cause damage to plants and animals (including humans). It has been

The concentration unit "ppm" stands for *parts per million.*

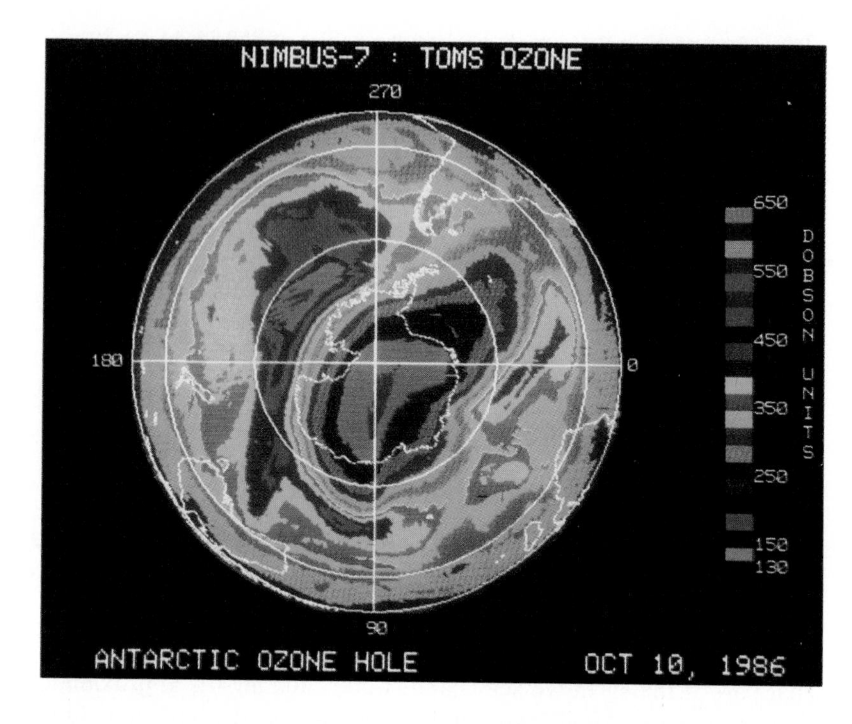

Figure 6-7
A computer map of ozone concentrations over the southern hemisphere. Dobson units are used to show the concentration of ozone. Measurements were made by the Total Ozone Mapping Spectrometer (TOMS) aboard the satellite Nimbus-7 on October 10, 1986. The depletion of the ozone layer by man-made pollutants is a major environmental concern. This map shows the hole (purple) in the ozone layer over Antarctica.

Small amounts of O_3 at the surface of the earth decompose rubber and plastic products by oxidation.

estimated that the incidence of skin cancer would increase by 2% for every 1% decrease in the concentration of ozone in the stratosphere (Figure 6-7). Although it decomposes rapidly in the upper atmosphere, the ozone supply is constantly replenished.

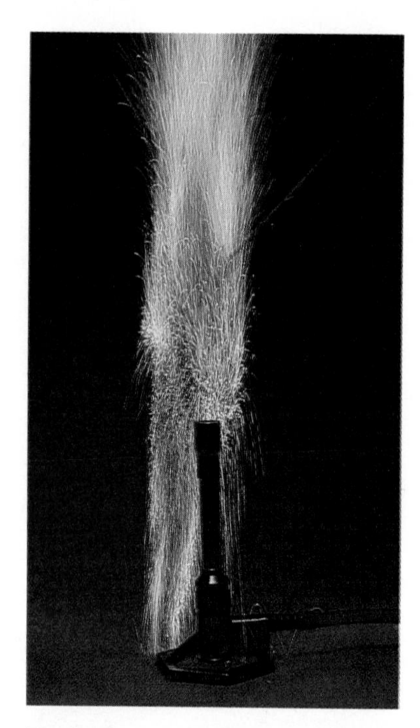

Figure 6-8
Iron powder burns brilliantly to form iron(III) oxide, Fe_2O_3.

2 Reactions of Oxygen and the Oxides

Oxygen combines directly with all other elements except the noble gases and noble (unreactive) metals (Au, Pd, Pt) to form **oxides**, binary compounds that contain oxygen. Although such reactions are generally very exothermic, many proceed quite slowly and require heating to supply the energy necessary to break the strong bonds in O_2 molecules. Once these reactions are initiated, most release more than enough energy to be self-sustaining and sometimes become "red hot."

Reactions of O_2 with Metals

In general, metallic oxides (and peroxides and superoxides) are ionic solids. The Group IA metals combine with oxygen to form three kinds of solid ionic products called oxides, peroxides, and superoxides. Lithium combines with oxygen to form lithium oxide:

$$4Li(s) + O_2(g) \longrightarrow 2Li_2O(s) \qquad \text{lithium oxide (mp > 1700°C)}$$

By contrast, sodium reacts with an excess of oxygen to form sodium peroxide, Na_2O_2, rather than sodium oxide, Na_2O, as the *major* product:

$$2Na(s) + O_2(g) \longrightarrow Na_2O_2(g) \qquad \text{sodium peroxide (decomposes at 460°C)}$$

Peroxides contain the $O—O^{2-}$, O_2^{2-} group, in which the oxidation number of oxygen is −1, whereas *normal oxides* such as lithium oxide, Li_2O, contain oxide ions, O^{2-}. The heavier members of the family (K, Rb, Cs) react with

Table 6-4
Oxygen Compounds of the IA and IIA Metals*

	IA					IIA				
	Li	Na	K	Rb	Cs	Be	Mg	Ca	Sr	Ba
oxide	Li_2O	Na_2O	K_2O	Rb_2O	Cs_2O	BeO	MgO	CaO	SrO	BaO
peroxide	Li_2O_2	Na_2O_2	K_2O_2	Rb_2O_2	Cs_2O_2			CaO_2	SrO_2	BaO_2
superoxide		NaO_2	KO_2	RbO_2	CsO_2					

* The shaded compounds represent the principal products of the direct reaction of the metal with oxygen.

excess oxygen to form **superoxides**. These contain the superoxide ion, O_2^-, in which the oxidation number of oxygen is $-\frac{1}{2}$. The reaction with K is

$$K(s) + O_2(g) \longrightarrow KO_2(s) \qquad \text{potassium superoxide (mp 430°C)}$$

The tendency of the Group IA metals to form oxygen-rich compounds increases as the group is descended. This is because cation radii increase going down the group. A similar trend is observed in the reactions of the Group IIA metals with oxygen. You can recognize these classes of compounds as follows:

Class	Contains Ions	Oxidation No. of Oxygen
oxide	O^{2-}	-2
peroxide	O_2^{2-}	-1
superoxide	O_2^-	$-\frac{1}{2}$

Beryllium reacts with oxygen only at elevated temperatures and forms only the normal oxide, BeO.

With the exception of Be, the Group IIA metals react with oxygen at moderate temperatures to form normal ionic oxides, MO, and at high pressures of oxygen the heavier ones form ionic peroxides, MO_2 (Table 6-4).

$$2M(s) + O_2(g) \longrightarrow 2(M^{2+}, O^{2-})(s) \qquad M = \text{Be, Mg, Ca, Sr, Ba}$$

$$M(s) + O_2(g) \longrightarrow (M^{2+}, O_2^{2-})(s) \qquad M = \text{Ca, Sr, Ba}$$

For example, the equations for the reactions of calcium and oxygen are

$$2Ca(s) + O_2(g) \longrightarrow 2CaO(s) \qquad \text{calcium oxide (mp 2580°C)}$$

$$Ca(s) + O_2(g) \longrightarrow CaO_2(s) \qquad \text{calcium peroxide (decomposes at 275°C)}$$

The other metals, with the exceptions noted previously (Au, Pd, and Pt), react with oxygen to form solid metal oxides. Many metals to the right of Group IIA show variable oxidation states, so they may form several oxides. For example, iron combines with oxygen in the following series of reactions to form three different oxides (Figure 6-8).

$$2Fe(s) + O_2(g) \xrightarrow{\Delta} 2FeO(s) \qquad \text{iron(II) oxide } \textit{or} \text{ ferrous oxide}$$

$$6FeO(s) + O_2(g) \xrightarrow{\Delta} 2Fe_3O_4(s) \qquad \text{magnetic iron oxide (a mixed oxide)}$$

$$4Fe_3O_4(s) + O_2(g) \xrightarrow{\Delta} 6Fe_2O_3(s) \qquad \text{iron(III) oxide } \textit{or} \text{ ferric oxide}$$

Increasing acidic character →

Increasing base character ↓

IV	IIA	IIIA	IVA	VA	VIA	VIIA
Li_2O	BeO	B_2O_3	CO_2	N_2O_5		F_2O
Na_2O	MgO	Al_2O_3	SiO_2	P_4O_{10}	SO_3	Cl_2O_7
K_2O	CaO	Ga_2O_3	GeO_2	As_2O_5	SeO_3	Br_2O_7
Rb_2O	SrO	In_2O_3	SnO_2	Sb_2O_5	TeO_3	I_2O_7
Cs_2O	BaO	Tl_2O_3	PbO_2	Bi_2O_5	PoO_3	At_2O_7

Figure 6-9
The normal oxides of the representative elements in their maximum oxidation states. Acidic oxides (acid anhydrides) are shaded red, amphoteric oxides are shaded purple, and basic oxides (basic anhydrides) are shaded blue. An amphoteric oxide is one that shows some acidic and some basic properties.

Copper reacts with a limited amount of oxygen to form red Cu_2O, whereas with excess oxygen it forms black CuO.

$$4Cu(s) + O_2(g) \xrightarrow{\Delta} 2Cu_2O(s) \quad \text{copper(I) oxide } or \text{ cuprous oxide}$$

$$2Cu(s) + O_2(g) \xrightarrow{\Delta} 2CuO(s) \quad \text{copper(II) oxide } or \text{ cupric oxide}$$

Metals that exhibit variable oxidation states react with a limited amount of oxygen to give lower oxidation state oxides (such as FeO and Cu_2O). They react with an excess of oxygen to give higher oxidation state oxides (such as Fe_2O_3 and CuO).

Reactions of Metal Oxides with Water

Oxides of metals are called **basic anhydrides** because many of them combine with water to form bases with no change in oxidation state of the metal (Figure 6-9). "Anhydride" means "without water"; in a sense, the metal oxide is a hydroxide base with the water "removed." Metal oxides that are soluble in water react to produce the corresponding hydroxides.

Metal Oxide + Water ⟶ Metal Hydroxide (base)

sodium oxide	$Na_2O(s) + H_2O(\ell) \longrightarrow 2\,NaOH(aq)$		sodium hydroxide
calcium oxide	$CaO(s) + H_2O(\ell) \longrightarrow Ca(OH)_2(aq)$		calcium hydroxide
barium oxide	$BaO(s) + H_2O(\ell) \longrightarrow Ba(OH)_2(aq)$		barium hydroxide

The oxides of the Group IA metals and the heavier Group IIA metals dissolve in water to give solutions of strong soluble bases. Most other metal oxides are insoluble in water.

Reactions of O_2 with Nonmetals

Oxygen combines with many nonmetals to form molecular oxides. For example, carbon burns in oxygen to form carbon monoxide or carbon dioxide, depending on the relative amounts of carbon and oxygen, as the following equations show:

$$2C(s) + O_2(g) \longrightarrow 2\overset{+2}{C}O(s) \quad \text{(excess C and limited } O_2)$$

$$C(s) + O(g) \longrightarrow \overset{+4}{C}O_2(g) \quad \text{(limited C and excess } O_2)$$

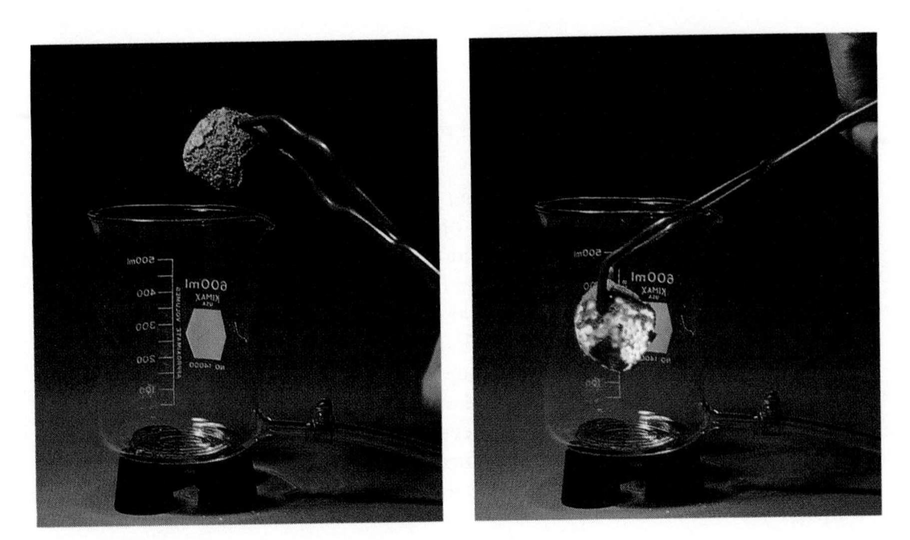

Carbon burns brilliantly in pure O_2 to form CO_2.

Carbon monoxide is also produced by the incomplete combustion of carbon-containing compounds such as gasoline and diesel fuel. It is a very poisonous gas because it forms a stronger bond than oxygen molecules form with the iron atom in hemoglobin. Attachment of the CO molecule to the iron atom destroys the ability of hemoglobin to pick up oxygen in the lungs and carry it to the brain and muscle tissues. Carbon monoxide poisoning is particularly insidious because the gas has no odor and because the victim first becomes drowsy.

Unlike carbon monoxide, carbon dioxide is not toxic. It is one of the products of the respiratory process. It is used to make carbonated beverages, which are mostly saturated solutions of carbon dioxide in water; a small amount of the carbon dioxide combines with the water to form carbonic acid (H_2CO_3), a very weak acid.

Phosphorus reacts with a limited amount of oxygen to form tetraphosphorus hexoxide, P_4O_6,

$$P_4(s) + 3O_2(g) \longrightarrow \overset{\small(+3)}{P_4}O_6(s) \qquad \text{tetraphosphorus hexoxide}$$

while an excess of oxygen reacts with phosphorus to form tetraphosphorus decoxide, P_4O_{10}:

$$P_4(s) + 5O_2(g) \longrightarrow \overset{\small(+5)}{P_4}O_{10} \qquad \text{tetraphosphorus decoxide}$$

Figure 6-10
Sulfur burns in oxygen to form sulfur dioxide.

Sulfur burns in oxygen to form primarily sulfur dioxide (Figure 6-10) and only very small amounts of sulfur trioxide.

$$S_8(s) + 8O_2(g) \longrightarrow 8\overset{\small(+4)}{S}O_2(g) \qquad \text{sulfur dioxide (mp } -73°\text{C)}$$

$$S_8(s) + 12O_2(g) \longrightarrow 8\overset{\small(+6)}{S}O_3(g) \qquad \text{sulfur trioxide (mp 32.5°C)}$$

The production of SO_3 at a reasonable rate requires the presence of a catalyst.

Oxidation States of Nonmetals

Nonmetals exhibit more than one oxidation state in their compounds. In general, the *most* common oxidation states of a nonmetal are (1) its periodic group number, (2) its periodic group number minus two, and (3) its periodic group number minus eight. The reactions of nonmetals with a limited amount of oxygen usually give products that contain the nonmetals (other than oxygen) in lower oxidation states, usually case (2). Reactions with excess oxygen give products in which the nonmetals exhibit higher oxidation states, case (1). The examples we have cited are CO and CO_2, P_4O_6 and P_4O_{10}, and SO_2 and SO_3. The molecular formulas of the oxides are sometimes not easily predictable, but the *simplest* formulas are. For example, the two most common oxidation states of phosphorus in molecular compounds are +3 and +5. The simplest formulas for the corresponding phosphorus oxides therefore are P_2O_3 and P_2O_5, respectively. The molecular formulas are twice these, P_4O_6 and P_4O_{10}.

Reactions of Nonmetal Oxides with Water

Nonmetal oxides are called **acid anhydrides** because many of them dissolve in water to form acids *with no change in oxidation state of the nonmetal* (Figure 6-9). Several **ternary acids** can be prepared by reaction of the appropriate nonmetal oxides with water. Ternary acids contain three elements, usually H, O, and another nonmetal.

	Nonmetal Oxide	+ Water	⟶	Ternary Acid	
carbon dioxide	(+4) $CO_2(g)$	$+ H_2O(\ell)$	⟶	(+4) $H_2CO_3(aq)$	carbonic acid
sulfur dioxide	(+4) $SO_2(g)$	$+ H_2O(\ell)$	⟶	(+4) $H_2SO_3(aq)$	sulfurous acid
sulfur trioxide	(+6) $SO_3(g)$	$+ H_2O(\ell)$	⟶	(+6) $H_2SO_4(aq)$	sulfuric acid
dinitrogen pentoxide	(+5) $N_2O_5(s)$	$+ H_2O(\ell)$	⟶	(+5) $2HNO_3(aq)$	nitric acid
tetraphosphorus decoxide	(+5) $P_4O_{10}(s)$	$+ 6H_2O(\ell)$	⟶	(+5) $4H_3PO_4(aq)$	phosphoric acid

Nearly all oxides of nonmetals dissolve in water to give solutions of ternary acids. The oxides of boron and silicon, which are insoluble, are two exceptions.

Reactions of Metal Oxides with Nonmetal Oxides

Another common kind of reaction of oxides is the *combination of metal oxides (basic anhydrides) with nonmetal oxides (acid anhydrides), with no change in oxidation states, to form salts.*

	Metal Oxide	+ Nonmetal Oxide	$\longrightarrow$ Salt	
calcium oxide + sulfur trioxide	$\overset{(+2)}{Ca}O(s)$	$+ \overset{(+6)}{S}O_3(g)$	$\longrightarrow \overset{(+2)}{Ca}\overset{(+6)}{S}O_4(s)$	calcium sulfate
magnesium oxide + carbon dioxide	$\overset{(+2)}{Mg}O(s)$	$+ \overset{(+4)}{C}O_2(g)$	$\longrightarrow \overset{(+2)}{Mg}\overset{(+4)}{C}O_3(s)$	magnesium carbonate
sodium oxide + tetraphosphorus decoxide	$6\overset{(+1)}{Na_2}O(s)$	$+ \overset{(+5)}{P_4}O_{10}(s)$	$\longrightarrow 4\overset{(+1)}{Na_3}\overset{(+5)}{P}O_4(s)$	sodium phosphate

Example 6-8

Arrange the following oxides in order of increasing molecular (acidic) character: SO_3, Cl_2O_7, CaO, and PbO_2.

Plan

Molecular (acidic) character of oxides increases in the same direction as nonmetallic character of the element that is combined with oxygen (Figure 6-9).

$$\xrightarrow{\text{Increasing nonmetallic character}}$$
$$\text{Ca} < \text{Pb} < \text{S} < \text{Cl}$$

Periodic group: IIA IVA VIA VIIA

Solution

Thus, the order is

$$\xrightarrow{\text{Increasing molecular character}}$$
$$CaO < PbO_2 < SO_3 < Cl_2O_7$$

Example 6-9

Arrange the oxides in Example 6-8 in order of increasing basicity.

Plan

The greater the molecular character of an oxide, the more acidic it is. Thus, the most basic oxides have the least molecular (most ionic) character (Figure 6-9).

Solution

$$\xrightarrow{\text{Increasing basic character}}$$
$$\text{molecular} \quad Cl_2O_7 < SO_3 < PBO_2 < CaO \quad \text{ionic}$$

Example 6-10

Predict the products of the reactions involving the following reactants. Write a balanced formula unit equation for each one.

(a) $Cl_2O_7(\ell) + H_2O(\ell) \longrightarrow$
(b) $As_4(s) + O_2(g)$ (excess) $\xrightarrow{\Delta}$
(c) $Mg(s) + O_2(g)$ (low pressure) $\xrightarrow{\Delta}$

Plan

(a) The reaction of a nonmetal oxide (acid anhydride) with water forms a ternary acid in which the nonmetal (Cl) has the same oxidation state (+7) as in the oxide. Thus, the acid is perchloric acid, $HClO_4$.
(b) Arsenic, a nonmetal of Group VA, exhibits common oxidation states of +5 and $+5 - 2 = +3$. Reaction of arsenic with *excess* oxygen produces the higher-

oxidation-state oxide, As_2O_5. By analogy with the oxide of phosphorus in the +5 oxidation state, P_4O_{10}, we might write the formula as As_4O_{10}, but this oxide is usually represented as As_2O_5.
(c) The reaction of a Group IIA metal with oxygen (at low pressure) produces the normal metal oxide—MgO in this case.

Solution

(a) $Cl_2O_7(\ell) + H_2O(\ell) \longrightarrow 2HClO_4(aq)$

(b) $As_4(s) + 5O_2(g) \xrightarrow{\Delta} 2As_2O_5(s)$

(c) $2Mg(s) + O_2(g) \xrightarrow{\Delta} 2MgO(s)$

Example 6-11
Predict the products of the following pairs of reactants. Write a balanced formula unit equation for each reaction.
(a) $CaO(s) + H_2O(\ell) \longrightarrow$
(b) $Li_2O(s) + SO_3(g) \longrightarrow$

Plan
(a) The reaction of a metal oxide with water produces the metal hydroxide.
(b) The reaction of a metal oxide with a nonmetal oxide produces a salt containing the cation of the metal oxide and the anion of the acid for which the nonmetal oxide is the anhydride. SO_3 is the acid anhydride of sulfuric acid, H_2SO_4.

Solution
(a) Calcium oxide reacts with water to form calcium hydroxide.

CaO is called quicklime. $Ca(OH)_2$ is called slaked lime.

$$CaO(s) + H_2O(\ell) \longrightarrow Ca(OH)_2(aq)$$

(b) Lithium oxide reacts with sulfur trioxide to form lithium sulfate.

$$Li_2O(s) + SO_3(g) \longrightarrow Li_2SO_4(s)$$

EOC 81–83

3 Combustion Reactions

Hydrocarbons are compounds that contain only hydrogen and carbon.

Combustion, or burning, is an oxidation–reduction reaction in which oxygen combines rapidly with oxidizable materials in highly exothermic reactions, usually with a visible flame. The complete combustion of **hydrocarbons**, in fossil fuels for example, produces carbon dioxide and water (steam) as the major products:

$$\overset{-4\ +1}{CH_4(g)} + \underset{\text{excess}}{2\overset{0}{O_2}(g)} \xrightarrow{\Delta} \overset{+4\ -2}{CO_2(g)} + 2\overset{+1\ -2}{H_2O(g)} + \text{heat}$$

$$\underset{\text{cyclohexane}}{\overset{-2\ +1}{C_6H_{12}(g)}} + \underset{\text{excess}}{9\overset{0}{O_2}(g)} \xrightarrow{\Delta} 6\overset{+4\ -2}{CO_2(g)} + 6\overset{+1\ -2}{H_2O(g)} + \text{heat}$$

As we have seen, the origin of the term "oxidation" lies in just such reactions, in which oxygen "oxidizes" another species.

4 Combustion of Fossil Fuels and Air Pollution

Fossil fuels are mixtures of variable composition that consist primarily of hydrocarbons. We burn them because they release energy, rather than to obtain chemical products (Figure 6-11). The incomplete combustion of hydrocarbons yields undesirable products, carbon monoxide and elemental carbon (soot), which pollute the air. Unfortunately, all fossil fuels—natural gas, coal, gasoline, kerosene, oil, and so on—also have undesirable non-hydrocarbon impurities that undergo combustion to produce oxides that act as additional air pollutants. At this time it is not economically feasible to remove all of these impurities.

Carbon in the form of soot is one of many kinds of *particulate matter* in polluted air.

Fossil fuels result from the decay of animal and vegetable matter (Figure 6-12). All living matter contains some sulfur and nitrogen, so fossil fuels also contain sulfur and nitrogen impurities to varying degrees. Table 6-5 gives composition data for some common kinds of coal.

Combustion of sulfur produces sulfur dioxide, SO_2, probably the single most harmful pollutant:

$$\overset{(0)}{S_8}(s) + 8O_2(g) \xrightarrow{\Delta} 8\overset{(+4)}{S}O_2(g)$$

Large amounts of SO_2 are produced by the burning of sulfur-containing coal.

Many metals occur in nature as sulfides. The process of extracting the free (elemental) metals involves **roasting**—heating an ore in the presence of air. For many metal sulfides this produces a metal oxide and SO_2. The metal oxides are then reduced to the free metals. Consider lead sulfide, PbS, as an example:

$$2PbS(s) + 3O_2(g) \longrightarrow 2PbO(s) + 2SO_2(g)$$

Sulfur dioxide is corrosive; it damages plants, structural materials, and humans. It is a nasal, throat, and lung irritant. Sulfur dioxide is slowly oxidized to sulfur trioxide, SO_3, by oxygen in air:

$$2SO_2(g) + O_2(g) \longrightarrow 2SO_3(\ell)$$

Sulfur trioxide combines with moisture in the air to form the strong, corrosive acid, sulfuric acid:

$$SO_3 + H_2O(\ell) \longrightarrow H_2SO_4(\ell)$$

Oxides of sulfur are the main cause of acid rain.

Table 6-5
Some Typical Coal Compositions in Percent (dry, ash-free)

	C	H	O	N	S
lignite	70.59	4.47	23.13	1.04	0.74
subbituminous	77.2	5.01	15.92	1.30	0.51
bituminous	80.2	5.80	7.53	1.39	5.11
anthracite	92.7	2.80	2.70	1.00	0.90

Figure 6-11
Georgia Power Company's Plant Bowen at Taylorsville, Georgia. In 1986 it burned 8,376,726 tons of coal and produced 21,170,999 megawatt-hours of electricity, a national record.

Figure 6-12
The luxuriant growth of vegetation that occurred during the carboniferous age is the source of our coal deposits.

Chemistry in Use. . .
Acid Rain

During the 1980s, the phenomenon of acid deposition, commonly known as acid rain, gained considerable attention from the public and the scientific community. It is the subject of intense research for thousands of environmental scientists. It has also been a source of political conflict and debate within and between nations in North America and Europe.

The term "acid rain" is used to describe all naturally occurring precipitation, including rain, snow, sleet, and hail, that has become acidified. The *acidity* or *alkalinity* of a substance is determined by the relative concentrations of hydrogen ions and hydroxide ions that it contains. Acidity is expressed on a logarithmic scale called the pH scale. Pure water, which has equal concentrations of hydrogen ions and hydroxide ions, has a pH of 7.0 and is said to be *neutral*. A substance with a higher concentration of hydrogen ions than hydroxide ions is acidic and has a value less than 7.0 on the pH scale. Conversely, a substance with a higher concentration of hydroxide ions than hydrogen ions is alkaline, or *basic*, and has a pH value greater than 7.0. The farther a reading is from 7.0, below or above, the more acidic or basic (respectively) the substance is. Each full pH unit decrease represents a tenfold increase in acidity. For example, a solution whose pH value is 6.0 contains ten times more hydrogen ions than pure water and is ten times more acidic. A substance with a pH of 5.0 is 100 times more acidic than pure water.

Normal, uncontaminated precipitation is naturally slightly acidic, having a pH value of about 5.6. This natural acidity is the result of the combination of carbon dioxide in the atmosphere with water vapor to form carbonic acid, H_2CO_3, a very weak acid. Any rainfall with pH lower than 5.6 is considered excessively acidic. Unfortunately, as a result of human activities such as burning fossil fuels and smelting ores, reports of rain with pH values of 3.8 to 4.5 are common. The most acidic rainfall recorded in the United States to date (April 1991) had a pH value of 1.5; this is 12,600 times more acidic than normal uncontaminated rain, and almost one-third as acidic as battery acid, a very strong acid. Widespread acid rain has been known in northern Europe and eastern North America for some time. More recent work has led to the discovery of acid rain in western North America, Japan, China, the Soviet Union, and South America.

The main compounds that produce acid rain are the oxides of sulfur and nitrogen, which quickly react with water to form acids such as sulfurous (H_2SO_3), sulfuric

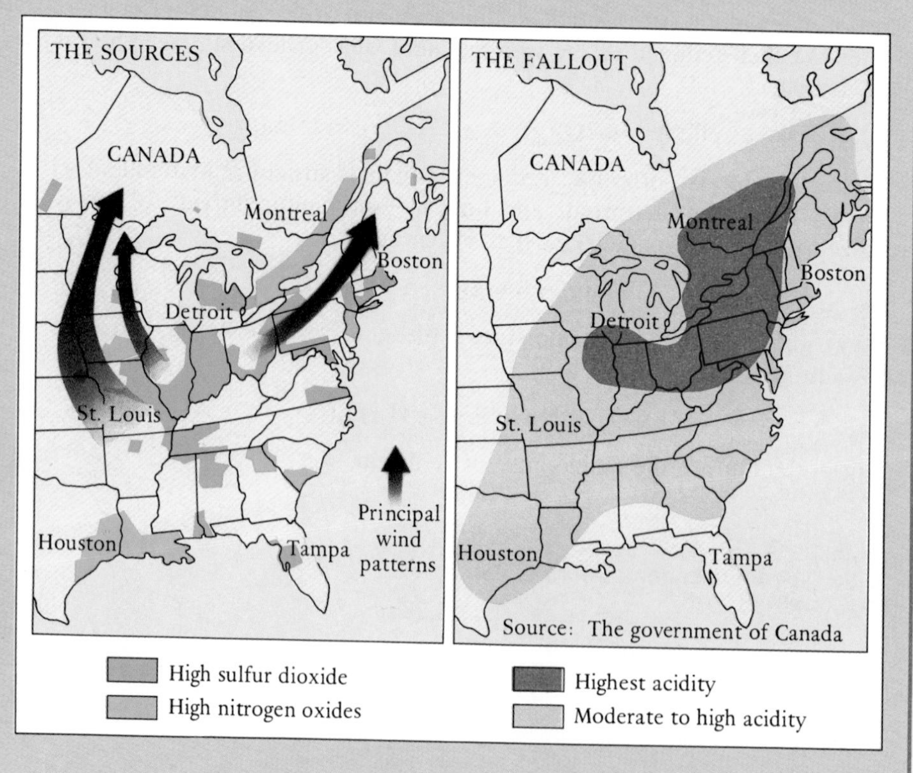

Most of the oxides of sulfur that contribute to acid rain in North America come from midwestern states. Prevailing winds carry the resulting acid droplets to the north and east, as far as Canada. Oxides of nitrogen also contribute to acid rain formation.

(H_2SO_4), nitrous (HNO_2), and nitric (HNO_3) acids. Sulfur oxides are produced mainly by the combustion of high-sulfur coal and oil and by the smelting of metal ores. Coal-burning utilities in the midwestern United States appear to be largely responsible for the presence of sulfur oxides in North American acid rain. Nitrogen oxides are by-products of gasoline combustion in automobile and airplane engines and of some processes that generate electricity. After these compounds are emitted into the atmosphere, they may be transported thousands of miles before returning to earth in solution with rain, sleet, or snow.

As a result of the great distances traveled by sulfur and nitrogen oxides before they result in acid rain, what was once assumed to be a harmless dispersal of contaminants has become an international pollution problem. For instance, about 50% of the sulfates falling in eastern Canada are believed to have originated in the United States, a fact which has been a source of controversy between the two countries. Similarly, much of the acidic precipitation in Scandinavia originates in industrialized areas of central Europe and the United Kingdom.

The consequences of acid rain depend in part on the characteristics of the soil and underlying rock upon which it falls. In areas where the principal rock is limestone (calcium carbonate), there is a natural buffer system that can prevent acidification of soil, lakes, and streams to some extent. In other areas where the soil and bodies of water do not contain such a natural buffer, the pH drops gradually as a result of acid precipitation. At times, it drops quite suddenly—for instance, when the spring melt causes all of the acids that have accumulated in the winter snows and ice to be released into soils and waters over a short period of time. Although the low pH resulting from the spring melt is usually temporary, it can be devastating to aquatic plant life and to many animals, such as small fish and frogs.

Although there is considerable debate about the extent to which acid deposition constitutes an environmental risk, the effects of acid rain on aquatic and terrestrial ecosystems have been amply documented. Many lakes are virtually sterile, devoid of fish as well as most plants and invertebrates. For example, in the Adirondacks region of the northeastern United States, more than 200 lakes have been rendered totally sterile by the effects of acid rain. Of the 100,000 lakes in Sweden, 4000 have become fishless. In acidified lakes in which fish still exist, the diversity of species has decreased and the life spans of the fish are greatly reduced. Some food chains have also been shortened due to the elimination of sensitive species at intermediate points in the chain. The disappearance of lower organisms may cause starvation of large predatory animals well before direct toxic action of hydrogen ions is evident. In terrestrial systems, trees and plants are suffering from nutrient deficiency, reduced efficiency of photosynthesis, and lowered resistance to disease. This is thought to be due mainly to the

Effects of acid rain on evergreen forests. The two photographs were taken at Camel's Hump, Vermont, 15 years apart.

Effects of acid rain on statues. The photo at the left was taken at the Lincoln Cathedral in England in 1910; the one at the right was taken in 1984.

leaching of important nutrients, such as calcium and magnesium, from the soils. Other elements, such as aluminum and trace metals, are leached from the soil and are subsequently washed into lakes and streams, where they are quite toxic to fishes and other organisms. The acidification of ground water also causes public water supplies to become acidified. Acidic water supplies dissolve metals from plumbing, creating a drinking-water hazard in some areas.

The effects of acid rain on pH can be offset by adding calcium carbonate, $CaCO_3$, to react with excess hydrogen ions. Here a helicopter sprays finely ground limestone (mostly $CaCO_3$) over forests affected by acid rain.

Acid rain has also done considerable damage to structures. Rates of corrosion of metal structures are greatly increased by acid rain. Exfoliation (flaking) of marble and limestone monuments and buildings is common, as is the pitting of granitic stonework. Ancient ruins such as the Acropolis in Greece are also showing signs of erosion by acid rain.

Acid rain is a serious worldwide pollution problem. Beyond its detrimental effect on other species, its potential consequences for humans are enormous: lowered crop yields, decreased timber production, and loss of important fishing and recreational areas as well as public water supplies. It is interesting to note that 21 European countries agreed in 1985 to lower their sulfur dioxide emissions by 30% or more over a 10-year period. By 1989, more than half of those countries had already reached that goal. In 1988, the Canadian government announced a goal of lowering its sulfur dioxide emissions by half by 1994. In the United States, progress has not been so rapid.

Measures to counteract the effects of acid rain have only limited effectiveness. The available processes that remove sulfur and nitrogen oxides *at the source,* before they enter the atmosphere to form acid rain, are expensive and so meet with resistance. But the monetary and ecological costs of allowing the conditions that create acid rain to continue are also very great. Although scientists can provide the information upon which to base important decisions regarding the acid rain problem, the ultimate decisions about alleviating the problem will have to be made in the political arena.

Beth A. Trust
Graduate student in chemistry
University of Texas Marine Sciences Institute

Figure 6-13
Photochemical pollution (a brown haze) enveloping a city.

Compounds of nitrogen are also impurities in fossil fuels, and they undergo combustion to form nitric oxide, NO. However, most of the nitrogen in the NO in exhaust gases from furnaces, automobiles, airplanes, and so on comes from the air that is mixed with the fuel:

Remember that "clean air" is *about* 80% N_2 and 20% O_2 by mass. This reaction does *not* occur at room temperature but does occur at the high temperatures of furnaces, internal combustion engines, and jet engines.

$$\overset{(0)}{N_2}(g) + O_2(g) \longrightarrow 2\overset{(+2)}{N}O(g)$$

NO can be further oxidized by oxygen to nitrogen dioxide, NO_2; this reaction is enhanced in the presence of ultraviolet light from the sun:

$$2\overset{(+2)}{N}O(g) + O_2(g) \xrightarrow[\text{light}]{\text{uv}} 2\overset{(+4)}{N}O_2(g) \qquad \text{(a reddish-brown gas)}$$

NO_2 is responsible for the reddish-brown haze that hangs over many cities on sunny afternoons (Figure 6-13) and probably for most of the respiratory problems associated with this kind of air pollution. It can react to produce other oxides of nitrogen and other secondary pollutants.

In addition to being a pollutant itself, nitrogen dioxide reacts with water in the air to form nitric acid, another major contributor to acid rain:

$$3NO_2(g) + H_2O(\ell) \longrightarrow 2HNO_3(\ell) + NO(g)$$

Key Terms

Acid anhydride The oxide of a nonmetal that reacts with water to form an acid.

Actinides Elements 90 through 103 (after *actinium*).

Amphoterism The ability to react with both acids and bases.

Angstrom (Å) 10^{-10} meter.

Atomic radius The radius of an atom.

Basic anhydride The oxide of a metal that reacts with water to form a base.

Catalyst A substance that speeds up a chemical reaction without itself being consumed in the reaction.

Combustion reaction The reaction of a substance with oxygen in a highly exothermic reaction, usually with a visible flame.

***d*-Transition elements (metals)** The B group elements except IIB in the periodic table; sometimes called simply transition elements.

Effective nuclear charge (Z_{eff}) The nuclear charge experienced by the outermost electrons of an atom; the actual nuclear charge minus the effects of shielding due to inner shell electrons.
Electron affinity The amount of energy absorbed in the process in which an electron is added to a neutral isolated gaseous atom to form a gaseous ion with a 1− charge; has a negative value if energy is released.
Electronegativity A measure of the relative tendency of an atom to attract electrons to itself when chemically combined with another atom.
f-Transition elements See *Inner transition elements*.
Hydride A binary compound of hydrogen.
Inner transition elements Elements 58 through 71 and 90 through 103; also called *f*-transition elements.
Ionic radius The radius of an ion.
Ionization energy The minimum amount of energy required to remove the most loosely held electron of an isolated gaseous atom or ion.
Isoelectronic Having the same electron configurations.
Lanthanides Elements 58 through 71 (after *lanthanum*).
Nanometer (nm) 10^{-9} meter.
Noble gases Elements of periodic Group 0; also called rare gases; formerly called inert gases.
Noble gas configuration The stable electronic configuration of a noble gas.
Normal oxide A metal oxide containing the oxide ion, O^{2-} (oxygen in the −2 oxidation state).
Nuclear shielding See *Shielding effect*.
Oxide A binary compound of oxygen.
Periodicity Regular periodic variations of properties of elements with atomic number (and position in the periodic table).
Periodic law The properties of the elements are periodic functions of their atomic numbers.
Peroxide A compound containing oxygen in the −1 oxidation state. Metal peroxides contain the peroxide ion, O_2^{2-}.
Radical A species containing one or more unpaired electrons; many radicals are very reactive.
Rare earths Inner transition elements.
Rare gases See *Noble gases*.
Representative elements A group elements in the periodic table.
Roasting Heating an ore of an element in the presence of air.
Shielding effect Electrons in filled sets of *s* and *p* orbitals between the nucleus and outer shell electrons shield the outer shell electrons somewhat from the effect of protons in the nucleus; also called screening effect.
Superoxide A compound containing the superoxide ion, O_2^- (oxygen in the $-\frac{1}{2}$ oxidation state).
Ternary acid An acid containing three elements, H, O, and (usually) another nonmetal.

Exercises

Classification of the Elements

1. Define and illustrate the following terms clearly and concisely: (a) representative elements, (b) *d*-transition elements, (c) inner transition elements.
2. Explain why Period 1 contains two elements and Period 2 contains eight elements.
3. Explain why Period 4 contains 18 elements.
4. The third major energy level ($n = 3$) has *s*, *p*, and *d* sublevels. Why does Period 3 contain only eight elements?
5. Account for the number of elements in Period 6.

*6. What would be the atomic number of the as-yet-undiscovered alkaline earth element of Period 8?

*7. How many elements are there in Period 7? Suppose that many new elements are discovered and it is found that Period 8 consists of *more* than 32 elements. How many would you predict? What would account for this number?

*8. In what periodic group would the as-yet-undiscovered element 116 be found? Would you classify it as a metal or a nonmetal? Suggest a reasonable electron configuration for the element.

9. Identify the group, family, and/or other periodic table location of each element with the outer electron configuration
   (a) ns^2np^3,
   (b) ns^1,
   (c) $ns^2(n-1)d^{0-2}(n-2)f^{1-14}$.
10. Repeat Exercise 9 for
    (a) ns^2np^5 (c) $ns^2(n-1)d^{1-10}$
    (b) ns^2 (d) ns^2np^1.
11. Write the outer electron configurations for the (a) alkaline earth metals, (b) *d*-transition metals, and (c) halogens.
12. Repeat Exercise 11 for the (a) noble gases, (b) alkali metals, (c) *f*-transition metals, and (d) vanadium family.
13. Identify the elements and the part of the periodic table in which the elements with the following configurations are found.
    (a) $1s^22s^22p^63s^23p^64s^2$
    (b) $[Kr]4d^85s^1$
    (c) $[Xe]4f^{14}5d^66s^2$
    (d) $[Xe]4f^{12}6s^2$
    (e) $[Kr]4d^{10}5s^25p^6$
    (f) $[Kr]4d^{10}4f^{14}5s^25p^65d^{10}6s^26p^2$
14. Which of the elements in the following periodic table

Hydrogen and the Hydrides

64. Summarize the physical properties of hydrogen.

65. Write balanced formula unit equations for (a) the reaction of iron with steam, (b) the reaction of calcium with hydrochloric acid, (c) the electrolysis of water, and (d) the "water gas" reaction.

66. Write a balanced formula unit equation for the preparation of (a) an ionic hydride and (b) a molecular hydride.

67. Classify the following hydrides as molecular or ionic: (a) NaH, (b) H_2S, (c) BaH_2, (d) KH, (e) NH_3.

68. Explain why NaH and H_2S are different kinds of hydrides.

69. Write formula unit equations for the reactions of (a) NaH and (b) BaH_2 with water.

70. Name the following (pure) compounds: (a) H_2S, (b) HF, (c) KH, (d) NH_3, (e) H_2Se, (f) MgH_2.

Oxygen and the Oxides

***71.** How are O_2 and O_3 similar? Different?

72. Briefly compare and contrast the properties of oxygen with those of hydrogen.

73. Write molecular equations to show how oxygen can be prepared from (a) mercury(II) oxide, HgO, (b) hydrogen peroxide, H_2O_2, and (c) potassium chlorate, $KClO_3$.

***74.** Which of the following elements form normal oxides as the *major* products of reactions with oxygen? (a) Li, (b) Na, (c) Rb, (d) Mg, (e) Zn (exhibits only one common oxidation state), (f) Al.

75. Write formula unit equations for the primary reactions of oxygen with the following elements: (a) Li, (b) Na, (c) K, (d) Ca.

76. Write formula unit equations for the reactions of the following elements with a *limited* amount of oxygen: (a) Sr, (b) Fe, (c) Mn, (d) Cu.

77. Write formula unit equations for the reactions of the following elements with an *excess* of oxygen: (a) Sr, (b) Fe, (c) Mn, (d) Cu.

78. Write formula unit equations for the reactions of the following elements with a *limited* amount of oxygen: (a) C, (b) As_4, (c) Ge.

79. Write formula unit equations for the reactions of the following elements with an *excess* of oxygen: (a) C, (b) As_4, (c) Ge.

***80.** Distinguish among normal oxides, peroxides, and superoxides. What is the oxidation state of oxygen in each case?

81. Which of the following can be classified as basic anhydrides? (a) SO_2, (b) Li_2O, (c) SeO_3, (d) CaO, (e) N_2O_5.

82. Write balanced formula unit equations for the following reactions and name the products:
(a) sulfur dioxide, SO_2, with water
(b) sulfur trioxide, SO_3, with water
(c) selenium trioxide, SeO_3, with water
(d) dinitrogen pentoxide, N_2O_5, with water
(e) dichlorine heptoxide, Cl_2O_7, with water

83. Write balanced formula unit equations for the following reactions and name the products:
(a) sodium oxide, Na_2O, with water
(b) calcium oxide, CaO, with water
(c) lithium oxide, Li_2O, with water
(d) magnesium oxide, MgO, with sulfur dioxide, SO_2
(e) barium oxide, BaO, with carbon dioxide, CO_2

***84.** Identify the acid anhydrides of the following ternary acids: (a) H_2SO_4, (b) H_2CO_3, (c) H_2SO_3, (d) H_3AsO_4, (e) HNO_2.

85. Identify the basic anhydrides of the following metal hydroxides: (a) NaOH, (b) $Mg(OH)_2$, (c) $Fe(OH)_2$ (d) $Al(OH)_3$.

Combustion Reactions

86. Define combustion. Why are all combustion reactions also redox reactions?

87. Write equations for the complete combustion of the following compounds: (a) ethane, $C_2H_6(g)$; (b) propane, $C_3H_8(g)$; (c) ethanol, $C_2H_5OH(\ell)$.

88. Write equations for the *incomplete* combustion of the following compounds to produce carbon monoxide: (a) ethane, $C_2H_6(g)$; (b) propane, $C_3H_8(g)$.

As we have seen, two substances may react to form different products when they are mixed in different proportions under different conditions. In Exercises 89 and 90, write balanced equations for the reactions described. Assign oxidation numbers.

89. (a) Methane burns in excess air to form carbon dioxide and water.
(b) Methane burns in a limited amount of air to form carbon monoxide and water.
(c) Methane burns (poorly) in a very limited amount of air to form elemental carbon and water.

90. (a) Butane (C_4H_{10}) burns in excess air to form carbon dioxide and water.
(b) Butane burns in a limited amount of air to form carbon monoxide and water.
(c) When heated in the presence of *very little* air, butane "cracks" to form acetylene, C_2H_2; carbon monoxide; and hydrogen.

***91.** (a) How much SO_2 would be formed by burning 1.00 ton of bituminous coal that is 5.11% sulfur by mass? Assume that all of the sulfur is converted to SO_2.
(b) If 27.0% of the SO_2 escaped into the atmosphere and 84.2% of it were converted to H_2SO_4, how many grams of H_2SO_4 would be produced in the atmosphere?

***92.** Write equations for the complete combustion of the following compounds. Assume that sulfur is converted to SO_2 and nitrogen is converted to NO. (a) $C_6H_5N(\ell)$, (b) $C_2H_5SH(\ell)$, (c) $C_7H_{10}NO_2S(\ell)$.

93. Describe the formation of the reddish-brown haze of some cities experiencing this kind of air pollution.

94. Account for the occurrence of acid rain.

7 Chemical Bonding and Inorganic Nomenclature

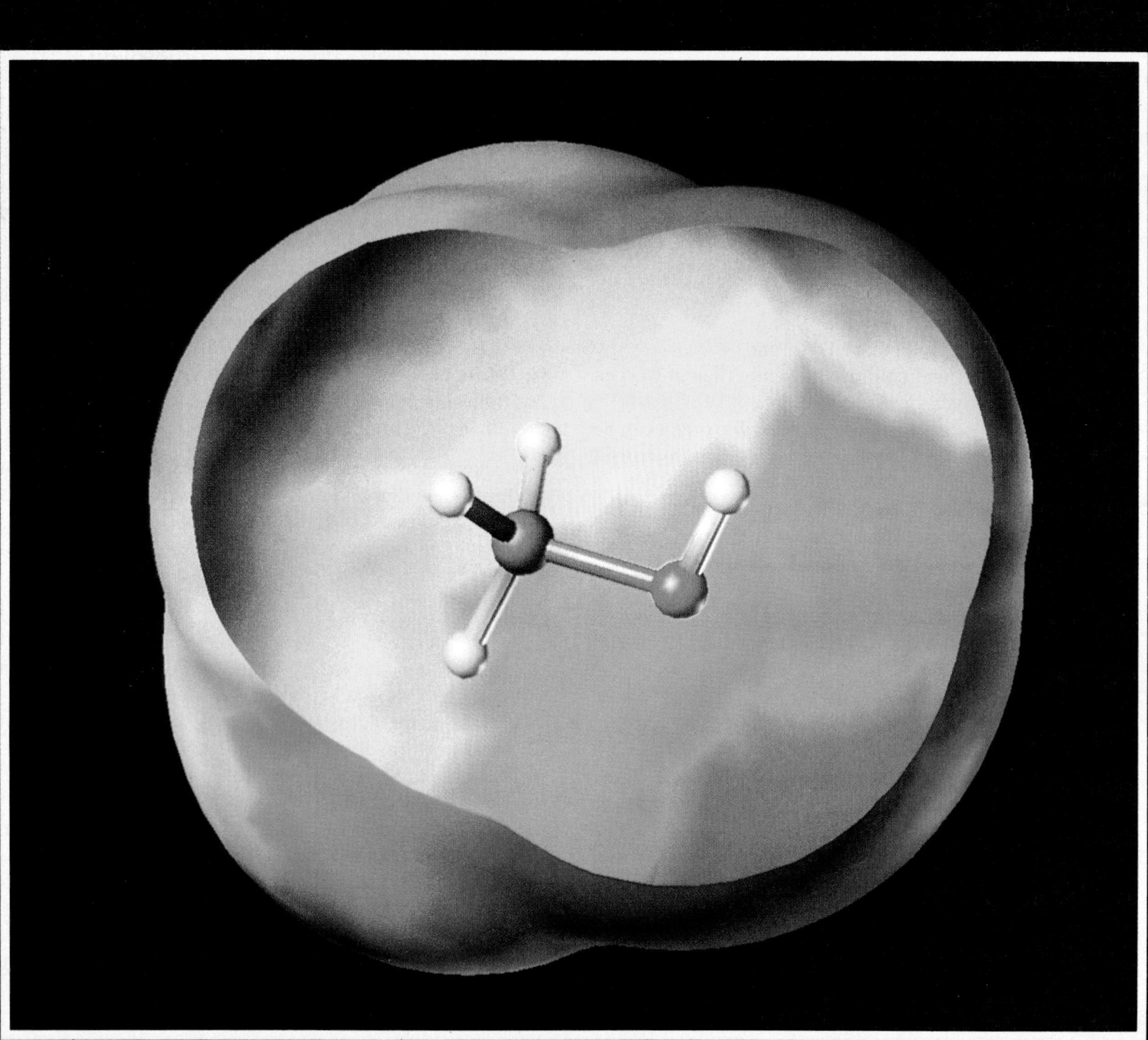

Outline

Objectives

As you study this chapter, you should learn

- ☐ To write Lewis dot representations of atoms
- ☐ To predict whether bonding will be primarily ionic, covalent, or polar covalent
- ☐ How the properties of compounds depend on their bonding
- ☐ How elements bond by electron transfer (ionic bonding)
- ☐ To predict the formulas of ionic compounds
- ☐ How elements bond by sharing electrons (covalent bonding)
- ☐ To write Lewis dot formulas for molecules and polyatomic ions
- ☐ To recognize exceptions to the octet rule
- ☐ To relate the nature of the bonding to electronegativity differences
- ☐ About resonance, when to write resonance structures, and how to do so
- ☐ How to write formal charges for atoms in covalent structures

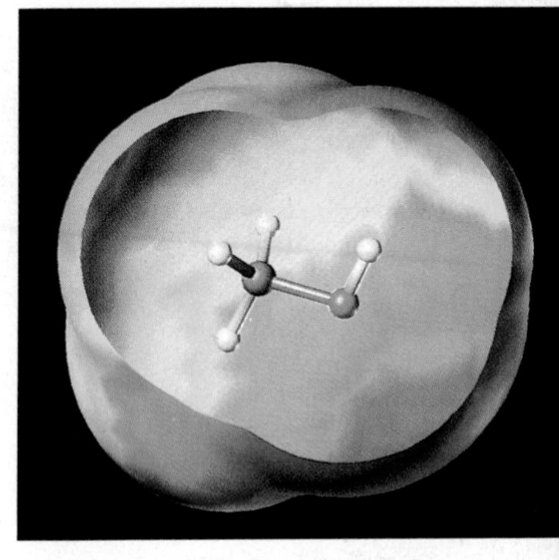

A computer model of wood alcohol, or methanol, CH_3OH (C gray, H white, O red). The ball-and-stick model is shown inside a computer-generated molecular surface. The surface is color-coded to show relative charge densities.

The attractive forces that hold atoms together in compounds are called **chemical bonds**. There are two major classes of bonding. (1) **Ionic bonding** results from electrostatic interactions among ions, which can take the form of the *transfer* of one or more electrons from one atom or group of atoms to another. (2) **Covalent bonding** results from *sharing* one or more electron pairs between two atoms. These two classes represent two extremes; all bonds have at least some degree of both ionic and covalent character. Compounds containing predominantly ionic bonding are called **ionic compounds**. Those containing predominantly covalent bonds are called **covalent compounds**. Some of the properties associated with many simple ionic and covalent compounds in the extreme cases are summarized below. The differences in properties can be accounted for by the differences in bonding between the atoms or ions.

Ionic Compounds	Covalent Compounds
1. They are solids with high melting points (typically >400°C).	1. They are gases, liquids, or solids with low melting points (typically <300°C).
2. Many are soluble in polar solvents such as water.	2. Many are insoluble in polar solvents.
3. Most are insoluble in nonpolar solvents, such as hexane, C_6H_{14}.	3. Most are soluble in nonpolar solvents, such as hexane, C_6H_{14}.
4. Molten compounds conduct electricity well because they contain mobile charged particles (ions).	4. Liquid and molten compounds do not conduct electricity.
5. Aqueous solutions conduct electricity well because they contain mobile charged particles (ions).	5. Aqueous solutions are *usually* poor conductors of electricity because most do not contain charged particles.

The distinction between polar and nonpolar molecules is made in Section 7-4.

As we saw in Section 4-2, aqueous solutions of some covalent compounds do conduct electricity.

7-1 Lewis Dot Representations of Atoms

The number and arrangements of electrons in the outermost shells of atoms determine the chemical and physical properties of the elements as well as the kinds of chemical bonds they form. We write **Lewis dot representations** (or **Lewis dot formulas**) as a convenient bookkeeping method for keeping track of these "chemically important electrons." We now introduce this

Table 7-1
Lewis Electron Dot Formulas for Representative Elements

Group	IA	IIA	IIIA	IVA	VA	VIA	VIIA	0
Number of electrons in outer shell	1	2	3	4	5	6	7	8 (except He)
Row 1	H·							He:
Row 2	Li·	Be:	B·	C·	·N·	·O:	·F:	:Ne:
Row 3	Na·	Mg:	Al·	Si·	·P·	·S:	·Cl:	:Ar:
Row 4	K·	Ca:	Ga·	Ge·	·As·	·Se:	·Br:	:Kr:
Row 5	Rb·	Sr:	In·	Sn·	·Sb·	·Te:	·I:	:Xe:
Row 6	Cs·	Ba:	Tl·	Pb·	·Bi·	·Po:	·At:	:Rn:
Row 7	Fr·	Ra:						

method for atoms of elements; in our discussion of chemical bonding in subsequent sections, we shall frequently use such formulas for atoms, molecules, and ions.

Chemical bonding usually involves only the outermost electrons of atoms, also called **valence electrons**. In Lewis dot representations, only the electrons in the outermost occupied *s* and *p* orbitals are shown as dots. Paired and unpaired electrons are also indicated. Table 7-1 shows Lewis dot formulas for the representative elements. All elements in a given group have the same outer shell configuration.

Because of the large numbers of dots, such formulas are not as useful for the transition and inner transition elements.

Ionic Bonding

7-2 Formation of Ionic Compounds

The first kind of chemical bonding we shall describe is the **ionic** or **electrovalent bonding**. Ionic bonding results from the *transfer of one or more electrons from one atom or group of atoms to another.* As our previous discussions of ionization energy, electronegativity, and electron affinity would indicate, ionic bonding occurs most easily when elements that have low ionization energies (metals) react with elements having high electronegativities and high electron affinities (nonmetals). Many metals are easily *oxidized*—that is, they lose electrons; and many nonmetals are readily *reduced*—that is, they gain electrons.

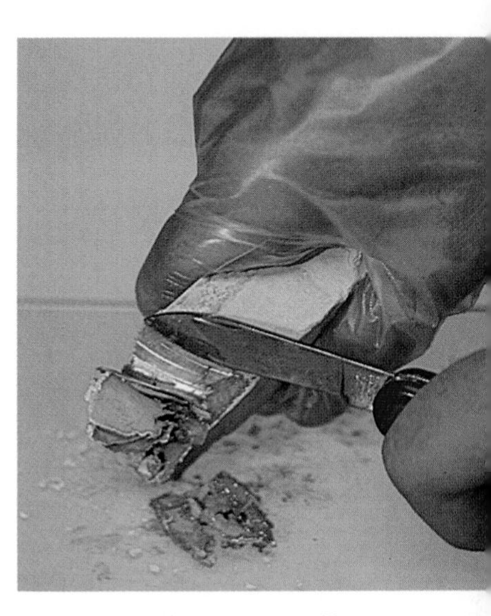

Freshly cut sodium has a metallic luster. A little while after being cut, the sodium metal surface turns white as it reacts with the air.

When the electronegativity difference, ΔEN, between two elements is large, the elements are likely to form a compound by ionic bonding (transfer of electrons).

Group IA Metals and Group VIIA Nonmetals

Consider the reaction of sodium (a Group IA metal) with chlorine (a Group VIIA nonmetal). Sodium is a soft silvery metal (mp 98°C), and chlorine is a yellowish-green corrosive gas at room temperature. Both sodium and chlorine react with water, sodium vigorously. By contrast, sodium chloride is a white solid (mp 801°C) that dissolves in water with no reaction and with the absorption of just a little heat. We can represent the reaction for its formation as

$$\underset{\text{sodium}}{2Na(s)} + \underset{\text{chlorine}}{Cl_2(g)} \longrightarrow \underset{\text{sodium chloride}}{2NaCl(s)}$$

We can understand this reaction in more detail by showing electron configurations for all species. We represent chlorine as individual atoms rather than molecules, for simplicity.

$$\left.\begin{array}{lll} & & \underline{3s} \quad \underline{\quad 3p \quad} \\ {}_{11}Na & [Ne] & \underline{\uparrow} \\ {}_{17}Cl & [Ne] & \underline{\uparrow\downarrow} \quad \underline{\uparrow\downarrow}\ \underline{\uparrow\downarrow}\ \underline{\uparrow} \end{array}\right\} \longrightarrow \left\{\begin{array}{llll} & & \underline{3s} \quad \underline{\quad 3p \quad} & \\ Na^+ & [Ne] & \underline{\quad} & 1e^- \text{ lost} \\ Cl^- & [Ne] & \underline{\uparrow\downarrow} \quad \underline{\uparrow\downarrow}\ \underline{\uparrow\downarrow}\ \underline{\uparrow\downarrow} & 1e^- \text{ gained} \end{array}\right.$$

Halite crystals (naturally occurring NaCl).

The loss of electrons is *oxidation* (Section 4-8). Na atoms are *oxidized* to form Na^+ ions.

The gain of electrons is *reduction* (Section 4-8). Cl atoms are *reduced* to form Cl^- ions.

In this reaction, Na atoms lose one electron each to form Na^+ ions, which contain only ten electrons, the same number as the *preceding* noble gas, neon. We say that sodium ions have the neon electronic structure; Na^+ is *isoelectronic* with Ne. In contrast, chlorine atoms gain one electron each to form chloride ions, Cl^-, which contain 18 electrons. This is the same number as the *next* noble gas, argon; Cl^- is *isoelectronic* with Ar. Similar observations apply to most ionic compounds formed by reactions between *representative metals and representative nonmetals.*

We can use Lewis dot formulas (Section 7-1) to represent the reaction.

$$Na\cdot + :\ddot{\underset{\cdot\cdot}{Cl}}\cdot \longrightarrow Na^+[:\ddot{\underset{\cdot\cdot}{Cl}}:]^-$$

The formula for sodium chloride, NaCl, indicates that the compound contains Na and Cl atoms in a 1:1 ratio. This is the formula we predict based on the fact that each Na atom contains only one electron in its highest energy level and each Cl atom needs only one electron to fill completely its outermost p orbitals.

The chemical formula NaCl does not explicitly indicate the ionic nature of the compound, only the ratio of atoms. So we must learn to recognize, from positions of elements in the periodic table and known trends in electronegativity, when the difference in electronegativity is large enough to favor ionic bonding.

The noble gases are excluded from this generalization.

The farther apart across the periodic table two elements are, the more likely they are to form an ionic compound.

The greatest difference in electronegativity occurs from lower left to upper right. Thus, CsF is more ionic than LiI.

All the Group IA metals (Li, Na, K, Rb, Cs) will react with the Group VIIA elements (F, Cl, Br, I) to form ionic compounds of the same general formula, MX. The resulting ions, M^+ and X^-, always have noble gas configurations. We can represent the general reaction of the IA metals with the VIIA elements as follows:

$$2M(s) + X_2 \longrightarrow 2MX(s) \qquad M = Li, Na, K, Rb, Cs;\ X = F, Cl, Br, I$$

The Lewis dot representation for the generalized reaction is

$$2M\cdot + :\ddot{\underset{\cdot\cdot}{X}}:\ddot{\underset{\cdot\cdot}{X}}: \longrightarrow 2\,(M^+[:\ddot{\underset{\cdot\cdot}{X}}:]^-)$$

Coulomb's Law is $F \propto \frac{q^+q^-}{d^2}$. The symbol $\propto$ means "is proportional to."

Because of the opposite charges on Na^+ and Cl^-, an attractive force is developed. According to Coulomb's Law, the force of attraction, F, between two oppositely charged particles of charge magnitudes q^+ and q^- is directly proportional to the product of the charges and inversely proportional to the square of the distance separating their centers, d. Thus, the greater the charges on the ions and the smaller the ions are, the stronger the resulting ionic bonding. Of course, like-charged ions repel each other, so the distances separating ions in solids are those at which the attractions exceed the repulsions by the greatest amount. The structure of common table salt, sodium chloride (NaCl), is shown in Figure 7-1. Like other simple ionic compounds,

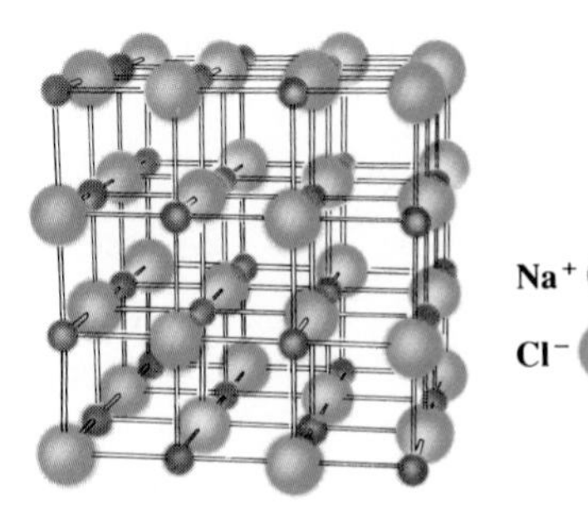

Figure 7-1
The crystal structure of NaCl, expanded for clarity. Each Cl^- (green) is surrounded by six sodium ions, and each Na^+ (gray) is surrounded by six chlorides. The crystal includes billions of ions in the pattern shown. Compare with Figure 2-7, a space-filling drawing of the NaCl structure.

Figure 7-1 has been expanded to show the spatial arrangement of ions. Adjacent ions actually are in contact with each other. The lines *do not* represent formal chemical bonds. They have been drawn to emphasize the spatial arrangement of ions.

NaCl(s) exists in a regular, extended array of positive and negative ions, Na^+ and Cl^-.

Distinct molecules of solid ionic substances do not exist, so we must refer to *formula units* (Section 2-3) instead of molecules. The sum of all the forces that hold all the particles in an ionic solid is quite large. This explains why such substances have quite high melting and boiling points (a topic that we will discuss more fully in Chapter 13). When an ionic compound is melted or dissolved in water, its charged particles are free to move in an electric field, so such a liquid shows high electrical conductivity (Section 4-2, part 1).

Group IA Metals and Group VIA Nonmetals

Next, consider the reaction of lithium (Group IA) with oxygen (Group VIA) to form lithium oxide, a solid ionic compound (mp >1700°C). We may represent the reaction as

$$\underset{\text{lithium}}{4Li(s)} + \underset{\text{oxygen}}{O_2(g)} \longrightarrow \underset{\text{lithium oxide}}{2Li_2O(s)}$$

The formula for lithium oxide, Li_2O, indicates that two atoms of lithium combine with one atom of oxygen. If we examine the structures of the atoms

Lithium is a metal, as the shiny surface of freshly cut Li shows. Where it has been exposed to air, the surface is covered with lithium oxide.

Each Li atom has 1 e^- in its valence shell. Each O atom has 6 e^- in its valence shell and needs 2 e^- more to give it a noble gas configuration. The Li^+ ions are formed by oxidation of Li atoms, and the O^{2-} ions are formed by reduction of O atoms.

before reaction, we can see why two lithium atoms react with one oxygen atom.

	1s	2s	2p			1s	2s	2p	
$_3Li$	⇅	↑			Li^+	⇅	_		1 e^- lost
$_3Li$	⇅	↑		⟶	Li^+	⇅	_		1 e^- lost
$_8O$	⇅	⇅	⇅ ↑ ↑		O^{2-}	⇅	⇅	⇅ ⇅ ⇅	2 e^- gained

The Lewis dot formulas for the atoms and ions are

$$2Li\cdot + :\ddot{\underset{\cdot}{O}}\cdot \longrightarrow 2Li^+[:\ddot{\underset{\cdot\cdot}{O}}:]^{2-}$$

Lithium ions, Li^+, are isoelectronic with helium atoms. Oxide ions, O^{2-}, are isoelectronic with neon atoms (10 e^-).

The very small size of the Li^+ ion gives it a much higher *charge density* than that of the larger Na^+ (Figure 6-1). Similarly, O^{2-} is smaller than Cl^-, so its doubly negative charge gives it a much higher charge density. These more concentrated charges and smaller sizes bring the Li^+ and O^{2-} ions closer together than the Na^+ and Cl^- ions are in NaCl. Consequently, the q^+q^- product in the numerator of Coulomb's Law is greater in Li_2O, and the d^2 term in the denominator is smaller. The net result is that the ionic bonding is much stronger in Li_2O than in NaCl. This accounts for the higher melting temperature of Li_2O (>1700°C) compared to NaCl (801°C).

Group IIA Metals and Group VIA Nonmetals

Calcium (Group IIA) reacts with oxygen (Group VIA) to form calcium oxide, a white solid ionic compound with a very high melting point, 2580°C.

$$\underset{\text{calcium}}{2Ca(s)} + \underset{\text{oxygen}}{O_2(g)} \longrightarrow \underset{\text{calcium oxide}}{2CaO(s)}$$

Again, we write out the electronic structures of the atoms and ions, representing the inner electrons by the symbol of the preceding noble gas (in brackets).

In writing equations in which electrons in different energy levels in different atoms are involved, atomic orbitals are more conveniently labeled *under* the lines that represent them.

$_{20}Ca$	[Ar]	⇅ (4s)		⟶	Ca^{2+}	[Ar]	_ (4s)		2 e^- lost
$_8O$	[He]	⇅ (2s)	⇅ ↑ ↑ (2p)		O^{2-}	[He]	⇅ (2s)	⇅ ⇅ ⇅ (2p)	2 e^- gained

The Lewis dot notation for the atoms and ions is

$$Ca: + :\ddot{\underset{\cdot}{O}}\cdot \longrightarrow Ca^{2+}[:\ddot{\underset{\cdot\cdot}{O}}:]^{2-}$$

Calcium ions, Ca^{2+}, are isoelectronic with argon (18 e^-), the preceding noble gas. Oxide ions, O^{2-}, are isoelectronic with neon (10 e^-), the following noble gas.

Ca^{2+} is about the same size as Na^+ (Figure 6-1) but carries twice the charge, so its charge density is higher. Because the attraction between the two small, highly charged ions Ca^{2+} and O^{2-} is quite high, the ionic bonding is very strong, accounting for the very high melting point of CaO, 2580°C.

Group IIA Metals and Group VA Nonmetals

As our final example of ionic bonding, consider the reaction of magnesium (Group IIA) with nitrogen (Group VA). At elevated temperatures, they form magnesium nitride, Mg_3N_2, a white, solid ionic compound that decomposes at 800°C. We can represent the reaction as

$$\underset{\text{magnesium}}{3Mg(s)} + \underset{\text{nitrogen}}{N_2(g)} \xrightarrow{\Delta} \underset{\text{magnesium nitride}}{Mg_3N_2(s)}$$

The formula Mg_3N_2 indicates that magnesium and nitrogen atoms combine in a 3:2 ratio. We could predict this from the fact that Mg atoms contain two electrons in their highest energy level, whereas N atoms need three electrons to fill theirs completely. Thus, three Mg atoms lose two electrons each for a total of 6 e^-, and two N atoms gain three electrons each, also for a total of 6 e^-. As before, examination of the electronic structures makes the picture clearer.

$$\left.\begin{matrix} 3 \times \left({}_{12}\text{Mg} \quad [\text{Ne}] \; \underset{3s}{\underline{\uparrow\downarrow}} \right) \\ 2 \times \left({}_{7}\text{N} \quad [\text{He}] \; \underset{2s}{\underline{\uparrow\downarrow}} \; \underset{2p}{\underline{\uparrow}\,\underline{\uparrow}\,\underline{\uparrow}} \right) \end{matrix}\right\} \xrightarrow{\Delta} \left\{\begin{matrix} 3 \times \left(\text{Mg}^{2+} \quad [\text{Ne}] \; \underset{3s}{\underline{\quad}} \right) & (2\,e^- \text{ lost}) \times 3 \\ 2 \times \left(\text{N}^{3-} \quad [\text{He}] \; \underset{2s}{\underline{\uparrow\downarrow}} \; \underset{2p}{\underline{\uparrow\downarrow}\,\underline{\uparrow\downarrow}\,\underline{\uparrow\downarrow}} \right) & (3\,e^- \text{ gained}) \times 2 \end{matrix}\right.$$

Using Lewis dot formulas for the atoms and ions, we have

$$3\text{Mg}: + 2\cdot\ddot{\underset{\cdot}{\text{N}}}\cdot \longrightarrow 3\text{Mg}^{2+}, 2[:\ddot{\underset{\cdot\cdot}{\text{N}}}:]^{3-}$$

Both Mg^{2+} and N^{3-} ions are isoelectronic with neon (10 e^-).

Table 7-2 summarizes the general formulas of binary ionic compounds formed by the representative elements. "M" represents metals and "X"

Binary compounds contain *two* elements.

Table 7-2
Simple Binary Ionic Compounds

Metal		Nonmetal		General Formula	Ions Present	Example	mp (°C)
IA*	+	VIIA	⟶	MX	(M^+, X^-)	LiBr	547
IIA	+	VIIA	⟶	MX_2	$(M^{2+}, 2X^-)$	$MgCl_2$	708
IIIA	+	VIIA	⟶	MX_3	$(M^{3+}, 3X^-)$	GaF_3	800 (subl)
IA*†	+	VIA	⟶	M_2X	$(2M^+, X^{2-})$	Li_2O	>1700
IIA	+	VIA	⟶	MX	(M^{2+}, X^{2-})	CaO	2580
IIIA	+	VIA	⟶	M_2X_3	$(2M^{3+}, 3X^{2-})$	Al_2O_3	2045
IA*	+	VA	⟶	M_3X	$(3M^+, X^{3-})$	Li_3N	840
IIA	+	VA	⟶	M_3X_2	$(3M^{2+}, 2X^{3-})$	Ca_3P_2	~1600
IIIA	+	VA	⟶	MX	(M^{3+}, X^{3-})	AlP	

* Hydrogen is considered a nonmetal. All binary compounds of hydrogen are covalent except certain metal hydrides such as NaH and CaH_2, which contain hydride, H^-, ions.

† As we saw in Section 6-8, part 2, the metals in Groups IA and IIA also commonly form peroxides (containing the O_2^{2-} ion) or superoxides (containing the O_2^- ion). See Table 6-4. The peroxide and superoxide ions contain atoms that are covalently bonded to one another.

represents nonmetals from the indicated groups. In these examples of ionic bonding, each of the metal atoms has lost one, two, or three electrons and each of the nonmetal atoms has gained one, two, or three electrons. *Simple (monatomic) ions rarely have charges greater than 3+ or 3−.* Ions with greater charges interact so strongly with the electron clouds of other ions in compounds that electron clouds are distorted severely, and considerable covalent character in the bonds results. (This point will be developed later.)

The distortion of the electron cloud of an anion by a small, highly charged cation is called *polarization.*

The *d*- and *f*-transition elements form many compounds that are essentially ionic in character. Most simple ions of the transition metals do not have noble gas configurations.

Enrichment

Introduction to Energy Relationships in Ionic Bonding

The following discussion may help you to understand why ionic bonding occurs between elements with low ionization energies and those with high electronegativities. There is a general tendency in nature to achieve stability. One way to do this is by lowering potential energy; *lower* energies generally represent *more stable* arrangements.

A more thorough discussion of these energy changes will be provided in Section 15-10.

Let us use energy relationships to describe why the ionic solid NaCl is more stable than a mixture of individual Na and Cl atoms. Consider a gaseous mixture of one mole of sodium atoms and one mole of chlorine atoms, Na(g) + Cl(g). The energy change associated with the loss of one mole of electrons by one mole of Na atoms to form one mole of Na^+ ions (Step 1 in Figure 7-2) is given by the *first ionization energy* of Na (Section 6-3).

$$Na(g) \longrightarrow Na^+(g) + e^- \qquad \text{first ionization energy} = 496 \text{ kJ/mol}$$

This is a positive value, so the mixture $Na^+(g) + e^- + Cl(g)$ is 496 kJ/mol higher in energy than the original mixture of atoms (the mixture $Na^+ + e^- + Cl$ is *less stable* than the mixture of atoms). The energy change for the gain of one mole of electrons by one mole of Cl atoms to form one mole of Cl^- (Step 2) is given by the *electron affinity* of Cl (Section 6-4).

$$Cl(g) + e^- \longrightarrow Cl^-(g) \qquad \text{electron affinity} = -348 \text{ kJ/mol}$$

This negative value, −348 kJ/mol, lowers the energy of the mixture, but the mixture of separated ions, $Na^+ + Cl^-$, is still *higher* in energy by (496 − 348) kJ/mol = 148 kJ/mol than the original mixture of atoms (the red arrow in Figure 7-2). Thus, just the formation of ions does not explain why the process occurs. The strong attractive force between ions of opposite charge draws the ions together into the regular array shown in Figure 7-1. The energy associated with this attraction (Step 3) is the *crystal lattice energy* of NaCl, −790 kJ/mol.

$$Na^+(g) + Cl^-(g) \longrightarrow NaCl(s) \qquad \text{crystal lattice energy} = -790 \text{ kJ/mol}$$

The crystal (solid) formation thus further *lowers* the energy to (148 − 790) kJ/mol = −642 kJ/mol. The overall result is that one mole of NaCl(s) is 642 kJ/mol lower in energy (more stable) than the original mixture of atoms (the blue arrow in Figure 7-2). Thus we see that a major driving force for the formation of ionic compounds is the large electrostatic stabilization due to the attraction of the ionic charges (Step 3).

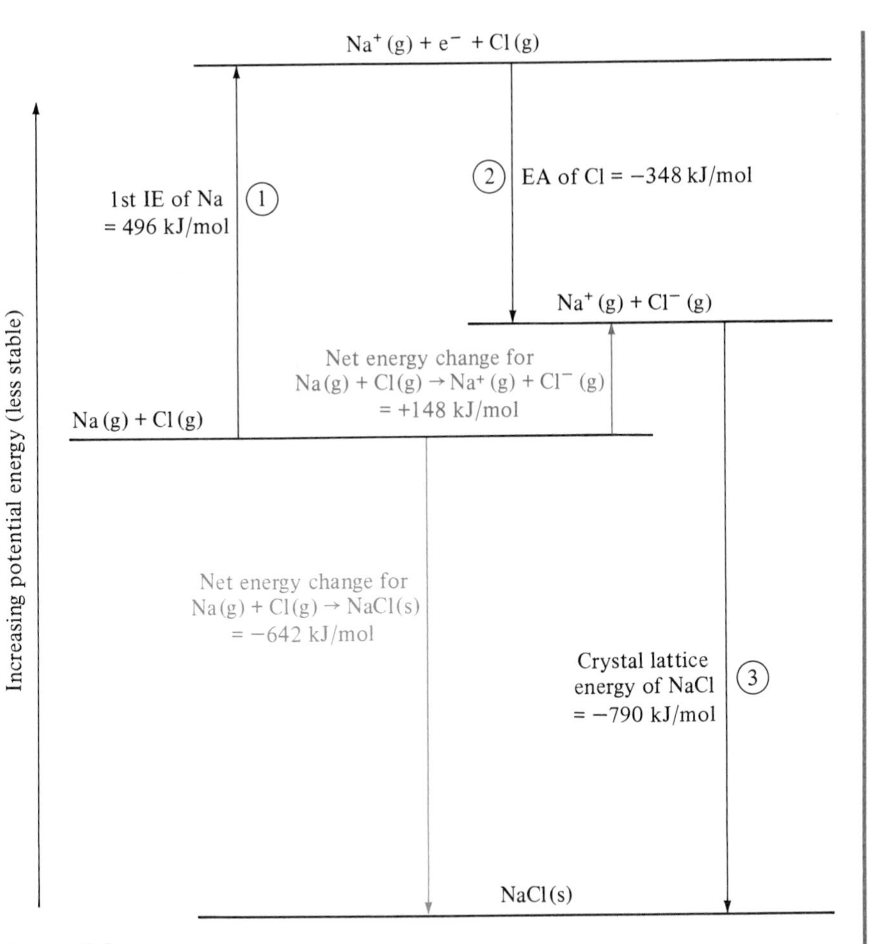

Figure 7-2
A schematic representation of the energy changes that accompany the process $Na^+(g) + Cl^-(g) \rightarrow NaCl(s)$. The red arrow represents the *positive* energy change (unfavorable) for the process of ion formation, $Na(g) + Cl(g) \rightarrow Na^+(g) + Cl^-(g)$. The blue arrow represents the *negative* energy change (favorable) for the overall process, including the formation of the ionic solid.

In this discussion we have not taken into account the fact that sodium is a solid metal or that chlorine actually exists as diatomic molecules. The additional energy changes involved when these are changed to gaseous Na and Cl atoms, respectively, are sufficiently small that the overall energy change starting from Na(s) and $Cl_2(g)$ is still negative. (This will be taken into account in the more thorough discussion in Section 15-10.)

Covalent Bonding

Ionic bonding cannot result from a reaction between two nonmetals, because their electronegativity difference is not great enough for electron transfer to take place. Instead, reactions between two nonmetals result in *covalent bonding*.

A **covalent bond** is formed when two atoms share one or more pairs of electrons. Covalent bonding occurs when the electronegativity difference, ΔEN, between elements (atoms) is zero or relatively small.

In predominantly covalent compounds the bonds between atoms *within* a molecule (*intra*molecular bonds) are relatively strong, but the forces of attraction *between* molecules (*inter*molecular forces) are relatively weak. As a result, covalent compounds have lower melting and boiling points than ionic compounds. By contrast, distinct molecules of solid ionic substances do not exist. The sum of the attractive forces of all the interactions in an ionic solid is substantial, and such a compound has high melting and boiling points. The relation of bonding types to physical properties of liquids and solids will be developed more fully in Chapter 13.

7-3 Formation of Covalent Bonds

Of course this one electron is *unpaired.*

Let us look at a simple case of covalent bonding, the reaction of two hydrogen atoms to form the diatomic molecule H_2. As you recall, an isolated hydrogen atom has the ground state electron configuration $1s^1$, with the probability density for this one electron spherically distributed about the hydrogen nucleus (Figure 7-3a). As two hydrogen atoms approach one another, the electron of each hydrogen atom is attracted by the nucleus of the *other* hydrogen atom as well as by its own nucleus (Figure 7-3b). If these two electrons have opposite spins so that they can occupy the same region (orbital), both electrons can now preferentially occupy the region *between* the two nuclei (Figure 7-3c), because they are attracted by both nuclei. The electrons are *shared* between the two hydrogen atoms, and a single covalent bond is formed. In other words, the $1s$ orbitals *overlap* so that both electrons are now in the orbitals of both hydrogen atoms. The closer together the atoms come, the more nearly this is true. In that sense, each hydrogen atom now has the helium configuration, $1s^2$.

The bonded atoms are at lower energy than the separated atoms. This is shown in the plot of energy versus distance in Figure 7-4. However, as the two atoms get closer together, the two nuclei, being positively charged, exert an increasing repulsion on one another. At some distance, a minimum energy, −435 kJ/mol, is reached; it corresponds to the most stable arrangement and occurs at 0.74 Å, the actual distance between two hydrogen nuclei in an H_2 molecule. At greater internuclear separation, the repulsive forces diminish, but the attractive forces decrease even faster. At smaller separations, repulsive forces grow more rapidly than attractive forces.

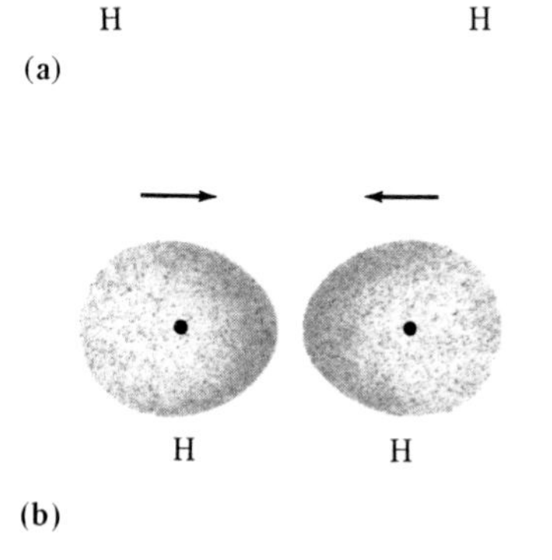

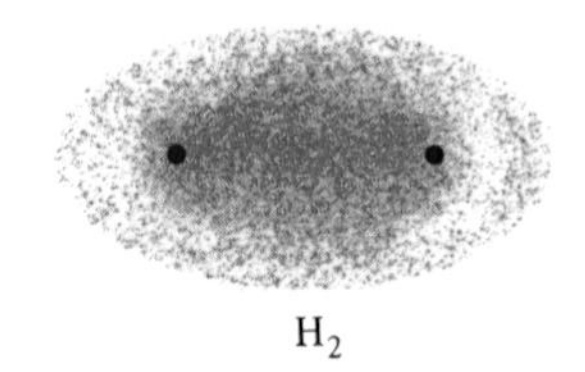

Figure 7-3
A representation of the formation of a covalent bond between two hydrogen atoms. The position of each positively charged nucleus is represented by a black dot. Electron density is indicated by the depth of shading. (a) Two hydrogen atoms separated by a large distance (essentially isolated). (b) As the atoms approach one another, the electron of each atom is attracted by the positively charged nucleus of the other atom, so the electron density begins to shift. (c) The two electrons can both occupy the region where the two $1s$ orbitals overlap; the electron density is highest in the region between the two atoms.

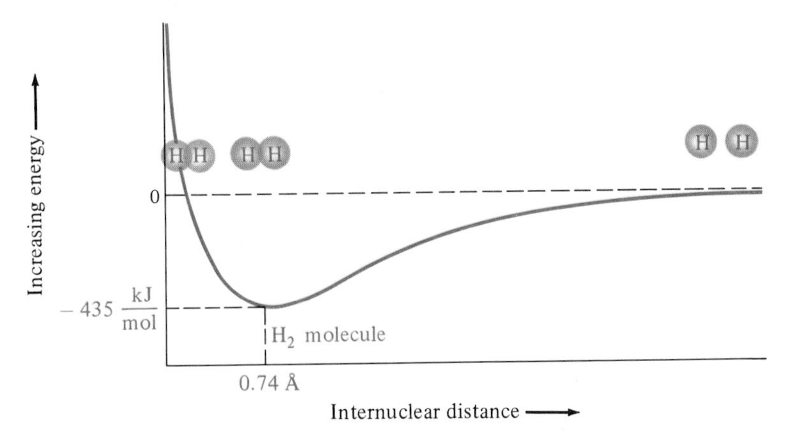

Figure 7-4
The potential energy of the H_2 molecule as a function of the distance between the two nuclei. The lowest point in the curve, −435 kJ/mol, corresponds to the internuclear distance actually observed in the H_2 molecule, 0.74 Å. (The minimum potential energy, −435 kJ/mol, corresponds to a value of -7.23×10^{-19} joule per H_2 molecule.) Energy is compared with that of two separated hydrogen atoms.

Other pairs of nonmetal atoms, and some metal atoms, share electron pairs to form covalent bonds. The result of this sharing is that each atom attains a more stable electron configuration—frequently the same as that of the nearest noble gas. (This is discussed in Section 7-8). Most covalent bonds involve two, four, or six electrons—that is, one, two, or three *pairs* of electrons. Two atoms form a **single covalent bond** when they share one pair of electrons, a **double covalent bond** when they share two electron pairs, and a **triple covalent bond** when they share three electron pairs. These are usually called simply *single, double,* and *triple* bonds. Covalent bonds involving one and three electrons are known, but are relatively rare.

We can represent the sharing of an electron pair by writing the dots in the Lewis formula between the two atom symbols. Thus, the formation of H_2 from two H atoms could be represented as

$$\text{H}\cdot + \cdot\text{H} \longrightarrow \text{H}:\text{H} \qquad \text{or} \qquad \text{H—H}$$

where the dash represents a single bond. Similarly, the combination of a hydrogen atom and a fluorine atom to form a hydrogen fluoride (HF) molecule can be shown as

$$\text{H}\cdot + \cdot\ddot{\underset{..}{\text{F}}}: \longrightarrow \text{H}:\ddot{\underset{..}{\text{F}}}: \qquad (\text{or H—}\ddot{\underset{..}{\text{F}}}:)$$

We shall see many more examples of this representation.

In our discussion, we have postulated that bonds form by the **overlap** of two atomic orbitals. This is the essence of the **valence bond theory**, which we will describe in more detail in the next chapter. Another theory, **molecular orbital theory**, is discussed in Chapter 9. For now, let us concentrate on the *number* of electron pairs shared and defer the discussion of *which* orbitals are involved in the sharing until the next chapter.

7-4 Polar and Nonpolar Covalent Bonds

Covalent bonds may be either *polar* or *nonpolar*. In a **nonpolar bond** such as that in the hydrogen molecule, H_2, the electron pair is *shared equally* between the two hydrogen nuclei. Recall (Section 6-6) that we defined electronegativity as the tendency of an atom to attract electrons to itself in a chemical bond. Both H atoms have the same electronegativity. This means

H:H or H—H

that the shared electrons are equally attracted to both hydrogen nuclei and therefore spend equal amounts of time near each nucleus. In this nonpolar covalent bond, the **electron density** is symmetrical about a plane that is perpendicular to a line between the two nuclei. This is true for all homonuclear **diatomic molecules**, such as H_2, O_2, N_2, F_2, and Cl_2, because the two identical atoms have identical electronegativities. We can generalize:

A *homonuclear* molecule contains only one kind of atom. A molecule that contains two or more kinds of atoms is described as *heteronuclear.*

> The covalent bonds in all homonuclear diatomic molecules must be nonpolar.

Let us now consider **heteronuclear diatomic molecules**. Start with the fact that hydrogen fluoride, HF, is a gas at room temperature. This tells us that it is a covalent compound. We also know that the H—F bond has some degree of polarity because H and F are not identical atoms and therefore do not attract the electrons equally. But how polar will this bond be?

Remember that ionic compounds are solids at room temperature.

H:F: (with two additional lone pairs on F)

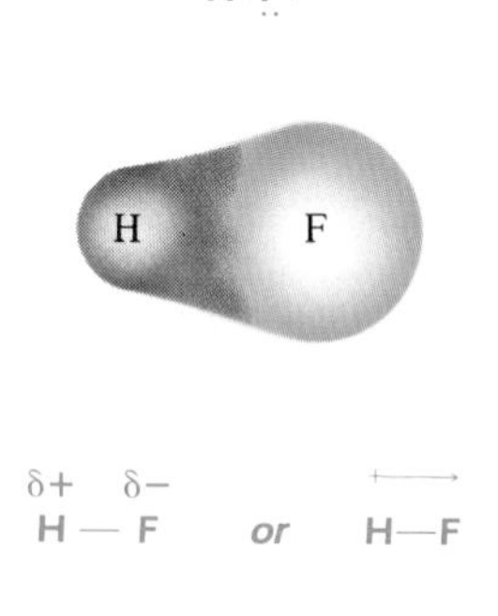

The electronegativity of hydrogen is 2.1, and that of fluorine is 4.0 (Table 6-3). Clearly, the F atom, with its higher electronegativity, attracts the shared electron pair much more strongly than does H. We represent the structure of HF as shown in the margin. Notice the unsymmetrical distribution of electron density; the electron density is distorted in the direction of the more electronegative F atom. This small shift of electron density leaves H somewhat positive.

$\overset{\delta+}{H} — \overset{\delta-}{F}$ *or* $\overset{\longmapsto}{H—F}$

Covalent bonds, such as the one in HF, in which the *electron pairs are shared unequally* are called **polar covalent bonds**. Two kinds of notation used to indicate polar bonds are shown in the margin.

The $\delta-$ over the F atom indicates a "partial negative charge." This means that the F end of the molecule is negative *with respect* to the H end. The $\delta+$ over the H atom indicates a "partial positive charge," or that the H end of the molecule is positive *with respect to* the F end. Please note that we are *not* saying that H has a charge of 1+ or that F has a charge of 1−! A second way to indicate the polarity is to draw an arrow so that the head points toward the negative end (F) of the bond and the crossed tail indicates the positive end (H).

The separation of charge in a polar covalent bond creates an electric **dipole**. We expect the dipoles in the covalent molecules HF, HCl, HBr, and HI to be different because F, Cl, Br, and I have different electronegativities. Therefore, atoms of these elements have different tendencies to attract electron pairs shared with hydrogen. We indicate this difference as shown below, where Δ(EN) is the difference in electronegativity between two atoms that are bonded together.

The word "dipole" means "two poles." Here it refers to the positive and negative poles that result from the separation of charge within a molecule.

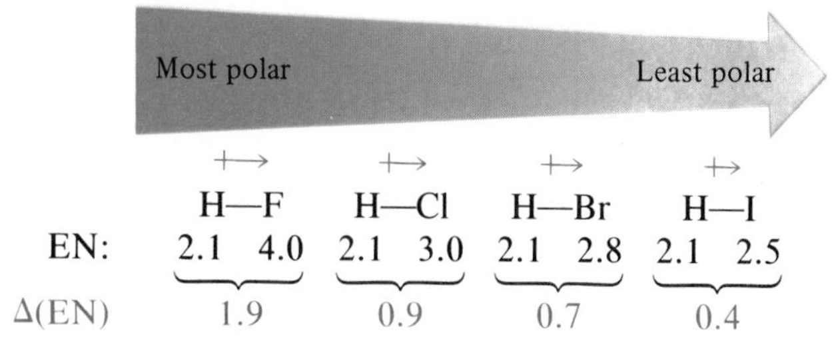

The values of electronegativity are obtained from Table 6-3.

Table 7-3
Dipole Moments and Δ(EN) Values for Some Pure (Gaseous) Substances

Substance	Dipole Moment (μ)*	Δ(EN)
HF	1.91 D	1.9
HCl	1.03 D	0.9
HBr	0.79 D	0.7
HI	0.38 D	0.4
H—H	0 D	0

* The magnitude of a dipole moment is given by the product of charge × distance of separation. Molecular dipole moments are usually expressed in debyes (D).

For comparison, the Δ(*EN*) values for some typical 1:1 ionic compounds are NaBr, 1.8; RbF, 3.1; and KCl, 2.1.

The longest arrow indicates the largest dipole, or greatest separation of electron density in the molecule (see Table 7-3).

7-5 The Continuous Range of Bonding Types

Let us now clarify our classification of bonding types. The degree of electron sharing or transfer depends on the electronegativity difference between the bonding atoms. Nonpolar covalent bonding (involving *equal sharing* of electron pairs) is one extreme, occurring when the atoms are identical (ΔEN is zero). Ionic bonding (involving *complete transfer* of electrons) represents the other extreme, and occurs when two elements with very different electronegativities interact (ΔEN is large).

Polar covalent bonds may be thought of as intermediate between pure (nonpolar) covalent bonds and pure ionic bonds. In fact, bond polarity is sometimes described in terms of **partial ionic character**. This usually increases with increasing difference in electronegativity between bonded atoms. Calculations based on the measured dipole moment (see the next section) of gaseous HCl indicate about 17% "ionic character."

When cations and anions interact strongly, some amount of electron sharing takes place; in such cases we can consider the ionic compound as having some **partial covalent character**. For instance, the high charge density of the very small Li^+ ion causes it to distort large anions that it approaches. The distortion attracts electron density from the anion to the region between it and the Li^+ ion, giving lithium compounds a higher degree of covalent character than in other alkali metal compounds.

Almost all bonds have both ionic and covalent character. By experimental means, a given type of bond can usually be identified as being "closer" to one or the other extreme type. We find it useful and convenient to use the labels for the major classes of bonds to describe simple substances, keeping in mind that they represent ranges of behavior.

In summary, we can describe chemical bonding as a continuum that may be represented as

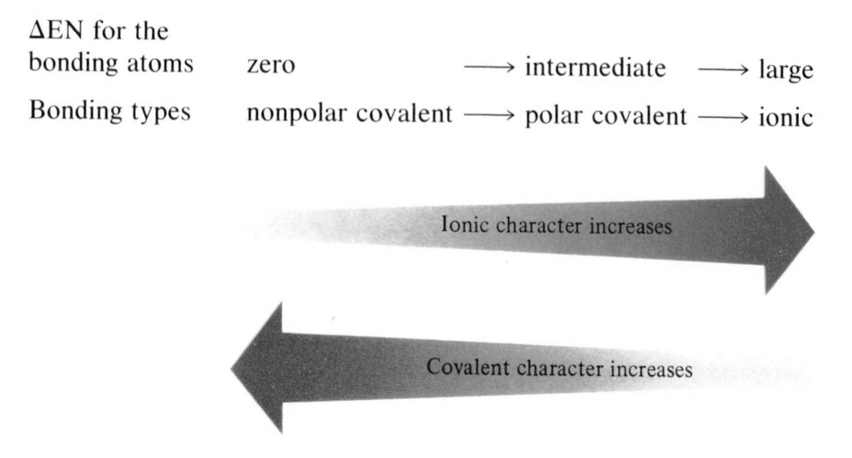

Above all, we must recognize that any classification of a compound that we might suggest based on electronic properties *must* be consistent with the physical properties of ionic and covalent substances described at the beginning of the chapter. For instance, HCl has a rather large electronegativity difference (0.9), and its aqueous solutions conduct electricity. But we know that we cannot view it as an ionic compound because it is a gas, and not a solid, at room temperature. Liquid HCl is a nonconductor.

HCl *ionizes* in aqueous solution.

Let us point out another aspect of the classification of compounds as ionic or covalent. Not all ions consist of single charged atoms. Many are small groups of atoms that are covalently bonded together, yet they still have excess positive or negative charge. Examples of such *polyatomic ions* are ammonium ion, NH_4^+, sulfate ion, SO_4^{2-}, and nitrate ion, NO_3^-. A compound such as potassium sulfate, K_2SO_4, contains potassium ions, K^+, and sulfate ions, SO_4^{2-}, in a 2:1 ratio. We should recognize that this compound contains both covalent bonding (electron sharing *within* the sulfate ions) and ionic bonding (electrostatic attractions *between* potassium and sulfate ions). However, we classify this compound as *ionic,* because it is a high-melting solid (mp 1069°C), it conducts electricity both in molten form and in aqueous solution, and it displays the properties that we generally associate with ionic compounds. Put another way, while covalent bonding holds a part of this substance together (the sulfate ions), the forces that hold the *entire* substance together are ionic.

7-6 Dipole Moments

$\mu = d \times q$

It is convenient to express bond polarities on a numerical scale. We indicate the polarity of a molecule by its dipole moment, which measures the separation of charge within the molecule. The **dipole moment**, μ, is defined as the product of the distance, d, separating charges of equal magnitude and opposite sign, and the magnitude of the charge, q. A dipole moment is measured by placing a sample of the substance between two plates and applying a voltage. This causes a small shift in electron density of any molecule, so the applied voltage is diminished very slightly. However, diatomic molecules that contain polar bonds, such as HF, HCl, and CO, tend to orient themselves in the electric field (Figure 7-5). This causes the mea-

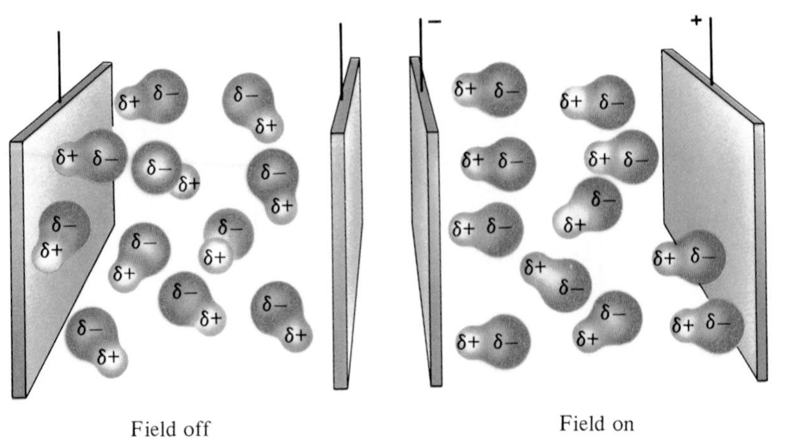

Figure 7-5
If polar molecules, such as HF, are subjected to an electric field, they tend to line up very slightly in a direction opposite to that of the field. This minimizes the electrostatic energy of the molecules. Nonpolar molecules are not oriented by an electric field. The effect is greatly exaggerated in this drawing.

sured voltage between the plates to decrease more markedly for these substances, and so we say that these molecules are *polar*. Molecules such as F_2 or N_2 do not reorient, so the change in voltage between the plates remains slight; we say that these molecules are *nonpolar*.

Generally, as electronegativity differences increase in diatomic molecules, the measured dipole moments increase. This can be seen clearly from the data for the hydrogen halides (Table 7-3).

Unfortunately, the dipole moments associated with *individual bonds* can be measured only in simple diatomic molecules. *Entire molecules* rather than selected pairs of atoms must be subjected to measurement. Measured values of dipole moments reflect the *overall* polarities of molecules. For polyatomic molecules they are the result of all the bond dipoles in the molecules. Later we shall see that structural features, such as molecular geometry and the presence of lone (unshared) pairs of electrons, also affect the polarity of a molecule. After we have discussed these additional aspects of covalent bonding, we shall return to the important topic of dipole moments (Section 8-3).

7-7 Lewis Dot Formulas for Molecules and Polyatomic Ions

In Sections 7-1 and 7-2 we drew *Lewis dot formulas* for atoms and monatomic ions. We can also use Lewis dot formulas to show the *valence electrons* in atoms that are covalently bonded in a molecule or a polyatomic ion. A water molecule can be represented by either of the following diagrams.

A polyatomic ion is an ion that contains more than one atom.

H : Ö :
　　H
Lewis dot formula

H—Ö :
　　|
　　H
dash formula

An H_2O molecule has two shared electron pairs, i.e., two single covalent bonds. The O atom also has two lone pairs.

In H_2O, the O atom contributes six valence electrons, and each H atom contributes one.

In *dash formulas,* a shared pair of electrons is indicated by a dash. There are two *double* bonds in carbon dioxide, and its Lewis dot formula is

:O: : C : :O:　　:O=C=O:

A CO_2 molecule has four shared electron pairs, i.e., two double bonds. The central atom (C) has no lone pairs.

In CO_2, the C atom contributes four valence electrons, and each O atom contributes six.

The covalent bonds in a polyatomic ion can be represented in the same way. The Lewis dot formula for the ammonium ion, NH_4^+, shows only eight electrons, even though the N atom has five electrons in its valence shell and each H atom has one, for a total of 5 + 4(1) = 9 electrons. The NH_4^+ ion, with a charge of 1+, has one less electron than the original atoms.

The NH_3 molecule, like the NH_4^+ ion, has eight valence electrons about the N atom.

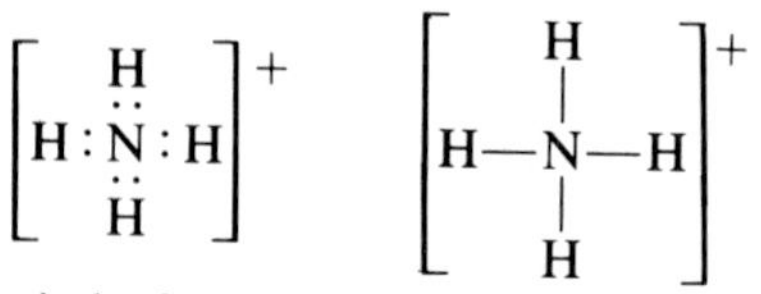

H:N̈:H or H—N̈—H (with H below N)

$$\left[\begin{array}{c} \text{H} \\ \text{H:}\ddot{\text{N}}\text{:H} \\ \text{H} \end{array}\right]^+ \qquad \left[\begin{array}{c} \text{H} \\ | \\ \text{H—N—H} \\ | \\ \text{H} \end{array}\right]^+$$

Lewis dot formula — dash formula

The writing of Lewis formulas is an electron bookkeeping method that is useful as a first approximation to suggest bonding schemes. It is important to remember that Lewis dot formulas only show the number of valence electrons, the number and kinds of bonds, and the order in which the atoms are connected. *They are not intended to depict the three-dimensional shapes of molecules and polyatomic ions.* We will show in Chapter 8, however, that the three-dimensional geometry of a molecule can be predicted from its Lewis structure.

7-8 The Octet Rule

Representative elements usually attain stable noble gas electron configurations when they share electrons. In the water molecule the O has a share in eight outer shell electrons, giving the neon configuration, while H shares two electrons in the helium configuration. Likewise, the C and O of CO_2 and the N of NH_3 and the NH_4^+ ion each have a share in eight electrons in their outer shells. The H atoms in NH_3 and NH_4^+ each share two electrons. Many Lewis formulas are based on the idea that

In some compounds, the central atom does not achieve a noble gas configuration. Such exceptions to the octet rule are discussed later in this chapter.

the representative elements achieve noble gas configurations in *most* of their compounds.

This statement is usually called the **octet rule**, because the noble gas configurations have 8 e^- in their outermost shells (except for He, which has 2 e^-).

For now, we shall restrict our discussion to compounds of the *representative elements*. The octet rule alone does not let us write Lewis formulas. We still need to know how to place the electrons around the bonded atoms—that is, how many of the available valence electrons are **bonding electrons** (shared) and how many are **unshared electrons** (associated with only one atom). A pair of unshared electrons in the same orbital is called a **lone pair**. A simple mathematical relationship is helpful here:

$$S = N - A$$

S is the total number of electrons *shared* in the molecule or polyatomic ion.

N is the number of valence shell electrons *needed* by all the atoms in the molecule or ion to achieve noble gas configurations ($N = 8 \times$ number of atoms not including H, plus $2 \times$ number of H atoms).

A is the number of electrons *available* in the valence shells of all of the (representative) atoms. This is equal to the sum of their periodic group numbers.

For example, in CO_2, A for each O atom is 6 and A for the carbon atom is 4, so A for CO_2 is $4 + 2(6) = 16$. The following general steps describe the use of this relationship in constructing dot formulas for molecules and polyatomic ions.

Writing Lewis Formulas

1. Select a reasonable (symmetrical) "skeleton" for the molecule or polyatomic ion.
   a. The *least electronegative element* is usually the central element, except that H is never the central element. The least electronegative element is usually the one that needs the most electrons to fill its octet. Example: CS_2 has the skeleton S C S.
   b. Oxygen atoms do not bond to each other except in (1) O_2 and O_3 molecules; (2) the peroxides, which contain the O_2^{2-} group; and (3) the rare superoxides, which contain the O_2^- group. Example: The sulfate ion, SO_4^{2-}, has the skeleton

$$\left[\begin{array}{ccc} & O & \\ O & S & O \\ & O & \end{array}\right]^{2-}$$

   c. In *ternary acids* (oxyacids), hydrogen usually bonds to an O atom, *not* to the central atom. Example: Nitrous acid, HNO_2, has the skeleton H O N O. However, there are a few exceptions to this rule, such as H_3PO_3 and H_3PO_2.

A ternary acid contains *three* elements—H, O, and another element, often a nonmetal.

   d. For ions or molecules that have more than one central atom, the most symmetrical skeletons possible are used. Examples: C_2H_4 and $P_2O_7^{4-}$ have the following skeletons:

$$\begin{array}{cc} H & H \\ C & C \\ H & H \end{array} \quad \text{and} \quad \left[\begin{array}{ccccc} & O & & O & \\ O & P & O & P & O \\ & O & & O & \end{array}\right]^{4-}$$

2. Calculate N, *the number of outer* (*valence*) *shell electrons* needed by all atoms in the molecule or ion to achieve noble gas configurations. Examples:

For compounds containing only representative elements, N is equal to $8 \times$ number of atoms *not* including H, plus $2 \times$ number of H atoms.

For H_2SO_4,

$$N = 1 \times 8 \text{ (S atom)} + 4 \times 8 \text{ (O atoms)} + 2 \times 2 \text{ (H atoms)}$$
$$= 8 + 32 + 4 = 44\ e^- \text{ needed}$$

For SO_4^{2-},

$$N = 8 + 32 = 40\ e^- \text{ needed}$$

For the representative elements, the number of valence shell electrons in an atom is equal to its periodic group number. Exceptions: 1 for an H atom and 8 for a noble gas (except 2 for He).

3. Calculate A, *the number of electrons available* in the outer (valence) shells of all the atoms. For negatively charged ions, add to the total the number of electrons equal to the charge on the anion; for positively charged ions, subtract the number of electrons equal to the charge on the cation. Examples:

   For H_2SO_4,

   $$A = 2 \times 1 \text{ (H atoms)} + 1 \times 6 \text{ (S atom)} + 4 \times 6 \text{ (O atoms)}$$
   $$= 2 + 6 + 24 = 32\ e^- \text{ available}$$

   For SO_4^{2-},

   $$A = 1 \times 6 \text{ (S atom)} + 4 \times 6 \text{ (O atoms)} + 2 \text{ (for 2− charge)}$$
   $$= 6 + 24 + 2 = 32\ e^- \text{ available}$$

4. Calculate S, *total number of electrons shared* in the molecule or ion, using the relationship $S = N - A$. Examples:

   For H_2SO_4,

   $$S = N - A = 44 - 32$$
   $$= 12 \text{ electrons shared (6 pairs of } e^- \text{ shared)}$$

   For SO_4^{2-},

   $$S = N - A = 40 - 32$$
   $$= 8 \text{ electrons shared (4 pairs of } e^- \text{ shared)}$$

C, N, and O often form double and triple bonds. S and Se can form double bonds with C, N, and O.

5. Place the S electrons into the skeleton as *shared pairs*. Use double and triple bonds only when necessary. Structures may be shown either by Lewis dot formulas or by dash formulas, in which a dash represents a shared pair of electrons.

Formula	Skeleton	Dot Formula ("bonds" in place, but incomplete)	Dash Formula ("bonds" in place, but incomplete)
H_2SO_4	O H O S O H O	O H:O:S:O:H O	O H—O—S—O—H O
SO_4^{2-}	[O O S O O]$^{2-}$	[O O:S:O O]$^{2-}$	[O O—S—O O]$^{2-}$

Please note that a Lewis dot (dash) formula *does not* indicate the geometry of a molecule or an ion. We will discuss the actual geometries of molecules and ions in Chapter 8.

6. Place the additional electrons into the skeleton as *unshared (lone) pairs* to fill the octet of every A group element (except H, which can share only 2 e^-). Check that the total number of electrons is equal to A, from Step 3. Examples:

   For H_2SO_4,

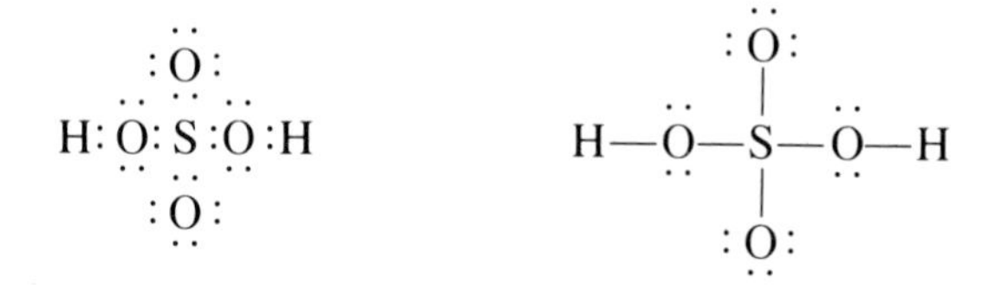

Check: 16 pairs of e^- have been used. $2 \times 16 = 32\ e^-$ available.

For SO_4^{2-},

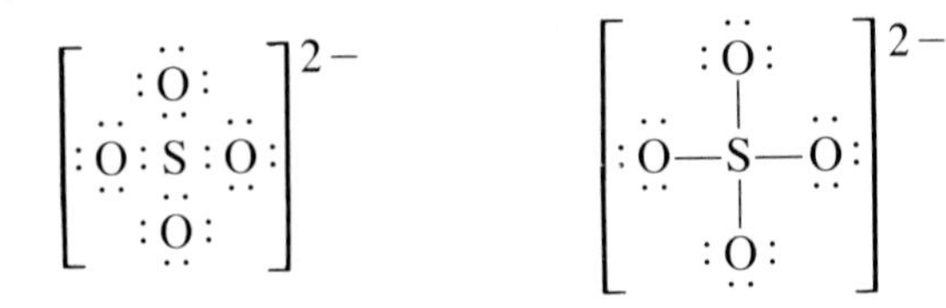

Check: 16 pairs of electrons have been used. $2 \times 16 = 32\ e^-$ available.

Example 7-1

Write the Lewis dot formula and the dash formula for the nitrogen molecule, N_2.

Plan

We follow the stepwise procedure for writing Lewis formulas that was just presented.

Solution

Step 1: The skeleton is N N.

Step 2: $N = 2 \times 8 = 16\ e^-$ needed (total) by both atoms

Step 3: $A = 2 \times 5 = 10\ e^-$ available (total) for both atoms

Step 4: $S = N - A = 16\ e^- - 10\ e^- = 6\ e^-$ shared

Step 5: N:::N 6 e^- (3 pairs) are shared; a *triple* bond.

Step 6: The additional 4 e^- are accounted for by a lone pair on each N. The complete Lewis diagram is

:N:::N: or :N≡N:

Check: 10 e^- (5 pairs) have been used.

Example 7-2

Write the Lewis dot and dash formula for carbon disulfide, CS_2, an ill-smelling liquid.

Plan

Again, we follow the stepwise procedure to apply the relationship $S = N - A$.

Solution

Step 1: The skeleton is S C S.

Step 2: $N = 1 \times 8$ (for C) $+ 2 \times 8$ (for S) $= 24\ e^-$ needed by all atoms

Step 3: $A = 1 \times 4$ (for C) $+ 2 \times 6$ (for S) $= 16\ e^-$ available

Step 4: $S = N - A = 24\ e^- - 16\ e^- = 8\ e^-$ shared

Step 5: S::C::S 8 e^- (4 pairs) are shared; two *double* bonds.

Step 6: C already has an octet, so the remaining $8e^-$ are distributed as lone pairs on the S atoms to give each S an octet. The complete Lewis formula is

:S::C::S: or :S=C=S:

Check: 16 e^- (8 pairs) have been used. The bonding picture is similar to that of CO_2; this is not surprising, S is below O in Group VIA.

EOC 42, 44

C is the central atom, or the element in the middle of the molecule. It needs four more electrons to acquire an octet, while each S atom needs only two more electrons.

A number of minerals contain the carbonate ion. A very common one is calcium carbonate, $CaCO_3$, the main constituent of limestone and of stalactites and stalagmites.

Example 7-3

Write the Lewis dot formula for the carbonate ion, CO_3^{2-}.

Plan

The same stepwise procedure can be applied to ions. We must remember to adjust A, the total number of electrons, to account for the charge shown on the ion.

Solution

Step 1: The skeleton is O C O (with a third O above C; charge 2−)

Step 2: $N = 1 \times 8$ (for C) $+ 3 \times 8$ (for O) $= 8 + 24 = 32\ e^-$ needed by all atoms

Step 3: $A = 1 \times 4$ (for C) $+ 3 \times 6$ (for O) $+ 2$ (for the 2− charge) $= 4 + 18 + 2 = 24\ e^-$ available

Step 4: $S = N - A = 32\ e^- - 24\ e^- = 8\ e^-$ (4 pairs) shared

Step 5: O:C::O (with O above C; charge 2−) (Four pairs are shared. At this point it doesn't matter which O is doubly bonded.)

Step 6: The Lewis formula is

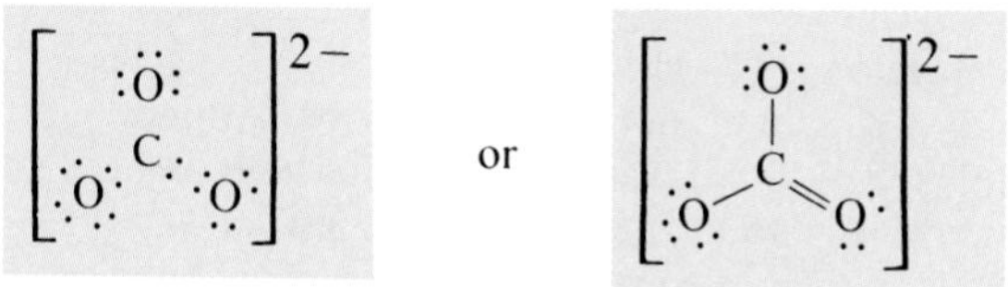

Check: $24\ e^-$ (12 pairs) have been used.

EOC 46

7-9 Resonance

In addition to the one shown in Example 7-3, two other Lewis formulas with the same skeleton for the CO_3^{2-} ion are equally acceptable. In these formulas, $4\ e^-$ could be shared between the carbon atom and either of the other two oxygen atoms.

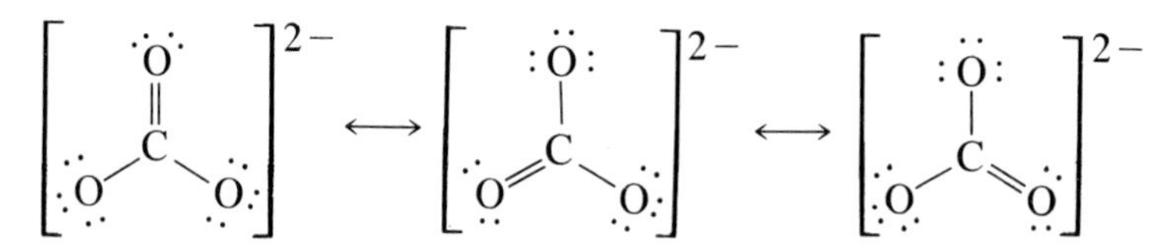

A molecule or polyatomic ion for which two or more dot formulas with the same arrangements of atoms can be drawn to describe the bonding is said to exhibit **resonance**. The three structures above are **resonance structures** of the carbonate ion. The relationship among them is indicated by the double-headed arrows, ↔. This symbol *does not mean* that the ion flips back and forth among these three structures. The true structure is like an average of the three.

The C—O bonds in CO_3^{2-} are really *neither* double nor single bonds, but are intermediate in bond length (and strength). This has been verified experimentally. Based on measurements in many compounds, the typical C—O single bond length is 1.43 Å, and the typical C=O double bond length is 1.22 Å. The C—O bond length for each bond in the CO_3^{2-} ion is intermediate at 1.29 Å. Another way to represent this situation is by **delocalization** of bonding electrons:

When electrons are shared among more than two atoms, the electrons are said to be *delocalized.* The concept of delocalization is important in molecular orbital theory (Chapter 9).

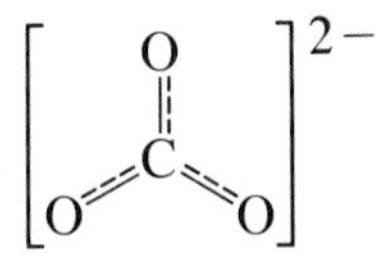

(lone pair on O atoms not shown)

The dashed lines indicate that some of the electrons shared between C and O atoms are *delocalized* among all four atoms; that is, the four pairs of shared electrons are equally distributed among three C—O bonds.

Example 7-4

Draw two resonance structures for the sulfur dioxide molecule, SO_2.

Plan

The stepwise procedure presented in Section 7-8 can be used to write each resonance structure.

Solution

$$N = 1(8) + 2(8) = 24\ e^-$$

$$A = 1(6) + 2(6) = 18\ e^-$$

$$S = N - A = 6\ e^- \text{ shared}$$

(The 1(...) terms are for S and the 2(...) terms are for O.)

The resonance structures are

:Ö::S:Ö: ⟷ :Ö:S::Ö: or :Ö=S—Ö: ⟷ :Ö—S=Ö:

We could show delocalization of electrons as follows:

O=S=O (with a lone pair on S; the S—O bonds drawn as dashed double bonds) (lone pairs on O not shown)

Remember that dot and dash formulas *do not necessarily show shapes.* SO_2 molecules are angular, not linear.

EOC 52, 53, 54

Trees killed by acid rain in a North Carolina forest. Combustion of sulfur-containing fossil fuels and smelting operations produce sulfur dioxide, SO_2. When this SO_2 is released into the atmosphere, it is one of the major contributors to acid rain. (Carolina Biological Supply Company)

Formal Charges

Enrichment

An experimental determination of the structure of a molecule or polyatomic ion is necessary to establish unequivocally its correct structure. However, we often do not have these results available. The concept of *formal charges* helps us to write Lewis formulas correctly in most cases. This bookkeeping system counts bonding electrons as though they were equally shared between the two bonded atoms. Generally, the most energetically favorable formula for a molecule is usually one in which the formal charge on each atom is zero or as near zero as possible.

Consider the reaction of NH_3 with hydrogen ion, H^+, to form the ammonium ion, NH_4^+.

$$\mathrm{H:\ddot{N}:H} + \mathrm{H^+} \longrightarrow \left[\begin{array}{c}\mathrm{H}\\ \mathrm{H:\ddot{N}:H}\\ \mathrm{H}\end{array}\right]^+$$

The unshared pair of electrons on the N atom in the NH_3 molecule is shared with the H^+ ion to form the NH_4^+ ion, in which the N atom has four covalent bonds. Because N is a Group VA element, we expect it to form three covalent bonds to complete its octet. How can we describe the fact that N has four covalent bonds in species like NH_4^+? The answer is obtained by calculating the *formal charge* on each atom in NH_4^+ by the following rules:

Rule for Assigning Formal Charges to Atoms of A Group Elements

1. a. In a molecule, the sum of the formal charges is zero.
   b. In a polyatomic ion, the sum of the formal charges is equal to the charge.
2. The formal charge, abbreviated FC, on an atom in a Lewis formula is given by the relationship

   $$\text{FC} = (\text{group number}) - [(\text{number of bonds}) + (\text{number of unshared } e^-)]$$

   The group number of the noble gases is taken as VIIIA, rather than zero, in calculating formal charges. Formal charges are represented by ⊕ and ⊖ to distinguish between formal charges and real charges on ions.
3. In a Lewis formula, an atom that has the same number of bonds as its periodic group number has no formal charge.
4. When difficulty arises in deciding which atom should be assigned a negative formal charge, the negative formal charge is assigned to the more electronegative element.
5. Atoms that are bonded to each other should not be assigned formal charges with the same sign (the *adjacent charge rule*) if this can be avoided. Lewis formulas in which adjacent atoms have formal charges of the same sign are usually *not* accurate representations.

Let us apply these rules to the ammonia molecule, NH_3, and to the ammonium ion, NH_4^+. Because N is a Group VA element, its group number is 5.

$$\mathrm{H:\ddot{N}:H} \qquad\qquad \left[\begin{array}{c}\mathrm{H}\\ \mathrm{H:\ddot{N}:H}\\ \mathrm{H}\end{array}\right]^+$$

In NH_3 the N atom has 3 bonds and 2 unshared e^-, and so for N,

$$\text{FC} = (\text{group number}) - [(\text{number of bonds}) + (\text{number of unshared } e^-)]$$
$$= 5 - (3 + 2) = 0 \text{ (for N)}$$

For H,

$$FC = (\text{group number}) - [(\text{number of bonds}) + (\text{number of unshared } e^-)]$$

$$= 1 - (1 + 0) = 0 \text{ (for H)}$$

The formal charges of N and H are both zero in NH_3, so the sum of the formal charges is $0 + 3(0) = 0$, consistent with Rule 1a.

In NH_4^+ the N atom has 4 bonds and no unshared e^-, and so for N,

$$FC = (\text{group number}) - [(\text{number of bonds}) + (\text{number of unshared } e^-)]$$

$$= 5 - (4 + 0) = 1+ \text{ (for N)}$$

Calculation of the FC for H atoms gives zero, as above. The sum of the formal charges in NH_4^+ is $(1+) + 4(0) = 1+$. This is consistent with Rule 1b.

The absence of a sign on a number means that the number is *positive*. Therefore, we attached a "+" sign to the 1.

FCs are indicated by ⊕ and ⊖. The sum of the formal charges in a polyatomic ion is equal to the charge on the ion—1+ in NH_4^+.

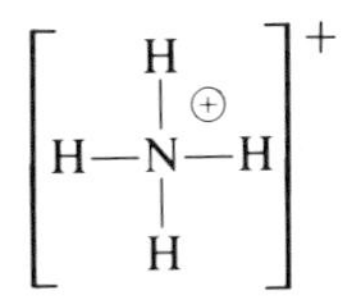

Thus we see that the octet rule is obeyed in both NH_3 and NH_4^+. The sum of the formal charges in each case is that predicted by Rule 1, even though nitrogen has four covalent bonds in the NH_4^+ ion.

Let us now write a Lewis formula for, and assign formal charges to, the atoms in thionyl chloride, $SOCl_2$, a compound often used in organic synthesis. Both Cl atoms and the O atom are bonded to the S atom. The Lewis formula is

:C̈l:S̈:C̈l:
  :Ö:

Formal charges on the atoms are calculated by the usual relationship.

$$FC = (\text{group number}) - [(\text{number of bonds}) + (\text{number of unshared } e^-)]$$

For Cl: $FC = 7 - (1 + 6) = 0$

For S: $FC = 6 - (3 + 2) = 1+$

For O: $FC = 6 - (1 + 6) = 1-$

⊕
:C̈l:S̈:C̈l:
  :Ö:
  ⊖

Dinitrogen oxide (N_2O, also called nitrous oxide) is used as a mild anesthetic. Let us write some possible Lewis formulas for N_2O and assign formal charges to each atom. It is known from experimental molecular structure studies that the oxygen is at the end of the three-atom molecule. We can write three Lewis formulas with O at the end.

(i) :N≡N—Ö:

Left-hand N, $FC = 5 - (3 + 2) = 0$

N in middle, $FC = 5 - (4 + 0) = 1+$

O, $FC = 6 - (1 + 6) = 1-$

(ii) $\dot{\underset{\cdot}{N}}=N=\ddot{\underset{\cdot\cdot}{O}}$ Left-hand N, FC = 5 − (2 + 4) = 1−

N in middle, FC = 5 − (4 + 0) = 1+

O, FC = 6 − (2 + 4) = 0

(iii) $:N-\ddot{\underset{\cdot\cdot}{N}}\equiv O:$ Left-hand N, FC = 5 − (1 + 6) = 2−

N in middle, FC = 5 − (4 + 0) = 1+

O, FC = 6 − (3 + 2) = 1+

Formula (iii) is less likely than the other two, because it contains one atom that has an unfavorably large formal charge (2− on left-hand N). Thus we write two resonance structures for N_2O:

$$:N\equiv N-\ddot{\underset{\cdot\cdot}{O}}: \longleftrightarrow \dot{\underset{\cdot}{N}}=N=\ddot{\underset{\cdot\cdot}{O}}$$

Suppose we did not know that the oxygen should be written at the end of the Lewis formula. We might proceed as follows:

(iv) $:N\equiv O-\ddot{\underset{\cdot\cdot}{N}}:$ Left-hand N, FC = 5 − (3 + 2) = 0

O, FC = 6 − (4 + 0) = 2+

Right-hand N, FC = 5 − (1 + 6) = 2−

(v) $\dot{\underset{\cdot}{N}}=O=\dot{\underset{\cdot}{N}}$ Left-hand N, FC = 5 − (2 + 4) = 1−

O, FC = 6 − (4 + 0) = 2+

Right-hand N, FC = 5 − (2 + 4) = 1−

Both of these formulas have formal charges that are unnecessarily large. In addition, we should expect that the more electronegative O would have a lower formal charge than any N atom to which it is bonded. On these bases, we might have expected they would be very unlikely formulas. This is consistent with experimental results.

7-10 Limitations of the Octet Rule for Lewis Formulas

Recall that representative elements achieve noble gas electronic configurations in *most* of their compounds. But when the octet rule is not applicable, the relationship $S = N - A$ is not valid without modification. The following are general cases for which the procedure in Section 7-8 *must be modified*—i.e., cases in which there are limitations of the octet rule.

1. Most covalent compounds of beryllium, Be. Because Be contains only two valence shell electrons, it usually forms only two covalent bonds when it bonds to two other atoms. Therefore, we use *four electrons* as the number *needed* by Be in Step 2, Section 7-8. In Steps 5 and 6 we use only two pairs of electrons for Be.
2. Most covalent compounds of the Group IIIA elements, especially boron, B. The IIIA elements contain only three valence shell electrons, so they

often form three covalent bonds when they bond to three other atoms. Therefore, we use *six electrons* as the number *needed* by the IIIA elements in Step 2; and in Steps 5 and 6 we use only three pairs of electrons for the IIIA elements.

3. Compounds or ions containing an odd number of electrons. Examples are NO, with 11 valence shell electrons, and NO_2, with 17 valence shell electrons.
4. Compounds or ions in which the central element needs a share in more than eight valence shell electrons to hold all the available electrons, A. Extra rules are added to Steps 4 and 6 when this is encountered.
   Step 4a: If S, the number of electrons shared, is less than the number needed to bond all atoms to the central atom, then S is increased to the number of electrons needed.
   Step 6a: If S must be increased in Step 4a, then the octets of all the atoms might be satisfied before all A of the electrons have been added. Place the extra electrons on the central element.

Lewis formulas are not normally written for compounds containing *d*- and *f*-transition metals. The *d*- and *f*-transition metals utilize *d* and/or *f* orbitals in bonding as well as *s* and *p* orbitals. Thus, they can accommodate more than eight valence electrons.

Many species that violate the octet rule are quite reactive. For instance, compounds containing atoms with only four valence electrons (limitation 1 above) or six valence electrons (limitation 2 above) frequently react with other species that supply electron pairs. Compounds such as these that accept a share in a pair of electrons are called *Lewis acids;* a *Lewis base* is a species that makes available a share in a pair of electrons. (This kind of behavior will be discussed in detail in Section 10-10.) Molecules with an odd number of electrons often *dimerize* (combine in pairs) to give products that do satisfy the octet rule. Examples are the dimerization of NO to form N_2O_2 (Section 26-8) and of NO_2 to form N_2O_4 (Section 26-10). Examples 7-5 through 7-8 illustrate some limitations and show how such Lewis formulas are constructed.

Example 7-5

Draw the Lewis dot formula and the dash formula for gaseous beryllium chloride, $BeCl_2$, a covalent compound.

Plan

This is an example of limitation 1. So, as we follow the steps in writing the Lewis formula, we must remember to use *four electrons* as the number *needed* by Be in Step 2. Steps 5 and 6 should show only two pairs of electrons for Be.

Solution

Step 1: The skeleton is Cl Be Cl.

Step 2: $N = 2 \times 8$ (for Cl) $+ 1 \times$ 4 (for Be) $= 20\ e^-$ needed (see limitation 1)

Step 3: $A = 2 \times 7$ (for Cl) $+ 1 \times 2$ (for Be) $= 16\ e^-$ available

Step 4: $S = N - A = 20\ e^- - 16\ e^- = 4\ e^-$ shared

Step 5: Cl : Be : Cl

Step 6: $:\ddot{\underset{..}{Cl}}:Be:\ddot{\underset{..}{Cl}}:$ or $:\ddot{\underset{..}{Cl}}—Be—\ddot{\underset{..}{Cl}}:$

Calculation of formal charges shows that

for Be, FC = 2 − (2 + 0) = 0 and for Cl, FC = 7 − (1 + 6) = 0

The chlorine atoms achieve the argon configuration, [Ar], while the beryllium atom has a share of only four electrons. Compounds such as $BeCl_2$, in which the central atom shares fewer than 8 e^-, are sometimes referred to as **electron deficient** compounds. This "deficiency" refers only to satisfying the octet rule for the central atom. The term does not imply that there are fewer electrons than there are protons in the nuclei, as in the case of a cation, because the molecule is neutral.

:Cl=Be=Cl:

A Lewis formula can be written for $BeCl_2$ that *does* satisfy the octet rule (see margin). Let us evaluate the formal charges for that formula:

for Be, FC = 2 − (4 + 0) = 2− and for Cl, FC = 7 − (2 + 4) = 1+

As mentioned earlier, the most favorable structure for a molecule is one in which the formal charge on each atom is zero, if possible. In case some atoms did have nonzero formal charges, we would expect that the more electronegative atoms (Cl) would be the ones with lowest formal charge. Thus, we prefer the Lewis structure shown in Example 7-5 over the one in the margin.

One might expect a similar situation for compounds of the other IIA metals, Mg, Ca, Sr, Ba, and Ra. However, these elements have *lower ionization energies* and *larger radii* than Be, so they usually form ions by losing two electrons.

BF_3 and BCl_3 are gases at room temperature. Liquid BBr_3 and solid BI_3 are shown here.

Example 7-6

Draw the Lewis dot formula and the dash formula for boron trichloride, BCl_3, a covalent compound.

Plan

This covalent compound of boron is an example of limitation 2. As we follow the steps in writing the Lewis formula, we use *six electrons* as the number *needed* by B in Step 2. Steps 5 and 6 should show only three pairs of electrons for B.

Solution

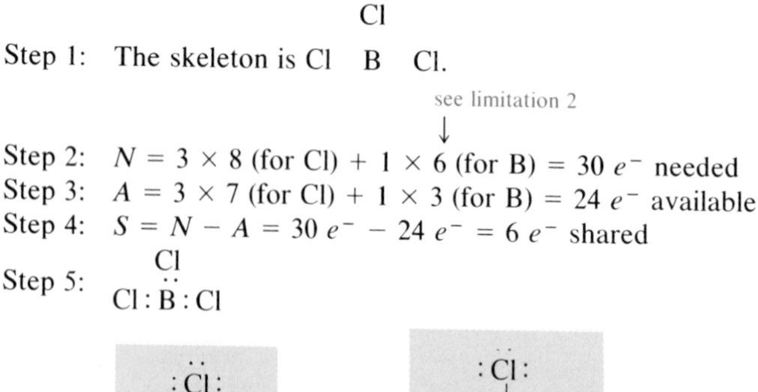

Step 1: The skeleton is Cl B Cl, with a third Cl above B.

see limitation 2 ↓

Step 2: $N = 3 \times 8$ (for Cl) $+ 1 \times 6$ (for B) $= 30\ e^-$ needed

Step 3: $A = 3 \times 7$ (for Cl) $+ 1 \times 3$ (for B) $= 24\ e^-$ available

Step 4: $S = N - A = 30\ e^- - 24\ e^- = 6\ e^-$ shared

Step 5: Cl : B : Cl, with Cl above B bonded by a pair

Step 6:

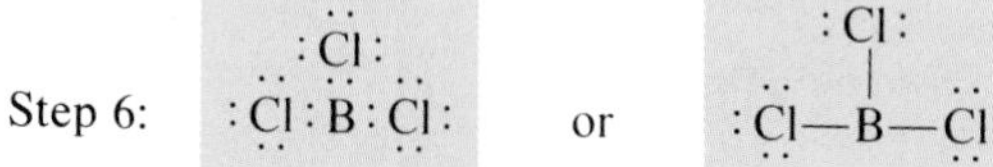

or

Each chlorine atom achieves the Ne configuration. The boron (central) atom acquires a share of only six valence shell electrons. Calculation of formal charges shows that

for B, FC = 3 − (3 + 0) = 0 and for Cl, FC = 7 − (1 + 6) = 0

Example 7-7

Write the Lewis dot formula and the dash formula for the covalent compound phosphorus pentafluoride, PF_5.

Plan

We apply the usual stepwise procedure to write the Lewis formula. In PF_5, all five F atoms are bonded to P. This requires the sharing of a minimum of 10 e^-, so this is an example of limitation 4. Therefore we add the extra Step 4a, and increase S from the calculated value of 8 e^- to 10 e^-.

Solution

Step 1: Skeleton is

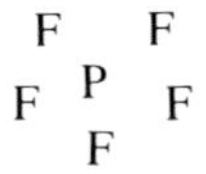

```
  F   F
F   P   F
    F
```

Step 2: $N = 5 \times 8$ (for F) + 1×8 (for P) = 48 e^- needed

Step 3: $A = 5 \times 7$ (for F) + 1×5 (for P) = 40 e^- available

Step 4: $S = N - A = 8\ e^-$ shared

Five F atoms are bonded to P. This requires the sharing of a minimum of 10 e^-. But only 8 e^- have been calculated in Step 4. Therefore, this is an example of limitation 4.

Step 4a: Increase S from 8 e^- to 10 e^-. The number of electrons available, 40, does not change.

Step 5:

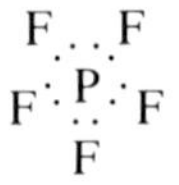

```
 F . . F
F : P : F
     F
```

Step 6:

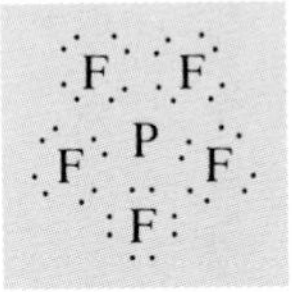

```
 :F: :F:
:F: P :F:   or   :F—P—F:  (with F—P bonds to all five F)
   :F:
```

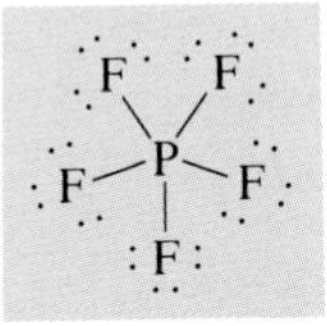

When the octets of the five F atoms have been satisfied, all 40 of the available electrons have been added. The phosphorus (central) atom has a share of ten electrons.

Calculation of formal charges shows that

for P, FC = 5 − (5 + 0) = 0 and for F, FC = 7 − (1 + 6) = 0

EOC 55

When an atom has a share of more than eight electrons, as does P in PF_5, we say that it exhibits an *expanded valence shell*. The electronic basis of the octet rule is that one s and three p orbitals in the valence shell of an atom can accommodate a maximum of eight electrons. The valence shell of phosphorus has $n = 3$, so it also has available d orbitals that can be involved in bonding. It is for this reason that phosphorus (and many other represen-

This is sometimes referred to as *hypervalence.*

tative elements of period 3 and beyond) can exhibit expansion of valence. By contrast, elements in the *second row* of the periodic table can *never* exceed eight electrons in their valence shells, because each atom has only one *s* and three *p* orbitals in that shell. Thus, we understand why NF_3 can exist but NF_5 cannot.

Example 7-8

Write the Lewis dot formula and the dash formula for the triiodide ion, I_3^-.

Plan

We apply the usual stepwise procedure. The calculation of $S = N - A$ in Step 4 shows only 2 e^- shared, but a minimum of 4 e^- are required to bond two I atoms to the central I. Limitation 4 applies, and we proceed accordingly.

Solution

Step 1: The skeleton is [I I I]$^-$.

Step 2: $N = 3 \times 8$ (for I) $= 24\ e^-$ needed

Step 3: $A = 3 \times 7$ (for I) $+ 1$ (for the 1− charge) $= 22\ e^-$ available

Step 4: $S = N - A = 2\ e^-$ shared. Two I atoms are bonded to the central I. This requires a minimum of 4 e^-, but only 2 e^- have been calculated in Step 4. Therefore, this is an example of limitation 4.

Step 4a: Increase S from 2 e^- to 4 e^-.

Step 5: [I : I : I]$^-$

Step 6: [: $\ddot{\underset{..}{I}}$: $\ddot{\underset{..}{I}}$: $\ddot{\underset{..}{I}}$:]$^-$

Step 6a: Now we have satisfied the octets of all atoms using only 20 of the 22 e^- available. We place the other two electrons on the central I atom.

The central iodine atom in I_3^- has an expanded valence shell.

[: I : I : I :]$^-$ or [: I—I—I :]$^-$

Calculation of formal charge shows that

for I on ends, FC = 7 − (1 + 6) = 0

for I in middle, FC = 7 − (2 + 6) = 1−

We have seen that *atoms attached to the central atom nearly always attain noble gas configurations*, even when the central atom does not.

Naming Inorganic Compounds

Because millions of compounds are known, it is important to be able to associate names and formulas in a systematic way.

The rules for naming inorganic compounds were set down in 1957 by the Committee on Inorganic Nomenclature of the International Union of Pure and Applied Chemistry (IUPAC). The concept of oxidation numbers (review Section 4-7) is essential in naming compounds.

7-11 Naming Binary Compounds

Binary compounds consist of two elements; they may be either ionic or covalent. The rule is to name the less electronegative element first and the more electronegative element second. The *more* electronegative element is named by adding an "-ide" suffix to the element's *unambiguous* stem. Stems for the nonmetals follow.

The stem for each element is derived from the name of the element.

IIIA		IVA		VA		VIA		VIIA	
								H	hydr
B	bor	C	carb	N	nitr	O	ox	F	fluor
		Si	silic	P	phosph	S	sulf	Cl	chlor
				As	arsen	Se	selen	Br	brom
				Sb	antimon	Te	tellur	I	iod

Binary ionic compounds contain metal cations and nonmetal anions. The cation is named first and the anion second according to the rule described.

Formula	Name	Formula	Name
KBr	potassium bromide	Rb_2S	rubidium sulfide
$CaCl_2$	calcium chloride	Al_2Se_3	aluminum selenide
NaH	sodium hydride	SrO	strontium oxide

The preceding method is sufficient for naming binary ionic compounds containing metals that exhibit *only one oxidation number* other than zero (Section 4-7). Most transition elements, and a few of the more electronegative representative metals, exhibit more than one oxidation number. These metals may form two or more binary compounds with the same nonmetal. To distinguish among all the possibilities, the oxidation number of the metal is indicated by a Roman numeral in parentheses following its name. This method can be applied to any binary compound of a metal and a nonmetal, whether the compound is ionic or covalent.

Roman numerals are *not* necessary for metals that commonly exhibit only one oxidation number in their compounds.

Formula	Ox. No. of Metal	Name	Formula	Ox. No. of Metal	Name
Cu_2O	+1	copper(I) oxide	$SnCl_2$	+2	tin(II) chloride
CuF_2	+2	copper(II) fluoride	$SnCl_4$	+4	tin(IV) chloride
FeS	+2	iron(II) sulfide	PbO	+2	lead(II) oxide
Fe_2O_3	+3	iron(III) oxide	PbO_2	+4	lead(IV) oxide

The advantage of the IUPAC system is that if you know the formula you can write the exact and unambiguous name; if you are given the name you can write the formula at once.

An older method, still in use but not recommended by the IUPAC, uses "-ous" and "-ic" suffixes to indicate lower and higher oxidation numbers, respectively. This system can distinguish between only two different oxidation numbers for a metal. Therefore, it is not as useful as the Roman numeral system.

Familiarity with the older system is still necessary. It is still widely used in many scientific, engineering, and medical fields.

Formula	Ox. No. of Metal	Name	Formula	Ox. No. of Metal	Name
CuCl	+1	cuprous chloride	SnF_2	+2	stannous fluoride
$CuCl_2$	+2	cupric chloride	SnF_4	+4	stannic fluoride
FeO	+2	ferrous oxide	Hg_2Cl_2	+1	mercurous chloride
$FeBr_3$	+3	ferric bromide	$HgCl_2$	+2	mercuric chloride

Pseudobinary ionic compounds contain more than two elements. In these compounds one or more of the ions consist of more than one element but behave as simple ions. Some common examples of such anions are the hydroxide ion, OH^-; the cyanide ion, CN^-; and the thiocyanate ion, SCN^-. As before, the name of the anion ends in "-ide." The ammonium ion, NH_4^+, is the common cation that behaves like a simple metal cation.

The Common Pseudobinary Anions

hydroxide	OH^-
cyanide	CN^-
thiocyanate	SCN^-

The Common Pseudobinary Cation

ammonium	NH_4^+

Formula	Name	Formula	Name
NH_4I	ammonium iodide	NH_4CN	ammonium cyanide
$Ca(CN)_2$	calcium cyanide	$Cu(OH)_2$	copper(II) hydroxide or cupric hydroxide
NaOH	sodium hydoxide	$Fe(OH)_3$	iron(III) hydroxide or ferric hydroxide

Nearly all **binary covalent compounds** involve two *nonmetals* bonded together. Although many nonmetals can exhibit different oxidation numbers, their oxidation numbers are *not* properly indicated by Roman numerals or suffixes. Instead, elemental proportions in binary covalent compounds are indicated by using a *prefix* system for both elements. The Greek and Latin prefixes used are mono, di, tri, tetra, penta, hexa, hepta, octa, nona, and deca. The prefix "mono-" is omitted for both elements except in the common name for CO, carbon monoxide. We use the minimum number of prefixes needed to name a compound unambiguously.

If you don't already know them, you should learn these prefixes.

Number	Prefix
2	di
3	tri
4	tetra
5	penta
6	hexa
7	hepta
8	octa
9	nona
10	deca

Formula	Name	Formula	Name
SO_2	sulfur dioxide	Cl_2O_7	dichlorine heptoxide
SO_3	sulfur trioxide	CS_2	carbon disulfide
N_2O_4	dinitrogen tetroxide	As_4O_6	tetraarsenic hexoxide

Chemists sometimes name binary covalent compounds that contain two nonmetals by the system used to name compounds of metals that show variable oxidation numbers; i.e., the oxidation number of the less electronegative element is indicated by a Roman numeral in parentheses. We do not recommend this procedure because it is incapable of naming compounds *unambiguously,* which is the principal requirement for a system of naming compounds. For example, both NO_2 and N_2O_4 are called nitrogen(IV) oxide by this system, but the name does not distinguish between the two compounds. The compound P_4O_{10} is tetraphosphorus decoxide, which indicates clearly its composition. Using the Roman numeral system, it would be called phosphorus(V) oxide, which could suggest the incorrect formula, P_2O_5. The simplest formula for P_4O_{10} is P_2O_5, but the name for a covalent compound must indicate clearly the composition of its molecules, not just its simplest formula.

Binary acids are compounds in which H is bonded to the more electronegative nonmetals; they act as acids when dissolved in water. The pure compounds are named as typical binary compounds. Their aqueous solutions are named by modifying the characteristic stem of the nonmetal with the prefix "hydro-" and the suffix "-ic" followed by the word "acid." The stem for sulfur in this instance is "sulfur" rather than "sulf."

Formula	Name of Compound	Name of Aqueous Solution
HCl	hydrogen chloride	hydrochloric acid, HCl(aq)
HF	hydrogen fluoride	hydrofluoric acid, HF(aq)
H_2S	hydrogen sulfide	hydrosulfuric acid, H_2S(aq)
HCN	hydrogen cyanide	hydrocyanic acid, HCN(aq)

A list of common cations and anions appears in Table 7-4. It will enable you to name many of the ionic compounds you encounter. In later chapters we shall learn the systematic rules for naming some other types of compounds.

Table 7-4
Formulas, Ionic Charges, and Names for Some Common Ions

Common Cations		
Formula	**Charge**	**Name**
Li^+	1+	lithium ion
Na^+	1+	sodium ion
K^+	1+	potassium ion
NH_4^+	1+	ammonium ion
Ag^+	1+	silver ion
Mg^{2+}	2+	magnesium ion
Ca^{2+}	2+	calcium ion
Ba^{2+}	2+	barium ion
Cd^{2+}	2+	cadmium ion
Zn^{2+}	2+	zinc ion
Cu^{2+}	2+	copper(II) ion or cupric ion
Hg_2^{2+}	2+	mercury(I) ion or mercurous ion
Hg^{2+}	2+	mercury(II) ion or mercuric ion
Mn^{2+}	2+	manganese(II) ion or manganous ion
Co^{2+}	2+	cobalt(II) ion or cobaltous ion
Ni^{2+}	2+	nickel(II) ion or nickelous ion
Pb^{2+}	2+	lead(II) ion or plumbous ion
Sn^{2+}	2+	tin(II) ion or stannous ion
Fe^{2+}	2+	iron(II) ion or ferrous ion
Fe^{3+}	3+	iron(III) ion or ferric ion
Al^{3+}	3+	aluminum ion
Cr^{3+}	3+	chromium(III) ion or chromic ion

Common Anions		
Formula	**Charge**	**Name**
F^-	1−	fluoride ion
Cl^-	1−	chloride ion
Br^-	1−	bromide ion
I^-	1−	iodide ion
OH^-	1−	hydroxide ion
CN^-	1−	cyanide ion
ClO^-	1−	hypochlorite ion
ClO_2^-	1−	chlorite ion
ClO_3^-	1−	chlorate ion
ClO_4^-	1−	perchlorate ion
CH_3COO^-	1−	acetate ion
MnO_4^-	1−	permanganate ion
NO_2^-	1−	nitrite ion
NO_3^-	1−	nitrate ion
SCN^-	1−	thiocyanate ion
O^{2-}	2−	oxide ion
S^{2-}	2−	sulfide ion
HSO_3^-	1−	hydrogen sulfite ion or bisulfite ion
SO_3^{2-}	2−	sulfite ion
HSO_4^-	1−	hydrogen sulfate ion or bisulfate ion
SO_4^{2-}	2−	sulfate ion
HCO_3^-	1−	hydrogen carbonate ion or bicarbonate ion
CO_3^{2-}	2−	carbonate ion
CrO_4^{2-}	2−	chromate ion
$Cr_2O_7^{2-}$	2−	dichromate ion
PO_4^{3-}	3−	phosphate ion
AsO_4^{3-}	3−	arsenate ion

7-12 Naming Ternary Acids and Their Salts

A ternary compound consists of three elements. Ternary acids **(oxyacids)** are compounds of hydrogen, oxygen, (usually) and a nonmetal. Nonmetals that exhibit more than one oxidation state form more than one ternary acid. These ternary acids differ in the number of oxygen atoms they contain. The suffixes "-ous" and "-ic" following the stem name of the central element indicate lower and higher oxidation states, respectively. One common ternary acid of each nonmetal is (somewhat arbitrarily) designated as the "-ic" acid." That is, it is named "stem*ic* acid." The common ternary "-ic acids" are shown below. There are no common "-ic" ternary acids for the omitted nonmetals.

The oxyacid with the central element in the highest oxidation state usually contains more O atoms. Oxyacids with their central elements in lower oxidation states usually have fewer O atoms.

It is important to learn the names and formulas of these acids, because the names of all other ternary acids and salts are derived from them.

Periodic Group of Central Elements				
IIIA	IVA	VA	VIA	VIIA
(+3) H_3BO_3 boric acid	(+4) H_2CO_3 carbonic acid	(+5) HNO_3 nitric acid		
	(+4) H_4SiO_4 silicic acid	(+5) H_3PO_4 phosphoric acid	(+6) H_2SO_4 sulfuric acid	(+5) $HClO_3$ chloric acid
		(+5) H_3AsO_4 arsenic acid	(+6) H_2SeO_4 selenic acid	(+5) $HBrO_3$ bromic acid
			(+6) H_6TeO_6 telluric acid	(+5) HIO_3 iodic acid

Note that the oxidation state of the central atom is equal to its periodic group number, except in the halogens.

Acids containing *one fewer oxygen atom* per central atom are named in the same way except that the "-ic" suffix is changed to "-ous." The oxidation number of the central element is *lower by 2* in the "-ous" acid than in the "-ic" acid.

Formula	Ox. No.	Name	Formula	Ox. No.	Name
H_2SO_3	+4	sulfur*ous* acid	H_2SO_4	+6	sulfur*ic* acid
HNO_2	+3	nitr*ous* acid	HNO_3	+5	nitr*ic* acid
H_2SeO_3	+4	selen*ous* acid	H_2SeO_4	+6	selen*ic* acid
$HBrO_2$	+3	brom*ous* acid	$HBrO_3$	+5	brom*ic* acid

Ternary acids that have one fewer O atom than the "-ous" acids (two fewer O atoms than the "-ic" acids) are named using the prefix "hypo-" and the suffix "-ous." These are acids in which the oxidation state of the central nonmetal is lower *by 2* than that of the central nonmetal in the "-ous acids."

Formula	Ox. No.	Name
$HClO$	+1	*hypo*chlor*ous* acid
H_3PO_2	+1	*hypo*phosphor*ous* acid
HIO	+1	*hypo*iod*ous* acid
$H_2N_2O_2$	+1	*hypo*nitr*ous* acid

Notice that $H_2N_2O_2$ has a 1:1 ratio of nitrogen to oxygen, as would the hypothetical HNO.

Acids containing *one more oxygen atom* per central nonmetal atom than the normal "-ic acid" are named "*per*stem*ic*" acids.

Formula	Ox. No.	Name
$HClO_4$	+7	*per*chlor*ic* acid
$HBrO_4$	+7	*per*brom*ic* acid
HIO_4	+7	*per*iod*ic* acid

The oxyacids of chlorine follow.

Formula	Ox. No.	Name
$HClO$	+1	*hypo*chlor*ous* acid
$HClO_2$	+3	chlor*ous* acid
$HClO_3$	+5	chlor*ic* acid
$HClO_4$	+7	*per*chlor*ic* acid

Ternary salts are compounds that result from replacing the hydrogen in a ternary acid with another ion. They usually contain metal cations or the ammonium ion. As with binary compounds, the cation is named first. The name of the anion is based on the name of the ternary acid from which it is derived.

An anion derived from a ternary acid with an "-ic" ending is named by dropping the "-ic acid" and replacing it with "-ate." An anion derived from an "-ous acid" is named by replacing the suffix "-ous acid" with "-ite." The "per-" and "hypo-" prefixes are retained.

Formula	Name
$(NH_4)_2SO_4$	ammonium sulfate (SO_4^{2-}, from H_2SO_4)
KNO_3	potassium nitrate (NO_3^-, from HNO_3)
$Ca(NO_2)_2$	calcium nitrite (NO_2^-, from HNO_2)
$LiClO_4$	lithium perchlorate (ClO_4^-, from $HClO_4$)
$FePO_4$	iron(III) phosphate (PO_4^{3-}, from H_3PO_4)
$NaClO$	sodium hypochlorite (ClO^-, from $HClO$)

Acidic salts contain anions derived from ternary acids in which one or more acidic hydrogen atoms remain. These salts are named as if they were the usual type of ternary salt, with the word "hydrogen" or "dihydrogen" inserted after the name of the cation to show the number of acidic hydrogen atoms.

Formula	Name	Formula	Name
$NaHSO_4$	sodium hydrogen sulfate	KH_2PO_4	potassium dihydrogen phosphate
$NaHSO_3$	sodium hydrogen sulfite	K_2HPO_4	potassium hydrogen phosphate
		$NaHCO_3$	sodium hydrogen carbonate

An older, commonly used method (which is not recommended by the IUPAC) involves the use of the prefix "bi-" attached to the name of the anion to indicate the presence of an acidic hydrogen. According to this system, $NaHSO_4$ is called sodium bisulfate and $NaHCO_3$ is called sodium bicarbonate.

Key Terms

Binary acid A binary compound in which H is bonded to one of the more electronegative nonmetals.

Binary compound A compound consisting of two elements; may be ionic or covalent.

Bonding pair A pair of electrons involved in a covalent bond.

Chemical bonds Attractive forces that hold atoms together in elements and compounds.

Covalent bond A chemical bond formed by the sharing of one or more electron pairs between two atoms.

Covalent compound A compound containing predominantly covalent bonds.

Debye The unit used to express dipole moments.

Delocalization of electrons Refers to bonding electrons distributed among more than two atoms that are bonded together; occurs in species that exhibit resonance.

Dipole Refers to the separation of charge between two covalently bonded atoms.

Dipole moment (μ) The product of the distance separating opposite charges of equal magnitude and the magnitude of the charge; a measure of the polarity of a bond or molecule. A measured dipole moment refers to the dipole moment of an entire molecule.

Double bond A covalent bond resulting from the sharing of four electrons (two pairs) between two atoms.

Electron deficient compound A compound containing at least one atom (other than H) that shares fewer than eight electrons.

Formal charge A method of counting electrons in a covalently bonded molecule or ion; bonding electrons are counted as though they were shared equally between the two atoms.

Heteronuclear Consisting of different elements.

Homonuclear Consisting of only one element.

Ionic bonding Chemical bonding resulting from the transfer of one or more electrons from one atom or group of atoms to another.

Ionic compound A compound containing predominantly ionic bonding.

Isoelectronic Having the same electronic configurations.

Lewis acid A substance that accepts a share in a pair of electrons from another species.

Lewis base A substance that makes available a share in an electron pair.

Lewis dot formula The representation of a molecule, ion, or formula unit by showing atomic symbols and only outer shell electrons; does not show shape.

Lone pair A pair of electrons residing on one atom and not shared by other atoms; unshared pair.

Nonpolar bond A covalent bond in which electron density is symmetrically distributed.

Octet rule Many representative elements attain at least a share of eight electrons in their valence shells when they form molecular or ionic compounds; there are some limitations.

Oxidation numbers Arbitrary numbers that can be used as mechanical aids in writing formulas and balancing equations. For single-atom ions they correspond to the charge on the ion; more electronegative atoms are assigned negative oxidation numbers.

Oxidation states See *Oxidation numbers*.

Polar bond A covalent bond in which there is an unsymmetrical distribution of electron density.

Pseudobinary ionic compound A compound that contains more than two elements but is named like a binary compound.

Resonance A concept in which two or more equivalent dot formulas for the same arrangement of atoms (resonance structures) are necessary to describe the bonding in a molecule or ion.

Single bond A covalent bond resulting from the sharing of two electrons (one pair) between two atoms.

Ternary acid A ternary compound containing H, O, and another element, often a nonmetal.

Ternary compound A compound consisting of three elements; may be ionic or covalent.

Triple bond A covalent bond resulting from the sharing of six electrons (three pairs) between two atoms.

Unshared pair See *Lone pair*.

Exercises

Chemical Bonding—Basic Ideas

1. Give a suitable definition of chemical bonding. What type of force is responsible for chemical bonding?
2. List the two basic types of chemical bonding. Give an example of a substance with each type of bonding. What are the differences between these types? What are some of the general properties associated with the two main types of bonding?
3. Why are covalent bonds called directional bonds, whereas ionic bonding is termed nondirectional?
4. The outermost electron configuration of any alkali metal atom, Group IA, is ns^1. How can an alkali metal atom attain a noble gas configuration?
5. The outermost electron configuration of any halogen atom, Group VIIA, is ns^2np^5. How can a halogen atom attain a noble gas configuration?
6. (a) What do Lewis dot representations for atoms show? (b) Draw Lewis dot representations for the following atoms: He, C, S, Ar, Br, Sr.
7. Draw Lewis dot representations for the following atoms: Li, B, P, K, Kr, As.
8. Describe the types of bonding in calcium hydroxide, $Ca(OH)_2$.

   Ca^{2+} 2[:Ö—H]$^-$

9. Describe the types of bonding in sodium chlorate, $NaClO_3$.

   Na^+ [:Ö—Cl(—:Ö:)—Ö:]$^-$

10. Based on the positions in the periodic table of the following pairs of elements, predict whether bonding between the two would be primarily ionic or covalent. Justify your answers. (a) Ba and S, (b) P and O, (c) Cl and Br, (d) Li and I, (e) Si and Cl, (f) Ca and F.

11. Predict whether the bonding between the following pairs of elements would be ionic or covalent. Justify your answers. (a) K and S, (b) N and O, (c) Ca and Br, (d) P and S, (e) C and Br, (f) K and N.
12. Classify the following compounds as ionic or covalent: (a) $MgSO_4$, (b) SO_2, (c) KNO_3, (d) $NiCl_2$, (e) H_2CO_3, (f) NCl_3, (g) Li_2O, (h) H_3PO_4, (i) $SOCl_2$.

Ionic Bonding

13. Describe what happens to the valence electron(s) as a metal atom and a nonmetal atom combine by ionic bonding.

*14. Describe an ionic crystal. What factors might determine the geometrical arrangement of the ions?

15. Why are most solid ionic compounds rather poor conductors of electricity? Why does conductivity increase when an ionic compound is melted or dissolved in water?
16. Write chemical equations for reactions between the following pairs of elements. Draw electronic structures of the atoms before reaction as well as electronic structures of the ions formed in the reactions, using both the orbital and ns^xnp^y notations, where n refers to the outermost occupied shell.
    (a) calcium and fluorine
    (b) barium and oxygen
    (c) sodium and sulfur
17. Write chemical equations for reactions between the following pairs of elements. Draw electronic structures of the atoms before reaction as well as electronic structures of the ions formed in the reactions, using both the orbital and ns^xnp^y notations, where n refers to the outermost occupied shell.
    (a) lithium and fluorine
    (b) magnesium and sulfur
    (c) calcium and chlorine
18. Write the formula for the ionic compound that forms between each of the following pairs of elements: (a) Cs and Cl_2, (b) Ba and F_2, (c) K and S.
19. Write the formula for the ionic compound that forms between each of the following pairs of elements: (a) Mg and Cl_2, (b) Sr and Cl_2, (c) Na and Se.
20. When a *d*-transition metal undergoes ionization, it loses its outer *s* electrons before it loses any *d* electrons. Using [noble gas]$(n-1)d^x$ representations, write the outer electron configurations for the following ions: (a) Co^{2+}, (b) Mn^{2+}, (c) Zn^{2+}, (d) Fe^{2+}, (e) Cu^{2+}, (f) Sc^{3+}, (g) Cu^+.
21. Which of the following do not accurately represent stable binary ionic compounds? Why? $BaCl_2$, KF_2, AlF_4, SrS_2, Ca_2O_3, $NaBr_2$, Li_2S.
22. Which of the following do not accurately represent stable binary ionic compounds? Why? MgI, BaO, InF_2, Al_2O_3, $RbCl_2$, CsSe, Be_2O.
23. (a) Write Lewis formulas for the positive and negative ions in these salts: $SrBr_2$, Li_2S, Ca_3P_2, PbF_2, Bi_2O_3. (b) Which ions do not have a noble gas configuration?
24. (a) What are isoelectronic species? List three pairs of isoelectronic species. (b) All but one of the following species are isoelectronic. Which one is not isoelectronic with the others? Ne, Al^{3+}, O^{2-}, Na^+, Mg^+, F^-.
25. All but one of the following species are isoelectronic. Which one is not isoelectronic with the others? S^{2-}, Ga^{2+}, Ar, K^+, Ca^{2+}, Sc^{3+}.
26. Write formulas for two cations and two anions that are isoelectronic with neon.
27. Write formulas for two cations and two anions that are isoelectronic with argon.
28. Write formulas for two cations that have the following electron configurations in their *highest* occupied energy level: (a) $4s^24p^6$, (b) $5s^25p^6$.
29. Write formulas for two anions that have the electron configurations listed in Exercise 28.

Covalent Bonding—General Concepts

30. What does Figure 7-4 tell us about the attractive and repulsive forces in a hydrogen molecule?
31. Distinguish between heteronuclear and homonuclear diatomic molecules.
32. How many electrons are shared between two atoms in (a) a single covalent bond, (b) a double covalent bond, and (c) a triple covalent bond?
33. What is the maximum number of covalent bonds that a second-period element could form? How can the representative elements beyond the second period form more than this number of covalent bonds?

Lewis Dot Formulas for Molecules and Polyatomic Ions

34. What information about chemical bonding can a Lewis formula for a compound or ion give? What information about bonding is not directly represented by a Lewis formula?
35. What is the octet rule? Is it generally applicable to compounds of the transition metals? Why?
36. (a) What is the simple mathematical relationship that is useful in writing Lewis dot formulas? (b) What does each term in the relationship represent?
37. Write Lewis dot formulas for the following: H_2, N_2, Cl_2, HCl, HBr.
38. Write Lewis dot formulas for the following: H_2O, NH_3, OH^-, F^-.
39. Use Lewis formulas to represent the covalent molecules formed by these pairs of elements. Write only structures that satisfy the octet rule. (a) P and H, (b) Se and Br, (c) C and F, (d) Si and F.
40. Use Lewis formulas to represent the covalent molecules formed by these pairs of elements. Write only structures that satisfy the octet rule. (a) O and F, (b) As and Cl, (c) I and Cl, (d) N and F.

41. Find the total number of valence electrons for (a) $SnCl_4$, (b) NH_2^-, (c) CH_5O^+ (d) XeO_3, (e) CN_2H_2.

42. Write Lewis formulas for the following covalent molecules: (a) H_2S, (b) PCl_3, (c) BCl_3, (d) SiH_4, (e) SF_4.

43. Write Lewis formulas for the following molecules: (a) CH_4S, (b) CH_2ClBr, (c) S_2Cl_2, (d) NH_2Cl, (e) C_2H_6S (two possibilities).

44. Write Lewis formulas without formal charges for these multiply bonded compounds: (a) CS_2, (b) ClNO, (c) C_2H_2O (two possibilities), (d) C_3H_4 (two possibilities).

45. Write Lewis formulas for (a) ClO_4^-, (b) NOF, (c) SeF_4, (d) $COCl_2$, (e) ClF_3.

46. Write Lewis structures for (a) H_2O_2, (b) IO_4^-, (c) BeH_2, (d) NCl_3, (e) HClO, (f) XeF_4.

47. Write Lewis formulas for (a) H_2NOH (i.e., one H bonded to O), (b) S_8 (a ring of eight atoms), (c) SiH_4, (d) F_2O_2 (O atoms in center, F atoms on outside), (e) CO, (f) $SeCl_6$.

48. Write Lewis formulas for (a) BBr_3, (b) SF_6, (c) CN^-, (d) AlH_4^-, (e) N_2H_4, (f) H_2SO_3.

***49.** (a) Write the Lewis formula for $AlCl_3$, a molecular compound. Note that in $AlCl_3$, the aluminum atom is an exception to the octet rule. (b) In the gaseous phase, two molecules of $AlCl_3$ join together (dimerize) to form Al_2Cl_6. (The two molecules are joined by two "bridging" Al—Cl—Al bonds.) Write the Lewis formula for this molecule.

***50.** (a) Write the Lewis formula for molecular ClO_2. There is a single unshared electron on the chlorine atom in this molecule. (b) Two of these molecules dimerize to form Cl_2O_4. (The two molecules are joined by a Cl—Cl bond.) Write the Lewis formula for this molecule.

51. What do we mean by the term "resonance"? Do the resonance structures that we draw actually represent the bonding in the substance? Explain your answer.

***52.** We can write two resonance structures for toluene, $C_6H_5CH_3$:

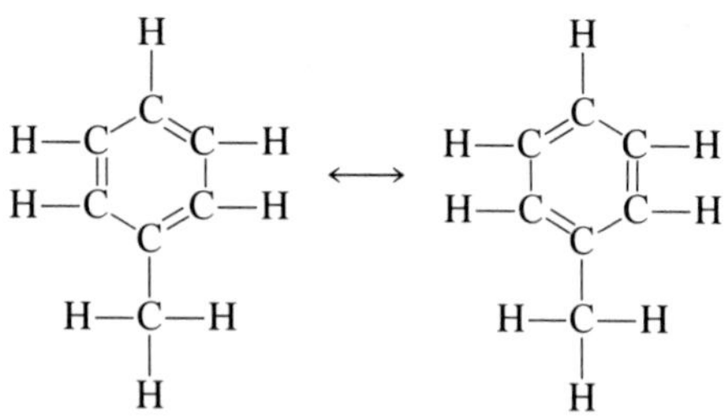

How would you expect the carbon–carbon bond length in the six-membered ring to compare with the carbon–carbon bond length between the CH_3 group and the carbon atom on the ring?

53. Write resonance structures for the formate ion, $HCOO^-$.

54. Write resonance structures for the nitrate ion, NO_3^-.

55. Which of the following species contain at least one atom that violates the octet rule?

(a) :F—F: (b) :O—Cl—O: (c) :F—Xe—F:

(d) [:O—S—O: (with :O: above and :O: below S)]$^{2-}$

56. Which of the following species contain at least one atom that violates the octet rule?

(a) P_4 (tetrahedron of P atoms: :P, P, P:, P)

(b) O=C=O

(c) [:O—Cl—O: (with :O: above Cl)]$^-$

(d) :Cl—B—Cl: (with :Cl: below B)

***57.** None of the following is known to exist. What is wrong with each one?

(a) :Cl—S=O

(b) H—H—O—P—Cl: (with :Cl: below P)

(c) :O≡N—O:$^-$

(d) Na—S: (with Na below S)

***58.** None of the following is known to exist. What is wrong with each one?

(a) :F—B—F: (with H above B and :F: below B)

(b) :O—Cl:$^{2-}$ (with :O: below O)

(c) H—O—H—C≡N:

(d) :Br—F: (with :F: below Br)

***59.** "El" is the general symbol for a representative element. In each case, in which periodic group is El located? Justify your answers and cite a specific example for each one.

(a) [:O—El—O: (with :O: below El)]$^-$

(b) H—O—El—O: (with :O: above and :O: below El)

(c) H—O—El=O

***60.** "El" is the general symbol for a representative element. In each case, in which periodic group is El located? Justify your answers and cite a specific example for each one.

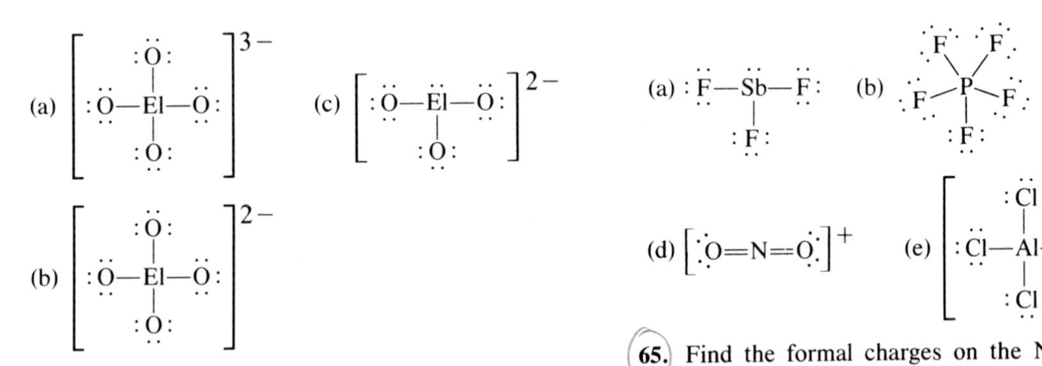

61. Many common stains, such as those of chocolate and other fatty foods, can be removed by dry-cleaning solvents such as tetrachloroethylene, C_2Cl_4. Is C_2Cl_4 ionic or covalent? Draw its Lewis dot formula.

*62. Draw acceptable dot formulas for the following common air pollutants: (a) SO_2, (b) NO_2, (c) CO, (d) O_3 (ozone), (e) SO_3, (f) $(NH_4)_2SO_4$. Which one is a solid? Which ones exhibit resonance?

Formal Charges

63. Assign a formal charge to each atom in the following:

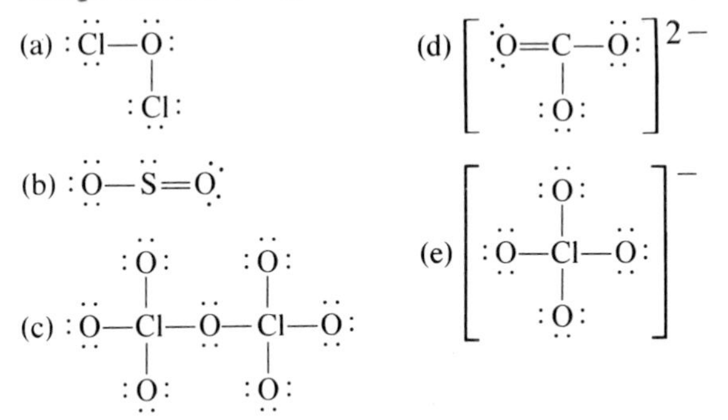

64. Assign a formal charge to each atom in the following:

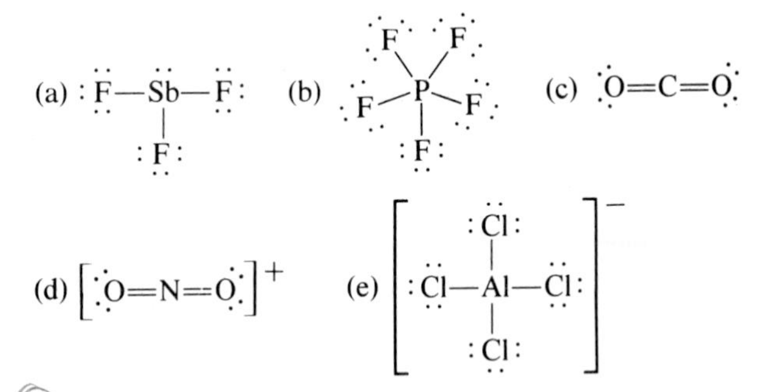

65. Find the formal charges on the N and B atoms in (a) NH_3, (b) BF_3, (c) NH_4^+, (d) BF_4^-, (e) NH_2^-, (f) $H_3B—NH_3$.

66. Find the formal charge of each atom, other than H and halogen atoms, in the following:
(a) $Cl_3P—O$

(b) $\left[\begin{matrix} CH_3O—H \\ | \\ H \end{matrix} \right]^+$ (c) $\left[\begin{matrix} O \\ | \\ O—S—O \\ | \\ O \end{matrix} \right]^{2-}$

*67. With the aid of formal charges, explain which Lewis formula is more likely to be correct for each given molecule.

(a) For Cl_2O, :C̤̈l—Ö̤—C̤̈l: or :C̤̈l—C̤̈l—Ö̤:

(b) For HN_3, H—N̈=N=N̤̈ or H—N≡N—N̤̈:

(c) For N_2O, N̤̈=O=N̤̈ or :N≡N—Ö̤:

68. Write Lewis formulas for three different atomic arrangements with the molecular formula HCNO. Indicate all formal charges. Predict which arrangement is likely to be the least stable and justify your selection.

Ionic versus Covalent Character and Bond Polarities

69. Distinguish between polar and nonpolar covalent bonds.

70. Why is an HCl molecule polar while a Cl_2 molecule is nonpolar?

71. Why do we show only partial charges, and not full charges, on the atoms of a polar molecule?

72. (a) Which two of the following pairs of elements are most likely to form ionic bonds? Te and H, C and F, Ba and F, N and F, K and O. (b) Of the remaining three pairs, which one forms the least polar, and which the most polar, covalent bond?

73. (a) List three reasonable nonpolar covalent bonds between dissimilar atoms. (b) List three pairs of elements whose compounds should exhibit extreme ionic character.

74. The elements X, Y, and Z are in the same period of the periodic table and are in groups IIA, VIA, and VIIA, respectively. (a) Write the Lewis formula for

the compound most likely to be formed between X and Z. Will this compound most likely be ionic or covalent? (b) Write the Lewis formula for the most probable compound of Y and Z. Will this compound be ionic or covalent?

75. Classify the bonding between the following pairs of atoms as ionic, polar covalent, or nonpolar covalent. (a) Si and O, (b) N and O, (c) Sr and F, (d) As and As.

76. Classify the bonding between the following pairs of atoms as ionic, polar covalent, or nonpolar covalent. (a) Li and O, (b) Br and I, (c) Ca and H, (d) O and O, (e) H and O.

77. State whether the structure of each substance is covalent or ionic, and give a reason for each answer. (a) HBr, mp −85.5°C, bp−67°C; (b) $MgCl_2$, mp 708°C, bp 1418°C (the molten substance conducts electricity).

78. Look up the properties of NaCl and PCl_3 in a handbook of chemistry. Why is NaCl classified as an ionic compound and PCl_3 as a covalent compound?

79. The following properties can be found in a handbook of chemistry:

camphor, $C_{10}H_{16}O$—colorless crystals; specific gravity 0.990 at 25°C; sublimes 204°C; insoluble in water; very soluble in alcohol and ether.

praseodymium chloride, $PrCl_3$—blue-green needle crystals; specific gravity 4.02; melting point 786°C; boiling point 1700°C; solubility in cold water, 103.9 g/100 mL H_2O; very soluble in hot water.

Would you classify each of these as ionic or covalent? Why?

80. (a) Write Lewis formulas for atoms of strontium, chlorine, and silicon. (b) Use the appropriate pairs of these atoms to show formation of (i) an ionic compound and (ii) a covalent molecule. (c) Indicate the polarity of the bond in the covalent molecule.

Oxidation Numbers (review)

81. What are oxidation numbers? How can they be useful?

82. What oxidation numbers are the following elements expected to exhibit in simple binary ionic compounds? Cl, O, Sr, K, Al, Se.

83. What oxidation numbers are the following elements expected to exhibit in simple binary ionic compounds? Ba, Li, Na, S, I, P.

84. Evaluate the oxidation number of N in NO, NO_2, N_2O_3, N_2O_5, N_2H_4, and H_2NOH.

85. Evaluate the oxidation number of (a) C in CH_4, CH_3OH, H_2CO, CO, and CO_2; (b) Cl in Cl_2, HClO, $HClO_2$, $HClO_3$, and $HClO_4$.

86. Evaluate the oxidation number of the underlined element in (a) $\underline{C}O_2$, (b) $\underline{S}OCl_2$, (c) $Na_2\underline{S}$, (d) $K_3\underline{P}O_4$, (e) $K_2\underline{P}HO_3$, (f) $K\underline{P}H_2O_2$, (g) $\underline{P}OCl_3$, (h) $K_2\underline{Mn}O_4$, and (i) $H\underline{C}OOH$. Assume that the other elements have their common oxidation numbers.

Naming Inorganic Compounds

87. Name the following monatomic cations, using the IUPAC system of nomenclature. (a) Li^+, (b) Cd^{2+}, (c) Fe^{2+}, (d) Mn^{2+}, (e) Al^{3+}.

88. Name the following monatomic cations, using the IUPAC system of nomenclature. (a) Au^+, (b) Au^{3+}, (c) Ba^{2+}, (d) Zn^{2+}, (e) Ag^+.

89. Write the chemical symbol for each of the following: (a) sodium ion, (b) zinc ion, (c) silver ion, (d) mercury(II) ion, (e) iron(III) ion.

90. Write the chemical symbol for each of the following: (a) lithium ion, (b) bismuth(III) ion, (c) iron(II) ion, (d) chromium(III) ion, (e) potassium ion.

91. Name the following ions: (a) N^{3-}, (b) O^{2-}, (c) Se^{2-}, (d) F^-, (e) Br^-.

92. Write the chemical symbol for each of the following: (a) chloride ion, (b) sulfide ion, (c) telluride ion, (d) iodide ion, (e) phosphide ion.

93. Name the following ionic compounds: (a) Li_2S, (b) SnO_2, (c) RbI, (d) Li_2O, (e) Ba_3N_2.

94. Name the following ionic compounds: (a) NaI, (b) Hg_2S, (c) Li_3N, (d) $MnCl_2$, (e) CuF_2, (f) FeO.

95. Write the chemical formula for each of the following compounds: (a) sodium fluoride, (b) zinc oxide, (c) barium oxide, (d) magnesium bromide, (e) hydrogen iodide, (f) copper(I) chloride.

96. Write the chemical formula for each of the following compounds: (a) sodium oxide, (b) calcium phosphide, (c) iron(II) oxide, (d) iron(III) oxide, (e) manganese(IV) oxide, (f) silver fluoride.

97. Name the following compounds: (a) NH_4CN, (b) $K_2Cr_2O_7$, (c) $Ca_3(PO_4)_2$, (d) $CaCO_3$, (e) $NaNO_3$.

98. Name the following compounds: (a) $(NH_4)_2SO_4$, (b) $Al(NO_3)_3$, (c) $Fe(ClO_4)_2$, (d) Li_2CO_3, (e) BaO_2.

99. Write the chemical formula for each of the following compounds: (a) potassium sulfite, (b) calcium permanganate, (c) sodium peroxide, (d) ammonium dichromate, (e) ammonium acetate.

100. Write the chemical formula for each of the following compounds: (a) iron(II) chlorate, (b) potassium nitrite, (c) barium phosphate, (d) copper(I) sulfate, (e) sodium carbonate.

101. Name the following common acids: (a) HCl, (b) H_3PO_4, (c) $HClO_4$, (d) HNO_3, (e) H_2SO_3.

102. Write the chemical formula for each of the following acids and bases: (a) nitrous acid, (b) sulfuric acid, (c) bromic acid, (d) sodium hydroxide, (e) calcium hydroxide.

103. What is the name of the acid with the formula H_2CO_3? Write the formulas of the two anions derived from it and name these ions.

104. What is the name of the acid with the formula H_3PO_3? What is the name of the HPO_3^{2-} ion?

105. Name the following binary molecular compounds: (a) CO, (b) CO_2, (c) SF_6, (d) $SiCl_4$, (e) IF.

106. Name the following binary molecular compounds: (a) AsF_3, (b) Br_2O, (c) BrF_5, (d) CSe_2, (e) Cl_2O_7.

107. Write the chemical formula for each of the following compounds: (a) iodine bromide, (b) silicon dioxide, (c) phosphorus trichloride, (d) tetrasulfur dinitride, (e) bromine trifluoride, (f) hydrogen telluride, (g) xenon tetrafluoride.

108. Write the chemical formula for each of the following compounds: (a) diboron trioxide, (b) dinitrogen pentasulfide, (c) phosphorus triiodide, (d) sulfur tetrachloride, (e) silicon sulfide, (f) hydrogen sulfide, (g) diphosphorus trioxide.

109. Write formulas for the compounds that are expected to be formed by the following pairs of ions:

	A. F^-	B. OH^-	C. SO_3^{2-}	D. PO_4^{3-}	E. NO_3^-
1. NH_4^+		Omit – see note			
2. K^+					
3. Mg^{2+}					
4. Cu^{2+}					
5. Fe^{3+}					
6. Ag^+					

NOTE: The compound NH_4OH does not exist. The solution commonly labeled "NH_4OH" is aqueous ammonia, $NH_3(aq)$.

110. Write the names for the compounds of Exercise 109.

111. Write the formula for each of the following substances and balance each equation.

(a) aluminum + iron(III) oxide → aluminum oxide + iron

(b) potassium hydroxide + zinc chlorate → zinc hydroxide + potassium chlorate

(c) silver nitrate + hydrogen sulfide → silver sulfide + nitric acid

(d) sodium carbonate + hydrogen chloride → sodium chloride + carbon dioxide + water

Mixed Exercises

112. Write a name for each formula or a formula for each name: (a) $Al(OH)_3$, (b) nitrogen trichloride, (c) tin(IV) oxide, (d) chromium(VI) oxide, (e) PbS, (f) $NaNO_2$.

113. Find the charge of the ion formed in each of the following ionization reactions, and name each of the resulting ions:

(a) Fe → Fe ion + 2 e^- (c) P + 3 e^- → P ion

(b) S + 2 e^- → S ion (d) Al → Al ion + 3 e^-

114. The following properties can be found in a handbook of chemistry:

barium chloride, $BaCl_3$—colorless cubic crystals; specific gravity 3.92; melting point 963°C; boiling point 1560°C; very soluble in water; very slightly soluble in alcohol.

Would you classify this substance as ionic or covalent? Why?

115. Describe the types of bonding in sodium sulfite, Na_2SO_3.

$2[Na]^+$ [:O—S—O:]$^{2-}$ (with :O: bonded below S)

116. Write the formula for the compound that forms between (a) calcium and nitrogen, (b) aluminum and oxygen, (c) potassium and selenium, (d) strontium and bromine. Name each compound.

117. Write Lewis formulas for the covalent molecules (a) silicon tetrahydride, (b) iodine chloride, (c) phosphorus pentafluoride, (d) carbon disulfide.

118. Draw the Lewis formulas for the nitric acid molecule (HNO_3) that are consistent with the following bond length data: 1.405 Å for the bond between the nitrogen atom and the oxygen atom that is attached to the hydrogen atom; 1.206 Å for the bonds between the nitrogen atom and each of the other oxygen atoms.

119. Which of the following species contain(s) at least one atom that violates the octet rule?

(a) Cl and Cl bonded to C=O (b) :F—P—F: with F above P and two F below P (c) [H—O—H with H above O]$^+$

120. Name the following compounds: (a) K_2CrO_4, (b) Na_2SO_3, (c) $FeCO_3$, (d) $Fe_2(SO_4)_3$, (e) $FeSO_4$.

121. Write the chemical formula for each of the following compounds: (a) silver nitrate, (b) uranium(IV) sulfate, (c) aluminum acetate, (d) manganese(II) phosphate.

122. Write the formula of each substance and balance each equation:

(a) ammonium sulfate + sodium hydroxide →
ammonia + sodium sulfate + water

(b) sulfur tetrafluoride + water →
hydrogen fluoride + sulfur dioxide

(c) phosphorus trichloride + chlorine →
phosphorus pentachloride

(d) hydrogen chloride + manganese(IV) oxide →
manganese(II) chloride + chlorine + water

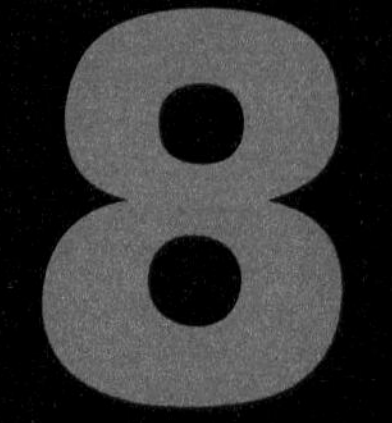

Molecular Structure and Covalent Bonding Theories

The primary genetic material of all cells is deoxyribonucleic acid, DNA. A DNA molecule is a double helix of nucleotides, each of which contains one molecule of a base, one molecule of deoxyribose (a sugar), and one molecule of phosphoric acid. A model of a small portion of this helix is shown here. Genetic information is encoded in DNA in the order in which pairs of bases (shown here in red and yellow) are stacked.

Outline

Objectives

As you study this chapter, you should learn

- ☐ The basic ideas of the valence shell electron pair repulsion (VSEPR) theory
- ☐ To use the VSEPR theory to predict electronic geometry of polyatomic molecules and ions
- ☐ To use the VSEPR theory to predict molecular geometry of polyatomic molecules and ions
- ☐ The relationships between molecular shapes and molecular polarities
- ☐ To predict whether a molecule is polar or nonpolar
- ☐ The basic ideas of the valence bond (VB) theory
- ☐ To analyze the hybrid orbitals used in bonding in polyatomic molecules and ions
- ☐ To use hybrid orbitals to describe the bonding in double and triple bonds

We know a great deal about the molecular structures of many thousands of compounds, all based on reliable experiments. In our discussion of theories of covalent bonding, we must keep in mind that the theories represent *an attempt to explain and organize experimental observations*. For bonding theories to be valid, they must be consistent with the large body of experimental observations about molecular structure. In this chapter, we shall study two theories of covalent bonding, which allow us to predict structures and properties that are usually accurate. Like any simplified theories, they are not entirely satisfactory in describing *every* known structure; however, their successful application to many thousands of structures justifies their continued use.

8-1 An Overview of the Chapter

The electrons in the outer shell, or **valence shell**, of an atom are the electrons involved in bonding. In most of our discussion of covalent bonding, we will focus attention on these electrons. Valence shell electrons may be thought of as those that were not present in the *preceding* noble gas, ignoring *filled* sets of *d* and *f* orbitals. Lewis dot formulas show the number of valence shell electrons in a polyatomic molecule or ion (Sections 7-7 through 7-10). We shall draw Lewis dot formulas for each molecule or polyatomic ion we discuss. The theories introduced in this chapter apply equally well to polyatomic molecules and to ions.

Two theories go hand in hand in a discussion of covalent bonding. The *valence shell electron pair repulsion (VSEPR) theory* helps us to understand and predict the spatial arrangement of atoms in a polyatomic molecule or ion. However, it does not explain *how* bonding occurs, just *where* it occurs, as well as where lone pairs of valence shell electrons are directed. The *valence bond (VB) theory* describes *how* the bonding takes place, in terms of *overlapping atomic orbitals*. In this theory, the atomic orbitals discussed in Chapter 5 are often "mixed," or *hybridized*, to form new orbitals with different spatial orientations. These two simple ideas, used together, enable us to understand the bonding, molecular shapes, and properties of a wide variety of polyatomic molecules and ions.

In Chapter 7 we used valence bond terminology to discuss the bonding in H_2 although we did not name the theory there.

We shall first discuss the basic ideas and application of these two theories. Then we shall learn how an important molecular property, *polarity*, depends on molecular shape. The major part of this chapter will then be devoted to studying how these ideas are applied to various types of polyatomic molecules and ions.

Important Note

Different instructors prefer to cover these two theories in different ways. We believe that one of these two alternative approaches is most effective. Your instructor will tell you the order in which you should study the material in this chapter. However you study this chapter, Tables 8-1, 8-2, 8-3, and 8-4 are important summaries, and you should refer to them often.

A. One approach is to discuss both the VSEPR theory and the VB theory together, emphasizing how they complement one another. If your instructor prefers this parallel approach, you should study the chapter in the order in which it is presented.

B. An alternative approach is to first master the VSEPR theory and the related topic of molecular polarity for many different structures, and then learn how the VB theory describes the overlap of bonding orbitals in these structures. If your instructor takes this approach, you should study this chapter in the following order:

1. Read the summary material under the main heading "Molecular Shapes and Bonding" preceding Section 8-5.
2. *VSEPR theory, molecular polarity.* Study Sections 8-2 and 8-3; then in Sections 8-5 through 8-12, study only the subsections marked A and B.
3. *VB theory.* Study Section 8-4; then in Sections 8-5 through 8-12, study the valence bond subsections, marked C; then study Sections 8-13 and 8-14.

Learn this procedure and use it as a mental "checklist." Trying to do this reasoning in a different order often leads to confusion or wrong answers.

No matter which order your instructor prefers, the following procedure will help you to analyze the structure and bonding in any compound.

1. Draw the Lewis formula for the molecule or polyatomic ion, and identify a *central atom*—an atom that is bonded to more than one other atom (Section 8-2).
2. Count the *number of regions of high electron density* on the central atom (Section 8-2).
3. Apply the VSEPR theory to determine the arrangement of the *regions of high electron density* (the *electronic geometry*) about the central atom (Section 8-2; Tables 8-1 and 8-4).
4. Using the Lewis formula as a guide, determine the arrangement of the *bonded atoms* (the *molecular geometry*) about the central atom, as well as the location of the unshared valence electron pairs on that atom (parts B of Sections 8-5 through 8-12; Tables 8-3 and 8-4). This description includes predicted bond angles.
5. If there are lone pairs of electrons on the central atom, consider how their presence might modify somewhat the *ideal* molecular geometry and bond angles deduced in Step 4 (Section 8-2; parts B of Sections 8-8 through 8-12).
6. Use the VB theory to determine the *hybrid orbitals* utilized by the central atom; describe the overlap of these orbitals to form bonds and the orbitals that contain unshared valence shell electron pairs on the central atom (parts C of Sections 8-5 through 8-12; Sections 8-13; 8-14; Tables 8-2 and 8-4).
7. If more than one atom can be identified as a central atom, repeat Steps 2 through 6 for each central atom, to build up a picture of the geometry and bonding in the entire molecule.
8. When all central atoms have been accounted for, use the entire molecular geometry, electronegativity differences, and the presence of lone pairs of valence shell electrons on the central atom to predict *molecular polarity* (Section 8-3; parts B of Sections 8-5 through 8-12).

Never try to skip to Step 5 until you have done Step 4. The electronic geometry and the molecular geometry may or may not be the same; knowing the electronic geometry first will enable you to find the correct molecular geometry.

The following diagram summarizes this procedure.

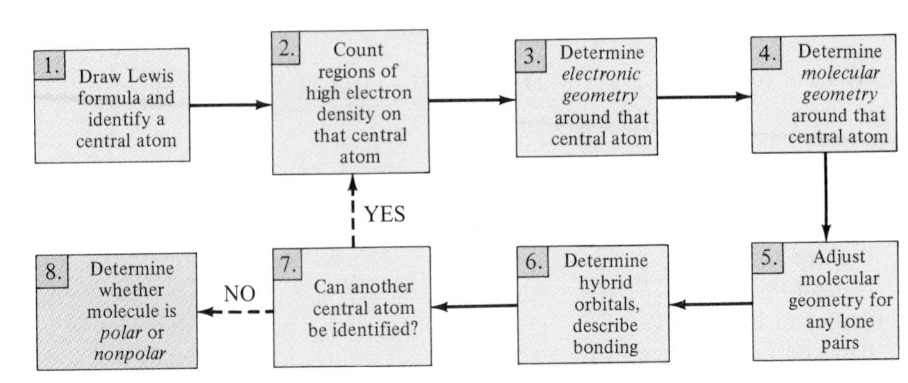

In Section 7-7 we showed that Lewis dot formulas of polyatomic ions can be constructed in the same way as dot formulas of neutral molecules. We must take into account the "extra" electrons on anions and the "missing" electrons of cations. Once the Lewis formula of an ion is known, we use the VSEPR and VB theories to deduce its electronic geometry, shape, and hybridization, just as for neutral molecules.

8-2 Valence Shell Electron Pair Repulsion (VSEPR) Theory

The basic ideas of the **valence shell electron pair repulsion (VSEPR) theory** follow:

> Each set of valence shell electrons on a central atom is significant. The sets of valence shell electrons on the *central atom* repel one another. They are arranged about the *central atom* so that repulsions among them are as small as possible.

This results in maximum separation among regions of high electron density about the central atom.

A **central atom** is any atom that is bonded to more than one other atom. In some molecules, more than one central atom may be present. In such cases, we determine the arrangement around each in turn, to build up a picture of the overall shape of the molecule or ion. We first count the number of **regions of high electron density** around the *central atom,* as follows:

> 1. Each bonded atom is counted as *one* region of high electron density, *whether the bonding is single, double, or triple.*
> 2. Each unshared pair of valence electrons on the central atom is counted as *one* region of high electron density.

As examples of this way of counting, consider the following molecules and polyatomic ions.

Formula:	CH_4	NH_3	CO_2	SO_4^{2-}
Lewis dot formula:	H | H—C—H | H	H | :N—H | H	$\ddot{O}$::C::$\ddot{O}$	[:Ö: | :Ö—S—Ö: | :Ö:]$^{2-}$
Central atom:	C	N	C	S
Number of atoms bonded to *central atom:*	4	3	2	4
Number of lone pairs on *central atom:*	0	1	0	0
Total number of regions of high electron density on *central atom:*	4	4	2	4

According to VSEPR theory, the structure is as stable as possible when the regions of high electron density on the central atom are as far apart as possible. For instance, two regions of high electron density are most stable on opposite sides of the central atom (the linear arrangement). Three regions are most stable when they are arranged at the corners of an equilateral triangle (the trigonal planar arrangement). The arrangement of these *regions of high electron density* around the central atom is referred to as the **electronic geometry** of the central atom.

Table 8-1 shows the relationship between the common numbers of regions of high electron density and the corresponding electronic geometries. After we know the electronic geometry (and *only then*), we consider how many of these regions of high electron density connect (bond) the central atom to other atoms. This lets us deduce the arrangement of *atoms* around the central atom, called the **molecular geometry**. If necessary, we repeat this procedure for each central atom in the molecule or ion. These procedures are illustrated in parts B of Sections 8-5 through 8-12.

Although the terminology is not as precise as we might wish, we use "molecular geometry" to describe the arrangement of atoms in *polyatomic ions* as well as in molecules.

8-3 Polar Molecules—The Influence of Molecular Geometry

In Chapter 7 we saw that the unequal sharing of electrons between two atoms with different electronegativities, $\Delta EN > 0$, results in a *polar bond*. For heteronuclear diatomic molecules such as HF, this bond polarity results in a *polar molecule*. Then the entire molecule acts as a dipole, and we would measure the *dipole moment* of such a molecule to be nonzero. When a molecule consists of more than two atoms joined by polar bonds, we must also take into account the *arrangement* of the resulting bond dipoles. For such a case, we first use VSEPR theory to deduce the atomic arrangement (molecular geometry), as described in the preceding section and exemplified in parts A and B of Sections 8-5 through 8-12. Then we determine whether

Table 8-1
Number of Regions of High Electron Density about a Central Atom

Number of Regions of High Electron Density	Electronic Geometry*	Angles†
2	linear	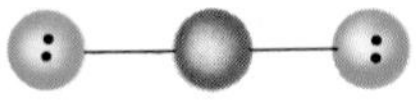 180°
3	trigonal planar	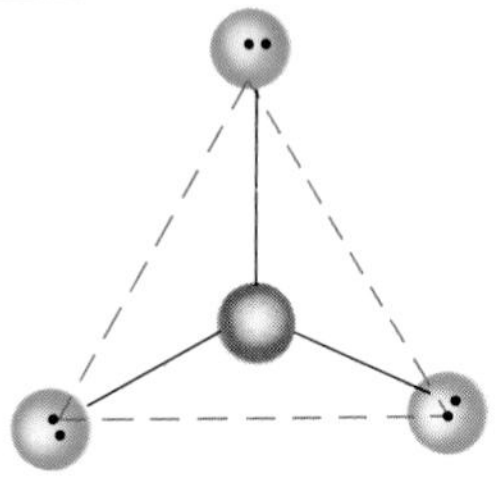 120°
4	tetrahedral	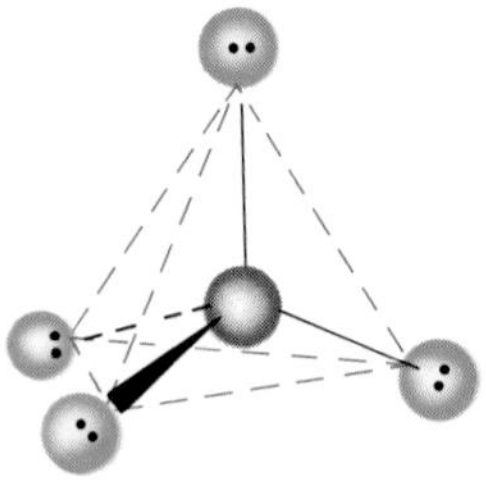 109.5°
5	trigonal bipyramidal	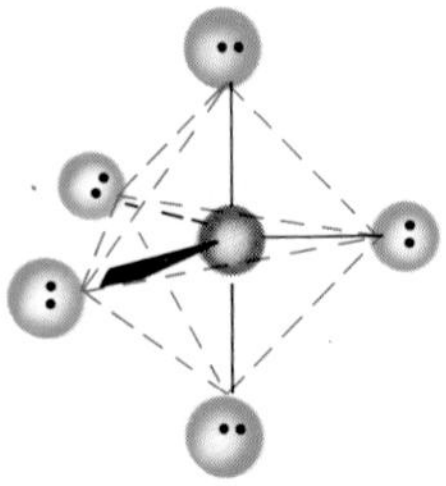 90°, 120°, 180°
6	octahedral	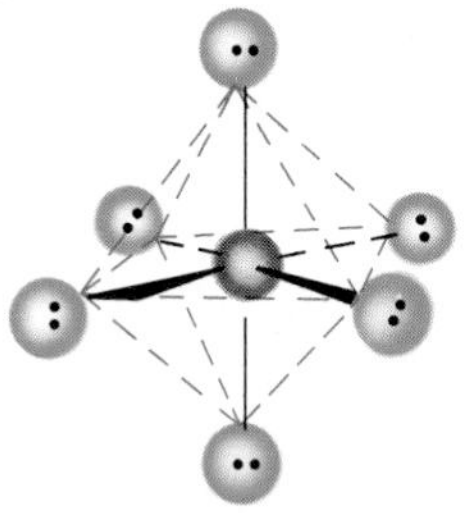 90°, 180°

* Electronic geometries are illustrated here using only single pairs of electrons as regions of high electron density. The symbol ⊙ represents the regions of high electron density about the central atom ●. By convention, a line in the plane of the drawing is represented by a solid line ______, a line behind this plane is shown as a dashed line ---, and a line in front of this plane is shown as a wedge ◀ with the fat end of the wedge nearest the viewer. Each shape is outlined in blue dashed lines to help you visualize it.

† Angles made by imaginary lines through the nucleus and the centers of regions of high electron density.

the bond dipoles are arranged in such a way that they cancel (so that the resulting molecule is *nonpolar*) or do not cancel (so that the resulting molecule is *polar*).

In this section, we shall discuss the ideas of cancellation of dipoles in general terms, using general atomic symbols X and Y. These ideas will be applied to specific molecular geometries and molecular polarities in parts B of Sections 8-5 through 8-12.

Let us consider a heteronuclear molecule with the formula XY_2 (X is the central atom). Such a molecule must have one of the following two molecular geometries:

The angular form could have different angles, but either the molecule is linear or it is not.

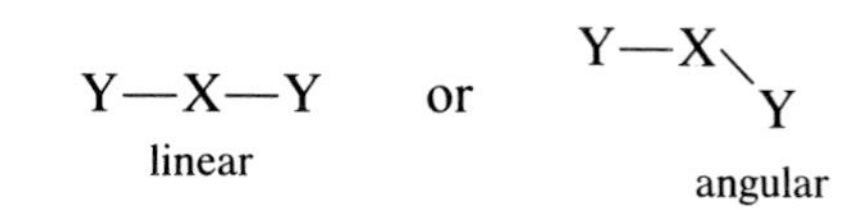

Suppose that atom Y has a higher electronegativity than atom X. Then each X—Y bond is polar, with the negative end of the bond dipole pointing toward Y. Each bond dipole can be viewed as an *electronic vector,* with a *magnitude* and a *direction.* In the linear XY_2 arrangement, the two bond dipoles are *identical* in magnitude and *opposite* in direction. Therefore, they cancel to give a nonpolar molecule (dipole moment equal to zero).

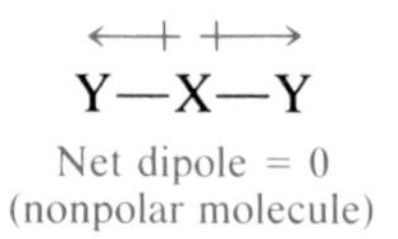

In the case of the angular arrangement, the two equal dipoles *do not cancel,* but add to give a nonzero result. The angular molecular arrangement represents a polar molecule (dipole moment greater than zero).

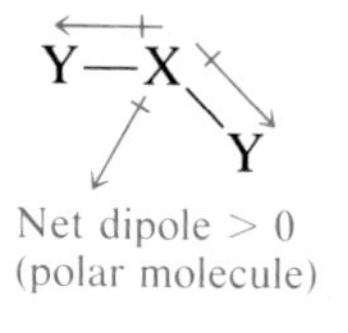

If the electronegativity differences were reversed in this Y—X—Y molecule—that is, if X were more electronegative than Y—the directions of all

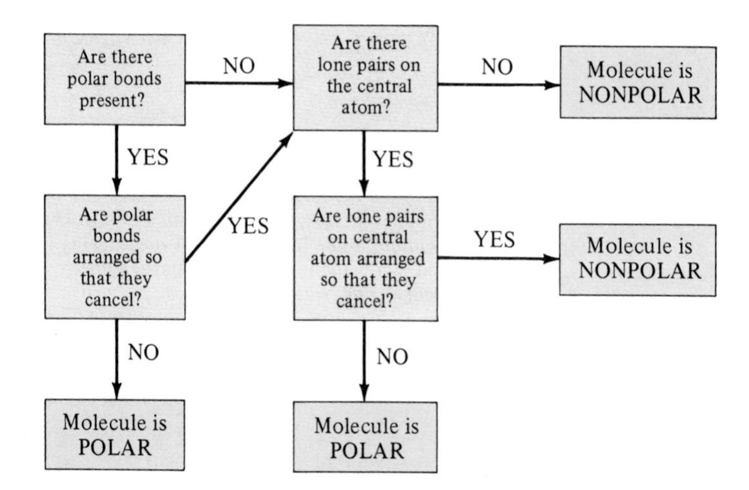

Figure 8-1
A guide to determining whether a polyatomic molecule is polar or nonpolar. Study the more detailed presentation in the text.

bond polarities would be reversed. But the bond polarities would still cancel in the linear arrangement, to give a nonpolar molecule. In the angular arrangement, bond polarities would still add to give a polar molecule, but with the net dipole pointing in the opposite direction from that described above.

Similar arguments based on addition of bond dipoles can be made for other arrangements. As we shall see in Section 8-8, lone pairs on the central atom also affect the direction and the magnitude of the net molecular dipole, so the presence of lone pairs on the central atom must always be taken into account.

For a molecule to be polar, *both* of the following conditions must be met:

1. There must be at least one polar bond or one lone pair on the central atom.

and

2. a. The polar bonds, if there are more than one, must not be so symmetrically arranged that their bond polarities cancel.

or

b. If there are two or more lone pairs on the central atom, they must not be so symmetrically arranged that their polarities cancel.

Put another way, if there are no polar bonds or lone pairs of electrons on the central atom, the molecule *cannot* be polar. Even if polar bonds or lone pairs are present, they might be arranged so that their polarities cancel one another, resulting in a nonpolar molecule.

For instance, carbon dioxide, CO_2, is a three-atom molecule in which each carbon–oxygen bond is *polar* because of the electronegativity difference between C and O. But the molecule *as a whole* is shown by experiment to be nonpolar. This tells us that the polar bonds are arranged in such a way that the bond polarities cancel. Water, H_2O, on the other hand, is a very polar molecule; this tells us that the H—O bond polarities do not cancel one another. Molecular shapes clearly play a crucial role in determining molecular dipole moments. We must develop a better understanding of molecular shapes in order to understand molecular polarities.

:O=C=O:

linear molecule; bond dipoles cancel; molecule is nonpolar

H—O: —H

angular molecule; bond dipoles do not cancel; molecule is polar

The logic used in deducing whether a molecule is polar or nonpolar is outlined in Figure 8-1. The approach described in this section will be applied to various electronic and molecular geometries in parts B of Sections 8-5 through 8-12.

8-4 Valence Bond (VB) Theory

In Chapter 7 we described covalent bonding as electron pair sharing that results from the overlap of orbitals from two atoms. This is the basic idea of the **valence bond (VB) theory**—it describes *how* bonding occurs. In many examples throughout this chapter, we shall first use the VSEPR theory to describe the *orientations* of the regions of high electron density. Then we shall use the VB theory to describe the atomic orbitals that overlap to produce the bonding with that geometry. We shall also assume that each

VSEPR theory describes the locations of bonded atoms around the central atom, as well as where its lone pairs of valence shell electrons are directed.

Table 8-2
Relation between Electronic Geometries and Hybridization

Regions of High Electron Density	Electronic Geometry	Hybridization	Atomic Orbitals Mixed from Valence Shell of Central Atom
2	linear	sp	one s, one p
3	trigonal planar	sp^2	one s, two p's
4	tetrahedral	sp^3	one s, three p's
5	trigonal bipyramidal	sp^3d	one s, three p's, one d
6	octahedral	sp^3d^2	one s, three p's, two d's

lone pair occupies a separate orbital. Thus, the two theories work together to give a fuller description of the bonding.

The atomic orbitals that we described in Chapter 5 refer to isolated atoms. Usually these "pure" atomic orbitals do not have the correct energies or the correct orientations to describe where the electrons are when an atom is bonded to other atoms. To explain the experimentally observed geometry, we usually need to invoke the concept of **hybrid orbitals**, which are formed by the combination, or **hybridization**, of the atom's valence shell atomic orbitals. The number of regions of high electron density about a central atom in a molecule or polyatomic ion (Table 8-1) suggests the kind of hybridization of that atom's valence atomic orbitals that occurs. The designation (label) given to a set of hybridized orbitals reflects the *number and kind* of atomic orbitals that hybridize to produce the set (Table 8-2). Further details about hybridization and hybrid orbitals appear in the following sections. Throughout the text, hybrid orbitals and, when appropriate, the atomic orbitals to which they are related are shaded in green.

Molecular Shapes and Bonding

We are now ready to study the structures of some simple molecules. In the sections that follow, we use generalized chemical formulas in which "A" represents the central atom, "B" represents an atom bonded to A, and "U" represents an unshared valence shell electron pair (lone pair) on the central atom A. For instance, AB_3U (Section 8-8) would represent any molecule with three B atoms bonded to a central atom A, with one unshared valence pair on A. For each type of bonding arrangement, we shall follow the eight steps of analysis outlined in Section 8-1. We shall first give the known (experimentally determined) facts about polarity and shape, and draw the Lewis dot formula (part A of each section). Then we shall explain these facts in terms of the VSEPR and VB theories. The simpler VSEPR theory will be used to explain (or predict) first the *electronic geometry* and then the *molecular geometry* in the molecule (part B). We shall then show how the molecular polarity of each molecule is a result of bond polarities, lone pairs, and molecular geometry. Finally, we shall use the VB theory to describe the bonding in molecules in more detail, usually using hybrid orbitals (part C). As you study each section, refer frequently to the summaries that appear in Table 8-3 on pages 324–325.

See the "Important Note" in Section 8-1 and consult your instructor for guidance on the order in which you should study Sections 8-5 through 8-12.

8-5 Linear Electronic Geometry—AB_2 Species (No Unshared Pairs of Electrons on A)

A. Experimental Facts and Lewis Formulas

Several linear molecules consist of a central atom plus two atoms of another element, abbreviated as AB_2. These compounds include $BeCl_2$, $BeBr_2$, and BeI_2, as well as CdX_2 and HgX_2, where X = Cl, Br, or I. All of these are known to be linear (bond angle = 180°), nonpolar, covalent compounds, although the individual bonds are polar.

Let's focus on *gaseous* $BeCl_2$ molecules (mp 405°C). The electronic structures of Be and Cl *atoms* in their ground states are

The high melting point of $BeCl_2$ is due to its polymeric nature in the solid state.

	1s	2s				1s	2s	2p			3s	3p		
Be	⇅	⇅	and		Cl	⇅	⇅	⇅	⇅	⇅	⇅	⇅	⇅	↑

We drew the Lewis dot formula for $BeCl_2$ in Example 7-5. It shows two single covalent bonds, with Be and Cl each contributing one electron to each bond:

$$:\ddot{\underset{..}{Cl}}:Be:\ddot{\underset{..}{Cl}}: \quad \text{or} \quad :\ddot{\underset{..}{Cl}}—Be—\ddot{\underset{..}{Cl}}:$$

B. VSEPR Theory

Valence Shell Electron Pair Repulsion theory places the two electron pairs on Be 180° apart, i.e., with *linear electronic geometry*. Both electron pairs are bonding pairs, so VSEPR predicts a linear atomic arrangement, or *linear molecular geometry*, for $BeCl_2$.

VSEPR theory assumes that regions of high electron density (electron pairs) on the central atom will be as far from one another as possible.

$$:\ddot{\underset{..}{Cl}}—\overset{180°}{Be}—\ddot{\underset{..}{Cl}}:$$

A model of a linear AB_2 molecule, e.g., $BeCl_2$.

If we examine the bond dipoles, we see that the electronegativity difference (see Table 6-3) is large (1.5 units) and the bonds are quite polar:

	Cl—Be—Cl
EN =	3.0 1.5 3.0
Δ(EN) =	1.5 1.5

$$\overset{\longleftarrow\!\!+}{:\ddot{\underset{..}{Cl}}—}Be\overset{+\!\!\longrightarrow}{—\ddot{\underset{..}{Cl}}:}$$

Net dipole = 0

The two bond dipoles are *identical* in magnitude and *opposite* in direction. Therefore, they cancel to give nonpolar molecules.

It is important to distinguish between *nonpolar bonds* and *nonpolar molecules*.

The difference in electronegativity between Be and Cl is so large that we might expect ionic bonding. However, the radius of Be^{2+} is so small (0.31 Å) and its **charge density** (ratio of charge to size) is so high that most simple beryllium compounds are covalent rather than ionic. The high charge density of Be^{2+} causes it to attract and distort the electron cloud of monatomic anions of all but the most electronegative elements. As a result, electrons are shared rather than being localized on ions. Two exceptions are BeF_2 and BeO. They are ionic compounds that contain very electronegative elements.

We say that the Be^{2+} ion *polarizes* the anions, Cl^-.

C. Valence Bond Theory

Cl ground state configuration:
[Ne] $\underline{\uparrow\downarrow}$ $\underline{\uparrow\downarrow}\ \underline{\uparrow\downarrow}\ \underline{\uparrow}$
3s 3p

Consider the ground state electronic configuration for Be. There are two electrons in the 1*s* orbital, but these nonvalence (inner) electrons are *not* involved in bonding. There are two more electrons *paired* in the 2*s* orbital. How, then, will two Cl atoms bond to Be? The Be atom must somehow make available one orbital for each bonding Cl electron (the unpaired *p* electrons). The following *ground state* electron configuration for Be is the configuration for an isolated Be atom. Another configuration may be more stable in a bonding environment. Suppose that the Be atom "promoted" one of the paired 2*s* electrons to one of the 2*p* orbitals, the next higher energy orbitals.

Be [He] $\underset{2s}{\underline{\uparrow\downarrow}}$ $\underset{2p}{\underline{\ }\ \underline{\ }\ \underline{\ }}$ $\xrightarrow{\text{promote}}$ Be [He] $\underset{2s}{\underline{\uparrow}}$ $\underset{2p}{\underline{\uparrow}\ \underline{\ }\ \underline{\ }}$

Then there would be two Be orbitals available for bonding, but we find a discrepancy between this description and experimental fact. The Be 2*s* and 2*p* orbitals would not overlap a Cl 3*p* orbital with equal effectiveness. So this "promoted pure atomic" arrangement would predict two *nonequivalent* Be—Cl bonds. Yet we observe experimentally that the Be—Cl bonds are *identical* in bond length and bond strength. So we reject the idea of simple "promotion" as an explanation.

Hund's Rule is discussed in Section 5-16.

For these two orbitals on Be to become equivalent, they must hybridize to give two orbitals intermediate between the *s* and *p* orbitals. These are called ***sp* hybrid orbitals**. Consistent with Hund's Rule, the two valence electrons of Be would occupy each of these orbitals individually:

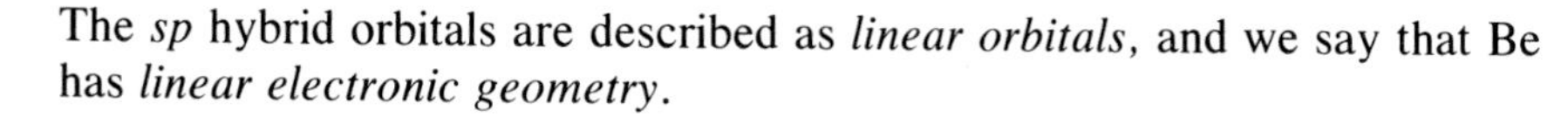

The *sp* hybrid orbitals are described as *linear orbitals*, and we say that Be has *linear electronic geometry*.

As we did for pure atomic orbitals, we often draw hybrid orbitals more slender than they actually are. Such drawings are intended to remind us of the orientations and general shapes of orbitals.

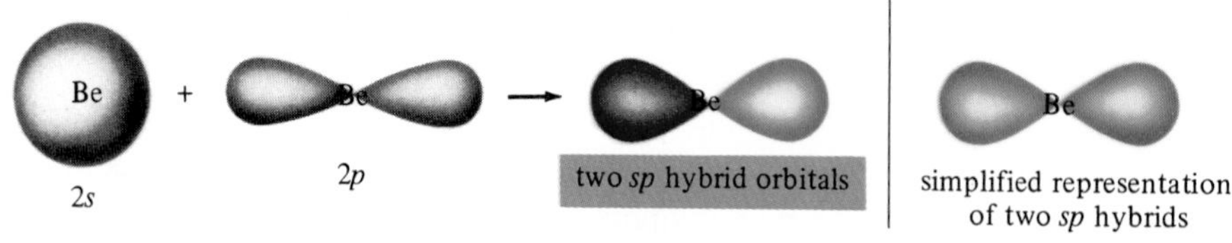

Recall that each Cl atom has a 3*p* orbital that contains only one electron, and so overlap with the *sp* hybrids of Be is possible. We picture the bonding in $BeCl_2$ in the following diagram, in which only the bonding electrons are represented:

Lone pairs of e^- on Cl atoms are not shown. The hybrid orbitals on the central atom are shown in green in this and subsequent drawings.

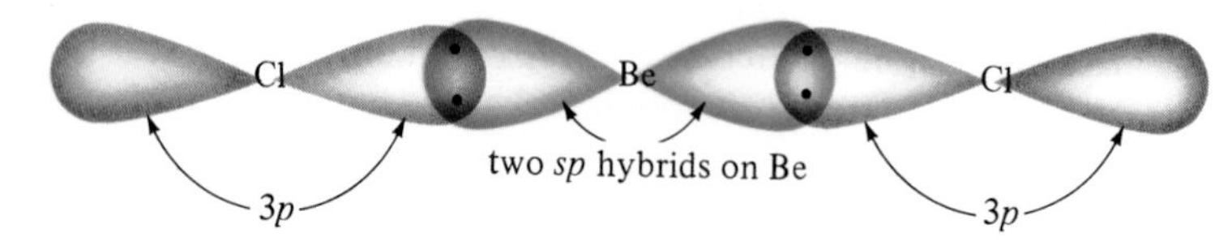

The Be and two Cl nuclei lie on a straight line. *This is consistent with the experimental observation that the molecule is linear.*

One additional idea about hybridization is worth special emphasis:

> The number of hybrid orbitals is always equal to the number of atomic orbitals that hybridize.

Hybrid orbitals are named by indicating the *number and kind* of atomic orbitals hybridized. Hybridization of *one s* orbital and *one p* orbital gives *two sp hybrid orbitals*. We shall see presently that hybridization of *one s* and *two p* orbitals gives *three* sp^2 hybrid orbitals; hybridization of *one s* orbital and *three p* orbitals gives *four* sp^3 hybrids, and so on (Table 8-2).

Hybridization usually involves orbitals from the same main shell (same *n*).

The structures of beryllium bromide, $BeBr_2$, and beryllium iodide, BeI_2, are similar to that of $BeCl_2$. The chlorides, bromides, and iodides of cadmium, CdX_2, and mercury, HgX_2, are also linear, covalent molecules (where X = Cl, Br, or I). A cadmium ion has two electrons in its 5*s* orbitals, and its 5*p* orbitals are vacant. Similarly, a mercury atom has two electrons in its 6*s* orbital, and its 6*p* orbitals are vacant. Thus, the possibility of *sp* hybridization exists in both metals. CdX_2 and HgX_2 are additional examples of this kind of covalent bonding.

The two X's within one structure are identical.

> *sp* hybridization occurs at the central atom whenever there are two regions of high electron density around the central atom. AB_2 molecules and ions with no lone pairs on the central atom have linear electronic geometry, linear molecular geometry, and *sp* hybridization on the central atom.

8-6 Trigonal Planar Electronic Geometry—AB_3 Species (No Unshared Pairs of Electrons on A)

A. Experimental Facts and Lewis Formulas

Boron is a Group IIIA element that forms many covalent compounds by bonding to three other atoms. Typical examples include boron trifluoride, BF_3 (mp −127°C); boron trichloride, BCl_3 (mp −107°C); boron tribromide, BBr_3 (mp −46°C); and boron triiodide, BI_3 (mp 50°C). All are trigonal planar nonpolar molecules.

A trigonal planar molecule is a flat molecule in which all three bond angles are 120°.

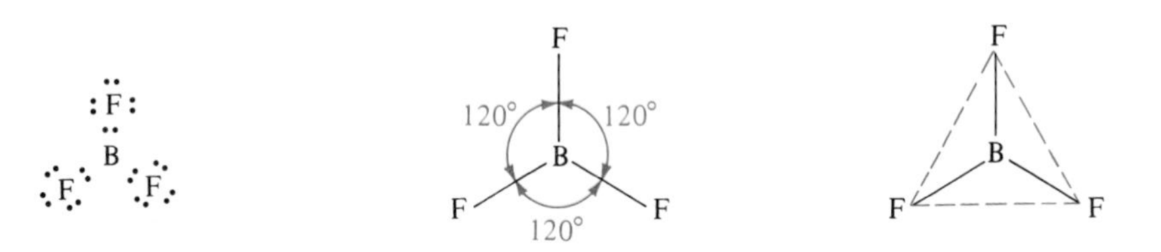

The solid lines represent bonds between B and F atoms. The dashed blue lines emphasize the shape of the molecule.

The Lewis dot formula for BF_3 is derived from the following: (a) each B atom has three electrons in its valence shell and (b) each B atom is bonded to three F (or Cl, Br, I) atoms. In Example 7-6 we drew the Lewis dot formula for BCl_3. F and Cl are both members of Group VIIA, and so the dot formula for BF_3 should be similar.

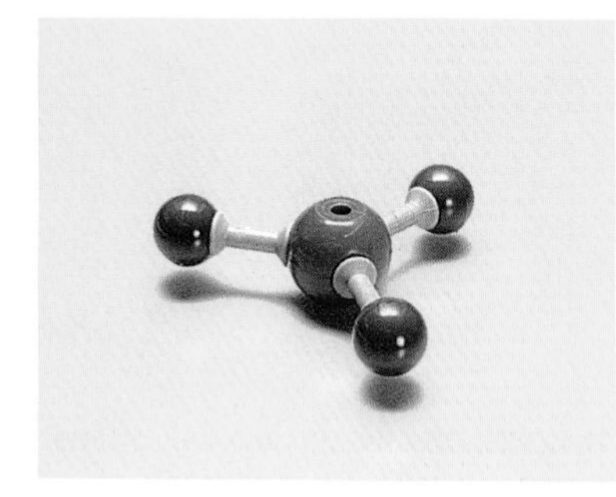

A model of a trigonal planar AB_3 molecule, e.g., BF_3.

We see that BF_3 and other similar molecules have central elements that do *not* attain a noble gas configuration by sharing electrons. Boron shares only six electrons.

B. VSEPR Theory

Boron, the central atom, has three regions of high electron density (three bonded atoms, no unshared pairs on B). VSEPR theory predicts **trigonal planar** *electronic geometry* for molecules such as BF_3 because this structure gives maximum separation among the three regions of high electron density. There are no lone pairs of electrons associated with the boron atom, so a fluorine atom is at each corner of the equilateral triangle, and the *molecular geometry* is also trigonal planar. The maximum separation of any three items (electron pairs) around a fourth item (B atom) is at 120° angles in a single plane. All four atoms are in the same plane. The three F atoms are at the corners of an equilateral triangle, with the B atom in the center. The structures of BCl_3, BBr_3, and BI_3 are similar.

The B^{3+} ion is so small (radius = 0.20 Å) that boron does not form simple ionic compounds.

Examination of the bond dipoles of BF_3 shows that the electronegativity difference (Table 6-3) is very large (2.0 units) and that the bonds are very polar.

$$\begin{array}{rcc} & \text{B—F} \\ \text{EN} = & \underbrace{2.0 \quad 4.0} \\ \Delta(\text{EN}) = & 2.0 \end{array}$$

Net molecular dipole = 0

However, the three bond dipoles are symmetrical and cancel to give nonpolar molecules.

C. Valence Bond Theory

To be consistent with experimental findings and the predictions of VSEPR theory, the VB theory must explain three *equivalent* B—F bonds. Again we invoke hybridization. Now the 2*s* orbital and two of the 2*p* orbitals of B hybridize to form a set of three degenerate ***sp*² hybrid orbitals**.

"Degenerate" refers to orbitals of the same energy.

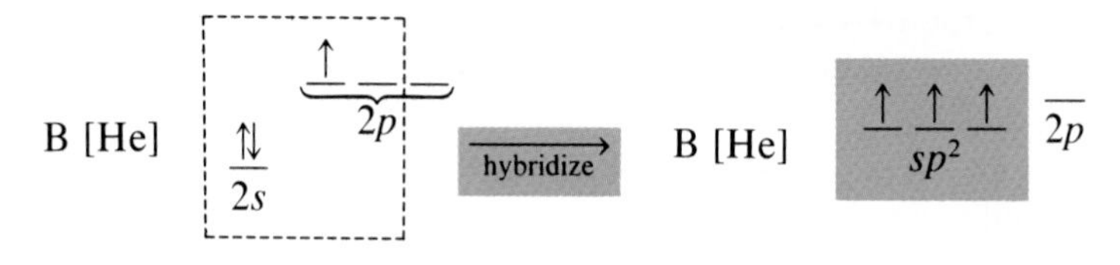

Three sp^2 hybrid orbitals point toward the corners of an equilateral triangle:

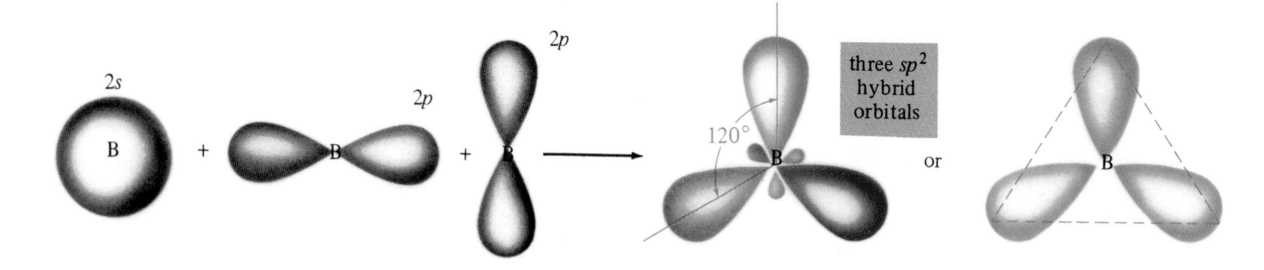

Each of the three F atoms has a $2p$ orbital with one unpaired electron. The $2p$ orbitals can overlap the three sp^2 hybrid orbitals on B. Three electron pairs are shared among one B and three F atoms:

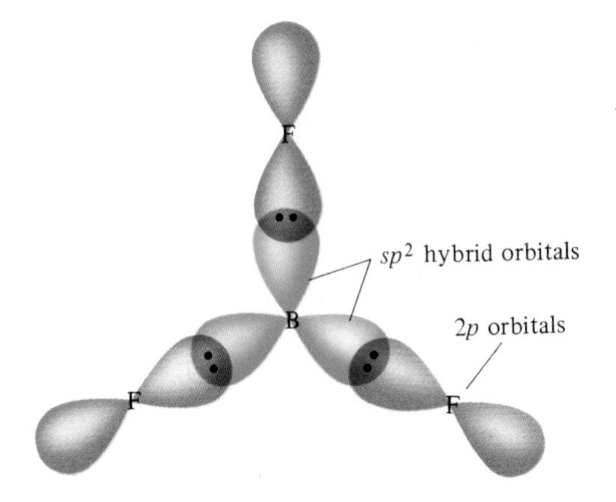

Lone pairs of e^- are not shown for the F atoms.

> sp^2 hybridization occurs at the central atom whenever there are three regions of high electron density around the central atom. AB_3 molecules and ions with no lone pairs on the central atom have trigonal planar electronic geometry, trigonal planar molecular geometry, and sp^2 hybridization on the central atom.

A molecule that has fewer than eight electrons in the valence shell of the central atom frequently reacts by accepting a share in an electron pair from another species. A substance that behaves in this way is called a **Lewis acid** (to be discussed more fully in Section 10-10). Both beryllium chloride, $BeCl_2$, and boron trichloride, BCl_3, react as Lewis acids. The fact that both compounds so readily take a share of additional pairs of electrons tells us that Be and B atoms do not have octets of electrons in these gaseous compounds.

8-7 Tetrahedral Electronic Geometry—AB_4 Species (No Unshared Pairs of Electrons on A)

A. Experimental Facts and Lewis Formulas

Each Group IVA element has four electrons in its highest occupied energy level. The Group IVA elements form many covalent compounds by sharing those four electrons with four other atoms. Typical examples include CH_4 (mp −182°C), CF_4 (mp −184°C), CCl_4 (mp −23°C), SiH_4 (mp −185°C), and SiF_4 (mp −90°C). All are tetrahedral, nonpolar molecules (bond angles = 109.5°). In each of them, the IVA atom is located in the center of a regular tetrahedron. The other four atoms are located at the four corners of the tetrahedron.

The Group IVA element contributes four electrons in a tetrahedral AB_4 molecule, and the other four atoms contribute one electron each. The Lewis dot formulas for methane, CH_4, and carbon tetrafluoride, CF_4, are typical:

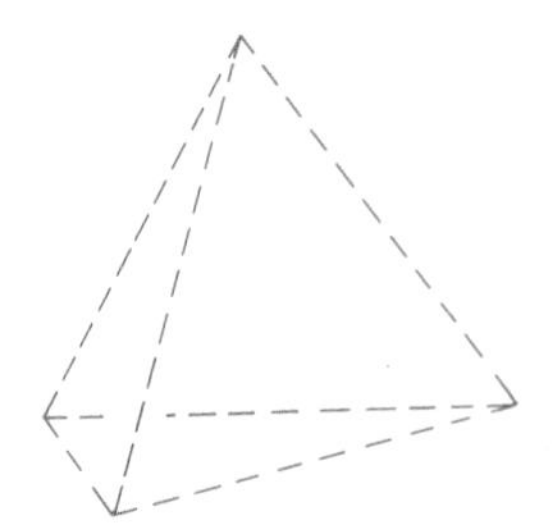

The names of many solid figures are based on the numbers of plane faces they have. A *regular* tetrahedron is a three-dimensional figure with four equal-sized equilateral triangular faces (the prefix *tetra-* means "four").

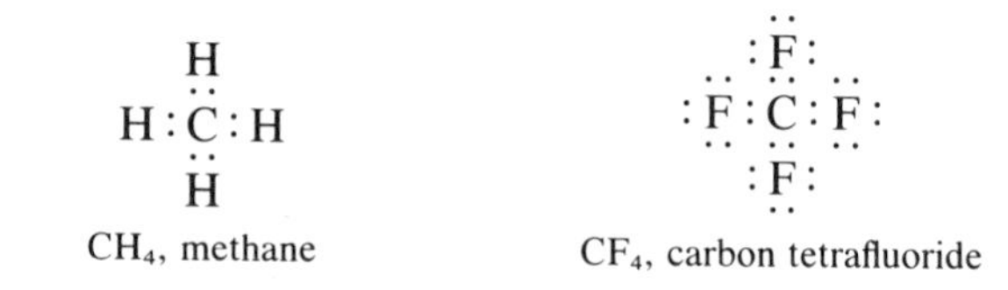

Ammonium ion, NH_4^+, and sulfate ion, SO_4^{2-}, are familiar examples of polyatomic ions of this type. In each of these ions, the central atom is located at the center of a regular tetrahedron with the other atoms at the corners (H—N—H and O—S—O bond angles = 109.5°).

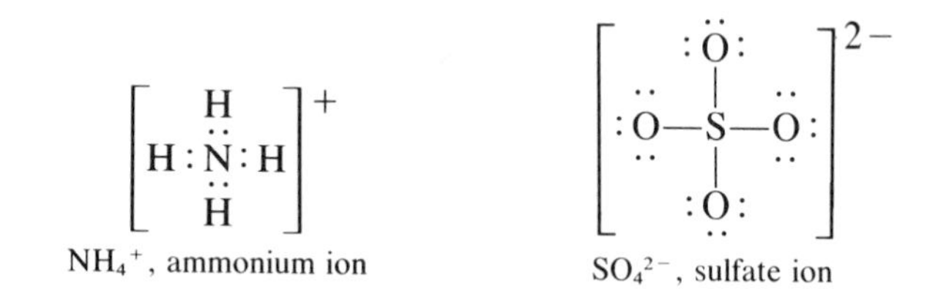

B. VSEPR Theory

VSEPR theory predicts that the four electron pairs are directed toward the corners of a regular tetrahedron. That shape gives the maximum separation for four electron pairs around one atom. Thus, VSEPR theory predicts a *tetrahedral electronic structure* for an AB_4 molecule that has no unshared electrons on A. Because there are no lone pairs of electrons on the central atom, another atom is at each corner of the tetrahedron. VSEPR theory predicts a *tetrahedral molecular geometry* of each of these molecules. Again, let us describe CH_4 and CF_4 molecules.

Models of two tetrahedral AB_4 molecules: CH_4 (left) and CF_4 (right).

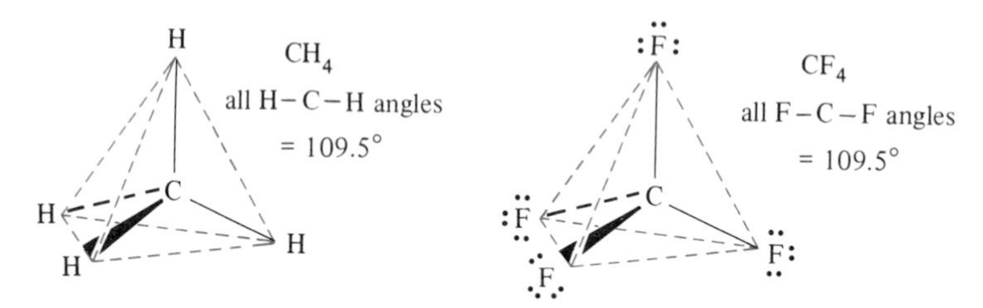

Examination of bond dipoles shows that in CH_4 the individual bonds are only slightly polar, whereas in CF_4 the bonds are quite polar. In CH_4 the bond dipoles are directed toward carbon, but in CF_4 they are directed away from carbon. Both molecules are quite symmetrical, so the bond dipoles cancel, and both molecules are nonpolar. This is true for *tetrahedral* AB_4 molecules in which there are *no unshared electron pairs on the central element* and all four B atoms are identical.

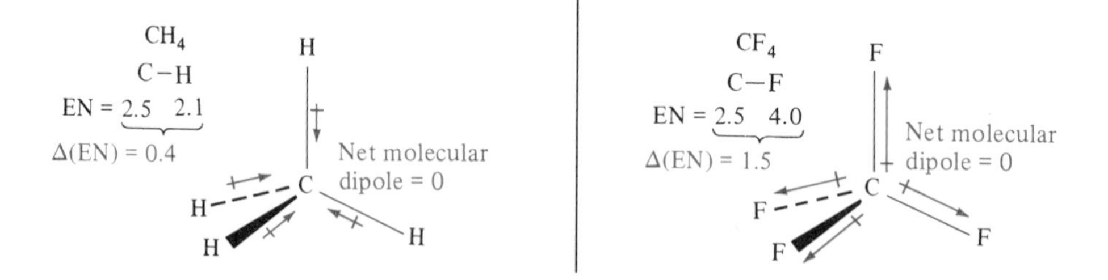

In some tetrahedral molecules, the atoms bonded to the central atom are not all the same. Such molecules may be polar, depending on the relative sizes of the bond dipoles present. In CH_3F or CH_2F_2, for example, the addition of unequal dipoles makes the molecule polar.

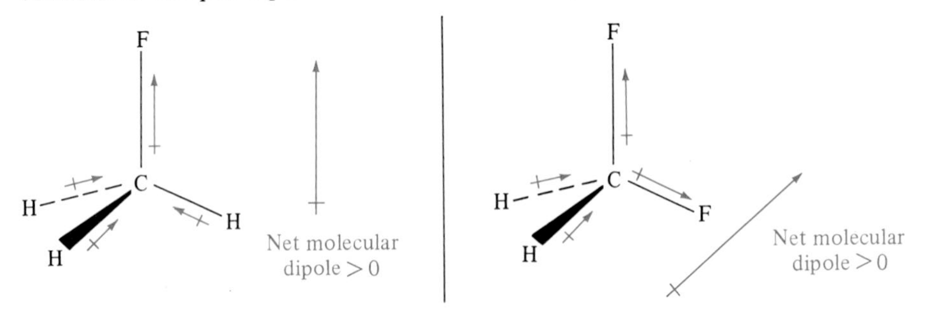

VSEPR theory predicts that NH_4^+ and SO_4^{2-} ions each have tetrahedral electronic geometry. Each region of high electron density bonds the central atom to another atom (H in NH_4^+, O in SO_4^{2-}) at the corner of the tetrahedral arrangement. We describe the molecular geometry of each of these ions as tetrahedral.

You may wonder whether square planar AB_4 molecules exist. They do, in compounds of some of the transition metals. All *simple* square planar AB_4 molecules have unshared electron pairs on A. The bond angles in square planar molecules are only 90°. Nearly all AB_4 molecules are tetrahedral, however, with larger bond angles (109.5°) and greater separation of valence electron pairs around A.

C. Valence Bond Theory

According to VB theory, each Group IVA atom (C in our example) must make four equivalent orbitals available for bonding. To do this, C forms four **sp^3 hybrid orbitals** by mixing the *s* and all three *p* orbitals in its outer shell. This results in four unpaired electrons:

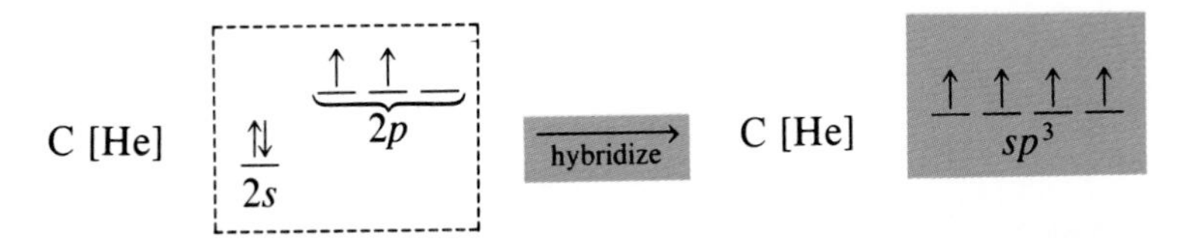

These sp^3 hybrid orbitals are directed toward the corners of a regular tetrahedron, which has a 109.5° angle from any corner to center to any other corner.

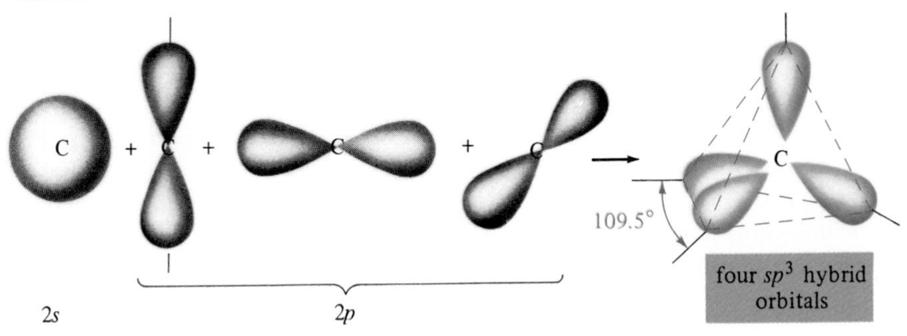

Each of the four atoms that bond to C possesses a half-filled atomic orbital that can overlap the half-filled sp^3 hybrids, as illustrated for CH_4 and CF_4.

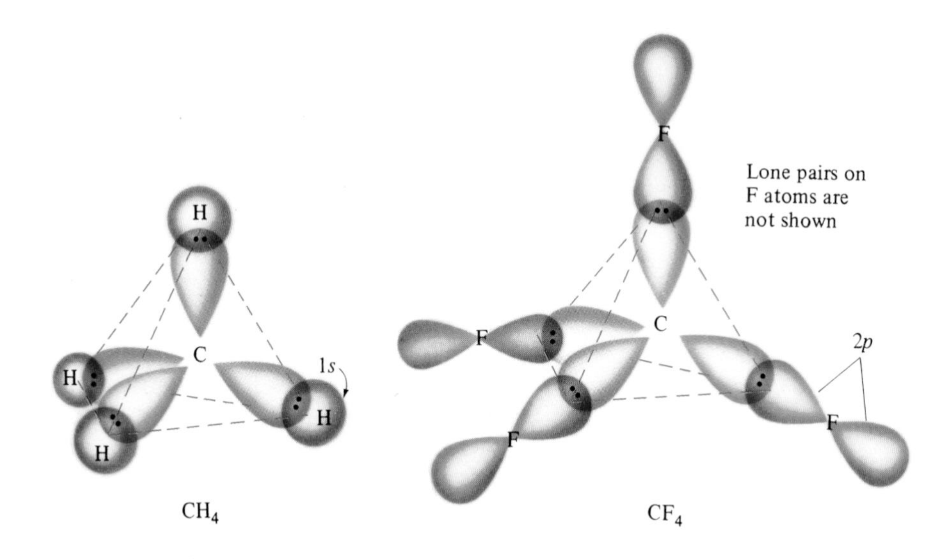

We can give the same VB description for the hybridization of the central atoms in polyatomic ions. In NH_4^+ and SO_4^{2-}, the N and S atoms, respectively, form four sp^3 hybrid orbitals directed toward the corners of a regular tetrahedron. Each of these sp^3 hybrid orbitals overlaps with an orbital on a neighboring atom (H in NH_4^+, O in SO_4^{2-}) to form a bond.

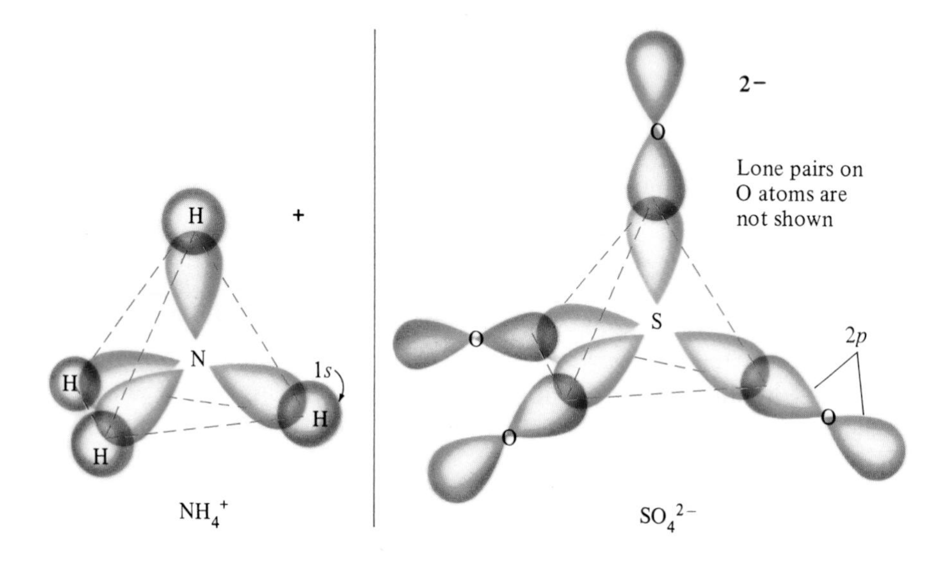

sp^3 hybridization occurs at the central atom whenever there are four regions of high electron density around the central atom. AB_4 molecules and ions with no lone pairs on the central atom have tetrahedral electronic geometry, tetrahedral molecular geometry, and sp^3 hybridization on the central atom.

8-8 Tetrahedral Electronic Geometry—AB_3U Species (One Unshared Pair of Electrons on A)

The symbol "U" in the generalized chemical formula stands for an unshared valence electron pair on atom A.

A. Experimental Facts and Lewis Formulas

Each Group VA element has five electrons in its valence shell. The Group VA elements form some covalent compounds by sharing three of those electrons with three other atoms. Let us describe two examples: ammonia, NH_3, and nitrogen trifluoride, NF_3. Each is a pyramidal, polar molecule with an unshared pair on the nitrogen atom. The Lewis dot formulas for NH_3 and NF_3 are

Some Group VA elements also form covalent compounds by sharing all five valence electrons (Sections 7-10 and 8-11).

NH_3 H : N̈ : H
         H

NF_3 :F̤̈ : N̤̈ : F̤̈:
          :F̤̈:

Sulfite ion, SO_3^{2-}, is an example of a polyatomic ion of the AB_3U type. It is a pyramidal ion with an unshared pair on the sulfur atom.

$$\left[:\ddot{\underset{..}{O}}-\ddot{S}-\ddot{\underset{..}{O}}: \atop \quad | \atop :\underset{..}{O}: \right]^{2-}$$

B. VSEPR Theory

As in Section 8-7, VSEPR theory predicts that the *four* regions of high electron density around a central atom will be directed toward the corners of a tetrahedron, because this gives maximum separation. Thus, in both molecules N has tetrahedral electronic geometry.

At this point let us reemphasize the distinction between electronic geometry and molecular geometry. *Electronic geometry* refers to the geometric arrangement of *regions of electron density* around the central atom. *Molecular geometry* refers to the arrangement of *atoms* (that is, nuclei), not just pairs of electrons, around the central atom. For example, CH_4, CF_4, NH_3, and NF_3 all have tetrahedral electronic geometry. But CH_4 and CF_4 have tetrahedral molecular geometry, whereas NH_3 and NF_3 have pyramidal molecular geometry.

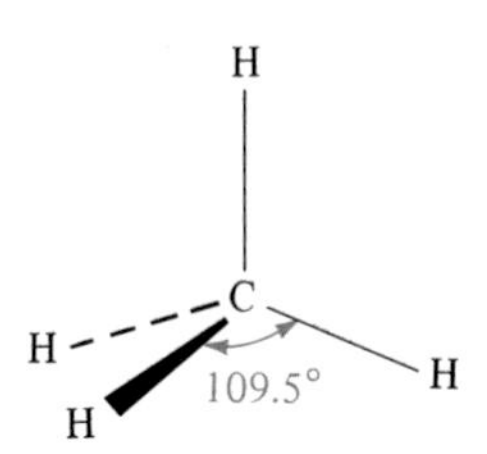

As we have seen, the term "lone pair" refers to a pair of valence electrons that is associated with only one nucleus. The known geometries of many molecules and polyatomic ions, based on measurements of bond angles, show that *lone pairs of electrons occupy more space than bonding pairs*. This is due to the fact that a lone pair has only one atom exerting strong attractive forces on it, so it resides closer to the nucleus than do bonding electrons. Additional observations indicate that the relative magnitudes of the repulsive forces between pairs of electrons on an atom are

$$lp/lp >> lp/bp > bp/bp$$

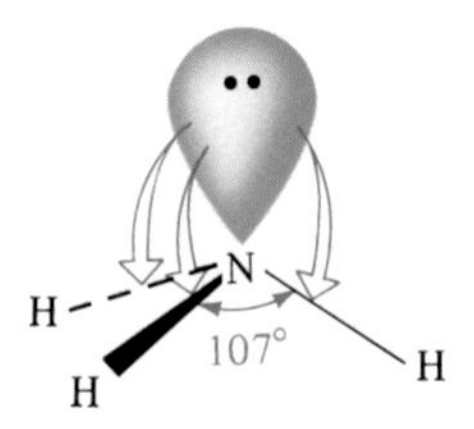

where *lp* refers to lone pairs and *bp* refers to bonding pairs of valence shell electrons. We are most concerned with the repulsions involving the electrons in the valence shell of the *central atom* of a molecule or polyatomic ion. The angles at which repulsive forces among valence shell electron pairs are exactly balanced are the angles at which the bonding pairs and lone pairs (and therefore nuclei) are found in covalently bonded molecules and polyatomic ions. Thus, the bond angles in NH_3 and NF_3 are *less* than the angles of 109.5° we observed in CH_4 and CF_4 molecules.

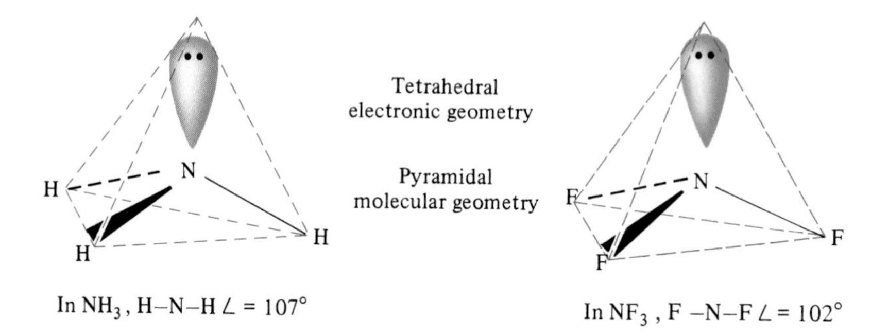

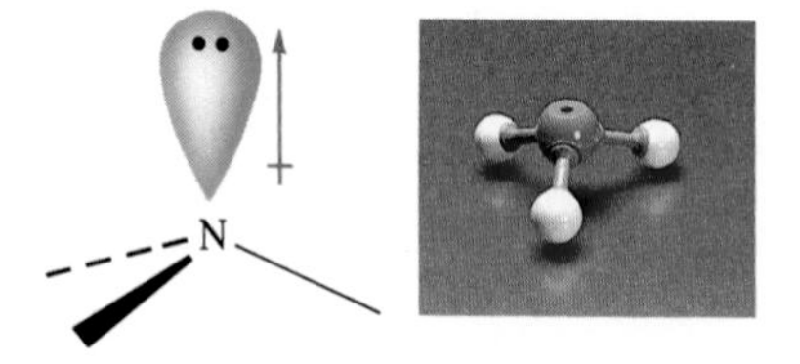

A drawing and a model of a pyramidal molecule (AB_3U).

The formulas are frequently written as $:NH_3$ and $:NF_3$ to emphasize the lone pairs of electrons. The lone pairs must be considered as the polarities of these molecules are examined; they are extremely important! The contribution of each lone pair to polarity can be depicted as shown in the margin.

The electronegativity differences in NH_3 and NF_3 are nearly equal, *but* the resulting nearly equal bond polarities are in opposite directions.

	N—H			N—F	
EN =	3.0 2.1	⟻ N—H	EN =	3.0 4.0	⟼ N—F
Δ(EN) =	0.9		Δ(EN) =	1.0	

Thus, we have

In NH_3 the bond dipoles *reinforce* the effect of the lone pair, so NH_3 is very polar (μ = 1.5 D). In NF_3 the bond dipoles *oppose* the effect of the lone pair, so NF_3 is only slightly polar (μ = 0.2 D).

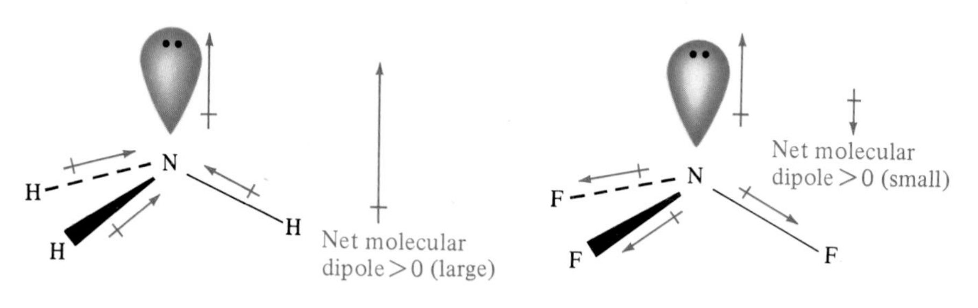

We can now use this information to explain the bond angles observed in NF_3 and NH_3. Because of the direction of the bond dipoles in NH_3, the electron-rich end of each N—H bond is at the central atom, N. In NF_3, on the other hand, the fluorine end of each bond is the electron-rich end. As a result, the lone pair can more closely approach the N in NF_3 than in NH_3. Therefore, in NF_3 the lone pair exerts greater repulsion toward the bonded pairs than in NH_3. In addition, the longer N—F bond length makes the *bp–bp* distance greater in NF_3 than in NH_3, so that the *bp–bp* repulsion in NF_3 is less than that in NH_3. The net effect is that the bond angles are reduced more in NF_3. We can represent this situation as follows:

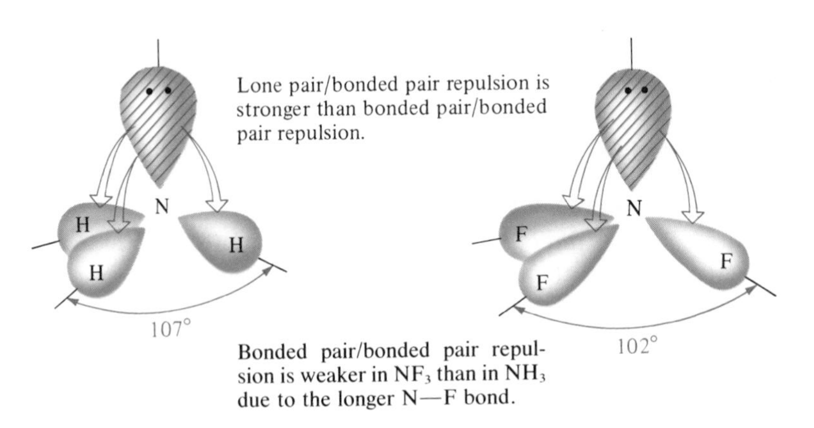

Lone pair/bonded pair repulsion is stronger than bonded pair/bonded pair repulsion.
N
H
H
H
107°
N
F
F
F
102°
Bonded pair/bonded pair repulsion is weaker in NF_3 than in NH_3 due to the longer N—F bond.

We might expect the larger F atoms (r = 0.64 Å) to repel each other more strongly than H atoms (r = 0.37 Å), leading to larger bond angles in NF_3 than in NH_3. This is not the case, however, because the N—F bond is longer than the N—H bond. The N—F bond density is farther from the N than the N—H bond density.

With the same kind of reasoning, VSEPR theory predicts that sulfite ion, SO_3^{2-}, has tetrahedral electronic geometry. One of these tetrahedral locations is occupied by the sulfur lone pair, and oxygen atoms are at the other three locations. The molecular geometry of this ion is trigonal pyramidal, the same as for other AB_3U species.

C. Valence Bond Theory

Experimental results suggest four nearly equivalent orbitals (three involved in bonding, a fourth to accommodate the lone pair), so we again need four sp^3 hybrid orbitals:

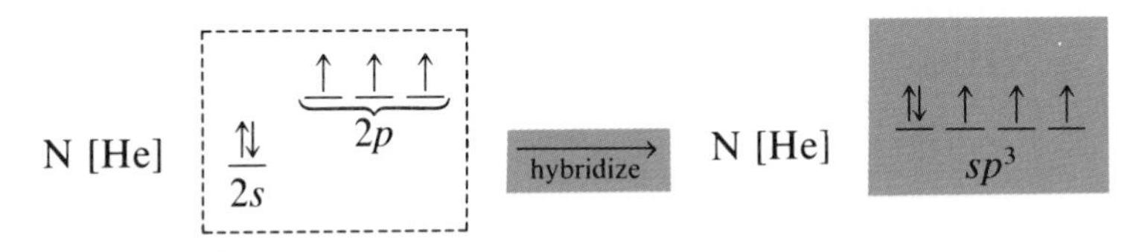

In both NH_3 and NF_3 the unshared pair of electrons occupies one of the sp^3 hybrid orbitals. Each of the other three sp^3 orbitals participates in bonding by sharing electrons with another atom. They overlap with half-filled H 1s orbitals and F 2p orbitals in NH_3 and NF_3, respectively.

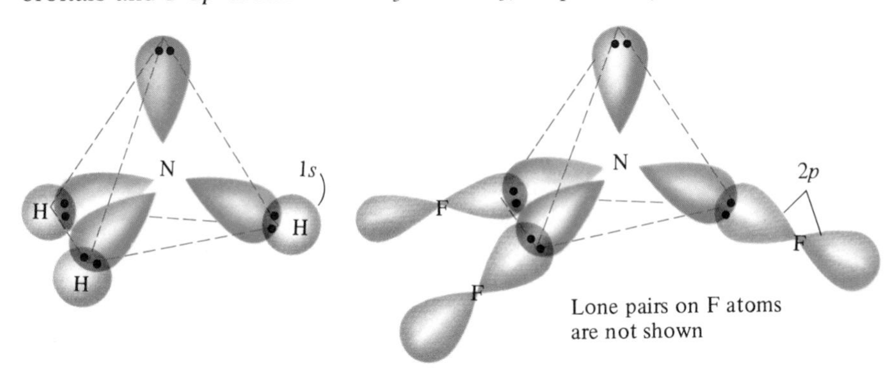

As we shall see in Section 10-10, many compounds that have an unshared electron pair on the central element are called Lewis bases. **Lewis bases** react by making available an electron pair that can be shared by other species.

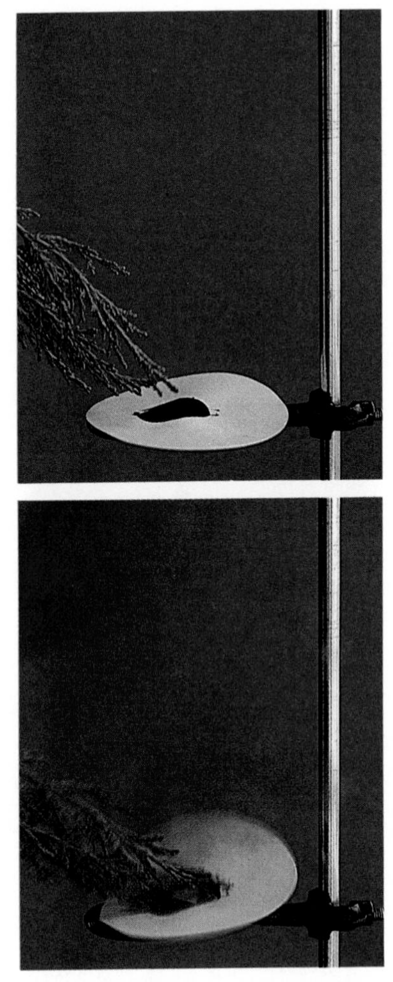

The structure of nitrogen triiodide, NI_3, is analogous to that of NF_3, described in the text. When dry, this dark brown compound is so sensitive that it explodes when touched by a small tree branch. The photo at the bottom was taken just as the compound exploded. *Caution! Do not attempt this experiment yourself.*

We must remember that *theory* (and its application) depends on fact, not the other way around. Sometimes the experimental facts are not consistent with the existence of hybrid orbitals. In such cases, we just use the "pure" atomic orbitals rather than hybrid orbitals. There appears to be no need to use hybridization to describe bonding in PH_3 and AsH_3. Each H—P—H bond angle is 93°, and each H—As—H bond angle is 92°. These angles very nearly correspond to three p orbitals at 90° to each other.

The sulfur atom in the sulfite ion, SO_3^{2-}, can be described as sp^3 hybridized. One of these hybrid orbitals contains the sulfur lone pair, and the remaining three overlap with oxygen orbitals to form bonds.

> AB_3U molecules and ions, each having four regions of high electron density around the central atom, *usually* have tetrahedral electronic geometry, trigonal pyramidal molecular geometry, and sp^3 hybridization on the central atom.

8-9 Tetrahedral Electronic Geometry—AB_2U_2 Species (Two Unshared Pairs of Electrons on A)

A. Experimental Facts and Lewis Formulas

A model of H_2O, an angular molecule (AB_2U_2).

Each Group VIA element has six electrons in its highest energy level. The Group VIA elements form many covalent compounds by acquiring a share in two additional electrons from two other atoms. Typical examples are H_2O, H_2S, and Cl_2O. All are angular, polar molecules. The Lewis dot formulas for these molecules are

H_2O H:Ö: (with H below O) H_2S H:S̈: (with H below S) Cl_2O :C̈l:Ö: (with :C̈l: below O)

Let us consider the structure of water in detail. The bond angle in water is 104.5°, and the molecule is very polar.

B. VSEPR Theory

Six electrons come from oxygen and two from the hydrogens.

VSEPR theory predicts that the four electron pairs around the oxygen atom in H_2O should be 109.5° apart in a tetrahedral arrangement. When we take into account increased repulsions between unshared pairs and bonding electron pairs, this theory satisfactorily explains the angular structure of water molecules and the observed bond angle of only 104.5°.

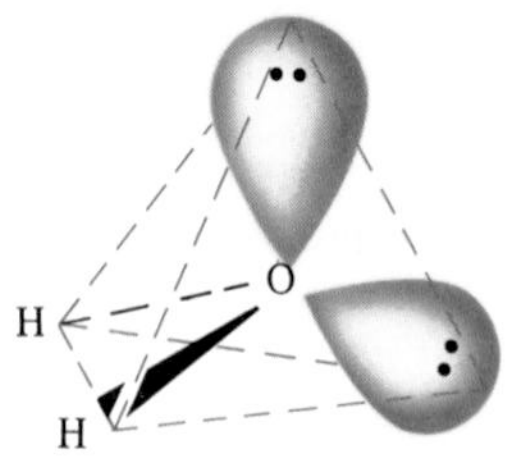

The electronegativity difference is large (1.4 units) and so the bonds are quite polar. Additionally, the bond dipoles *reinforce* the effect of the two lone pairs, so the H_2O molecule is very polar. Its dipole moment is 1.7 D. Water has unusual properties, which can be explained in large part by its high polarity.

	O—H	
EN =	3.5	2.1
Δ(EN) =	1.4	

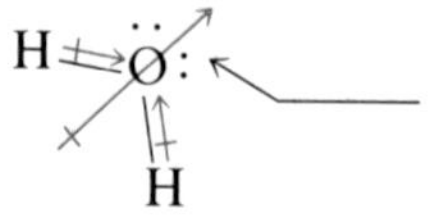

Molecular dipole; includes effect of two unshared electron pairs.

C. Valence Bond Theory

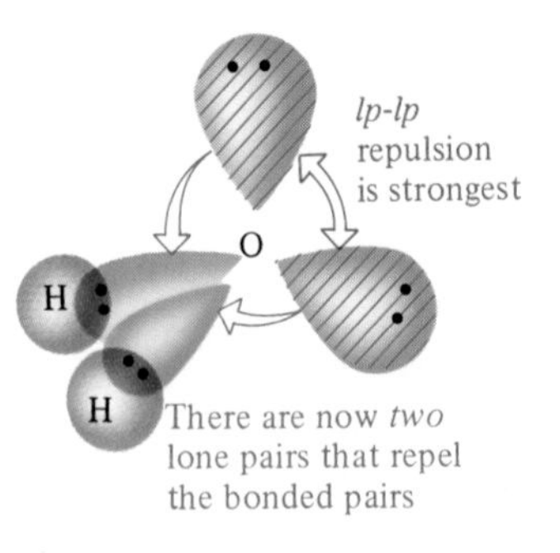

lp-lp repulsion is strongest
O
H
H
There are now two lone pairs that repel the bonded pairs

The bond angle in H_2O (104.5°) is near the tetrahedral value (109.5°). Valence bond theory postulates four sp^3 hybrid orbitals centered on the O atom: two to participate in bonding and two to hold the two lone pairs.

We can easily explain the observed bond angle of 104.5°. The expected bond angle for sp^3 hybridization (tetrahedral electronic geometry) is 109.5°. However, the two lone pairs strongly repel each other and the bonding pairs of electrons. These repulsions force the bonding pairs closer together and result in the decreased bond angle. The decrease in the H—O—H bond angle (from 109.5° to 104.5°) is greater than the corresponding decrease in the H—N—H bond angles in ammonia (from 109.5° to 107°).

Hydrogen sulfide, H_2S, is also an angular molecule, but the H—S—H bond angle is 92°. This is very close to the 90° angles between two unhybridized 3*p* orbitals of S. Therefore, we *do not* postulate hybrid orbitals. The two H atoms are able to exist at approximately right angles to each other when they are bonded to the larger S atom. Likewise, the bond angles in H_2Se and H_2Te are 91° and 89.5°, respectively.

Sulfur is located directly below oxygen in Group VIA.

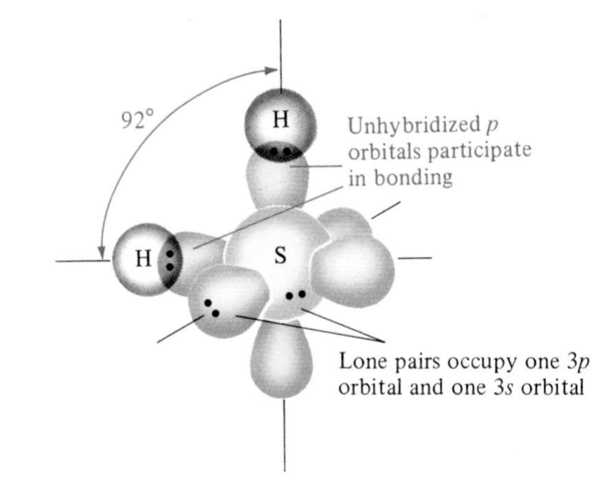

92°
H
H
S
Unhybridized p orbitals participate in bonding
Lone pairs occupy one 3p orbital and one 3s orbital

> AB_2U_2 molecules and ions, each having four regions of high electron density around the central atom, *usually* have tetrahedral electronic geometry, angular molecular geometry, and sp^3 hybridization on the central atom.

8-10 Tetrahedral Electronic Geometry—ABU_3 Species (Three Unshared Pairs of Electrons on A)

Each Group VIIA element has seven electrons in its highest occupied energy level. The Group VIIA elements form covalent molecules such as H—F, H—Cl, Cl—Cl, and I—I by sharing one of those electrons with another atom. The other atom contributes one electron to bonding. Lewis dot formulas for these molecules are shown in the margin. All diatomic molecules are of necessity linear. Neither VSEPR theory nor VB theory adds anything to what we already know about the molecular geometry of such molecules.

We represent the halogen as X.

HX, $H:\ddot{\underset{..}{X}}:$ X_2, $:\ddot{\underset{..}{X}}:\ddot{\underset{..}{X}}:$

In the latter case, either halogen may be considered the "A" atom of AB.

8-11 Trigonal Bipyramidal Electronic Geometry—AB_5, AB_4U, AB_3U_2, and AB_2U_3

A. Experimental Facts and Lewis Formulas

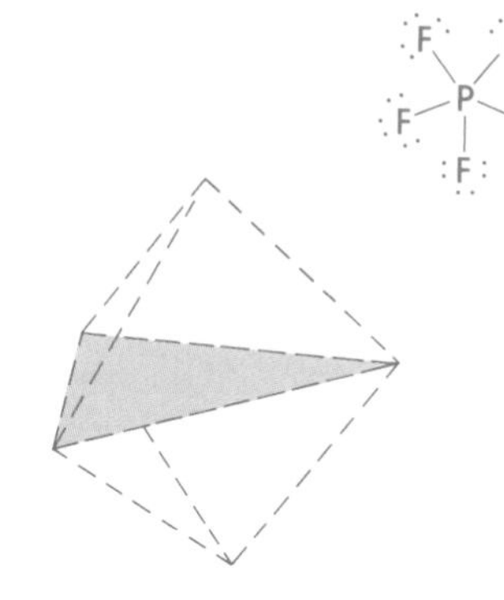

A trigonal bipyramid. The triangular base common to the two pyramids is shaded.

In Section 8-8 we saw that the Group VA elements have five electrons in their outermost occupied shells and form some covalent molecules by sharing only three of these electrons with other atoms (for example, NH_3 and NF_3). Other Group VA elements (P, As, and Sb) form some covalent compounds by sharing all five of their valence electrons with five other atoms (Section 7-10). Phosphorus pentafluoride, PF_5 (mp −83°C), is such a compound. Each P atom has five valence electrons to share with five F atoms. The Lewis formula for PF_5 (Example 7-7) is shown in the margin. PF_5 molecules are *trigonal bipyramidal* nonpolar molecules. A **trigonal bipyramid** is a six-sided polyhedron consisting of two pyramids joined at a common triangular (trigonal) base.

B. VSEPR Theory

A model of a trigonal bipyramidal AB_5 molecule, e.g., PF_5.

VSEPR theory predicts that the five regions of high electron density around the phosphorus atom in PF_5 should be as far apart as possible. Maximum separation of five items around a sixtn item is achieved when the five items (bonding pairs) are placed at the corners and the sixth item (P atom) is placed in the center of a trigonal bipyramid. This is in agreement with experimental observation.

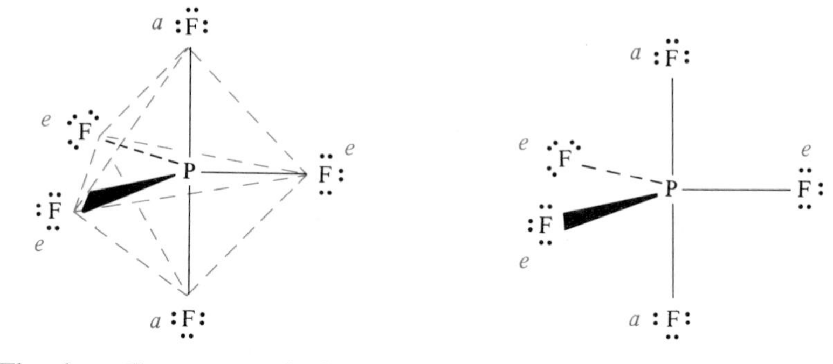

The three F atoms marked *e* are at the corners of the common base, in the same plane as the P atom. These are called *equatorial* F atoms (*e*). The other two F atoms, one above and one below the plane, are called *axial* F atoms (*a*). The axial P—F bonds are longer than the equatorial P—F bonds. The F—P—F bond angles are 90° (axial to equatorial), 120° (equatorial to equatorial), and 180° (axial to axial).

As an exercise in geometry, in how many different ways can five fluorine atoms be arranged *symmetrically* around a phosphorus atom? Compare the hypothetical bond angles in such arrangements with those in a trigonal bipyramidal arrangement.

The large electronegativity difference between P and F (1.9) indicates very polar bonds. Let's consider the bond dipoles in two groups, because there are two different kinds of P—F bonds in PF_5 molecules, axial and equatorial.

	P—F
EN =	2.1 4.0
Δ(EN) =	1.9

axial bonds: F—P—F, 180°

equatorial bonds: F, F, P, F, 120°

The two axial bond dipoles cancel each other, and the three equatorial bond dipoles cancel, so PF_5 molecules are nonpolar.

C. Valence Bond Theory

Because phosphorus is the central element in PF_5 molecules, it must have available five half-filled orbitals to form bonds with five F atoms. Hybridization is again the explanation. Now it involves one *d* orbital from the vacant set of 3*d* orbitals and the 3*s* and 3*p* orbitals of the P atom.

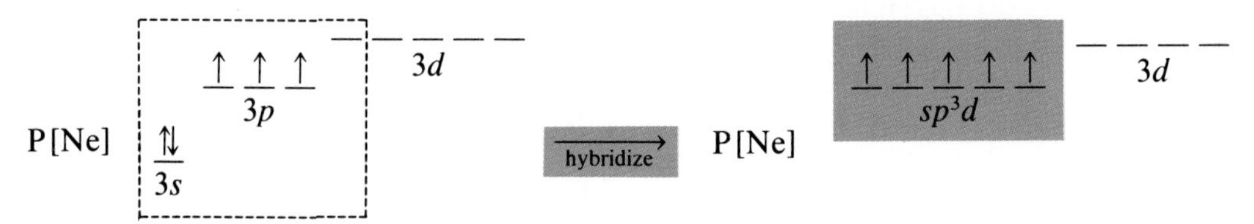

The five ***sp^3d* hybrid orbitals** point toward the corners of a trigonal bipyramid. Each is overlapped by a singly occupied 2*p* orbital of an F atom. The resulting pairing of P and F electrons forms five covalent bonds.

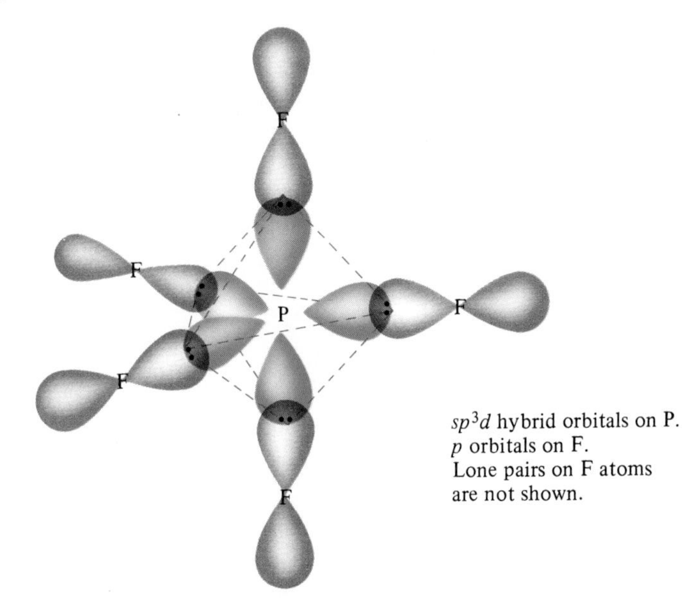

sp^3d hybrid orbitals on P.
p orbitals on F.
Lone pairs on F atoms are not shown.

> sp^3d hybridization occurs at the central atom whenever there are five regions of high electron density around the central atom. AB_5 molecules and ions with no lone pairs on the central atom have trigonal bipyramidal electronic geometry, trigonal bipyramidal molecular geometry, and sp^3d hybridization on the central atom.

We see that sp^3d hybridization uses an available *d* orbital in the outermost occupied shell of the central atom, P. The heavier Group VA elements—P, As, and Sb—can form *five-coordinate compounds* using this hybridization. But nitrogen, also in Group VA, cannot form such five-coordinate compounds. This is because the valence shell of N has only one *s* and three *p* orbitals (and no *d* orbitals). The set of *s* and *p* orbitals in a given energy level (and therefore any set of hybrids composed only of *s* and *p* orbitals)

The P atom is said to have an *expanded valence shell* (Section 7-10).

can accommodate a *maximum* of eight electrons and participate in a *maximum* of four covalent bonds. The same is true of all elements of the second period, because they have only *s* and *p* orbitals in their valence shells. No atoms before the third period can exhibit expanded valence.

D. Unshared Valence Electron Pairs in Trigonal Bipyramidal Electronic Geometry

Relative magnitudes of repulsive forces:

lp/lp >> *lp/bp* > *bp/bp*.

As we saw in Sections 8-8 and 8-9, lone pairs of electrons occupy more space than bonding pairs, resulting in increased repulsions from lone pairs. What happens when one or more of the five regions of high electron density on the central atom are unshared electron pairs? Let us first consider a molecule such as SF_4, for which the Lewis formula is

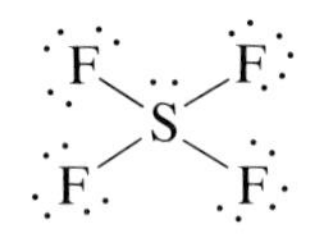

The central atom, S, is bonded to four atoms and has one unshared valence electron pair. This is an example of the general formula AB_4U. Sulfur has five regions of high electron density, so we know that the electronic geometry is trigonal bipyramidal and that the bonding orbitals are sp^3d hybrids. But now a new question arises: Is the lone pair more stable in an axial (*a*) or in an equatorial (*e*) position? If it were in an axial position, it would be 90° from the *three* closest other pairs (the pairs bonding three F atoms in equatorial positions) and 180° from the other axial pair. If it were in an equatorial position, only the *two* axial pairs would be at 90°, while the other two equatorial pairs would be less crowded at 120° apart.

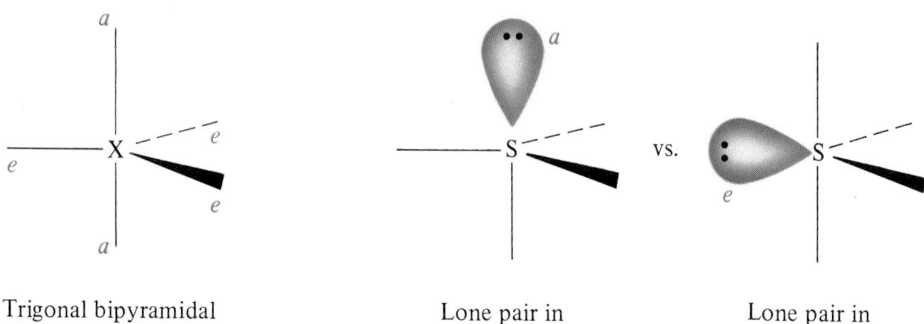

Trigonal bipyramidal electronic geometry

Lone pair in axial position

Lone pair in equatorial position (preferred arrangement)

We conclude that the lone pair would be less crowded in an *equatorial* position. The four F atoms then occupy the remaining four positions. We describe the resulting arrangement of *atoms* as a **seesaw arrangement**.

Imagine rotating the arrangement so that the line joining the two axial positions is the board on which the two seesaw riders sit, and the two bonded equatorial positions are the pivot of the seesaw.

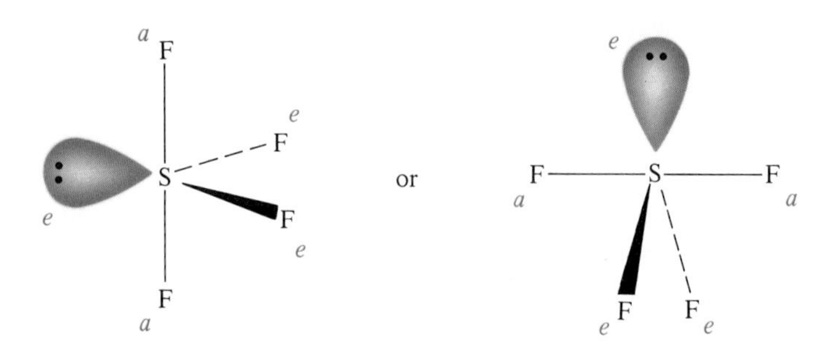

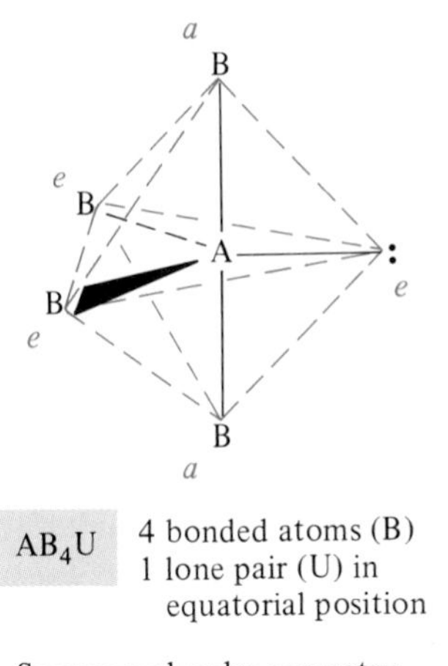

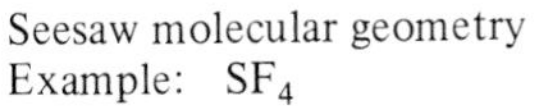

AB_4U 4 bonded atoms (B)
1 lone pair (U) in equatorial position

Seesaw molecular geometry
Example: SF_4

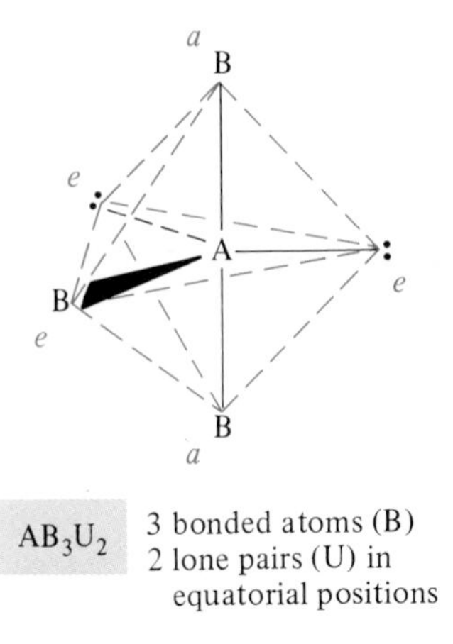

AB_3U_2 3 bonded atoms (B)
2 lone pairs (U) in equatorial positions

T-shaped molecular geometry
Examples: ICl_3, ClF_3

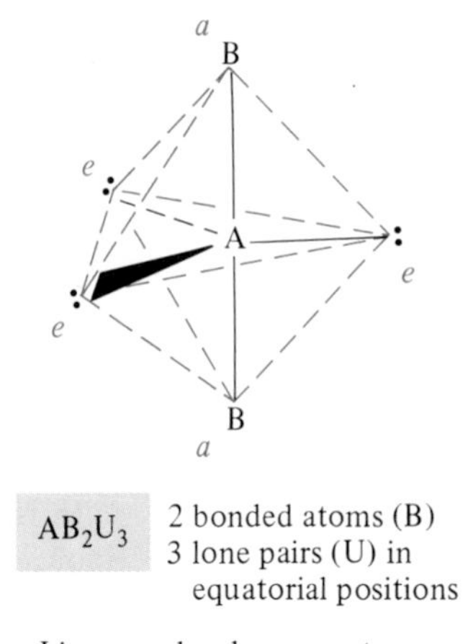

AB_2U_3 2 bonded atoms (B)
3 lone pairs (U) in equatorial positions

Linear molecular geometry
Examples: XeF_2, I_3^-

Figure 8-2
Arrangements of bonded atoms and lone pairs (five regions of high electron density—trigonal bipyramidal electronic geometry).

As we saw in Sections 8-8 and 8-9, the differing magnitudes of repulsions involving lone pairs and bonding pairs often result in actual bond angles that deviate slightly from idealized values. For instance, *bp*/*lp* repulsion in the seesaw molecule SF_4 causes distortion of the axial S—F bonds away from the lone pair, to an angle of 177°; the two equatorial S—F bonds, ideally at 120°, also move closer together to an angle of 104°.

By the same reasoning, we understand why additional lone pairs also take equatorial positions (AB_3U_2 with both lone pairs equatorial or AB_2U_3 with all three lone pairs equatorial). These arrangements are summarized in Figure 8-2.

8-12 Octahedral Electronic Geometry—AB_6, AB_5U, and AB_4U_2

A. Experimental Facts and Lewis Formulas

The heavier Group VIA elements form some covalent compounds of the AB_6 type by sharing their six valence electrons with six other atoms. Sulfur hexafluoride SF_6 (mp −51°C), an unreactive gas, is an example. Sulfur hexafluoride molecules are nonpolar octahedral molecules. Hexafluorophosphate ion, PF_6^-, is an example of a polyatomic ion of the type AB_6.

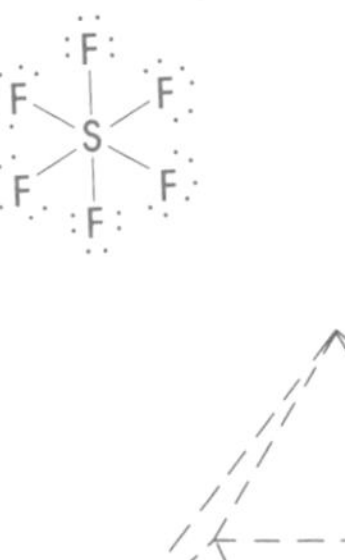

B. VSEPR Theory

In an SF_6 molecule we have six valence electron pairs and six F atoms surrounding one S atom. Because there are no lone pairs in the valence shell of sulfur, the electronic and molecular geometries are identical. The maximum separation possible for six electron pairs around one S atom is achieved when the electron pairs are at the corners and the S atom is at the center of a regular octahedron. Thus, VSEPR theory is consistent with the observation that SF_6 molecules are octahedral.

In a regular octahedron, each of the eight faces is an equilateral triangle.

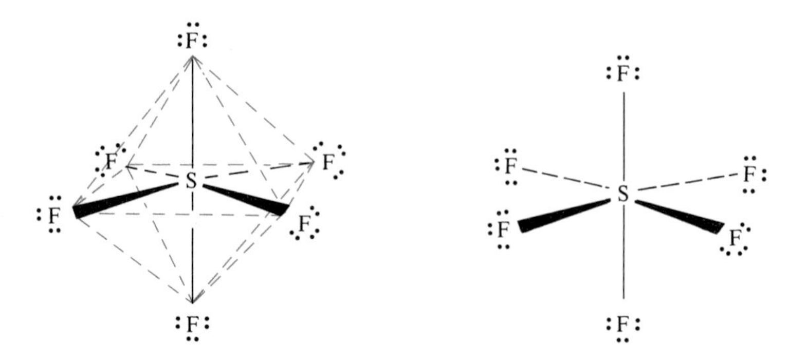

In this octahedral molecule the F—S—F bond angles are 90° and 180°. Each S—F bond is quite polar, but each bond dipole is cancelled by an equal dipole at 180° from it. So the large bond dipoles cancel and the SF_6 molecule is nonpolar.

By similar reasoning, VSEPR theory predicts octahedral electronic geometry and octahedral molecular geometry for the PF_6^- ion, which has six valence electron pairs and six F atoms surrounding one P atom.

C. Valence Bond Theory

Sulfur atoms can use one 3*s*, three 3*p*, and two 3*d* orbitals to form six hybrid orbitals that accommodate six electron pairs:

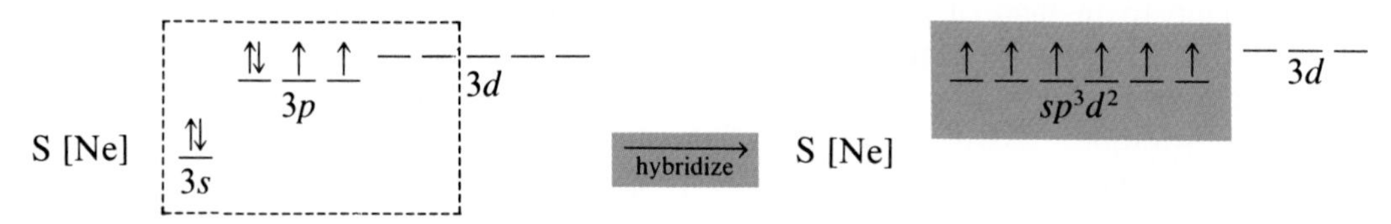

Se and Te, in the same group, form analogous compounds. O cannot do so, for the reasons discussed earlier for N (Section 8-11).

The six sp^3d^2 **hybrid orbitals** are directed toward the corners of a regular octahedron. Each sp^3d^2 hybrid orbital is overlapped by a half-filled 2*p* orbital from fluorine, to form a total of six covalent bonds.

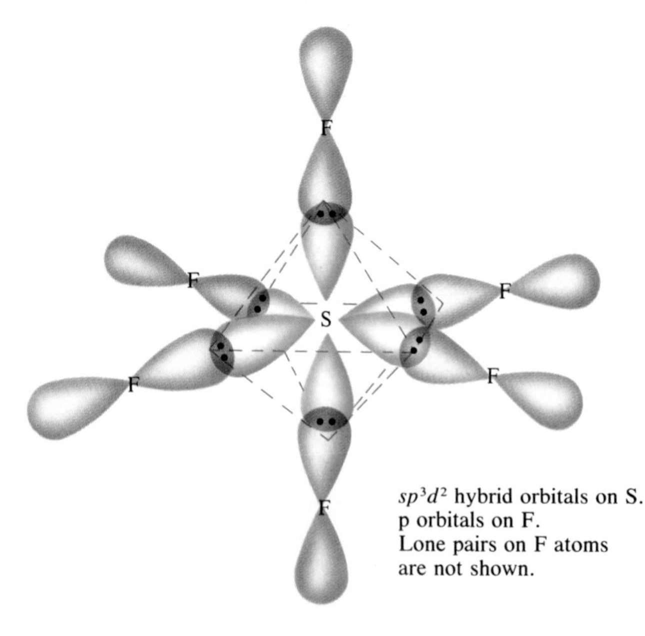

sp^3d^2 hybrid orbitals on S.
p orbitals on F.
Lone pairs on F atoms are not shown.

(Left) A model of an octahedral AB_6 molecule, e.g., SF_6. (Right) Some familiar octahedral toys.

An analogous picture could be drawn for the PF_6^- ion.

> sp^3d^2 hybridization occurs at the central atom whenever there are six regions of high electron density around the central atom. AB_6 molecules and ions with no lone pairs on the central atom have octahedral electronic geometry, octahedral molecular geometry, and sp^3d^2 hybridization on the central atom.

D. Unshared Valence Electron Pairs in Octahedral Electronic Geometry

We can reason along the lines of part D in Section 8-10 to predict the preferred locations of unshared electron pairs on the central atom in octahedral electronic geometry. Because of the high symmetry of the octahedral arrangement, all six positions are equivalent, so it does not matter in which position in the drawing we put the first lone pair. AB_5U molecules and ions are described as having square pyramidal molecular geometry. When a second lone pair is present, the two lone pairs (which repel one another more strongly than bonding pairs are repelled) are most stable when they are in two octahedral positions at 180° from one another. This leads to a square planar molecular geometry for AB_4U_2 species. These arrangements are shown in Figure 8-3. Table 8-3 summarizes a great deal of information—study this table carefully.

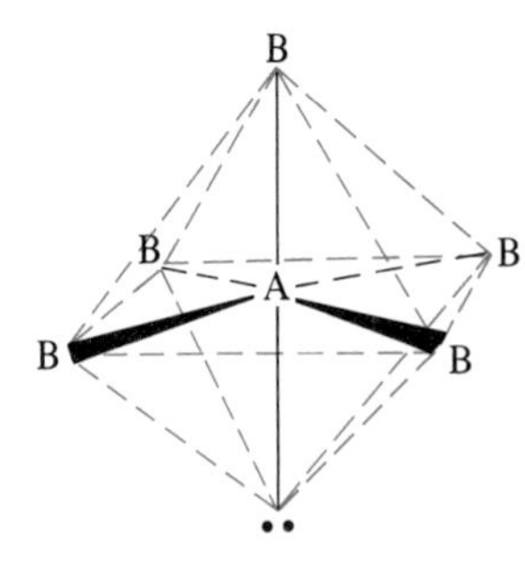

AB_5U 5 bonded atoms (B)
1 lone pair (U)

Square pyramidal molecular geometry
Examples: IF_5, BrF_5

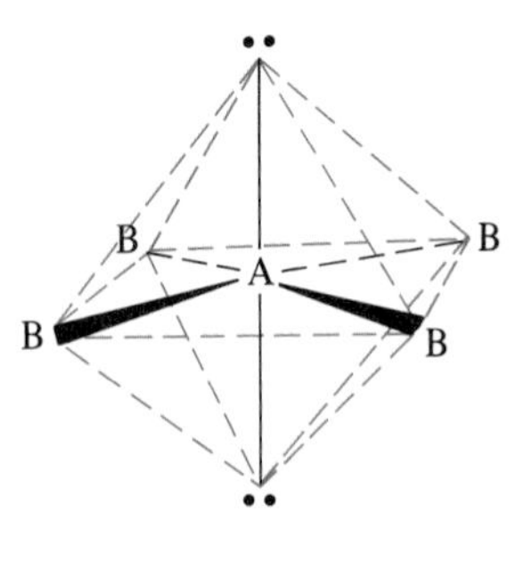

AB_4U_2 4 bonded atoms (B)
2 lone pairs (U)

Square planar molecular geometry
Examples: XeF_4, IF_4^-

Figure 8-3
Arrangements of bonded atoms and lone pairs (six regions of high electron density—octahedral electronic geometry).

Table 8-3
Molecular Geometry of Species with Lone Pairs (U) on the Central Atom

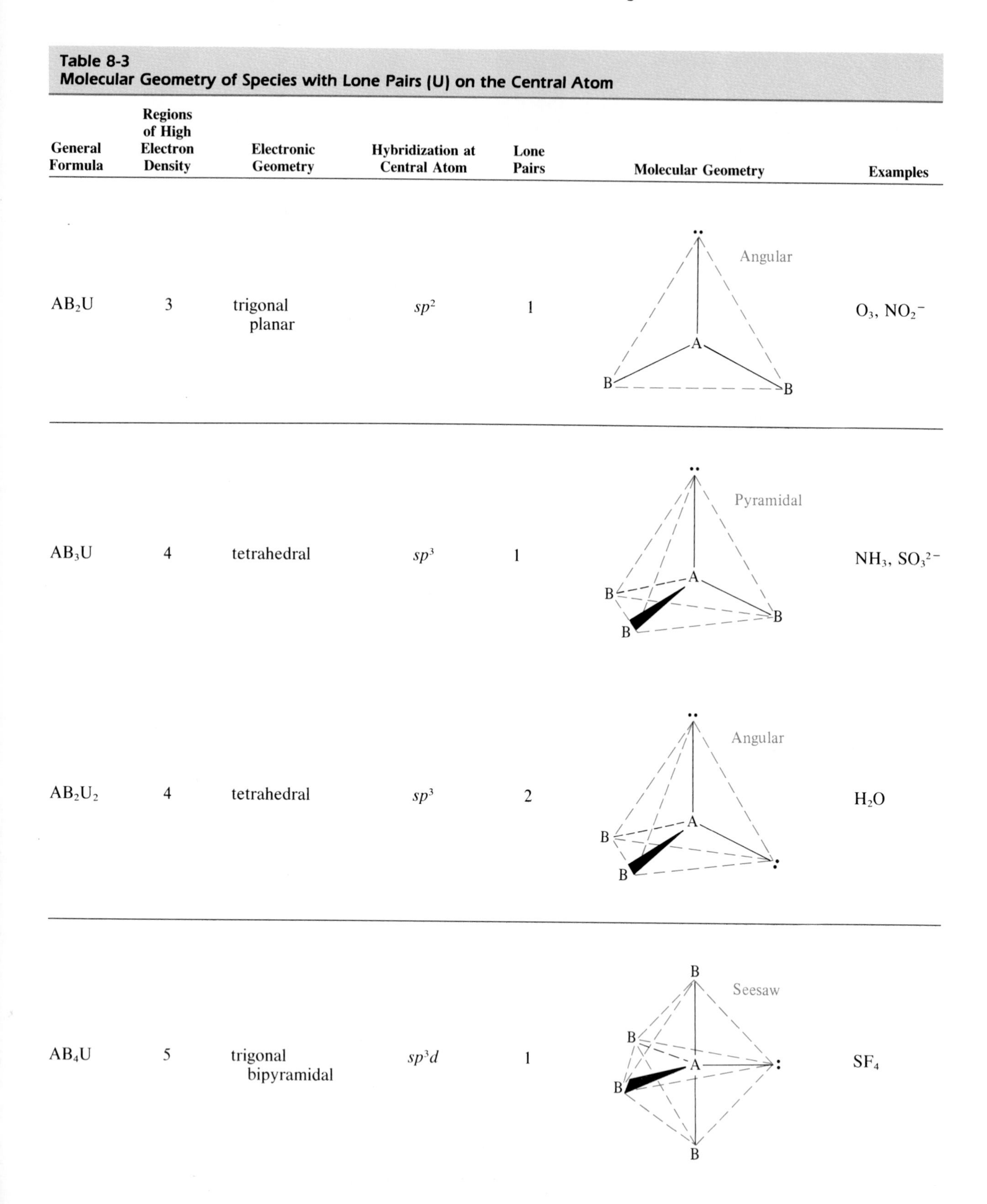

General Formula	Regions of High Electron Density	Electronic Geometry	Hybridization at Central Atom	Lone Pairs	Molecular Geometry	Examples
AB_2U	3	trigonal planar	*sp*^{2}	1	Angular	O_3, NO_2^-
AB_3U	4	tetrahedral	*sp*^{3}	1	Pyramidal	NH_3, SO_3^{2-}
AB_2U_2	4	tetrahedral	*sp*^{3}	2	Angular	H_2O
AB_4U	5	trigonal bipyramidal	*sp*^{3}*d*	1	Seesaw	SF_4

Table 8-3 (*continued*)

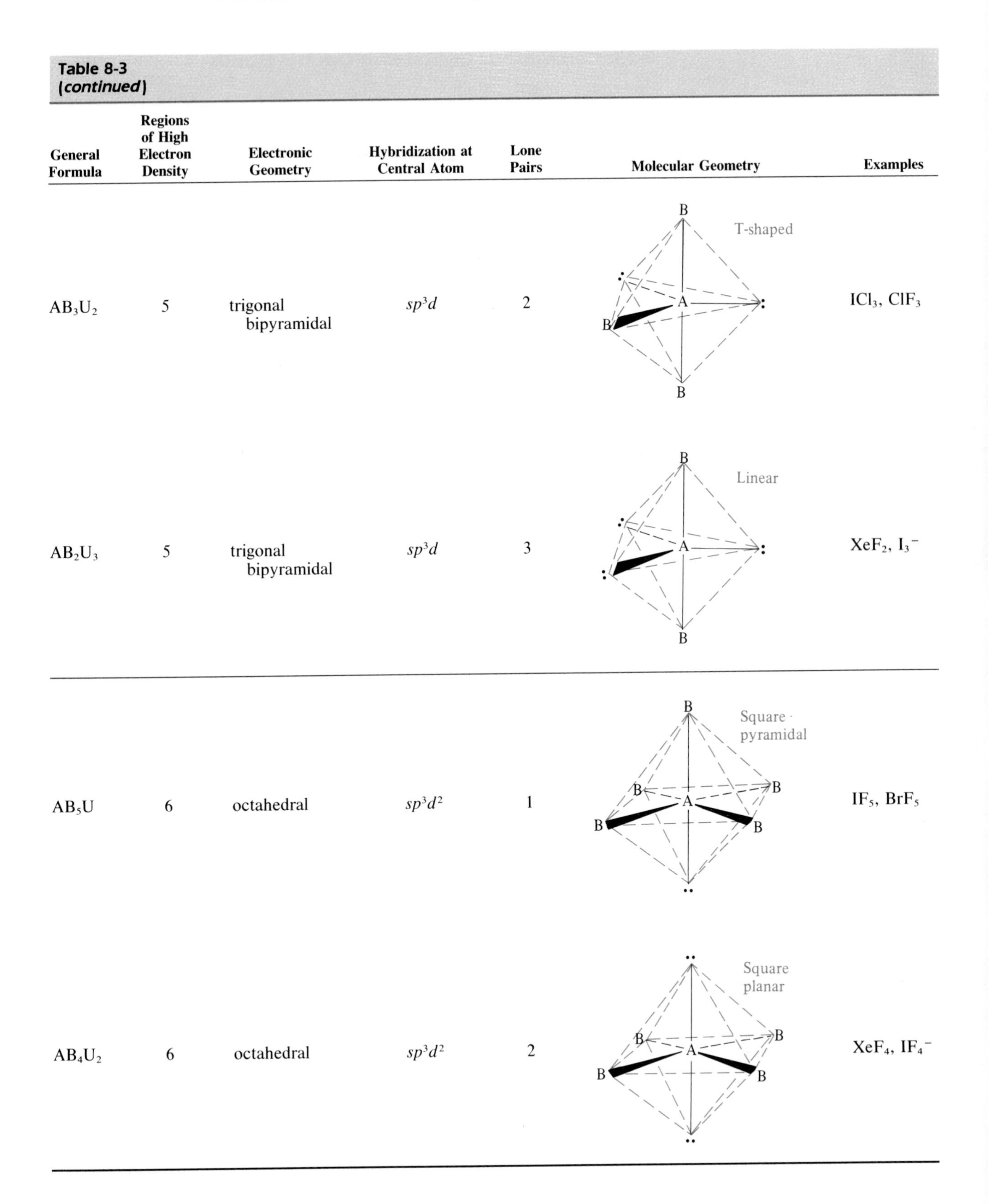

General Formula	Regions of High Electron Density	Electronic Geometry	Hybridization at Central Atom	Lone Pairs	Molecular Geometry	Examples
AB_3U_2	5	trigonal bipyramidal	sp^3d	2	T-shaped	ICl_3, ClF_3
AB_2U_3	5	trigonal bipyramidal	sp^3d	3	Linear	XeF_2, I_3^-
AB_5U	6	octahedral	sp^3d^2	1	Square pyramidal	IF_5, BrF_5
AB_4U_2	6	octahedral	sp^3d^2	2	Square planar	XeF_4, IF_4^-

8-13 Compounds Containing Double Bonds

In Chapter 7 we constructed dot formulas for some molecules and polyatomic ions that contain double and triple bonds. We have not yet considered bonding and shapes for such species. Let us consider ethene (ethylene), C_2H_4, as a specific example. Its dot formula is

$$S = N - A = 24 - 12 = \underline{12e^- \text{ shared}}$$

H.   .H
:C::C:
H   H

H   H
 C=C
H   H

Here each C atom is considered a central atom.

There are three regions of high electron density around each C atom. VSEPR theory tells us that each C atom is at the center of a trigonal plane.

Valence bond theory pictures each doubly bonded carbon atom as sp^2 hybridized with one electron in each sp^2 hybrid orbital and one electron in the unhybridized $2p$ orbital. This $2p$ orbital is perpendicular to the plane of the three sp^2 hybrid orbitals:

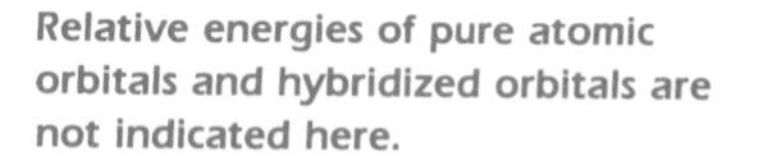

Relative energies of pure atomic orbitals and hybridized orbitals are not indicated here.

Recall that sp^2 hybrid orbitals are directed toward the corners of an equilateral triangle. Figure 8-4 shows top and side views of these hybrid orbitals.

The two C atoms interact by head-on (end-to-end) overlap of sp^2 hybrids pointing toward each other to form a *sigma* (σ) *bond* and by side-on overlap of the unhybridized $2p$ orbitals to form a *pi* (π) *bond*. A **sigma bond** is a bond resulting from head-on overlap of atomic orbitals. *The region of electron sharing is along and cylindrically around the imaginary line connecting the bonded atoms.* All single bonds are sigma bonds. We have seen that many kinds of pure atomic orbitals and hybridized orbitals can be involved in sigma bond formation. A **pi bond** is a bond resulting from side-on overlap of atomic orbitals. *The regions of electron sharing are on opposite sides of the imaginary line connecting the bonded atoms and parallel to this line.* A pi bond can form *only* if there is *also* a sigma bond between the same two atoms. The sigma and pi bonds together make a double bond (Figure 8-5).

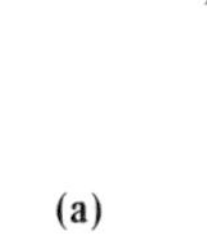

(a) (b)

Figure 8-4
(a) A top view of three sp^2 hybrid orbitals (green). The remaining unhybridized p orbital (not shown in this view) is perpendicular to the plane of the drawing. (b) A side view of a carbon atom in a trigonal planar (sp^2-hybridized) environment, showing the remaining p orbital (purple). This p orbital is perpendicular to the plane of the three sp^2 hybrid orbitals.

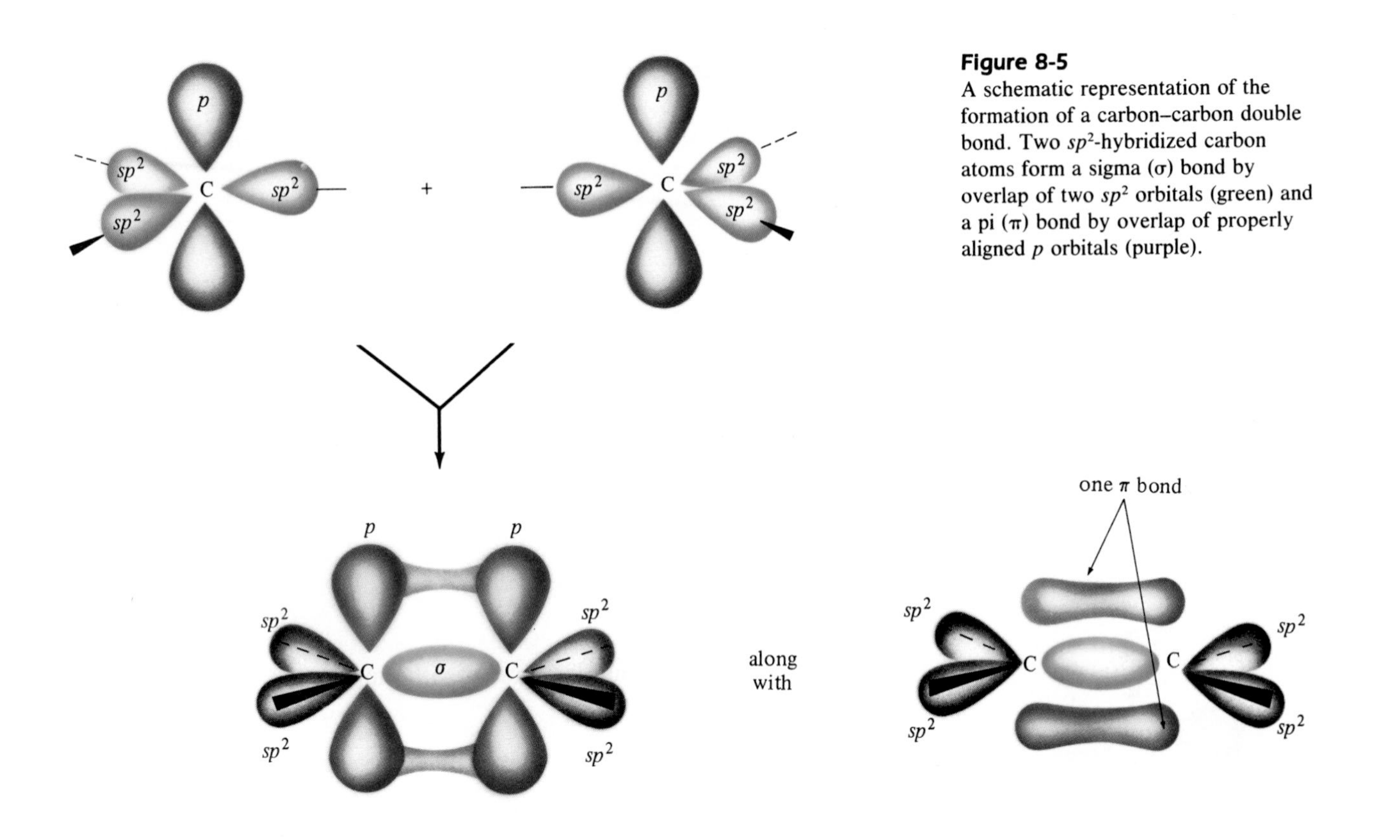

Figure 8-5
A schematic representation of the formation of a carbon–carbon double bond. Two sp^2-hybridized carbon atoms form a sigma (σ) bond by overlap of two sp^2 orbitals (green) and a pi (π) bond by overlap of properly aligned *p* orbitals (purple).

The 1*s* orbitals (with one e^- each) of four hydrogen atoms overlap the remaining four sp^2 orbitals (with one e^- each) on the carbon atoms to form four C—H sigma bonds (Figure 8-6).

A double bond consists of one sigma bond and one pi bond.

As a consequence of the sp^2 hybridization of C atoms in carbon–carbon double bonds, each carbon atom is at the center of a trigonal plane. The *p* orbitals that overlap to form the π bond must be parallel to each other for effective overlap to occur. This adds the further restriction that these trigonal planes (sharing a common corner) must also be *coplanar*. Thus, all four atoms attached to the doubly bonded C atoms lie in the same plane (Figure 8-6). Many other important organic compounds contain carbon–carbon double bonds. Several are described in Chapter 31.

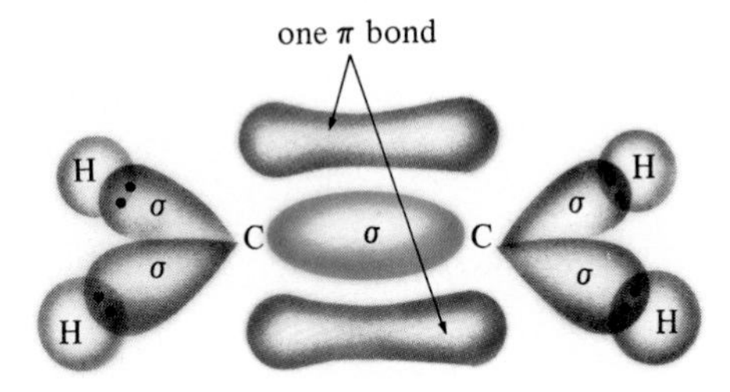

Figure 8-6
Four C—H σ bonds (gray), one C—C σ bond (green), and one C—C π bond (purple) in the planar C_2H_4 molecule.

8-14 Compounds Containing Triple Bonds

One compound that contains a triple bond is ethyne (acetylene), C_2H_2. This has the dot formula

$$S = N - A$$
$$= 20 - 10 = \underline{10e^-\text{shared}}$$

H:C⋮⋮C:H   H—C≡C—H

VSEPR theory predicts that the two regions of high electron density around each carbon atom are 180° apart.

Valence bond theory postulates that each triply bonded carbon atom is *sp*-hybridized (see Section 8-5) because each has two regions of high electron density. Let us designate the p_x orbitals as the ones involved in hybridization. Carbon has one electron in each *sp* hybrid orbital and one electron in each of the $2p_y$ and $2p_z$ orbitals (before bonding is considered). See Figure 8-7.

The three *p* orbitals in a set are indistinguishable. We label the one involved in hydridization as "p_x" to help us visualize the orientations of the two unhybridized *p* orbitals on carbon.

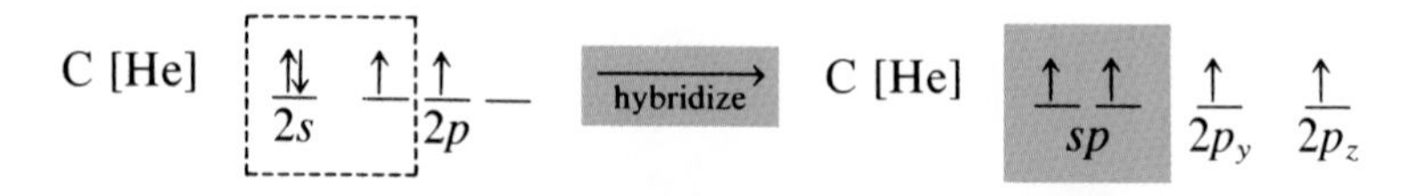

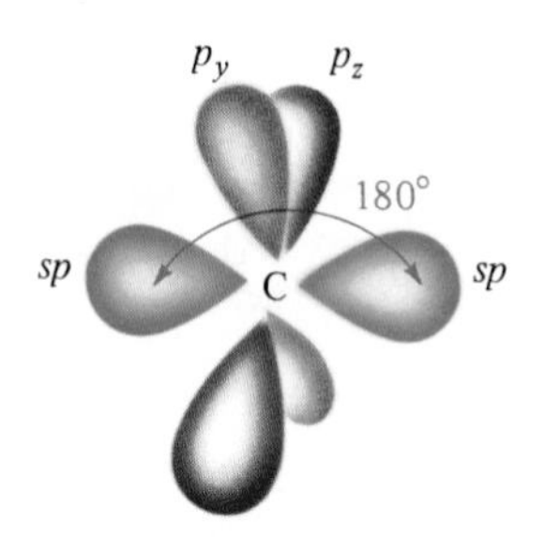

Figure 8-7
Diagram of the two linear hybridized *sp* orbitals (green) of an atom. These lie in a straight line, and the two unhybridized *p* orbitals p_y (orange) and p_z (purple) lie in the perpendicular plane.

Each carbon atom forms one sigma bond with the other C atom and a sigma bond with one H atom.

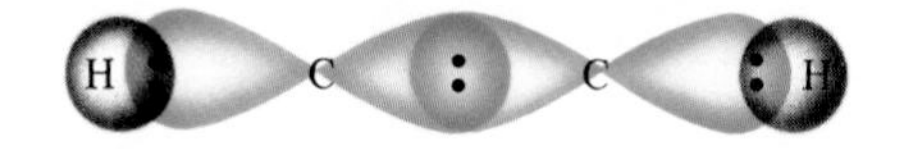

The unhybridized atomic $2p_y$ and $2p_z$ orbitals are perpendicular to each other and to a line through the centers of the two *sp* hybrid orbitals (Figure 8-8). The *sp* hybrids are 180° apart. The *sp* hybrids on the two C atoms overlap head-on. Thus, the entire molecule must be linear.

A triple bond consists of one sigma bond and two pi bonds.

Other molecules containing triply bonded atoms are nitrogen, :N≡N:, propyne, CH_3—C≡C—H, and hydrogen cyanide, H—C≡N:. In each case, both atoms involved in the triple bonds are *sp*-hybridized. In the triple bond, each atom participates in one sigma and two pi bonds. The C atom in carbon dioxide, :Ö=C=Ö:, must participate in two pi bonds (to two

In propyne, the C atom in the CH_3 group is sp^3-hybridized and at the center of a tetrahedral arrangement.

Acetylene (ethyne) is used in welding.

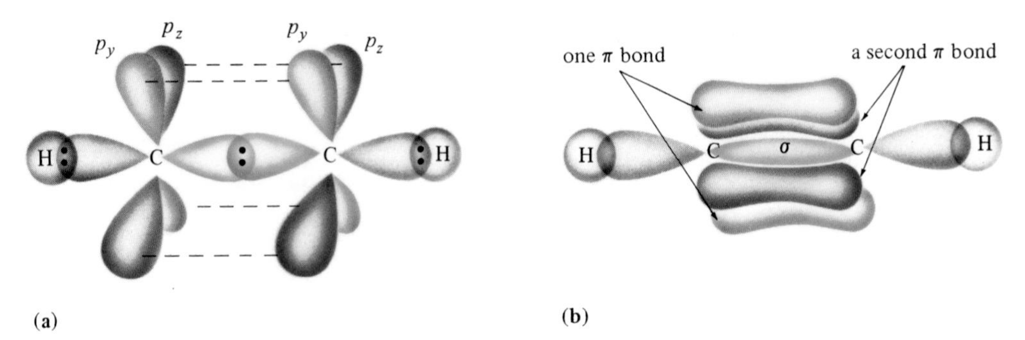

Figure 8-8
The acetylene molecule, C_2H_2. (a) The overlap diagram of two *sp*-hybridized carbon atoms and two *s* orbitals from two hydrogen atoms. One hybridized *sp* orbital on each C is shown in green, the other *sp* orbital on each atom is shown in gray, and the unhybridized *p* orbitals are shown in purple and orange. The dashed lines, each connecting two lobes, indicate the side-by-side overlap of the four unhybridized *p* orbitals to form two π bonds. There are two C—H σ bonds (gray), one C—C σ bond (green), and two C—C π bonds (purple and orange). This makes the net carbon–carbon bond a triple bond. (b) The π bonding orbitals (purple and orange) are positioned with one above and below the line of the σ bonds (green) and the other behind and in front of the line of the σ bonds.

different O atoms). It also participates in two sigma bonds, so it is also *sp*-hybridized and the molecule is linear.

8-15 A Summary of Electronic and Molecular Geometries

We have discussed several common types of polyatomic molecules and ions, and provided a reasonable explanation for the observed structures and polarities of these species. Table 8-4 provides a summation of the points developed in this chapter.

Our discussion of covalent bonding illustrates two important points:

1. Molecules and polyatomic ions have definite shapes.
2. The properties of molecules and polyatomic ions are determined to a great extent by their shapes. Incompletely filled electron shells and unshared pairs of electrons on the central element are very important.

Our ideas about chemical bonding have developed over many years. As experimental techniques for determining the *structures* of molecules have improved, our understanding of chemical bonding has improved also. Experimental observations on molecular geometry support our ideas about chemical bonding. The ultimate test for any theory is this: Can it correctly predict the results of experiments before they are performed? When the answer is *yes*, we have confidence in the theory. When the answer is *no*, the theory must be modified. Current theories of chemical bonding enable us to make predictions that are usually accurate.

Table 8-4
A Summary of Electronic and Molecular Geometries of Polyatomic Molecules and Ions

Regions of High Electron Density[a]	Electronic Geometry	Hybridization at Central Atoms (A)	Hybridized Orbital Orientation	Examples	Molecular Geometry
2	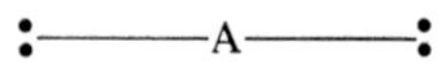 linear	*sp* (180°)		$BeCl_2$ $HgBr_2$ CdI_2 CO_2[b] C_2H_2[c]	linear linear linear linear linear
3	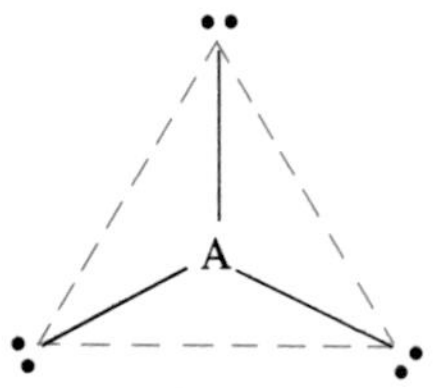 trigonal planar	*sp*^{2} (120°)	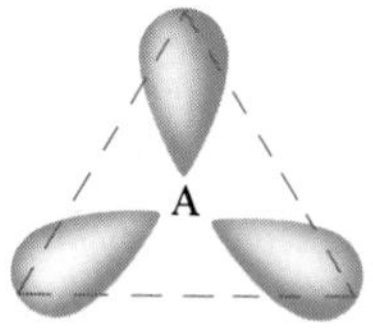	BF_3 BCl_3 NO_3^-[e] SO_2[d,e] NO_2^-[d,e] C_2H_4[f]	trigonal planar trigonal planar trigonal planar angular (AB_2U) angular (AB_2U) planar (trig. planar at each C)
4	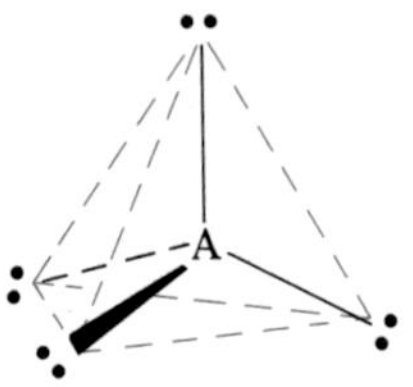 tetrahedral	*sp*^{3} (109.5°)	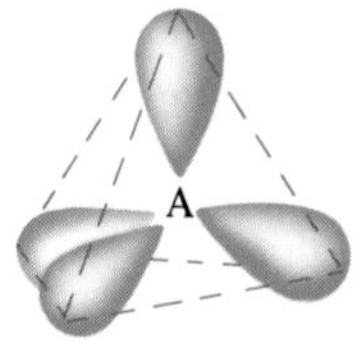	CH_4 CCl_4 NH_4^+ SO_4^{2-} $CHCl_3$ NH_3[d] SO_3^{2-}[d] H_3O^+[d] H_2O[d]	tetrahedral tetrahedral tetrahedral tetrahedral distorted tet. pyramidal (AB_3U) pyramidal (AB_3U) pyramidal (AB_3U) angular (AB_2U_2)
5	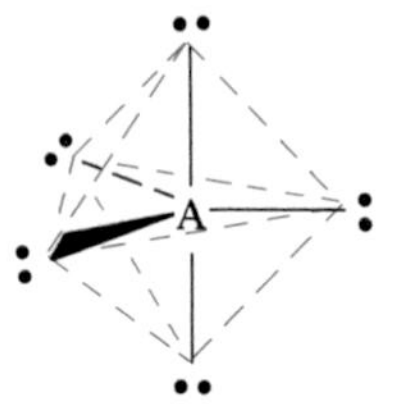 trigonal bipyramidal	*sp*^{3}*d* or *dsp*^{3} (90°, 120°, 180°)	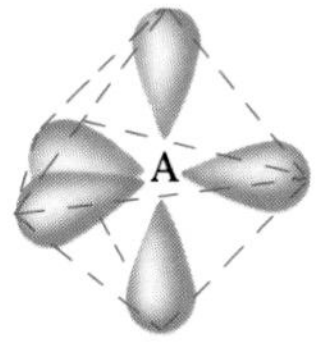	PF_5 $AsCl_5$ SF_4[d] ICl_3[d] XeF_2[d] I_3^-[d]	trigonal bipyramidal trigonal bipyramidal seesaw (AB_4U) T-shaped (AB_3U_2) linear (AB_2U_3) linear (AB_2U_3)
6	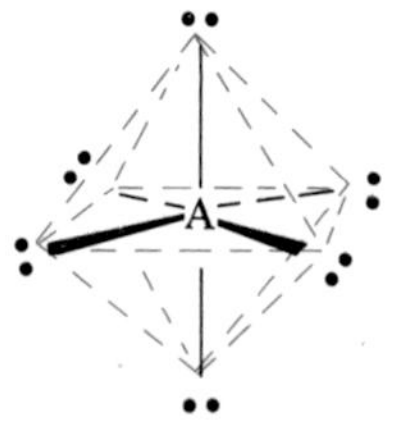 octahedral	*sp*^{3}*d*^{2} or *d*^{2}*sp*^{3} (90°, 180°)	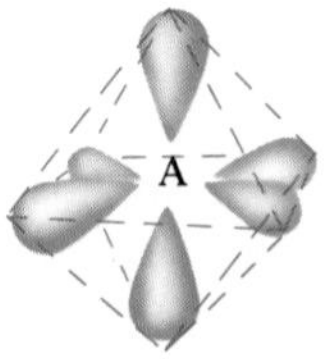	SF_6 SeF_6 PF_6^- BrF_5[d] XeF_4[d]	octahedral octahedral octahedral square pyramidal (AB_5U) square planar (AB_4U_2)

[a] The number of locations of high electron density around the central atom. A region of high electron density may be a single bond, a double bond, a triple bond, or a lone pair. These determine the electronic geometry, and thus hybridization at the central element.
[b] Contains two double bonds.
[c] Contains a triple bond.
[d] Central atom in molecule or ion has lone pair(s) of electrons.
[e] Contains a resonant double bond.
[f] Contains one double bond.

Key Terms

Central atom An atom in a molecule or polyatomic ion that is bonded to more than one other atom.

Electronic geometry The geometric arrangement of orbitals containing the shared and unshared electron pairs surrounding the central atom of a molecule or polyatomic ion.

Hybridization The mixing of a set of atomic orbitals to form a new set of atomic orbitals with the same total electron capacity and with properties and energies intermediate between those of the original unhybridized orbitals.

Ionic geometry The arrangement of atoms (not lone pairs of electrons) about the central atom of a polyatomic ion.

Lewis acid A substance that accepts a share in a pair of electrons from another species.

Lewis base A substance that makes available a share in an electron pair.

Lewis dot formula A method of representing a molecule or formula unit by showing atoms and only outer shell electrons; does not show shape.

Molecular geometry The arrangement of atoms (*not* lone pairs of electrons) around a central atom of a molecule or polyatomic ion.

Octahedral A term used to describe a molecule or polyatomic ion that has one atom in the center and six atoms at the corners of a regular octahedron.

Octahedron A polyhedron with eight equal-sized, equilateral triangular faces and six apices (corners).

Overlap of orbitals The interaction of orbitals on different atoms in the same region of space.

Pi (π) bond A bond resulting from the side-on overlap of atomic orbitals, in which the regions of electron sharing are on opposite sides of and parallel to the imaginary line connecting the bonded atoms.

Sigma (σ) bond A bond resulting from the head-on overlap of atomic orbitals, in which the region of electron sharing is along and (cylindrically) symmetrical to the imaginary line connecting the bonded atoms.

Square planar A term used to describe molecules and polyatomic ions that have one atom in the center and four atoms at the corners of a square.

Tetrahedral A term used to describe a molecule or polyatomic ion that has one atom in the center and four atoms at the corners of a regular tetrahedron.

Tetrahedron A polyhedron with four equal-sized, equilateral triangular faces and four apices (corners).

Trigonal bipyramid A six-sided polyhedron with five apices (corners), consisting of two pyramids sharing a common triangular base.

Trigonal bipyramidal A term used to describe a molecule or polyatomic ion that has one atom in the center and five atoms at the corners of a trigonal bipyramid.

Trigonal planar A term used to describe a molecule or polyatomic ion that has one atom in the center and three atoms at the corners of an equilateral triangle.

Valence bond (VB) theory Assumes that covalent bonds are formed when atomic orbitals on different atoms overlap and electrons are shared.

Valence shell electron pair repulsion (VSEPR) theory Assumes that electron pairs are arranged around the central element of a molecule or polyatomic ion so that there is maximum separation (and minimum repulsion) among regions of high electron density.

Exercises

VSEPR Theory—General Concepts

1. State in your own words the basic idea of the VSEPR theory.
2. (a) Distinguish between "lone pairs" and "bonding pairs" of electrons. (b) Which has the greater spatial requirement? How do we know this? (c) Indicate the order of increasing repulsions among lone pairs and bonding pairs of electrons.
3. Distinguish between electronic geometry and molecular geometry.
4. Under what conditions is molecular (or ionic) geometry identical to electronic geometry about a central atom?
5. What two shapes can a triatomic species have? How would the electronic geometries for the two shapes differ?
6. When using VSEPR theory to predict molecular geometry, how are double and triple bonds treated? How is a single nonbonding electron treated?
7. How does the presence of lone pairs of electrons on an atom influence the bond angles around that atom?
8. Sketch the three different possible arrangements of the two B atoms around the central atom A for the molecule AB_2U_3. Which of these structures correctly describes the molecular geometry? Why?
9. Sketch the three different possible arrangements of the three B atoms around the central atom A for the molecule AB_3U_2. Which of these structures correctly describes the molecular geometry? Why? What are the predicted ideal bond angles? How would the actual bond angles deviate from these values?

Valence Bond Theory—General Concepts

10. Describe the orbital overlap model of covalent bonding.
11. What are hybridized atomic orbitals? How is the theory of hybridized orbitals useful?
12. Prepare sketches of the overlaps of the following atomic orbitals: (a) *s* with *s*, (b) *s* with *p* along the bond axis, (c) *p* with *p* along the bond axis (head-on overlap), (d) *p* with *p* perpendicular to the bond axis (side-on overlap).
13. Prepare a sketch of the cross-section (through the atomic centers) taken between two atoms that have formed (a) a single σ bond, (b) a double bond consisting of a σ bond and a π bond, and (c) a triple bond consisting of a σ bond and two π bonds.
14. Prepare sketches of the orbitals around atoms that are (a) sp, (b) sp^2, (c) sp^3, (d) sp^3d, and (e) sp^3d^2 hybridized. Show in the sketches any unhybridized *p* orbitals that might participate in multiple bonding.
15. What form of hybridization is associated with these electronic geometries: trigonal planar, linear, tetrahedral, octahedral, trigonal bipyramidal?
16. What angles are associated with orbitals in the following hybridized sets of orbitals? (a) sp, (b) sp^2, (c) sp^3, (d) sp^3d, (e) sp^3d^2
17. What types of hybridization would you predict for molecules having the following general formulas? (a) AB_3, (b) AB_2U_2, (c) AB_3U, (d) ABU_4, (e) ABU_3
18. Repeat Exercise 17 for (a) ABU_5, (b) AB_2U_4, (c) AB_4, (d) AB_3U_2, (e) AB_5
19. What are the primary factors upon which we base a decision on whether the bonding in a molecule is better described in terms of simple orbital overlap or overlap involving hybridized atomic orbitals?

Electronic and Molecular Geometry

20. Draw a Lewis dot formula for each of the following species. Indicate the number of regions of high electron density and describe the electronic and molecular geometries. (a) H_2Be, (b) molecular $AlCl_3$, (c) SiH_4, (d) SF_6, (e) IO_4^-, (f) NCl_3
21. Draw a Lewis dot formula for each of the following species. Indicate the number of regions of high electron density and describe the electronic and molecular or ionic geometries. (a) IF_4^-, (b) CO_2, (c) AlH_4^-, (d) NH_4^+, (e) $AsCl_3$, (f) ClO_3^-
22. Draw a Lewis dot formula for each of the following species. Indicate the number of regions of high electron density and describe the electronic and molecular or ionic geometries. (a) SeF_6, (b) ONCl (N is the central atom), (c) Cl_2CO, (d) PCl_5, (e) BCl_3, (f) ClO_4^-
23. Draw a Lewis dot formula for each of the following species. Indicate the number of regions of high electron density and describe the electronic and molecular geometries. (a) molecular $HgCl_2$, (b) ClO_2, (c) XeF_2, (d) CCl_4, (e) $CdCl_2$, (f) $AsCl_5$
24. Draw a Lewis dot formula for each of the following species. Indicate the number of regions of high electron density and the electronic and molecular or ionic geometries. (a) H_2O, (b) $SnCl_4$, (c) SeF_6, (d) SbF_6^-
25. Draw a Lewis dot formula for each of the following species. Indicate the number of regions of high electron density and the electronic and molecular or ionic geometries. (a) BF_3, (b) NF_3, (c) BrO_3^-, (d) $SiCl_4$
26. (a) What would be the ideal bond angles in the species in Exercise 24, ignoring lone pair effects? (b) How do these differ, if at all, from the actual values? Why?
27. (a) What would be the ideal bond angles in the molecules or ions in Exercise 25, ignoring lone pair effects? (b) Are these values greater than, less than, or equal to the actual values? Why?
28. Carbon forms two common oxides, CO and CO_2, both of which are linear. It forms a third (very uncommon) oxide, carbon suboxide, C_3O_2, which is also linear. The structure has terminal oxygen atoms on both ends. Draw Lewis dot and dash formulas for C_3O_2. How many regions of high electron density are there about each of the three carbon atoms?
29. Pick the member of each pair that you would expect to have the smaller bond angles, if different, and explain why. (a) SF_2 and SO_2, (b) BF_3 and BCl_3, (c) CF_4 and SF_4, (d) NH_3 and H_2O.
30. Draw a Lewis dot formula, sketch the three-dimensional shape, and name the electronic and ionic geometries for the following polyatomic ions. (a) $AsCl_4^-$, (b) PCl_6^-, (c) PCl_4^-, (d) $AsCl_4^+$
31. As the name implies, the interhalogens are compounds that contain two halogens. Draw Lewis dot formulas and three-dimensional structures for the following. Name the electronic and molecular geometries of each. (a) IF_5, (b) IBr, (c) ICl_5
32. A number of ions derived from the interhalogens are known. Draw Lewis dot formulas and three-dimensional structures for the following ions. Name the electronic and ionic geometries of each. (a) IF_4^+, (b) ICl_2^-, (c) BrF_4^-.

*33. (a) Draw a Lewis dot formula for each of the following molecules: BF_3, NF_3, BrF_3. (b) Contrast the molecular geometries of these three molecules. Account for differences in terms of the VSEPR theory.

*34. (a) Draw a Lewis dot formula for each of the following molecules: GeF_4, SF_4, XeF_4. (b) Contrast the molecular geometries of these three molecules. Account for differences in terms of the VSEPR theory.

35. Draw the Lewis structures and predict the shapes of these very reactive carbon-containing species: H_3C^+ (a carbocation), $H_3C:^-$ (a carbanion), and $:CH_2$ (a carbene whose unshared electrons are paired).
36. Draw the Lewis structures and predict the shapes of (a) ICl_4^-, (b) $TeCl_4$, (c) XeO_3, (d) BrNO (N is the central atom), (e) $ClNO_2$ (N is the central atom), (f) Cl_2SO (S is the central atom).

37. Describe the shapes of these polyatomic ions: (a) BO_3^{3-}, (b) PO_4^{3-}, (c) SO_3^{2-}, (d) NO_2^-.
38. Describe the shapes of these polyatomic ions: (a) H_3O^+, (b) GeF_3^-, (c) ClF_3^{2-}, (d) $IO_2F_2^-$.
39. Which of the following molecules are polar? Why? (a) CH_4, (b) CH_3Br, (c) CH_2Br_2, (d) $CHBr_3$, (e) CBr_4
40. Which of the following molecules are polar? Why? (a) CdI_2, (b) BCl_3, (c) PCl_3, (d) H_2O, (e) SF_6
41. Which of the following molecules have dipole moments of zero? Justify your answer. (a) SO_3, (b) IF, (c) Cl_2O, (d) $AsCl_3$
42. The PF_3Cl_2 molecule has a dipole moment of zero. Use this information to sketch its three-dimensional shape. Justify your choice.
*43. In what two major ways does the presence of lone pairs of valence electrons affect the polarity of a molecule? Describe two molecules for which the presence of lone pairs on the central atom helps to make the molecules polar. Can you think of a bonding arrangement that has lone pairs of valence electrons on the central atom but that is nonpolar?

Valence Bond Theory

44. What is the hybridization on the central atom in each of the following? (a) H_2Be, (b) molecular $AlCl_3$, (c) SiH_4, (d) SF_6, (e) IO_4^-, (f) NCl_3
45. What is the hybridization on the central atom in each of the following? (a) ICl_4^-, (b) CO_2, (c) AlH_4^-, (d) NH_4^+, (e) $AsCl_3$, (f) ClO_3^-
46. What is the hybridization on the central atom in each of the following? (a) SeF_6, (b) ONCl (N is the central atom), (c) Cl_2CO, (d) PCl_5, (e) BCl_3, (f) ClO_4^-
47. What is the hybridization on the central atom in each of the following? (a) molecular $MgCl_2$, (b) ClO_2, (c) XeF_2, (d) CCl_4, (e) $CdCl_2$, (f) $AsCl_5$
48. (a) Describe the hybridization of the central atom in each of these covalent species. (1) $CHCl_3$, (2) BeH_2, (3) NCl_3, (4) ClO_3^-, (5) IF_6^+, (6) SiF_6^{2-}. (b) Give the shape of each species.
49. Describe the hybridization of the underlined atoms in $\underline{C}_2Cl_4$, $\underline{C}_2Cl_2$, $\underline{N}_2F_4$, and $(H_2\underline{N})_2\underline{C}O$.
50. (a) Describe the hybridization of N in NO_2^+ and NO_2^-. (b) Predict the bond angle in each case.
*51. After comparing experimental and calculated dipole moments, Charles A. Coulson suggested that the Cl atom in HCl is *sp*-hybridized. (a) Give the orbital electronic structure for an *sp*-hybridized Cl atom. (b) Which HCl molecule would have a larger dipole moment—one in which the chlorine uses pure *p* orbitals for bonding with the H atom or one in which *sp* hybrid orbitals are used?
*52. Predict the hybridization at each carbon atom in each of the following molecules.
(a) ethanol (ethyl alcohol or grain alcohol):

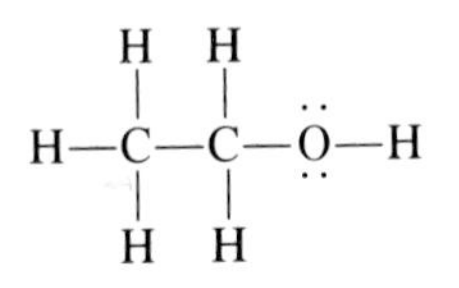

(b) alanine (an amino acid):

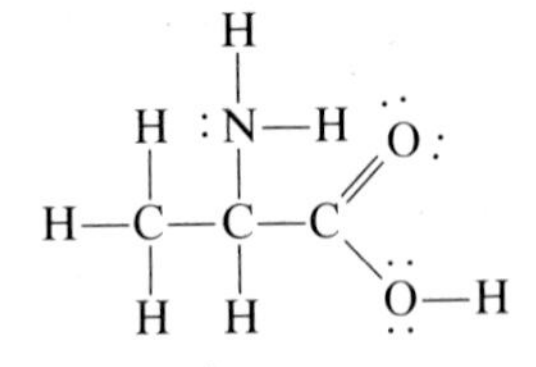

(c) tetracyanoethylene:

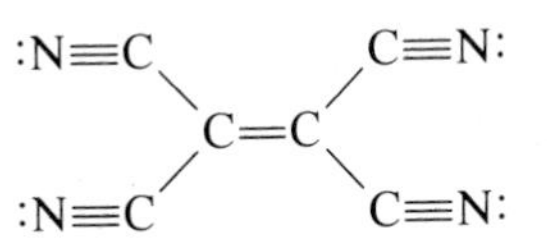

(d) chloroprene (used to make neoprene, a synthetic rubber):

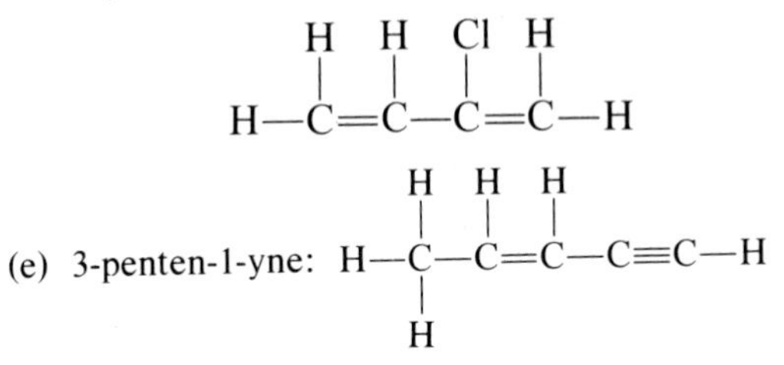

(e) 3-penten-1-yne:

*53. Predict the hybridization at the numbered atoms (①, ②, and so on) in the following molecules and predict the approximate bond angles at those atoms.
(a) diethyl ether, an anesthetic:

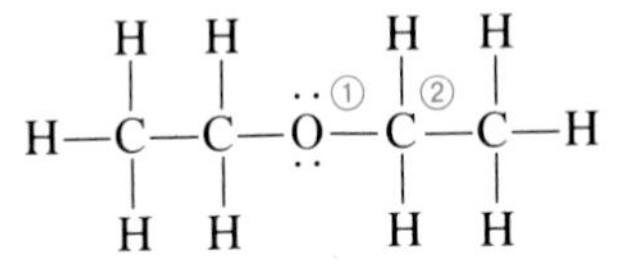

(b) caffeine, a stimulant in coffee and in many over-the-counter medicinals:*

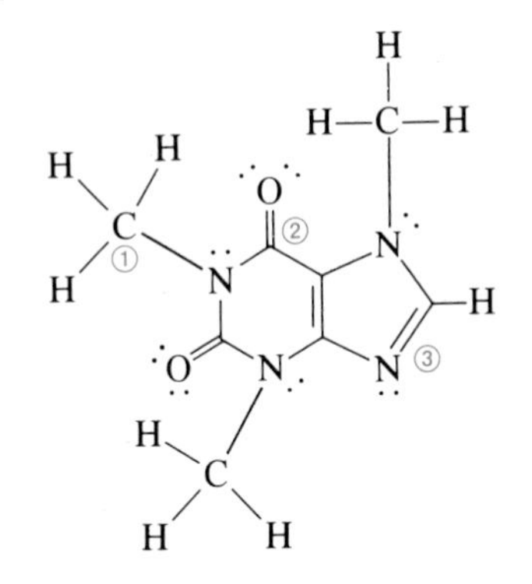

* In this kind of structural drawing, each intersection of lines represents a C atom; sometimes H atoms are not shown at all these intersections.

(c) acetylsalicylic acid (aspirin):

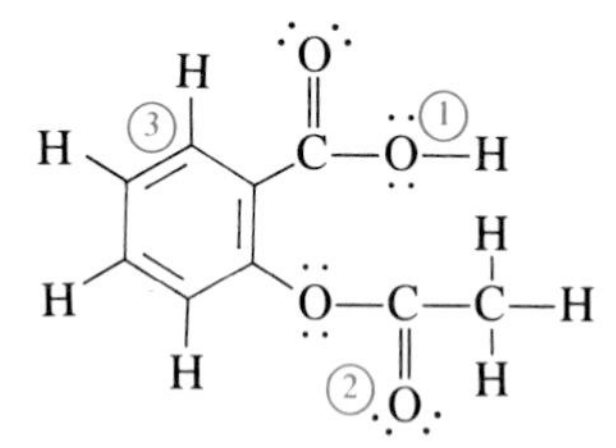

(d) nicotine:

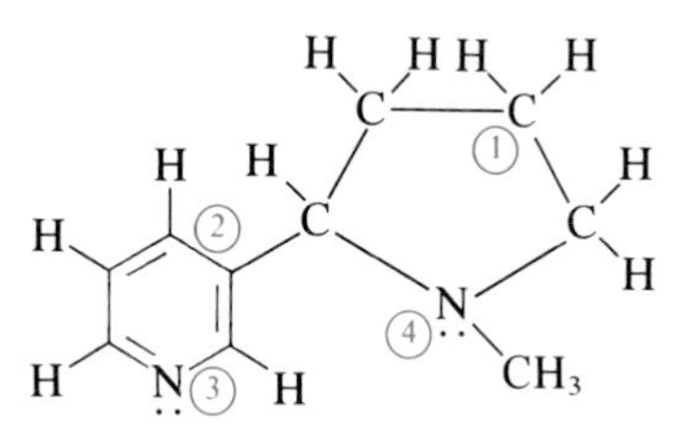

(e) ephedrine (used as a nasal decongestant):

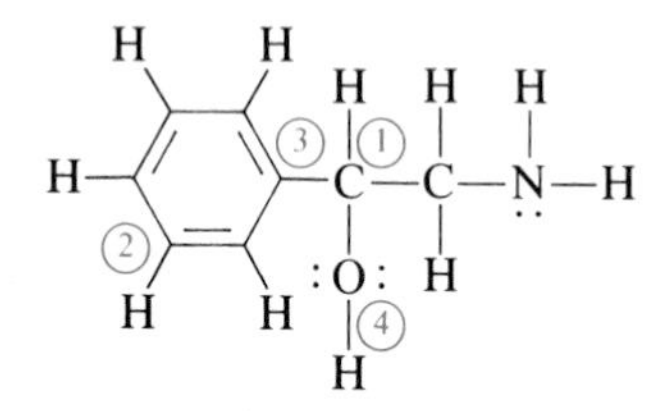

54. How many sigma and how many pi bonds are there in each of the following molecules?

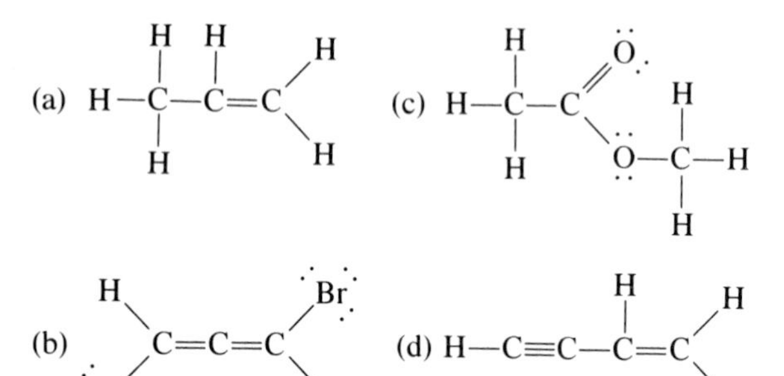

*55. How many sigma and how many pi bonds are there in each of the following molecules?

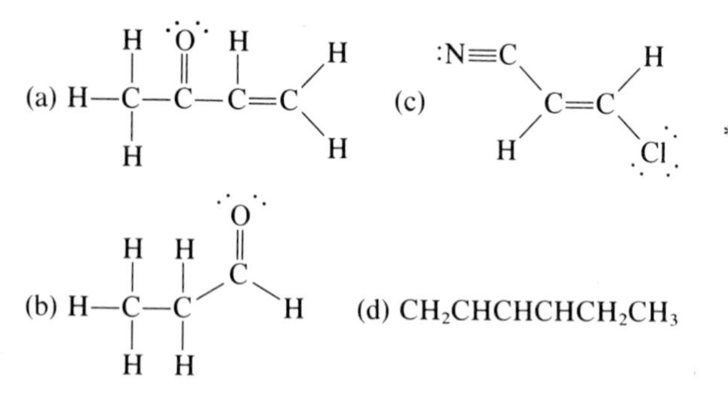

(d) $CH_2CHCHCHCH_2CH_3$

56. Describe the bonding in the N_2 molecule with a three-dimensional VB structure. Show the orbital overlap and label the orbitals.

57. Draw the Lewis structures for molecular oxygen and ozone. Assuming that all of the oxygen atoms are hybridized, what will be the hybridization of the oxygen atoms in each substance? Prepare sketches of the molecules.

*58. A water solution of cadmium bromide, $CdBr_2$, contains not only Cd^{2+} and Br^- ions, but also $CdBr^+$, $CdBr_2$, $CdBr_3^-$, and $CdBr_4^{2-}$. Describe the type of hybrid orbital used by Cd in each polyatomic species and describe the shape of the species.

*59. In their crystalline states, PCl_5 exists as $(PCl_4)^+(PCl_6)^-$, and PBr_5 exists as $PBr_4^+Br^-$. (a) Predict the shapes of all the polyatomic ions. (b) Indicate the hybrid orbital structure for the P atom in each of its different types of ions.

*60. Draw a dash formula and a three-dimensional structure for each of the following polycentered molecules. Indicate hybridizations and bond angles at each carbon atom. (a) Butane, C_4H_{10}; (b) 1-butene, $H_2C{=}CHCH_2CH_3$; (c) 1-butyne, $HC{\equiv}CCH_2CH_3$; (d) acetaldehyde, CH_3CHO

61. How many σ bonds and how many π bonds are there in each of the molecules of Exercise 60?

62. (a) Describe the hybridization of N in each of these species? (1) $:NH_3$, (2) NH_4^+, (3) $H\ddot{N}{=}\ddot{N}H$, (4) $HC{\equiv}N:$.
(b) Give an orbital description for each species, specifying the location of any unshared pairs and the orbitals used for the multiple bonds.

63. Draw the Lewis structures and predict the orbital types and the shapes of these polyatomic ions and covalent molecules: (a) $HgCl_2$, (b) BF_3, (c) BF_4^-, (d) SeF_2, (e) $AsCl_5$, (f) SbF_6^-.

64. (a) What is the hybridized state of each C in these molecules? (1) $Cl_2C{=}O$, (2) $HC{\equiv}N$, (3) CH_3CH_3, (4) ketene, $H_2C{=}C{=}O$. (b) Describe the shape of each molecule.

*65. The following fluorides of xenon have been well characterized: XeF_2, XeF_4, and XeF_6. (a) Draw Lewis structures for these substances and decide what type of hybridization of the Xe atomic orbitals has taken place. (b) Draw all of the possible atomic arrangements of XeF_2 and discuss your choice of molecular geometry. (c) What shape do you predict for XeF_4?

*66. Iodine and fluorine form a series of interhalogen molecules and ions. Among these are IF (minute quantities observed spectroscopically), IF_3, IF_4^-, IF_5, IF_6^-, and IF_7. (a) Draw Lewis structures for each of these species. (b) Identify the type of hybridization that the orbitals of the iodine atom have undergone in each substance. (c) Identify the shape of the molecule or ion.

Mixed Exercises

67. In the pyrophosphate ion, $P_2O_7^{4-}$, one oxygen atom is bonded to both phosphorus atoms. Draw a Lewis dot formula and sketch the three-dimensional shape of the ion. Describe the ionic geometry with respect to the central O atom and with respect to each P atom.

68. Briefly discuss the bond angles in the hydroxylamine molecule in terms of the ideal geometry and the small changes caused by electron pair repulsions.

$$H-\ddot{N}-\ddot{\underset{..}{O}}-H$$
$$\quad\;\, |$$
$$\quad\;\, H$$

69. Repeat Exercise 68 for nitric acid.

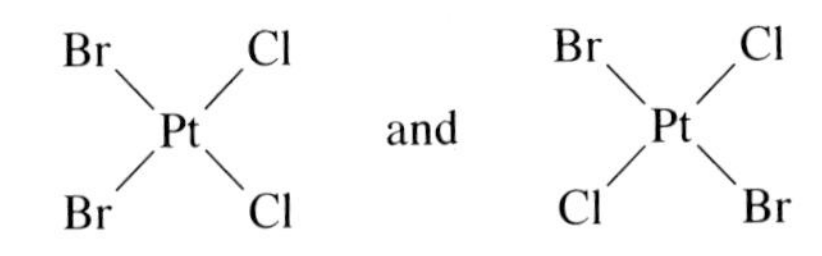

***70.** The methyl free radical $\cdot CH_3$ has bond angles of about 120°, whereas the methyl carbanion $:CH_3^-$ has bond angles of about 109°. What can you infer from these facts about the repulsive force exerted by an unpaired, unshared electron as compared to that exerted by an unshared pair of electrons?

***71.** Two Lewis structures can be written for the square-planar molecule $PtCl_2Br_2$:

Br Cl / Pt / Br Cl and Br Cl / Pt / Cl Br

Show how a difference in dipole moments can distinguish between these two possible structures.

72. Prepare a sketch of the molecule $CH_3CH{=}CH_2$ showing orbital overlaps. Identify the type of hybridization of atomic orbitals for each carbon atom.

***73.** The skeleton for the nitrous acid molecule, HNO_2, is shown. What are the hybridizations at the middle O and N atoms? Draw the dash formula. Is this consistent with your predictions?

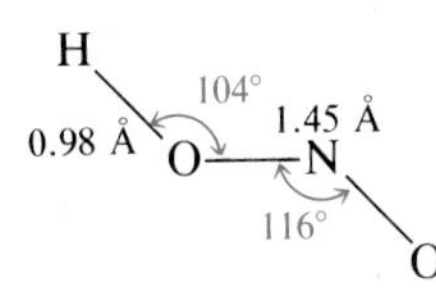

74. Describe the change in hybridization that occurs at the central atom of the reactant at the left in each of the following reactions.

(a) $PF_5 + F^- \longrightarrow PF_6^-$
(b) $2\ CO + O_2 \longrightarrow 2\ CO_2$
(c) $AlI_3 + I^- \longrightarrow AlI_4^-$
(d) What change in hybridization occurs in the following reaction?

$$:NH_3 + BF_3 \longrightarrow H_3N:BF_3$$

***75.** Consider the following proposed Lewis structures for ozone (O_3):

(i) $:\ddot{O}-\ddot{O}=\ddot{O} \longleftrightarrow \ddot{O}=\ddot{O}-\ddot{O}:$ (ii) cyclic $:O-O:$ with $:O:$ at the apex (triangle)

(iii) $:\ddot{O}-\ddot{O}-\ddot{O}:$

(a) Which of these structures correspond to a polar molecule?
(b) Which of these structures predict covalent bonds of equal lengths and strengths?
(c) Which of these structures predict a diamagnetic molecule?
(d) The properties listed in (a), (b), and (c) are those observed for ozone. Which structure correctly predicts all three?
(e) Which of these structures contains a considerable amount of "strain"?

76. What hybridizations are predicted for the central atoms in molecules having the formulas AB_2U_2 and AB_3U? What are the predicted bond angles for these molecules? The actual bond angles for representative substances are

H_2O	104.45°	NH_3	106.67°
H_2S	92.2°	PH_3	93.67°
H_2Se	91.0°	AsH_3	91.8°
He_2Te	89.5°	SbH_3	91.3°

What would be the predicted bond angle if no hybridization occurred? What conclusion can you draw concerning the importance of hybridization for molecules of compounds involving elements with higher atomic numbers?

77. Describe the orbitals (s, p, sp^2, and so on) of the central atom used for bond orbitals and unshared pairs in (a) H_2S (bond angle 91°), (b) CH_3OCH_3 (C—O—C angle 110°), (c) S_8 (S—S—S angle 105°), (d) $(CH_3)_2Sn^{2+}$ (C—Sn—C angle 180°), (e) $(CH_3)_3Sn^+$ (planar with respect to the Sn and 3 C atoms), (f) $SnCl_2$ (angle 95°).

9 Molecular Orbitals in Chemical Bonding

Outline

9-1 Molecular Orbitals
9-2 Molecular Orbital Energy-Level Diagrams
9-3 Bond Order and Bond Stability
9-4 Homonuclear Diatomic Molecules
9-5 Heteronuclear Diatomic Molecules
9-6 Delocalization and the Shapes of Molecular Orbitals

Objectives

As you study this chapter, you should learn to

- □ Understand the basic ideas of molecular orbital theory
- □ Relate the shapes and overlap of atomic orbitals to the shapes and energies of the resulting molecular orbitals
- □ Distinguish between bonding and antibonding orbitals
- □ Apply the Aufbau Principle to find the molecular orbital descriptions for homonuclear diatomic molecules and their ions
- □ Apply the Aufbau Principle to find the molecular orbital descriptions for heteronuclear diatomic molecules and their ions
- □ Find the bond order in diatomic molecules
- □ Relate bond order to bond stability
- □ Use the MO concept of delocalization for molecules in which VB theory would postulate resonance

An early triumph of molecular orbital theory was its ability to account for the observed paramagnetism of oxygen, O_2. According to earlier theories, O_2 was expected to be diamagnetic, that is, to have only paired electrons.

We have described bonding and molecular geometry in terms of valence bond theory. In valence bond theory, we postulate that bonds result from the sharing of electrons in overlapping orbitals of different atoms. These orbitals may be *pure atomic orbitals* or *hybridized atomic orbitals* of *individual* atoms. We describe electrons in overlapping orbitals of different atoms as being localized in the bonds between the two atoms involved, rather than delocalized over the entire molecule. We then invoke hybridization when it helps to account for the geometry of a molecule.

In **molecular orbital theory**, we postulate

> the combination of atomic orbitals on different atoms forms **molecular orbitals** (MOs), so that electrons in them belong to the molecule as a whole.

In some polyatomic molecules, a molecular orbital may extend over only a fraction of the molecule. Molecular orbitals also exist for polyatomic ions such as CO_3^{2-}, SO_4^{2-}, and NH_4^+.

The valence bond and molecular orbital approaches have strengths and weaknesses that are complementary. They are alternative descriptions of chemical bonding. Valence bond theory is descriptively attractive and it lends itself well to visualization. Molecular orbital (MO) theory gives better descriptions of electron cloud distributions, bond energies, and magnetic properties, but its results are not as easy to visualize.

The valence bond picture of bonding in the O_2 molecule, involves a double bond and no unpaired electrons. It predicts sp^2 hybridization at each oxygen because there are three sets of valence electrons on each O atom.

Although this structure is correct by the standards of valence bond theory, it is inconsistent with the observed paramagnetism of O_2.

$$\ddot{\underset{..}{O}}::\ddot{\underset{..}{O}}$$

However, experiments show that O_2 is paramagnetic; therefore, it has unpaired electrons. Thus, the VB structure is inconsistent with experiment and cannot be accepted as a suitable description of the bonding. Molecular orbital theory *predicts* that O_2 has two unpaired electrons. This ability of MO theory to explain the paramagnetism of O_2 brought it to the forefront as a major theory in bonding. In the following sections, we shall develop some of the notions of MO theory. Then we shall apply them to some molecules and ions.

9-1 Molecular Orbitals

We learned in Chapter 5 that each solution to the Schrödinger equation, called a wave function, represents an atomic orbital. The mathematical pictures of hybrid orbitals in valence bond theory can be generated by combining the wave functions that describe two or more atomic orbitals on a *single* atom. Similarly, combining wave functions that describe atomic orbitals on *separate* atoms generates mathematical descriptions of molecular orbitals.

An orbital has physical meaning only when we square its wave function to describe the electron density. Thus, the overall sign on the wave function that describes an atomic orbital is arbitrary. But when we *combine* two orbitals, the signs relative to one another do matter. When waves are combined, they may interact either constructively or destructively (Figure 9-1). Likewise, when two atomic orbitals overlap, they can be in phase (added) or out of phase (subtracted). When they overlap in phase, constructive interference occurs in the region between the nuclei, and a **bonding orbital** is produced. The energy of the bonding orbital is always lower (more stable) than the energies of the combining orbitals. When they overlap out of phase, destructive interference reduces the probability of finding electrons in the region between the nuclei, and an **antibonding orbital** is produced. This is higher in energy (less stable) than the original atomic orbitals. The overlap of two atomic orbitals always produces two MOs—one bonding and one antibonding.

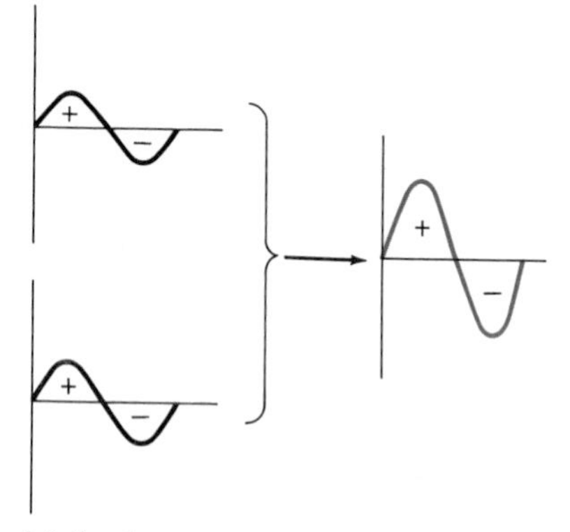

(a) In-phase overlap (add)

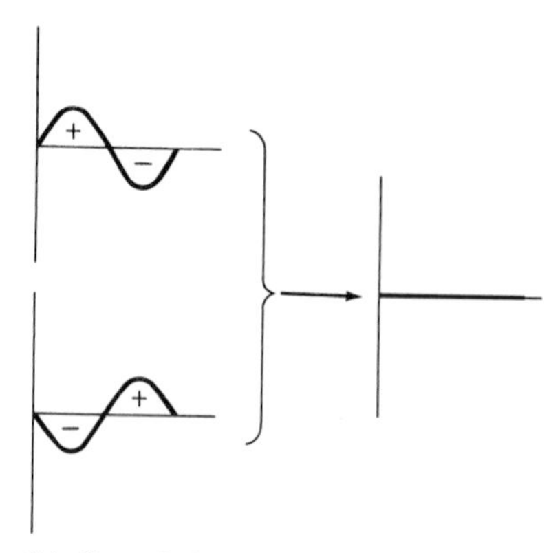

(b) Out-of-phase overlap (subtract)

Figure 9-1
An illustration of constructive and destructive interaction of waves. (a) If the two identical waves shown at the left are added, they interfere constructively to produce the more intense wave at the right. (b) Conversely, if they are subtracted, it is as if the phases (signs) of one wave were reversed and added to the first wave. This causes destructive interference, resulting in the wave at the right with zero amplitude.

We can illustrate this basic principle by considering the combination of the 1*s* atomic orbitals *on two different atoms* (Figure 9-2). When these orbitals are occupied by electrons, the shapes of the orbitals are really plots of electron density. These plots show the regions in molecules where the probabilities of finding electrons are greatest.

In the bonding orbital, the two 1*s* orbitals have reinforced one another in the region between the two nuclei by in-phase overlap, or addition of their electron waves. In the antibonding orbital, they have canceled one another in this region by out-of-phase overlap, or subtraction of their electron waves. We designate both molecular orbitals as *sigma* (σ) *orbitals* (which indicates that they are cylindrically symmetrical about the internuclear axis). We

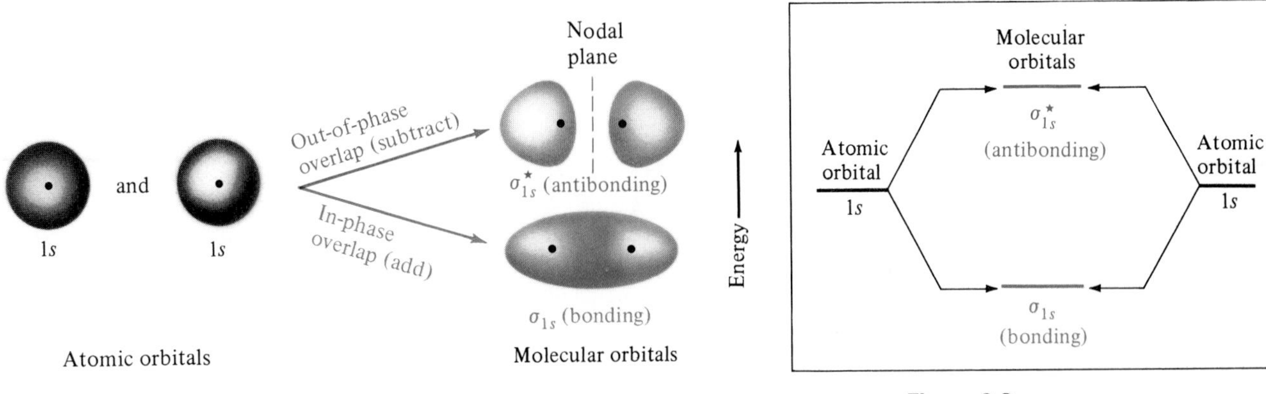

Figure 9-2
Molecular orbital (MO) diagram for the combination of the $1s$ atomic orbitals on two identical atoms (at the left) to form two MOs. One is a *bonding* orbital, σ_{1s} (blue), resulting from addition of the wave functions of the $1s$ orbitals. The other is an *antibonding* orbital, $\sigma^{\star}_{1s}$ (red), at higher energy resulting from subtraction of the waves that describe the combining $1s$ orbitals. In all σ-type MOs, the electron density is symmetrical about an imaginary line connecting the two nuclei. The terms "subtraction of waves," "out of phase," and "destructive interference in the region between the nuclei" all refer to the formation of an antibonding MO. Nuclei are represented by dots.

indicate with subscripts the atomic orbitals that have been combined. The star denotes an antibonding orbital. Thus, two $1s$ orbitals produce a σ_{1s} (read "sigma-$1s$") bonding orbital and a $\sigma^{\star}_{1s}$ (read "sigma-$1s$-star") antibonding orbital. The right-hand side of Figure 9-2 shows the relative energy levels of these orbitals. All sigma antibonding orbitals have nodal planes bisecting the internuclear axis. A **node** or **nodal plane** is a region in which the probability of finding electrons is zero.

Another way of viewing the relative stabilities of these orbitals follows. In a bonding molecular orbital, there is high electron density *between* the two atoms, where it stabilizes the arrangement by exerting a strong attraction for both nuclei. By contrast, an antibonding orbital has a node (a region of zero electron density) between the nuclei; this allows a strong net repulsion between the nuclei, destabilizing the arrangement. Electrons are *more* stable (have lower energy) in bonding molecular orbitals than in the individual atoms. Placing electrons in antibonding orbitals, on the other hand, requires raising their energy, which makes them *less* stable than in the individual atoms.

For any two sets of p orbitals on two different atoms, corresponding orbitals such as p_x orbitals can overlap *head-on*. This gives σ_p and $\sigma^{\star}_p$ orbitals, as shown in Figure 9-3 for the head-on overlap of $2p_x$ orbitals on two atoms. If the remaining p orbitals overlap (p_y with p_y and p_z with p_z), they must do so sideways, or *side-on*, forming *pi* (π) *molecular orbitals*. De-

How we name the axes is arbitrary. We shall designate the internuclear axis as the x direction.

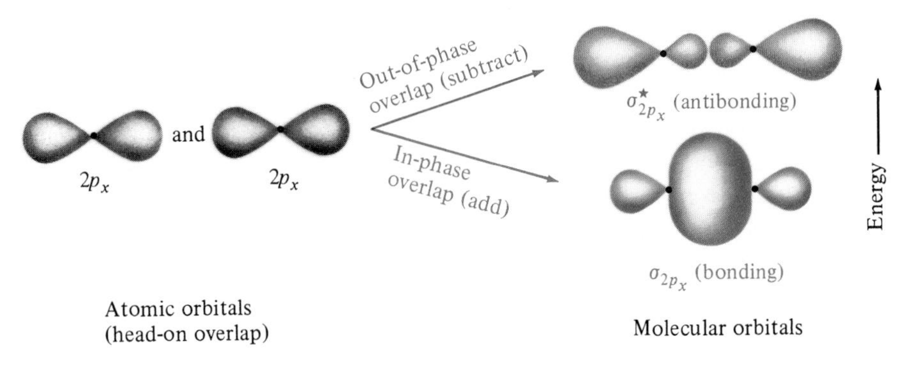

Figure 9-3
Production of σ_{2p_x} and $\sigma^{\star}_{2p_x}$ molecular orbitals by overlap of $2p_x$ orbitals on two atoms.

If we had chosen the *z*-axis as the axis of head-on overlap of the $2p$ orbitals in Figure 9-3, side-on overlap of the $2p_x$–$2p_x$ and $2p_y$–$2p_y$ orbitals would form the π-type molecular orbitals.

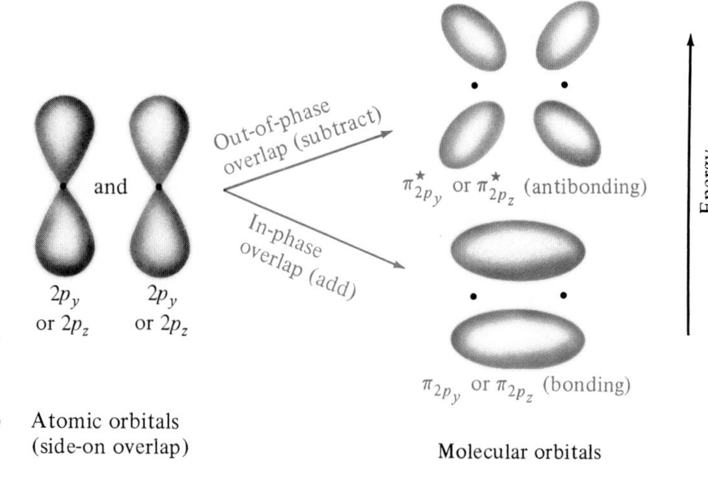

Figure 9-4
The π_{2p} and $\pi^\star_{2p}$ molecular orbitals from overlap of one pair of $2p$ atomic orbitals (for instance, $2p_y$ orbitals). There can be an identical pair of molecular orbitals at right angles to these, formed by another pair of p orbitals on the same two atoms (in this case, $2p_z$ orbitals).

pending on whether all p orbitals overlap, there can be as many as two π_p and two $\pi^\star_p$ orbitals. Figure 9-4 illustrates the overlap of two corresponding $2p$ orbitals on two atoms to form π_{2p} and $\pi^\star_{2p}$ molecular orbitals. There is a nodal plane along the internuclear axis for all pi molecular orbitals. If one views a sigma molecular orbital along the internuclear axis, it appears to be symmetrical around the axis like a pure *s* atomic orbital. A similar cross-sectional view of a pi molecular orbital looks like a pure *p* atomic orbital, with a node along the internuclear axis.

This would involve rotating Figures 9-2, 9-3, and 9-4 by 90° so that the internuclear axes are perpendicular to the plane of the pages.

The number of molecular orbitals (MOs) formed is equal to the number of atomic orbitals that are combined. When two atomic orbitals are combined, one of the resulting MOs is at a *lower* energy than the original atomic orbitals; this is a *bonding* orbital. The other MO is at a *higher* energy than the original atomic orbitals; this is an *antibonding* orbital.

9-2 Molecular Orbital Energy-Level Diagrams

In the same way that atomic orbitals can be arranged by increasing energy into an energy-level diagram (Figure 5-28), we can draw molecular orbital energy-level diagrams for simple molecules. The simplest examples, shown in Figure 9-5, apply to homonuclear diatomic molecules of elements in the first and second periods. Each diagram is an extension of the right-hand diagram in Figure 9-2, to which we have added the molecular orbitals formed from $2s$ and $2p$ atomic orbitals.

For the cases shown in Figure 9-5a, the two π_{2p} orbitals are lower in energy than the σ_{2p} orbital. However, molecular orbital calculations indicate that for O_2, F_2, and (hypothetical) Ne_2 molecules, the σ_{2p} orbital is lower in energy than the π_{2p} orbitals (Figure 9-5b).

Spectroscopic data support these orders.

Diagrams such as these are used to describe the bonding in a molecule in MO terms. We simply apply the Aufbau Principle and follow the same rules that we did for atomic energy-level diagrams (Section 5-16).

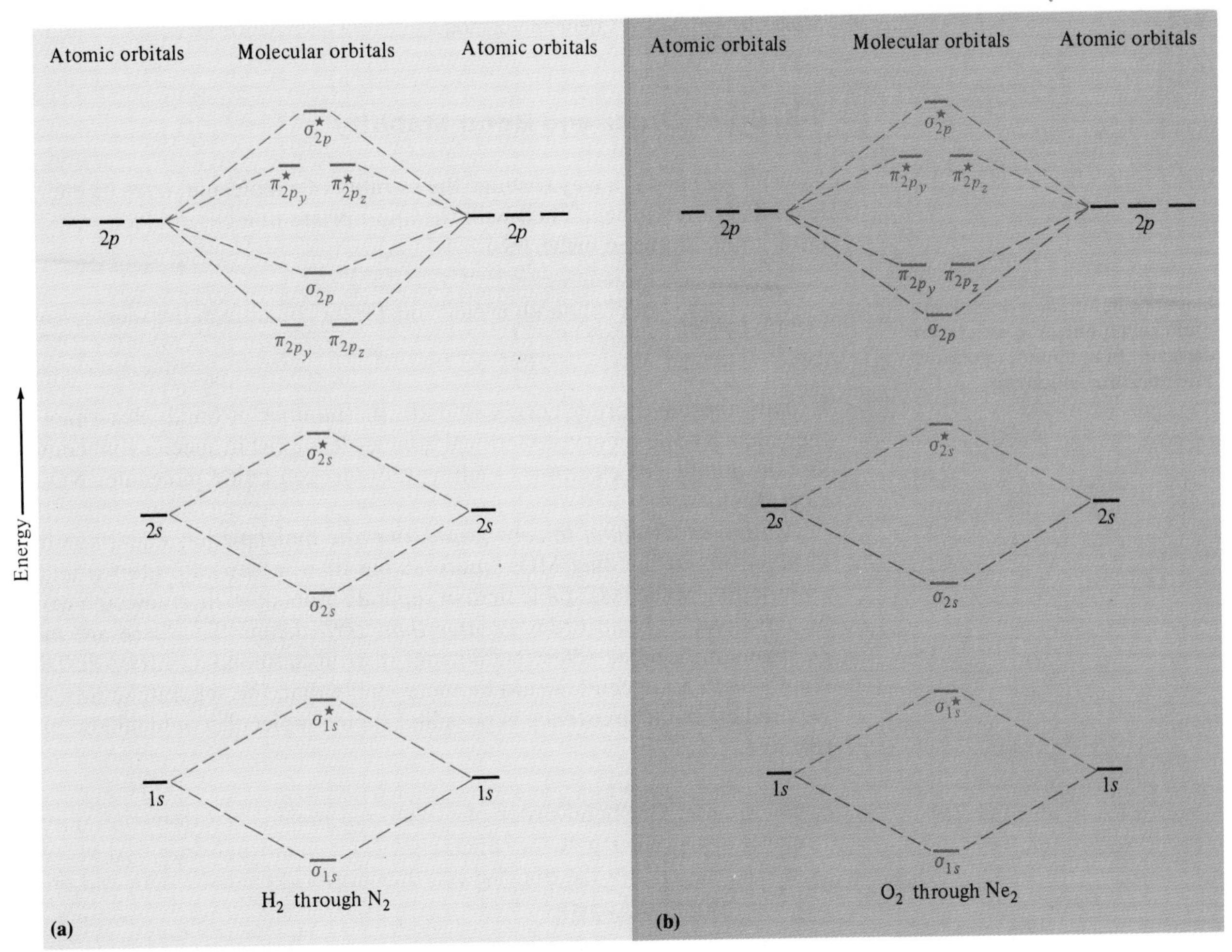

Figure 9-5
Energy-level diagrams for first- and second-period homonuclear diatomic molecules and ions. The solid lines represent the relative energies of the indicated atomic and molecular orbitals. (a) The diagram for H_2, He_2, Li_2, Be_2, B_2, C_2, and N_2 molecules and their ions. (b) The diagram for O_2, F_2, and Ne_2 molecules and their ions.

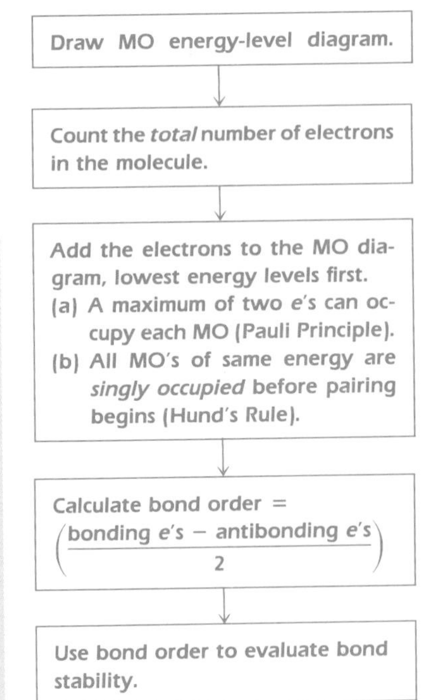

Draw MO energy-level diagram.
Count the *total* number of electrons in the molecule.
Add the electrons to the MO diagram, lowest energy levels first.
(a) A maximum of two e's can occupy each MO (Pauli Principle).
(b) All MO's of same energy are *singly occupied* before pairing begins (Hund's Rule).
Calculate bond order = (bonding e's − antibonding e's / 2)
Use bond order to evaluate bond stability.

1. Draw (or select) the appropriate molecular orbital energy-level diagram.
2. Determine the *total* number of electrons in the molecule. Note that in applying MO theory, we usually account for *all* electrons. This includes both the core electrons and the valence electrons.
3. Add these electrons to the energy-level diagram, putting each electron into the lowest energy level available.
   a. A maximum of *two* electrons can occupy any given molecular orbital, and then only if they have opposite spin (Pauli Exclusion Principle).
   b. Electrons must occupy all the orbitals of the same energy singly before pairing begins. These unpaired electrons have parallel spins (Hund's Rule).

9-3 Bond Order and Bond Stability

Now all we need is a way to judge the stability of a molecule, once its energy-level diagram has been filled with the appropriate number of electrons. This criterion is the **bond order** (bo):

Electrons in bonding orbitals are often called **bonding electrons**; electrons in antibonding orbitals, **antibonding electrons**.

$$\text{bond order} = \frac{(\text{no. of bonding electrons}) - (\text{no. of antibonding electrons})}{2}$$

Usually the bond order corresponds to the number of bonds described by the valence bond theory. Fractional bond orders exist in species that contain an odd number of electrons, such as the nitrogen oxide molecule, NO (15 electrons).

A bond order *equal to zero* means that the molecule has equal numbers of electrons in bonding MOs (more stable than in separate atoms) and in antibonding MOs (less stable than in separate atoms); such a molecule would be *unstable*. A bond order *greater than zero* means that there are more electrons in bonding MOs (stabilizing) than in antibonding MOs (destabilizing). Such a molecule would be more stable than the separate atoms, and we predict that its existence is possible; such a molecule could still be quite reactive.

The greater the bond order of a diatomic molecule or ion, the more stable we predict it to be. Likewise, for a bond between two given atoms, the greater the bond order, the shorter the bond length and the greater the bond energy.

The **bond energy** is the amount of energy necessary to break a mole of bonds (Section 15-9); it is therefore a measure of bond strength.

9-4 Homonuclear Diatomic Molecules

"Homonuclear" means consisting only of atoms of the same element. "Diatomic" means consisting of two atoms.

The electron distributions for the homonuclear diatomic molecules of the first and second periods are shown in Table 9-1 together with their bond orders, bond lengths, and bond energies. We shall now look at some of these as examples.

The Hydrogen Molecule, H_2

The overlap of the 1*s* orbitals of two hydrogen atoms produces σ_{1s} and $\sigma_{1s}^{\star}$ molecular orbitals. The two electrons of the molecule occupy the lower-energy σ_{1s} orbital (Figure 9-6).

$$H_2 \text{ bo} = \frac{2-0}{2} = 1$$

Because the two electrons in an H_2 molecule are in a bonding orbital, the bond order is one. We conclude that the H_2 molecule would be stable, and it is. The energy associated with two electrons in the H_2 molecule is less than that associated with the same two electrons in 1*s* atomic orbitals. The lower the energy of a system, the more stable it is.

Table 9-1
Molecular Orbitals for First and Second Row Diatomic Molecules[a]

Increasing energy (not to scale) ↑

	H_2	He_2[c]	Li_2[b]	Be_2[b]	B_2[b]	C_2[b]	N_2		O_2	F_2	Ne_2[c]
σ^*_{2p}	—	—	—	—	—	—	—		—	—	⇅
$\pi^*_{2p_y}, \pi^*_{2p_z}$	— —	— —	— —	— —	— —	— —	— —		↑ ↑	⇅ ⇅	⇅ ⇅
σ_{2p}	—	—	—	—	—	—	⇅	π_{2p_y}, π_{2p_z}	⇅ ⇅	⇅ ⇅	⇅ ⇅
π_{2p_y}, π_{2p_z}	— —	— —	— —	— —	↑ ↑	⇅ ⇅	⇅ ⇅	σ_{2p}	⇅	⇅	⇅
σ^*_{2s}	—	—	—	⇅	⇅	⇅	⇅		⇅	⇅	⇅
σ_{2s}	—	—	⇅	⇅	⇅	⇅	⇅		⇅	⇅	⇅
σ^*_{1s}	—	⇅	⇅	⇅	⇅	⇅	⇅		⇅	⇅	⇅
σ_{1s}	⇅	⇅	⇅	⇅	⇅	⇅	⇅		⇅	⇅	⇅
Paramagnetic?	no	no	no	no	yes	no	no		yes	no	no
Bond order	1	0	1	0	1	2	3		2	1	0
Observed bond length (Å)	0.74	—	2.67	—	1.59	1.31	1.09		1.21	1.43	—
Observed bond energy (kJ/mol)	435	—	110	9	~270	602	946		498	159	—

[a] Electron distribution in molecular orbitals, bond order, bond length, and bond energy of homonuclear diatomic molecules of the first- and second-row elements. Note that nitrogen molecules, N_2, have the highest bond energies listed; they have a bond order of three. The species C_2 and O_2, with a bond order of two, have the next highest bond energies.
[b] Exists only in the vapor state at elevated temperatures.
[c] Unknown species.

The Helium Molecule (hypothetical), He_2

The energy-level diagram for He_2 is similar to that for H_2 except that it has two more electrons. These occupy the antibonding σ^*_{1s} orbital (Figures 9-5a and 9-6b and Table 9-1), giving it a bond order of zero. That is, two electrons in He_2 would be *more stable* than in the separate atoms, but the other two would be *less stable* by the same amount. Thus, the molecule would be no more stable than the separate atoms and would not exist. In fact, He_2 is not known.

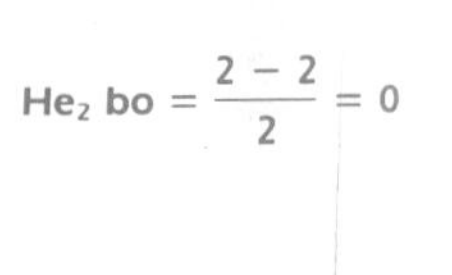
$$He_2 \text{ bo} = \frac{2-2}{2} = 0$$

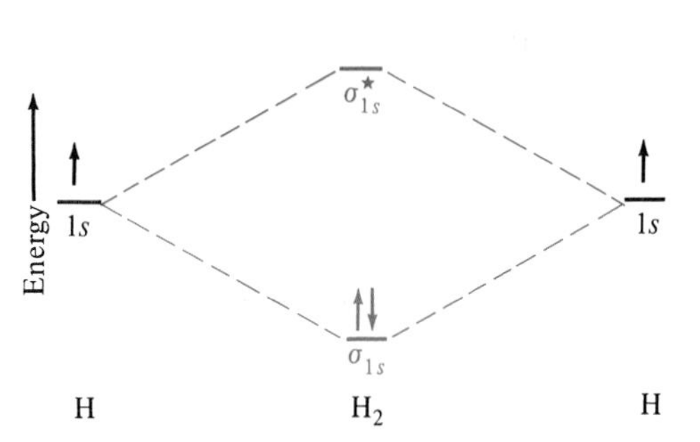

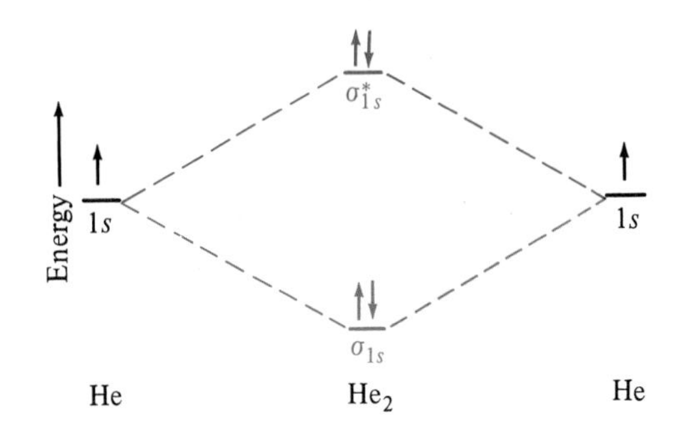

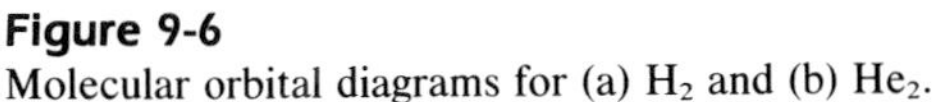

Figure 9-6
Molecular orbital diagrams for (a) H_2 and (b) He_2.

The Boron Molecule, B_2

B_2 bo $= \frac{6-4}{2} = 1$

Orbitals of equal energy are called *degenerate* orbitals. Hund's Rule for filling degenerate orbitals was discussed in Section 5-16.

The boron atom has the configuration $1s^22s^22p^1$. Now the *p* electrons do participate in bonding. Figure 9-5a and Table 9-1 show that the π_{p_y} and π_{p_z} molecular orbitals are lower in energy than the σ_{2p} for B_2. Thus, the electron configuration is $\sigma_{1s}{}^2\ \sigma^\star_{1s}{}^2\ \sigma_{2s}{}^2\ \sigma^\star_{2s}{}^2\ \pi_{2p_y}{}^1\ \pi_{2p_z}{}^1$. The unpaired electrons are consistent with the observed paramagnetism of B_2. Here we illustrate Hund's Rule in molecular orbital theory. The π_{2p_y} and π_{2p_z} orbitals are equal in energy and contain a total of two electrons. As a result, one electron occupies each orbital. The bond order is one. Experiments verify that the molecule exists in the vapor state.

The Nitrogen Molecule, N_2

N_2 bo $= \frac{10-4}{2} = 3$

In the valence bond representation, N_2 is shown as N≡N, with a triple bond.

Experimental thermodynamic data show that the N_2 molecule is stable, is diamagnetic, and has a very high bond energy, 946 kJ/mol. This is consistent with molecular orbital theory. Each nitrogen atom has seven electrons, so the diamagnetic N_2 molecule has 14 electrons, distributed as follows:

$$\sigma_{1s}{}^2\quad \sigma^\star_{1s}{}^2\quad \sigma_{2s}{}^2\quad \sigma^\star_{2s}{}^2\quad \pi_{2p_y}{}^2\quad \pi_{2p_z}{}^2\quad \sigma_{2p}{}^2$$

There are six more electrons in bonding orbitals than in antibonding orbitals, so the bond order is three. We see (Table 9-1) that N_2 has a very short bond length, only 1.09 Å, the shortest of any diatomic species except H_2.

The Oxygen Molecule, O_2

Among the homonuclear diatomic molecules, only N_2 and the very small H_2 have shorter bond lengths than O_2, 1.21 Å. Recall that VB theory predicts that O_2 is diamagnetic. However, experiments show that it is paramagnetic, with two unpaired electrons. MO theory predicts a structure consistent with this observation. For O_2, the σ_{2p} orbital is lower in energy than the π_{2p_y} and π_{2p_z} orbitals. Each oxygen atom has eight electrons, so the O_2 molecule has 16, distributed as follows:

$$\sigma_{1s}{}^2\quad \sigma^\star_{1s}{}^2\quad \sigma_{2s}{}^2\quad \sigma^\star_{2s}{}^2\quad \sigma_{2p}{}^2\quad \pi_{2p_y}{}^2\quad \pi_{2p_z}{}^2\quad \pi^\star_{2p_y}{}^1\quad \pi^\star_{2p_z}{}^1$$

O_2 bo $= \frac{10-6}{2} = 2$

The two unpaired electrons reside in the *degenerate* antibonding orbitals, $\pi^\star_{2p_y}$ and $\pi^\star_{2p_z}$. Because there are four more electrons in bonding orbitals than in antibonding orbitals, the bond order is two (Figure 9-5b and Table 9-1). We understand why the molecule is much more stable than two free O atoms.

Molecular orbital theory can also be used to predict the structures and stabilities of ions, as Example 9-1 shows.

Example 9-1

Predict the stabilities and bond orders of the ions (a) $O_2{}^+$ and (b) $O_2{}^-$.

Plan

(a) The $O_2{}^+$ ion is obtained by removing one electron from the O_2 molecule, given above. The electrons that are withdrawn most easily are those in the highest energy orbitals. (b) The superoxide ion, $O_2{}^-$, results from adding an electron to the O_2 molecule.

Solution

(a) We remove one of the $\pi^\star_{2p}$ electrons of O_2 to find the configuration of O_2^+:

$$\sigma_{1s}^2 \quad \sigma_{1s}^{\star 2} \quad \sigma_{2s}^2 \quad \sigma_{2s}^{\star 2} \quad \sigma_{2p}^2 \quad \pi_{2p_y}^2 \quad \pi_{2p_z}^2 \quad \pi_{2p_y}^{\star 1}$$

There are five more electrons in bonding orbitals than in antibonding orbitals, so the bond order is 2.5. We conclude that the ion would be reasonably stable relative to other diatomic ions, and it does exist.

In fact, the unusual ionic compound $[O_2^+][PtF_6^-]$ played an important role in the discovery of the first noble gas compound, $XePtF_6$ (Section 24-3).

(b) We add one electron to the appropriate orbital of O_2 to find the configuration of O_2^-. Following Hund's Rule, we add this electron into the $\pi^\star_{2p_y}$ orbital to form a pair:

$$\sigma_{1s}^2 \quad \sigma_{1s}^{\star 2} \quad \sigma_{2s}^2 \quad \sigma_{2s}^{\star 2} \quad \sigma_{2p}^2 \quad \pi_{2p_y}^2 \quad \pi_{2p_z}^2 \quad \pi_{2p_y}^{\star 2} \quad \pi_{2p_z}^{\star 1}$$

The bond order is 1.5 because there are three more bonding electrons than antibonding electrons. Thus, we can conclude that the ion should exist but be less stable than O_2.

The known superoxides of the heavier Group IA elements—KO_2, RbO_2, and CsO_2—contain the superoxide ion, O_2^-. These compounds are formed by combination of the free metals with oxygen (Section 6-8, part 2).

EOC 32

The Fluorine Molecule, F_2

Each fluorine atom has the $1s^22s^22p^5$ configuration. The 18 electrons of F_2 are distributed as follows:

$$\sigma_{1s}^2 \quad \sigma_{1s}^{\star 2} \quad \sigma_{2s}^2 \quad \sigma_{2s}^{\star 2} \quad \sigma_{2p}^2 \quad \pi_{2p_y}^2 \quad \pi_{2p_z}^2 \quad \pi_{2p_y}^{\star 2} \quad \pi_{2p_z}^{\star 2}$$

The bond order is one. Experiments show that F_2 exists. The F—F bond distance is longer (1.43 Å) than the bond distances in O_2 (1.21 Å), N_2 (1.09 Å), and even C_2 (1.31 Å) molecules. The bond energy of the F_2 molecules is quite low (159 kJ/mol), so F_2 molecules are very reactive.

F_2 bo $= \frac{10 - 8}{2} = 1$

Heavier Homonuclear Diatomic Molecules

It might appear reasonable to use the same types of molecular orbital diagrams to predict the stability or existence of homonuclear diatomic molecules of the third and subsequent periods. However, the heavier halogens, Cl_2, Br_2, and I_2, which contain only sigma (single) bonds, are the only well-characterized examples at room temperature. We would predict from both molecular orbital theory and valence bond theory that the other (nonhalogen) homonuclear diatomic molecules from below the second period would exhibit pi bonding and therefore multiple bonding.

Some heavier elements exist as diatomic species, such as S_2, in the vapor phase at elevated temperatures. These are neither prominent nor very stable.

The instability appears to be directly related to the inability of the heavier elements to form strong pi bonds with each other. For larger atoms, the sigma bond length is too great to allow the atomic *p* orbitals on different atoms to overlap very effectively. Therefore, the strength of pi bonding decreases rapidly with increasing atomic size. For example, N_2 is *much* more stable than P_2. This is because the 3*p* orbitals on one P atom do not overlap side by side in a pi-bonding manner with corresponding 3*p* orbitals on another P atom nearly as effectively as do the corresponding 2*p* orbitals on the smaller N atoms. MO theory does not predict multiple bonding for Cl_2, Br_2, or I_2, each of which has a bond order of one.

9-5 Heteronuclear Diatomic Molecules

Heteronuclear Diatomic Molecules of Second-Period Elements

The corresponding atomic orbitals of two different elements, such as the 2*s* orbitals of carbon and oxygen atoms, have different energies because their differently charged nuclei have different attractions for the electrons. The atomic orbitals of the *more electronegative element* are *lower* in energy than the corresponding orbitals of the less electronegative element. As a result, a molecular orbital diagram such as Figure 9-5 is inappropriate for *heter-*

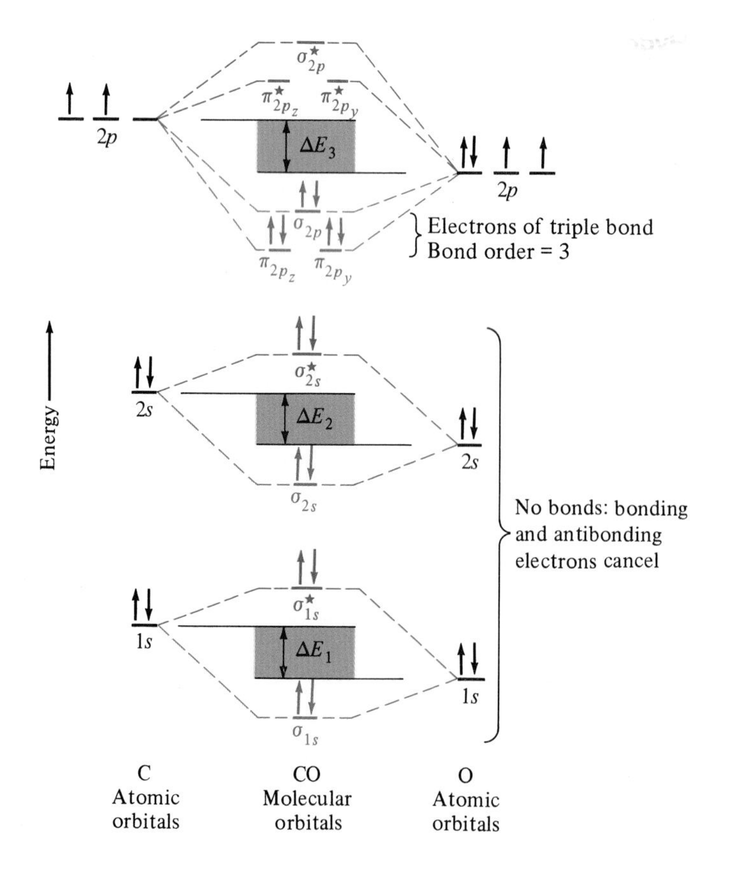

Figure 9-7
MO energy-level diagram for carbon monoxide, CO, a slightly polar heteronuclear diatomic molecule (μ = 0.11 D). The atomic orbitals of oxygen, the more electronegative element, are a little lower in energy than the corresponding atomic orbitals of carbon, the less electronegative element. For this molecule, the energy differences ΔE_1, ΔE_2, and ΔE_3 are not very large; the molecule is not very polar.

onuclear diatomic molecules. If the two elements are similar (as in CO, NO, or CN molecules, for example), we can modify the diagram of Figure 9-5 by skewing it slightly. Figure 9-7 shows the energy-level diagram and electron configuration for carbon monoxide, CO.

NOTE: CN is a reactive molecule, not the cyanide ion, CN^-.

The closer the energy of a molecular orbital is to the energy of one of the atomic orbitals of which it is composed, the more of the character of that atomic orbital it shows. Thus, in the CO molecule, the bonding MOs have more oxygen-like atomic orbital character, and the antibonding orbitals have more carbon-like atomic orbital character.

In general, the energy differences ΔE_1, ΔE_2, and ΔE_3 (shown in green in Figure 9-7) depend on the difference in electronegativities between the two atoms. The greater these energy differences, the more polar is the bond joining the atoms (the greater is its ionic character). On the other hand, the energy differences reflect the degree of overlap between atomic orbitals; the smaller these differences, the more the orbitals can overlap, and the greater is the covalent character of the bond.

We see that CO has a total of 14 electrons, making it isoelectronic with the stable N_2 molecule. Therefore, the distribution of electrons is the same in CO as in N_2, although we expect the energy levels of the MOs to be different. In accord with our predictions, carbon monoxide is a very stable molecule. It has a bond order of three, a short carbon–oxygen bond length of 1.13 Å, a low dipole moment of 0.11 D, and a very high bond energy of 1071 kJ/mol.

The Hydrogen Fluoride Molecule, HF

The electronegativity difference between hydrogen (EN = 2.1) and fluorine (EN = 4.0) is very large (ΔEN = 1.9). The hydrogen fluoride molecule contains a very polar bond (μ = 1.9 D). The bond in HF involves the 1*s* electron of H and an unpaired electron from a F 2*p* orbital. Figure 9-8 shows the overlap of the 1*s* orbital of H with a 2*p* orbital of F to form σ_{sp} and $\sigma^\star_{sp}$ molecular orbitals. The remaining two F 2*p* orbitals have no net overlap with H orbitals. They are called **nonbonding** orbitals. The same is true for the F 2*s* and 1*s* orbitals. These nonbonding orbitals retain the characteristics of the F atomic orbitals from which they are formed. The MO diagram of HF is shown in Figure 9-9.

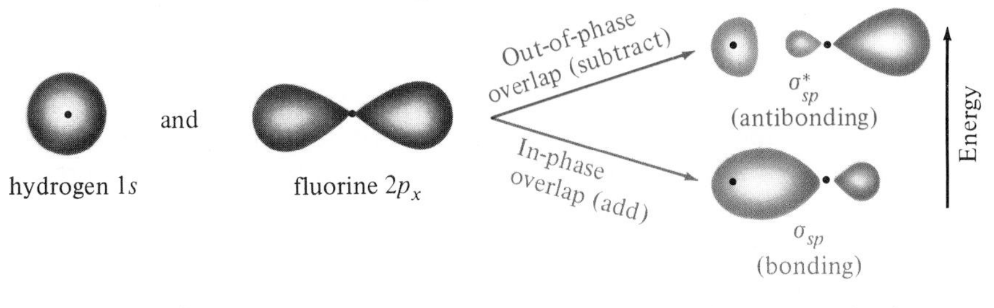

Figure 9-8
Formation of σ_{sp} and $\sigma^\star_{sp}$ molecular orbitals in HF by overlap of the 1*s* orbital of H with a 2*p* orbital of F.

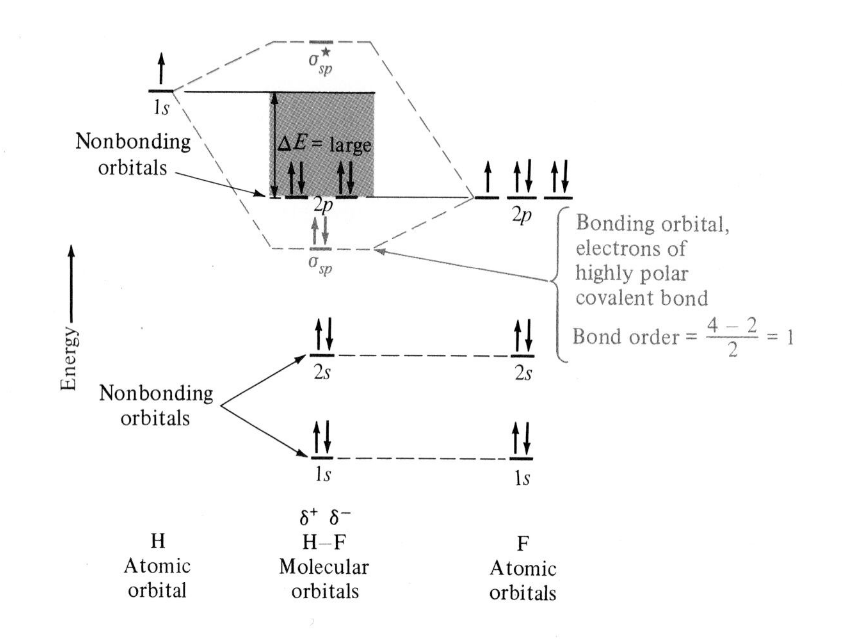

Figure 9-9
MO energy-level diagram for hydrogen fluoride, HF, a very polar molecule diagram ($\mu = 1.9$ D). Because of the large electronegativity difference, ΔE is large.

9-6 Delocalization and the Shapes of Molecular Orbitals

In Section 7-9 we described resonance formulas for molecules and polyatomic ions. Resonance is said to exist when two or more equivalent Lewis dot formulas can be drawn for the same species and a single such formula does not account for the properties of a substance. In molecular orbital terminology, a more appropriate description involves *delocalization* of electrons. The shapes of molecular orbitals for species in which electron delocalization occurs can be determined by combining all the contributing atomic orbitals.

The Carbonate Ion, CO_3^{2-}

The average carbon–oxygen bond order in the CO_3^{2-} ion is $1\frac{1}{3}$.

Consider the trigonal planar carbonate ion, CO_3^{2-}, as an example. All the carbon–oxygen bonds in the ion have equal bond length and equal bond energy, intermediate between those of typical C—O and C=O bonds. Valence bond theory describes the ion in terms of three contributing resonance structures (Figure 9-10a). No one of the three resonance forms adequately describes the bonding.

According to valence bond theory, the C atom is described as sp^2-hybridized, and it forms one sigma bond with each of the three O atoms. This leaves one unhybridized $2p$ atomic orbital on the C atom, say the $2p_z$ orbital. This orbital is capable of overlapping and mixing with the $2p_z$ orbital of any of the three O atoms. The sharing of two electrons in the resulting localized pi orbital would form a pi bond. Thus, three equivalent resonance structures can be drawn in valence bond terms (Figure 9-10b). We must emphasize that there is *no evidence* for the existence of these separate resonance structures.

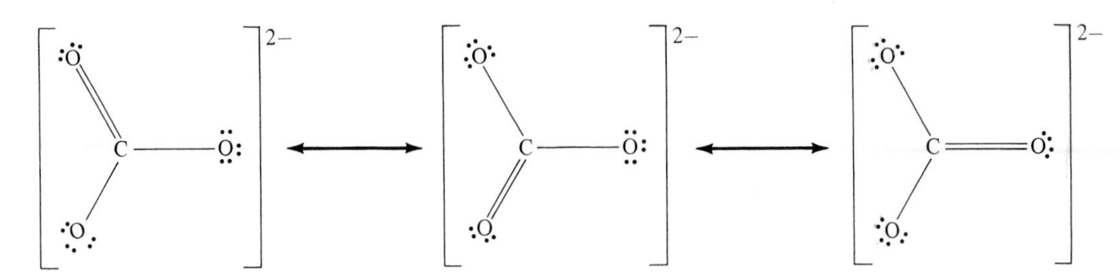

(a) Lewis formulas for valence bond resonance structures

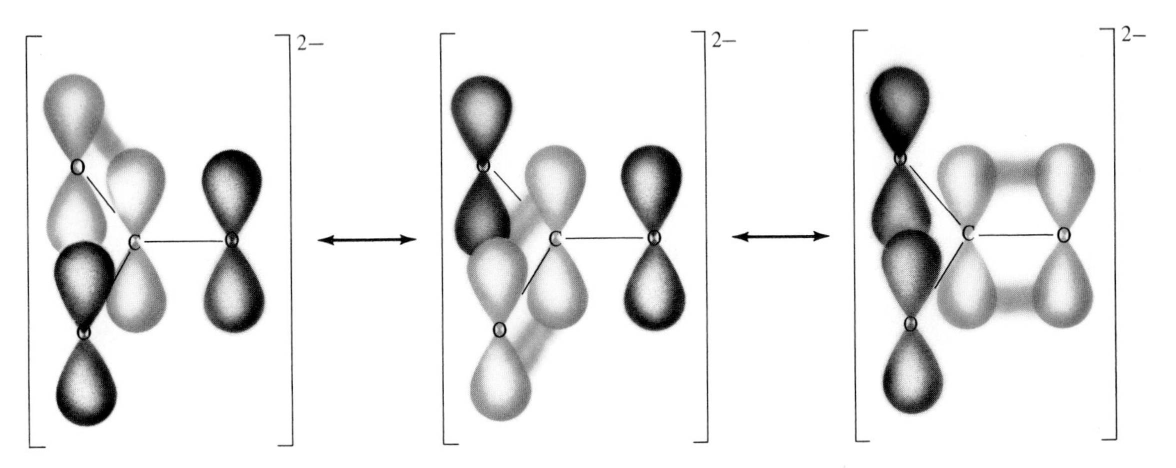

(b) *p*-orbital overlap in valence bond resonance structures

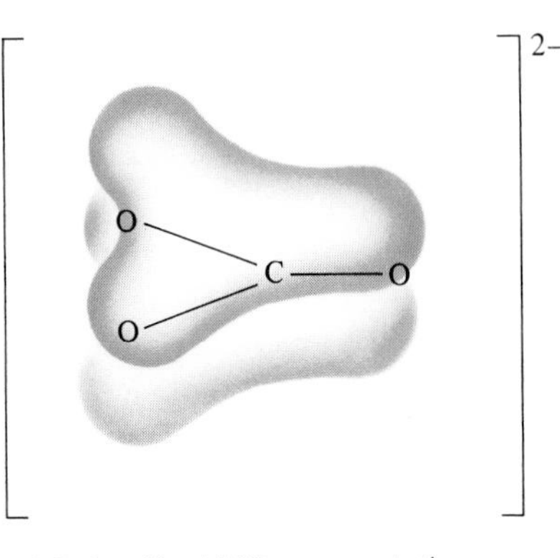

(c) Delocalized MO representation

Figure 9-10
Alternative representations of the bonding in the carbonate ion, CO_3^{2-}. (a) Lewis formulas of the three valence bond resonance structures. (b) Representation of the p orbital overlap in the valence bond resonance structures. In each resonance form, the p orbitals on two atoms would overlap to form the π components of the hypothetical double bonds. Each O atom has two additional sp^2 orbitals (not shown) in the plane of the nuclei. Each of these additional sp^2 orbitals contains an oxygen lone pair. (c) In the MO description, the electrons in the π-bonded region are spread out, or *delocalized,* over all four atoms of the CO_3^{2-} ion. This MO description is more consistent with the experimental observation of equal bond lengths and energies than are the valence bond pictures in (a) and (b).

The MO description of the pi bonding involves the simultaneous overlap and mixing of the carbon $2p_z$ orbital with the $2p_z$ orbitals of all three oxygen atoms. This forms a delocalized bonding pi molecular orbital system extending above and below the plane of the sigma system, as well as an antibonding pi orbital system. Electrons are said to occupy the entire set of bonding pi MOs, as depicted in Figure 9-10c. The shape is obtained by averaging the three contributing valence bond resonance structures. The bonding in such species as nitrate ion, NO_3^-, and sulfur dioxide, SO_2, can be described in a similar manner.

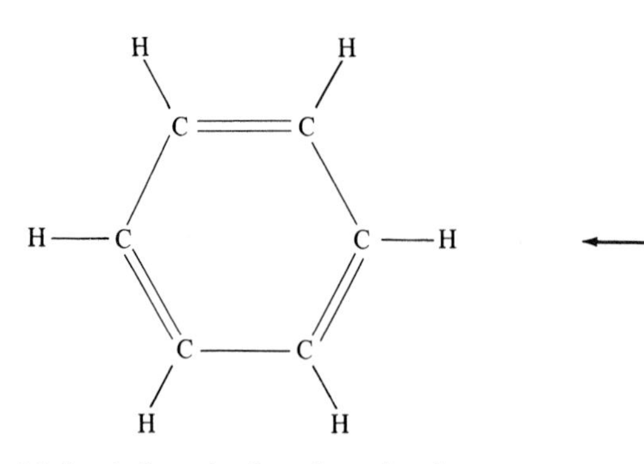

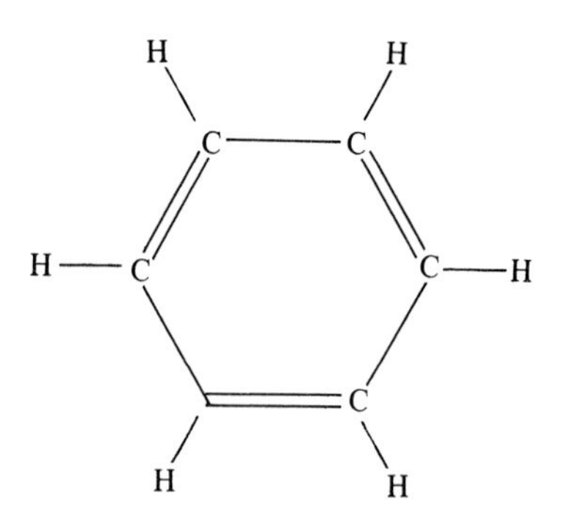

(a) Lewis formulas for valence bond resonance structures

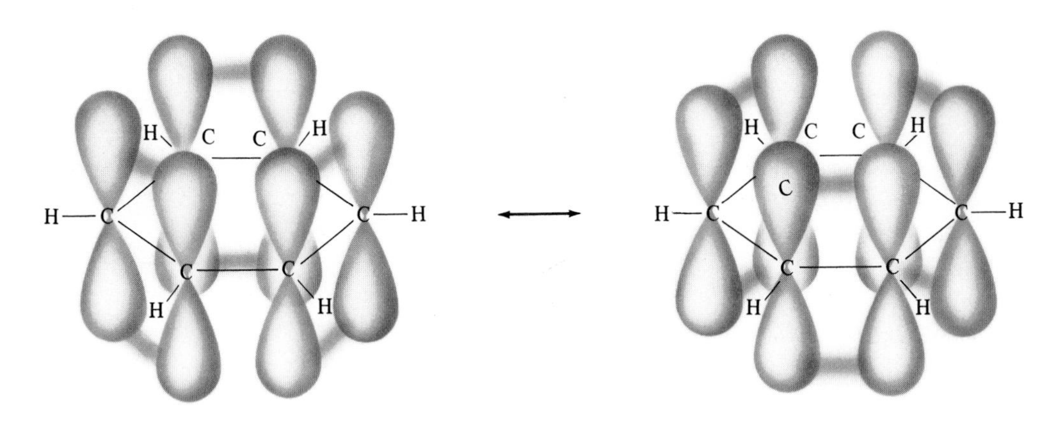

(b) p-orbital overlap in valence bond resonance structures

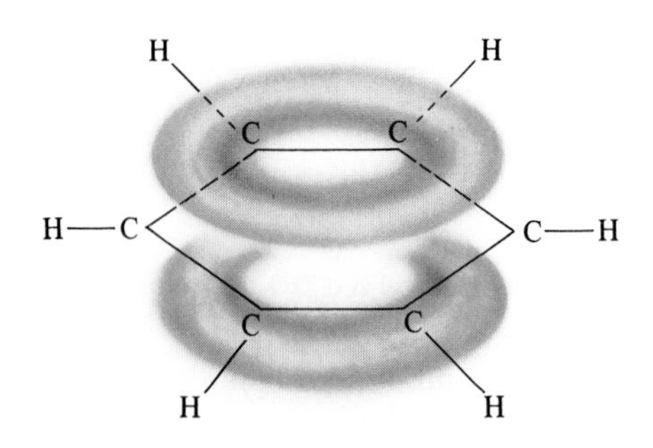

(c) Delocalized MO representation

Figure 9-11
Representations of the bonding in the benzene molecule, C_6H_6. (a) Lewis formulas of the two valence bond resonance structures. (b) The six p orbitals of the benzene ring, shown overlapping to form the (hypothetical) double bonds of the two resonance forms of valence bond theory. (c) In the MO description, the six electrons in the pi-bonded region are *delocalized,* occupying an extended pi-bonding region above and below the plane of the six C atoms.

The Benzene Molecule, C_6H_6

Now let us consider the benzene molecule, C_6H_6, whose two valence bond resonance forms are shown in Figure 9-11a. The valence bond description involves sp^2 hybridization at each C atom. Each C atom is at the center of a trigonal plane, and the entire molecule is known to be planar. There are sigma bonds from each C atom to the two adjacent C atoms and to one H atom. This leaves one unhybridized $2p_z$ orbital on each C atom and one remaining valence electron for each. According to valence bond theory, adjacent pairs of $2p_z$ orbitals and the six remaining electrons occupy the regions of overlap to form a total of three pi bonds in either of the two ways shown in Figure 9-11b.

There is no evidence for the existence of either of these forms of benzene. The MO description of benzene is far superior to the VB description.

Experimental studies of the C_6H_6 structure prove that it does *not* contain alternating single and double carbon–carbon bonds. The usual C—C single bond length is 1.54 Å, and the usual C=C double bond length is 1.34 Å. All six of the carbon–carbon bonds in benzene are the same length, 1.39 Å, intermediate between those of single and double bonds.

This is well explained by the MO theory, which predicts that the six $2p_z$ orbitals of the C atoms overlap and mix to form three pi bonding and three pi antibonding molecular orbitals. For instance, the most strongly bonding pi molecular orbital in the benzene pi–MO system is that in Figure 9-11c. The six pi electrons occupy three bonding MOs of this extended (delocalized) system. Thus, they are distributed throughout the molecule as a whole, above and below the plane of the sigma-bonded framework. This results in identical character for all carbon–carbon bonds in benzene. Each carbon–carbon bond has a bond order of $1\frac{1}{2}$. The MO representation of the extended pi system is the same as that obtained by averaging the two contributing valence bond resonance structures.

Key Terms

Antibonding orbital A molecular orbital higher in energy than any of the atomic orbitals from which it is derived; when populated with electrons, lends instability to a molecule or ion. Denoted with a star (★) superscript on its symbol.

Bond energy The amount of energy necessary to break one mole of bonds of a given kind (in the gas phase).

Bond order Half the number of electrons in bonding orbitals minus half the number of electrons in antibonding orbitals.

Bonding orbital A molecular orbital lower in energy than any of the atomic orbitals from which it is derived; when populated with electrons, lends stability to a molecule or ion.

Delocalization The formation of a set of molecular orbitals that extend over more than two atoms; important in species that valence bond theory describes in terms of *resonance*.

Heteronuclear Consisting of different elements.

Homonuclear Consisting of only one element.

Molecular orbital (MO) An orbital resulting from overlap and mixing of atomic orbitals on different atoms. An MO belongs to the molecule as a whole.

Molecular orbital theory A theory of chemical bonding based upon the postulated existence of molecular orbitals.

Nodal plane (node) A region in which the probability of finding an electron is zero.

Nonbonding orbital A molecular orbital derived only from an atomic orbital of one atom; lends neither stability nor instability to a molecule or ion when populated with electrons.

Pi (π) bond A bond resulting from electron occupation of a pi molecular orbital.

Pi (π) orbital A molecular orbital resulting from side-on overlap of atomic orbitals.

Sigma (σ) bond A bond resulting from electron occupation of a sigma molecular orbital.

Sigma (σ) orbital A molecular orbital resulting from head-on overlap of two atomic orbitals.

Exercises

MO Theory—General Concepts

1. Describe the main differences between the valence bond theory and the molecular orbital theory.
2. What is a molecular orbital? What two types of information can be obtained from molecular orbital calculations? How do we use such information to describe the bonding within a molecule?
3. What is the relationship between the maximum number of electrons that can be accommodated by a set of molecular orbitals and the maximum number that can be accommodated by the atomic orbitals from which the MOs are formed? What is the maximum number of electrons that one MO can hold?
4. Answer Exercise 3 after replacing "molecular orbitals" with "hybridized atomic orbitals."
5. What differences and similarities exist among (a) atomic orbitals, (b) localized hybridized atomic orbitals according to valence bond theory, and (c) molecular orbitals?
6. What is the relationship between the energy of a bonding molecular orbital and the energies of the original atomic orbitals? What is the relationship between the energy of an antibonding molecular orbital and the energies of the original atomic orbitals?
7. Draw an energy-level diagram for the formation of molecular orbitals from two atomic orbitals of equal energy. Identify the bonding and antibonding molecular orbitals.
8. In terms of overlapping AO's, describe three ways in which a σ MO can be formed.
9. Complete the following energy-level diagram for the formation of molecular orbitals:

$\overline{2p_x}\ \overline{2p_y}\ \overline{2p_z}$		$\overline{2p_x}\ \overline{2p_y}\ \overline{2p_z}$
atomic orbitals	*molecular orbitals*	*atomic orbitals*

Assume the bonding axis to be the x axis. Identify the bonding and antibonding molecular orbitals that are formed.

10. State the three rules for placing electrons in molecular orbitals.
11. What is meant by the term "bond order"? How is the value of the bond order calculated?
12. Compare and illustrate the differences between (a) atomic orbitals and molecular orbitals, (b) bonding and antibonding molecular orbitals, (c) σ bonds and π bonds, and (d) localized and delocalized molecular orbitals.
13. How are the electron occupancies of bonding, nonbonding, and antibonding orbitals related to the stability of a molecule or ion?
14. From memory, draw the energy-level diagram of molecular orbitals produced by the overlap of orbitals of two identical atoms from the second period. (Show the π_{2p_y} and π_{2p_z} orbitals below the σ_{2p} in energy.)
15. Is it possible for a molecule or complex ion in its ground state to have a negative bond order? Why?

Homonuclear Diatomic Species

16. What do we mean when we say that a molecule or ion is (a) *homonuclear,* (b) *heteronuclear,* or (c) *diatomic*?
17. Use the appropriate molecular orbital energy diagram to write the electron configuration for each of the following molecules or ions, calculate the bond order of each, and predict which would exist. (a) H_2^+, (b) H_2, (c) H_2^-, (d) H_2^{2-}
18. Repeat Exercise 17 for (a) He_2^+ and (b) He_2.
19. Repeat Exercise 17 for (a) N_2, (b) Ne_2, and (c) C_2.
20. Repeat Exercise 17 for (a) Li_2, (b) Li_2^+, and (c) F_2.
21. Determine the electron configurations of the following molecules and ions: (a) Be_2, Be_2^+, Be_2^-; (b) B_2, B_2^+, B_2^-; (c) C_2^+.
22. What is the bond order of each of the species in Exercise 21?
23. Which of the species in Exercise 21 are diamagnetic (D) and which are paramagnetic (P)?
24. Apply MO theory to predict relative stabilities of the species in Exercise 21. Comment on the validity of these predictions. What else *must* be considered in addition to electron occupancy of MOs?

*25. Which homonuclear diatomic molecules or ions of the second period have the following electron distributions in MOs? In other words, identify X in each of the following cases.
(a) X_2 $\sigma_{1s}{}^2\ \sigma^{\star}_{1s}{}^2\ \sigma_{2s}{}^2\ \sigma^{\star}_{2s}{}^2\ \pi_{2p_y}{}^2\ \pi_{2p_z}{}^2$
(b) X_2^+ $\sigma_{1s}{}^2\ \sigma^{\star}_{1s}{}^2\ \sigma_{2s}{}^2\ \sigma^{\star}_{2s}{}^2\ \sigma_{2p}{}^2\ \pi_{2p_y}{}^2\ \pi_{2p_z}{}^2\ \pi^{\star}_{2p_y}{}^2\ \pi^{\star}_{2p_z}{}^1$
(c) X_2^- $\sigma_{1s}{}^2\ \sigma^{\star}_{1s}{}^2\ \sigma_{2s}{}^2\ \sigma^{\star}_{2s}{}^2\ \pi_{2p_y}{}^2\ \pi_{2p_z}{}^2\ \sigma_{2p}{}^2\ \pi^{\star}_{2p_y}{}^1$

26. What is the bond order of each of the species in Exercise 25?
27. (a) Give the MO designations for O_2, O_2^-, O_2^{2-}, O_2^+, and O_2^{2+}. (b) Give the bond order in each case. (c) Match these species with the following observed bond lengths: 1.04 Å, 1.12 Å, 1.21 Å, 1.33 Å, and 1.49 Å.
28. (a) Give the MO designations for N_2, N_2^-, and N_2^+. (b) Give the bond order in each case. (c) Rank these three species by increasing predicted bond length.
29. Assuming that the σ_{2p} MO is lower in energy than the π_{2p_y} and π_{2p_z} MOs for the following species, write out electronic configurations for all of them. (a) F_2, F_2^+, F_2^-; (b) Ne_2, Ne_2^+.
30. (a) What is the bond order of each of the species in Exercise 29? (b) Are they diamagnetic or paramagnetic? (c) What would MO theory predict about the stabilities of these species?

Heteronuclear Diatomic Species

The following is the molecular orbital energy-level diagram for a heteronuclear diatomic molecule, XY, in which both X and Y are from Period 2 and Y is the more electronegative element. This diagram may be useful in answering questions in this section.

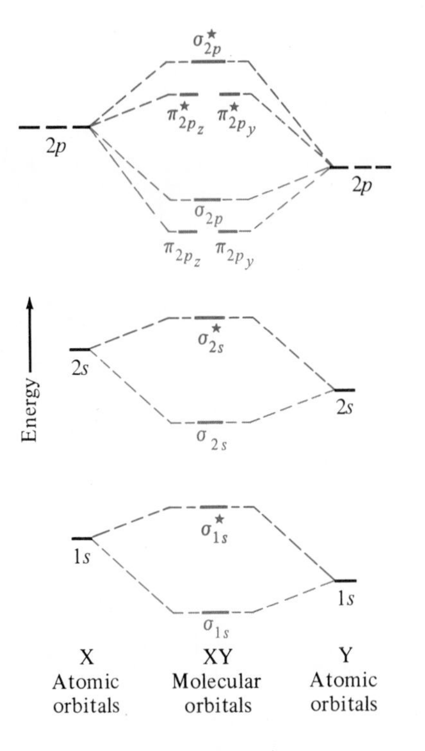

31. Use the preceding diagram to fill in an MO diagram for nitrogen monoxide, NO. What is the bond order of NO? Is it paramagnetic? How would you assess its stability?
32. Assuming that the preceding MO diagram is valid for CN, CN^+, and CN^-, write the MO descriptions for these species. Which would be most stable? Why?
33. For each of the two species OF and OF^+: (a) Draw MO energy-level diagrams. (b) Write out electronic configurations. (c) Determine bond orders and predict relative stabilities. (d) Predict diamagnetism or paramagnetism. Refer to the preceding diagram. Assume that the σ_{2p} MO is lower in energy than the π_{2p_y} and π_{2p_z} MOs.
34. For each of the two species NF and NF^-: (a) Draw MO energy-level diagrams. (b) Write out electronic configurations. (c) Determine bond orders and predict relative stabilities. (d) Predict diamagnetism or paramagnetism. Refer to the preceding diagram. Assume that the σ_{2p} MO is lower in energy than the π_{2p_y} and π_{2p_z} MOs.
35. Considering the shapes of MO energy-level diagrams for nonpolar covalent and polar covalent molecules, what would you predict about MO diagrams, and therefore about overlap of atomic orbitals, for ionic compounds?
36. To increase the strength of the bonding in the hypothetical compound BO, would you add or subtract an electron? Explain your answer with the aid of an MO electron structure.

Delocalization

37. Draw Lewis dot formulas depicting the resonance structures of the following species from the valence bond point of view, and then draw MOs for the delocalized π systems. (a) SO_2, sulfur dioxide; (b) HCO_3^-, hydrogen carbonate ion (H is bonded to O); (c) NO_3^-, nitrate ion.
38. Draw Lewis dot formulas depicting the resonance structures of the following species from the valence bond point of view, and then draw MOs for the delocalized π systems: (a) SO_3, sulfur trioxide; (b) O_3, ozone; (c) HCO_2^-, formate ion (H is bonded to C).

Mixed Exercises

39. Draw and label the complete MO energy-level diagrams for the following species. For each, determine the bond order, predict the stability of the species, and predict whether the species will be paramagnetic. (a) He_2^+; (b) CN; (c) HeH^+.
40. Draw and label the complete MO energy-level diagrams for the following species. For each, determine the bond order, predict the stability of the species, and predict whether the species will be paramagnetic. (a) O_2^{2+}; (b) HO^-; (c) HCl.
41. Which of these species would you expect to be paramagnetic? (a) He_2^+; (b) NO; (c) NO^+; (d) N_2^{2+}; (e) CO^{2-}; (f) F_2^+.
42. Rationalize the following observations in terms of the stabilities of σ and π bonds: (a) The most common form of nitrogen is N_2, whereas the most common form of phosphorus is P_4 (see the structure in Figure 2-3). (b) The most common forms of oxygen are O_2 and (less common) O_3, whereas the most common form of sulfur is S_8.
43. (a) Give the MO designations for (1) Be_2, (2) F_2, (3) HeH^+, (4) OF, (5) Ne_2. (b) Which of these species is/are unlikely to exist? Explain.

10 Reactions in Aqueous Solutions I: Acids, Bases, and Salts

Outline

- **10-1** The Arrhenius Theory
- **10-2** The Hydrated Hydrogen Ion
- **10-3** The Brønsted–Lowry Theory
- **10-4** Properties of Aqueous Solutions of Acids and Bases
- **10-5** Strengths of Binary Acids
- **10-6** Reactions of Acids and Bases
- **10-7** Acidic Salts and Basic Salts
- **10-8** Strengths of Ternary Acids and Amphoterism
- **10-9** The Preparation of Acids
- **10-10** The Lewis Theory

Objectives

As you study this chapter, you should learn

- ☐ To understand the Arrhenius theory
- ☐ About hydrated hydrogen ions
- ☐ To understand the Brønsted–Lowry theory
- ☐ The properties of aqueous solutions of acids
- ☐ The properties of aqueous solutions of bases
- ☐ How to predict the strengths of binary acids
- ☐ About acid–base reactions
- ☐ About acidic and basic salts
- ☐ To understand the strengths of ternary acids
- ☐ About amphoterism
- ☐ How to prepare acids
- ☐ To understand the Lewis theory

Many common household liquids are acidic, including soft drinks, vinegar, and fruit juices. Most cleaning materials are basic.

In highly developed societies, acids, bases, and salts are indispensable compounds. Table 4-10 lists the 18 such compounds that were included in the top 50 chemicals produced in the United States in 1990. The production of H_2SO_4 (number 1) was more than twice as great as the production of NH_3 (number 3). Sixty-five percent of the H_2SO_4 is used in the production of fertilizers.

Many acids, bases, and salts occur in nature and serve a wide variety of purposes. For instance, your "digestive juice" contains approximately 0.10 mole of hydrochloric acid per liter. The liquid in your automobile battery is approximately 40% H_2SO_4 by mass. Sodium hydroxide, a base, is used in the manufacture of soaps, paper, and many other chemicals. "Drāno" is solid NaOH that contains some aluminum turnings. Sodium chloride is used to season food and as a food preservative. Calcium chloride is used to melt ice on highways and in the emergency treatment of cardiac arrest. Several ammonium salts are used as fertilizers.

You will encounter many of these in your laboratory work.

Many organic acids (carboxylic acids) and their derivatives occur in nature. Amino acids are the building blocks of proteins, which are important materials in the bodies of animals, including humans. Amino acids are carboxylic acids that also contain basic groups derived from ammonia. The pleasant odors and flavors of ripe fruit are due in large part to the presence of esters (Section 31-13), which are formed from the acids that are present in unripe fruit.

10-1 The Arrhenius Theory

In 1680 Robert Boyle noted that acids (1) dissolve many substances, (2) change the colors of some natural dyes (indicators), and (3) lose their characteristic properties when mixed with alkalis (bases). By 1814 J. Gay-Lussac concluded that acids *neutralize* bases and that the two classes of substances can be defined only in terms of their reactions with each other.

This is an extremely important idea.

In 1884 Svante Arrhenius presented this theory of electrolytic dissociation, which resulted in the Arrhenius theory of acid–base reactions. In his view, an **acid** is a substance that contains hydrogen and produces H^+ in aqueous solution. A **base** is a substance that contains the OH group and produces hydroxide ions, OH^-, in aqueous solution. **Neutralization** is defined as the combination of H^+ ions with OH^- ions to form H_2O molecules:

We now know that all ions are hydrated in aqueous solution.

$$H^+(aq) + OH^+(aq) \longrightarrow H_2O(\ell) \qquad \text{(neutralization)}$$

The Arrhenius theory of acid–base behavior satisfactorily explained reactions of **protonic acids** (those containing acidic hydrogen atoms) with metal hydroxides (hydroxy bases). It was a significant contribution to chemical thought and theory in the latter part of the nineteenth century. We used this theory in introducing acids and bases and discussing some of their reactions. The Arrhenius model of acids and bases, although limited in scope, led to the development of other general theories of acid–base behavior. They will be considered in later sections.

You may wish to review Sections 4-2, 4-5, 6-7, and 6-8.

10-2 The Hydrated Hydrogen Ion

The most common isotope of hydrogen, ^1_1H, has no neutrons. Thus, $^1_1H^+$ is a bare proton.

Although Arrhenius described H^+ ions in water as bare ions (protons), we now know that they are hydrated in aqueous solution and exist as $H^+(H_2O)_n$ in which n is some small integer. This is due to the attraction of the H^+ ions, or protons, for the oxygen end ($\delta-$) of water molecules. While we do not know the extent of hydration of H^+ in most solutions, we often represent the hydrated hydrogen ion as the hydronium ion, H_3O^+, or $H^+(H_2O)_n$ in which $n = 1$. The hydrated hydrogen ion is the species that gives aqueous solutions of acids their characteristic acidic properties. Whether we use the designation $H^+(aq)$ or H_3O^+, we are always referring to the hydrated hydrogen ion.

$$H^+ + :\underset{\substack{\cdot\cdot \\ H}}{\overset{\cdot\cdot}{O}}:H \longrightarrow H:\underset{\substack{\cdot\cdot \\ H}}{\overset{\cdot\cdot}{O}}:H^+$$

10-3 The Brønsted–Lowry Theory

The Brønsted–Lowry theory is especially useful for reactions in aqueous solutions. It is widely used in medicine and in the biological sciences.

In 1923 J. N. Brønsted and T. M. Lowry independently presented logical extensions of the Arrhenius theory. Brønsted's contribution was more thorough than Lowry's, and the result is known as the **Brønsted theory** or the **Brønsted–Lowry theory**.

An **acid** is defined as a *proton donor*, H^+, and a **base** is defined as a *proton acceptor*. The definitions are sufficiently broad that any hydrogen-containing molecule or ion capable of releasing a proton, H^+, is an acid, while any molecule or ion that can accept a proton is a base. According to the Brønsted–Lowry theory,

> An acid–base reaction is the transfer of a proton from an acid to a base.

Thus, the complete ionization of hydrogen chloride, HCl, a *strong* acid, in water is an acid–base reaction in which water acts as a base, a proton acceptor.

Step 1:	$HCl(aq) \longrightarrow H^+(aq) + Cl^-(aq)$	(Arrhenius description)
Step 2:	$H^+(aq) + H_2O(\ell) \longrightarrow H_3O^+$	
Overall:	$HCl(aq) + H_2O(\ell) \longrightarrow H_3O^+ + Cl^-(aq)$	(Brønsted–Lowry description)

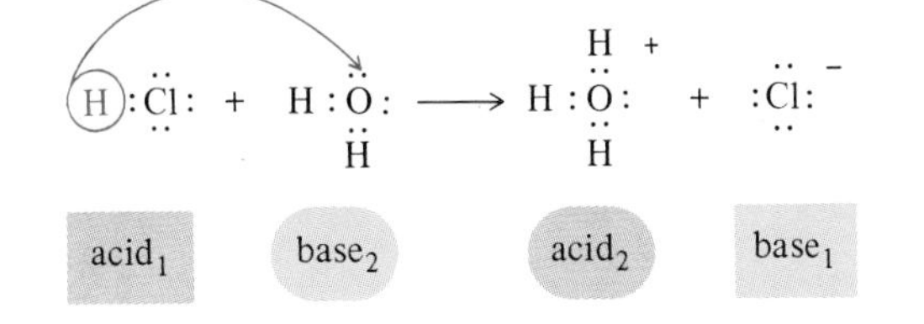

We use red to indicate acids and blue to indicate bases. We use rectangles to indicate one conjugate acid–base pair and ovals to indicate the other pair.

The ionization of hydrogen fluoride, a *weak* acid, is similar, but it occurs to only a slight extent. So we use double arrows to indicate that it is reversible.

Various measurements (electrical conductivity, freezing point depression, and so on) indicate that HF is only *slightly* ionized in water.

$$HF(g) + H_2O(\ell) \rightleftharpoons H_3O^+ + F^-(aq)$$

acid$_1$ base$_2$ acid$_2$ base$_1$

$$H\!:\!\ddot{\underset{..}{F}}\!: + H:\ddot{\underset{..}{O}}:H \rightleftharpoons [H:\ddot{\underset{..}{O}}:H\,H]^+ + [:\ddot{\underset{..}{F}}:]^-$$

We can describe Brønsted–Lowry acid–base reactions in terms of **conjugate acid–base pairs**. These are species that differ by a proton. In the preceding equation, HF (acid$_1$) and F^- (base$_1$) are one conjugate acid–base pair, and H_2O (base$_2$) and H_3O^+ (acid$_2$) are the other pair. The members of each conjugate pair are designated by the same numerical subscript. In the forward reaction, HF and H_2O act as acid and base, respectively. In the reverse reaction, H_3O^+ acts as the acid, or proton donor, and F^- acts as the base, or proton acceptor.

It makes no difference which conjugate acid–base pair, HF and F^- or H_3O^+ and H_2O, is assigned the subscripts 1 and 2.

When HF is dissolved in water, the HF molecules give up some H^+ ions that can be accepted by either of two bases, F^- and H_2O. The fact that HF is only slightly ionized tells us that F^- is a stronger base than H_2O. When HCl (a *strong* acid) is dissolved in water, the HCl molecules give up H^+

ions that can be accepted by either of two bases, Cl^- and H_2O. The fact that HCl is completely ionized in dilute aqueous solution tells us that Cl^- is a weaker base than H_2O. Thus, the stronger acid, HCl, has the weaker conjugate base, Cl^-. The weaker acid, HF, has the stronger conjugate base, F^-. We can generalize:

> The weaker an acid is, the greater is the base strength of its conjugate base. Likewise, the weaker a base is, the stronger is its conjugate acid.

"Strong" and "weak," like many other adjectives, are used in a relative sense. We do not mean to imply that the fluoride ion, F^-, is a strong base compared with species such as the hydroxide ion, OH^-. We mean that *relative to the anions of strong acids,* which are very weak bases, F^- is a much stronger base.

Ammonia is very *soluble* in water (~15 mol/L at 25°C). In 0.10 *M* solution, NH_3 is only 1.3% ionized and 98.7% nonionized.

Ammonia acts as a weak Brønsted–Lowry base, and water acts as an acid in the ionization of aqueous ammonia:

$$\underset{\text{base}_1}{NH_3(aq)} + \underset{\text{acid}_2}{H_2O(\ell)} \rightleftharpoons \underset{\text{acid}_1}{NH_4^+(aq)} + \underset{\text{base}_2}{OH^-(aq)}$$

$$H:\overset{H}{\underset{H}{\ddot{N}}}: \;+\; (H):\underset{H}{\ddot{\underset{..}{O}}}: \;\rightleftharpoons\; \left[H:\overset{H}{\underset{H}{\ddot{N}}}:H\right]^+ \;+\; \left[:\underset{H}{\ddot{\underset{..}{O}}}:\right]^-$$

As we see in the reverse reaction, ammonium ion, NH_4^+, is the conjugate acid of NH_3. The hydroxide ion, OH^-, is the conjugate base of water. In three dimensions, the molecular structures are

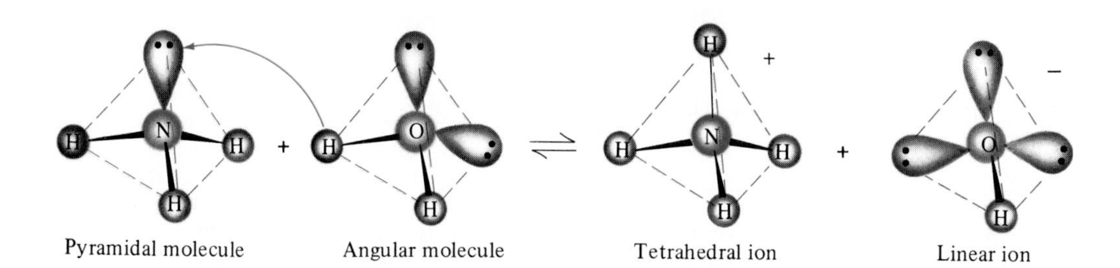

Pyramidal molecule — Angular molecule — Tetrahedral ion — Linear ion

Water acts as an acid (H^+ donor) in its reaction with NH_3, whereas it acts as a base (H^+ acceptor) in its reaction with HCl and HF.

> Whether water acts as an acid or as a base depends on the other species present.

Careful measurements show that pure water ionizes ever so slightly to produce equal numbers of hydrated hydrogen ions and hydroxide ions:

$$H_2O + H_2O \rightleftharpoons H_3O^+ + OH^-$$

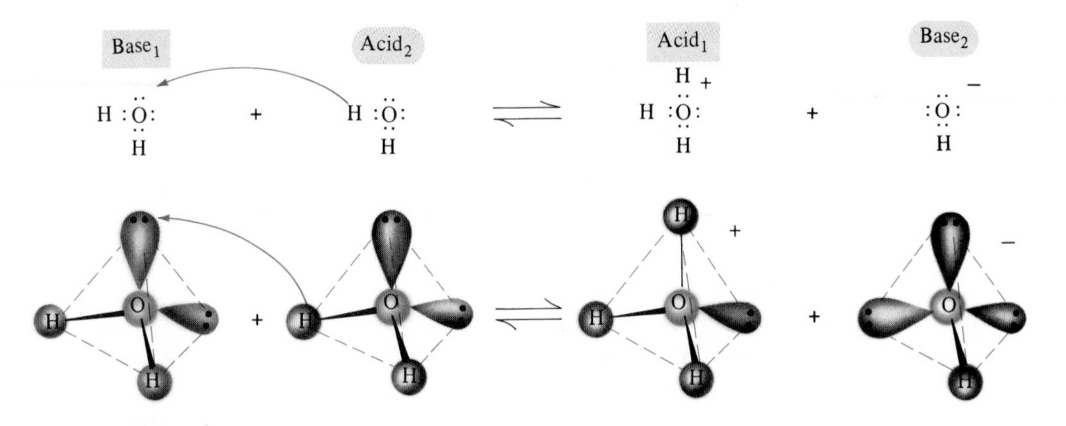

In simplified notation,

$$H_2O \rightleftharpoons H^+(aq) + OH^-(aq)$$

The **autoionization** (self-ionization) of water is an acid–base reaction according to the Brønsted–Lowry theory. One H_2O molecule (the acid) donates a proton to another H_2O molecule (the base). The H_2O molecule that donates a proton becomes an OH^- ion, the conjugate base of water. The H_2O molecule that accepts a proton becomes an H_3O^+ ion. Examination of the reverse reaction (right to left) shows that H_3O^+ (an acid) donates a proton to OH^- (a base) to form two H_2O molecules. One H_2O molecule behaves as an acid while the other acts as a base in the autoionization of water. Water is said to be **amphiprotic**; that is, H_2O molecules can accept and donate protons.

The prefix "amphi-" means "of both kinds." "Amphiprotism" refers to amphoterism by accepting and donating a proton in different reactions.

As we saw in Section 4-5, H_3O^+ and OH^- ions combine to form nonionized water molecules when strong acids and strong soluble bases react to form soluble salts and water. The reverse reaction, the autoionization of water, occurs only slightly, as expected.

10-4 Properties of Aqueous Solutions of Acids and Bases

Aqueous solutions of most protonic acids exhibit certain properties, which are properties of hydrated hydrogen ions.

1. They have a sour taste. Pickles are usually preserved in vinegar, a 5% solution of acetic acid. Many pickled condiments contain large amounts of sugar so that the taste of acetic acid is not so pronounced. Lemons contain citric acid, which is responsible for their characteristic sour taste.
2. They change the colors of many indicators (highly colored dyes). Acids turn blue litmus red, and bromthymol blue changes from blue to yellow in acids.
3. Nonoxidizing acids react with metals above hydrogen in the activity series (Section 4-6, part 2) to liberate hydrogen gas, H_2.

The indicator bromthymol blue is yellow in acidic solution and blue in basic solution.

4. They react with (neutralize) metal oxides and metal hydroxides to form salts and water (Section 4-5).
5. They react with salts of weaker or more volatile acids to form the weaker or more volatile acid and a new salt.
6. Aqueous solutions of protonic acids conduct an electric current because they are wholly or partly ionized.

Aqueous solutions of most bases also exhibit certain properties. These are due to the hydrated hydroxide ions that are present in aqueous solutions of bases:

1. They have a bitter taste.
2. They have a slippery feeling. Soaps are common examples; they are mildly basic. A solution of household bleach feels very slippery because it is strongly basic.
3. They change the colors of many indicators: litmus changes from red to blue, and bromthymol blue changes from yellow to blue, in bases.
4. They react with (neutralize) protonic acids to form salts and water.
5. Their aqueous solutions conduct an electric current because they are ionized or dissociated.

10-5 Strengths of Binary Acids

A *weak* acid may be very reactive. For example, HF dissolves sand and glass. The equation for its reaction with sand is

$$SiO_2(s) + 4HF(g) \longrightarrow SiF_4(g) + 2H_2O(\ell)$$

The reaction with glass and other silicates is similar. These reactions are not related to acid strength.

The ease of ionization of binary protonic acids depends on both (1) the ease of breaking H—X bonds and (2) the stability of the resulting ions in solution. Let us consider the relative strengths of the Group VIIA hydrohalic acids. Hydrogen fluoride ionizes only slightly in dilute aqueous solutions:

$$HF(aq) \xrightleftharpoons{H_2O} H^+(aq) + F^-(aq)$$

However, HCl, HBr, and HI ionize completely or nearly completely in dilute aqueous solutions because the H—X bonds are much weaker.

$$HX(aq) \xrightarrow{H_2O} H^+(aq) + X^-(aq) \qquad X = Cl, Br, I$$

The order of *bond strengths* for the hydrogen halides is

$$HF >> HCl > HBr > HI$$

Bond strength is shown by the bond energies introduced in Chapter 7 and tabulated in Section 15-9. The strength of the H—F bond is due largely to the very small size of the F atom.

To understand why HF is so much weaker an acid than the other hydrogen halides, let us consider the following factors:

1. In HF the electronegativity difference is 1.9, compared with 0.9 in HCl, 0.7 in HBr, and 0.4 in HI. We expect the very polar H—F bond in HF to ionize easily. The fact that HF is the *weakest* of these acids suggests that this effect must be of minor importance.
2. The bond strength is considerably greater in HF than in the other three molecules. This tells us that the H—F bond is harder to break than the H—Cl, H—Br, and H—I bonds.
3. The small, highly charged F^- ion, formed when HF ionizes, causes increased ordering of the water molecules. This increase is unfavorable to the process of ionization.

Table 10-1
Relative Strengths of Conjugate Acid–Base Pairs

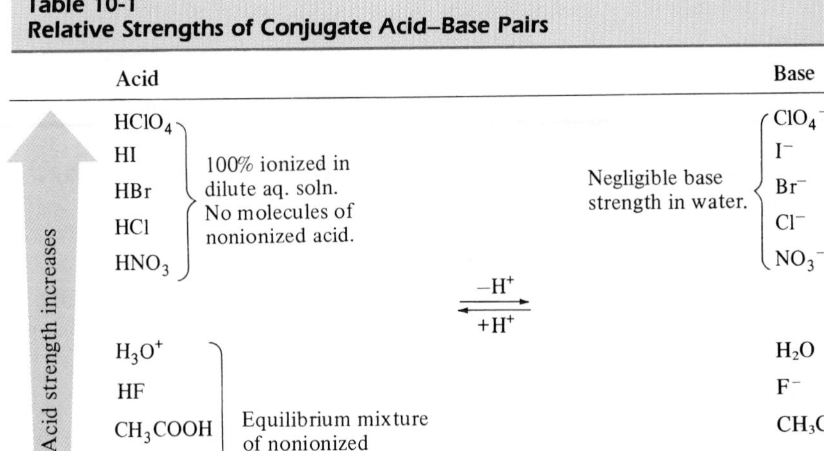

Acid				Base
$HClO_4$	100% ionized in dilute aq. soln. No molecules of nonionized acid.	$\underset{+H^+}{\overset{-H^+}{\rightleftharpoons}}$	Negligible base strength in water.	ClO_4^-
HI				I^-
HBr				Br^-
HCl				Cl^-
HNO_3				NO_3^-
H_3O^+	Equilibrium mixture of nonionized molecules of acid, conjugate base, and H^+(aq).			H_2O
HF				F^-
CH_3COOH				CH_3COO^-
HCN				CN^-
NH_4^+				NH_3
H_2O				OH^-
NH_3			Reacts completely with H_2O; cannot exist in aqueous solution.	NH_2^-

Acid strength increases (upward); Base strength increases (downward).

The net result of all factors is that the order of *acid strengths* is

$$HF << HCl < HBr < HI$$

In dilute aqueous solutions, hydrochloric, hydrobromic, and hydroiodic acids are completely ionized and all show the same apparent acid strength. Water is sufficiently basic that it does not distinguish among the acid strengths of HCl, HBr, and HI, and therefore it is referred to as a **leveling solvent**. It is not possible to determine the order of the strengths of these three acids in water because they are so nearly completely ionized.

When these compounds are dissolved in anhydrous acetic acid or other solvents less basic than water, however, significant differences in their acid strengths are observed.

The strengths of *ternary* acids are discussed in Section 10-8.

$$HCl < HBr < HI \qquad \text{(strongest acid)}$$

One more observation is appropriate to describe the leveling effect:

> The hydrated hydrogen ion is the strongest acid that can exist in aqueous solution.

Acids stronger than H^+(aq) react with water to produce H^+(aq) and their conjugate bases. For example, $HClO_4$ (see Table 10-1) reacts with H_2O completely to form H^+(aq) and ClO_4^-(aq).

Similar observations can be made for aqueous solutions of strong soluble bases such as NaOH and KOH. Both are completely dissociated in dilute aqueous solutions:

$$Na^+OH^-(aq) \xrightarrow{H_2O} Na^+(aq) + OH^-(aq)$$

The hydroxide ion is the strongest base that can exist in aqueous solution.

The amide ion, NH_2^-, is a stronger base than OH^-.

Bases stronger than OH^- react with H_2O to produce OH^- and their conjugate acids. When metal amides such as sodium amide, $NaNH_2$, are placed in H_2O, the amide ion, NH_2^-, reacts with H_2O completely, as shown by the following equation:

$$NH_2^- + H_2O \longrightarrow NH_3(aq) + OH^-(aq)$$

Thus, we see that H_2O is a leveling solvent for all bases stronger than OH^-.

The trends in binary acid strengths *across* a period (e.g., $CH_4 < NH_3 < H_2O < HF$) are *not* those predicted from trends in bond energies and electronegativity differences. The correlations used for *vertical* trends cannot be used for *horizontal* trends. This is because a "horizontal" series of compounds has different stoichiometries and different numbers of lone pairs of electrons on its central atoms.

Acid strengths for other *vertical* series of binary acids vary in the same way as those of the VIIA elements. The order of bond strengths for the VIA hydrides is

$$H_2O >> H_2S > H_2Se > H_2Te$$

H—O bonds are much stronger than the other H—El bonds. As we might expect, the order of acid strengths for these hydrides is just the reverse of the order of bond strengths.

$$H_2O << H_2S < H_2Se < H_2Te \qquad \text{(strongest acid)}$$

Table 10-1 displays relative acid and base strengths of a number of conjugate acid–base pairs.

10-6 Reactions of Acids and Bases

Common Strong Acids	
Binary	**Ternary**
HCl	$HClO_4$
HBr	$HClO_3$
HI	HNO_3
	H_2SO_4

Strong Soluble Bases	
LiOH	
NaOH	
KOH	$Ca(OH)_2$
RbOH	$Sr(OH)_2$
CsOH	$Ba(OH)_2$

In Section 4-5 we introduced classical acid–base reactions. We defined neutralization as the reaction of an acid with a base to form a salt and (in most cases) water. Most *salts* are ionic compounds that contain a cation other than H^+ and an anion other than OH^- or O^{2-}. The *strong acids* and *strong soluble bases* are listed in the margin. Recall that other common acids may be assumed to be weak. The other common metal hydroxides (bases) are insoluble in water.

Arrhenius and Brønsted–Lowry acid–base neutralization reactions all have one thing in common. They involve the reaction of an acid with a base to form a salt that contains the cation characteristic of the base and the anion characteristic of the acid. Water is also usually formed. This is indicated in the formula unit (molecular) equation. However, the general form of the net ionic equation and the essence of the reaction are different for different acid–base reactions. They depend upon the solubility and extent of ionization or dissociation of each reactant and product.

In writing ionic equations, we always write the formulas of the predominant forms of the compounds in, or in contact with, aqueous solution. Writing ionic equations from formula unit equations requires a knowledge of the lists of strong acids and strong soluble bases, as well as of the generalizations on solubilities of inorganic compounds. Please review carefully all of Sections 4-2 and 4-3. Study Tables 4-9 and 4-10 carefully because they summarize much information that you are about to use again.

In Section 4-5 we examined some reactions of strong acids with strong soluble bases to form soluble salts. Let us illustrate one additional example.

Perchloric acid, $HClO_4$, reacts with sodium hydroxide to produce sodium perchlorate, $NaClO_4$, a soluble ionic salt.

$$HClO_4(aq) + NaOH(aq) \longrightarrow NaClO_4(aq) + H_2O(\ell)$$

The total ionic equation for this reaction is

$$[H^+(aq) + ClO_4^-(aq)] + [Na^+(aq) + OH^-(aq)] \longrightarrow [Na^+(aq) + ClO_4^-(aq)] + H_2O(\ell)$$

Eliminating the spectator ions, Na^+ and ClO_4^-, gives the net ionic equation

$$H^+(aq) + OH^-(aq) \longrightarrow H_2O(\ell)$$

This is the same as $H_3O^+ + OH^- \rightarrow 2H_2O$.

This is the net ionic equation for the reaction of all strong acids with strong soluble bases to form soluble salts and water.

Many weak acids react with strong soluble bases to form soluble salts and water. For example, acetic acid, CH_3COOH, reacts with sodium hydroxide, NaOH, to produce sodium acetate, $NaCH_3COO$.

$$CH_3COOH(aq) + NaOH(aq) \longrightarrow NaCH_3COO(aq) + H_2O(\ell)$$

The total ionic equation for this reaction is

$$CH_3COOH(aq) + [Na^+(aq) + OH^-(aq)] \longrightarrow [Na^+(aq) + CH_3COO^-(aq)] + H_2O(\ell)$$

Elimination of Na^+ from both sides gives the net ionic equation

$$CH_3COOH(aq) + OH^-(aq) \longrightarrow CH_3COO^-(aq) + H_2O(\ell)$$

In general terms, the reaction of a *weak monoprotic acid* with a *strong soluble base* to form a *soluble salt* may be represented as

$$HA(aq) + OH^-(aq) \longrightarrow A^-(aq) + H_2O(aq) \qquad \text{(net ionic equation)}$$

Monoprotic acids contain one, diprotic acids contain two, and triprotic acids contain three acidic (ionizable) hydrogen atoms per formula unit. Polyprotic acids (those that contain more than one ionizable hydrogen atom) are discussed in detail in Chapter 18.

Example 10-1

Write (a) formula unit, (b) total ionic, and (c) net ionic equations for the neutralization of aqueous ammonia with nitric acid.

Plan

(a) The salt contains the cation of the base, NH_4^+, and the anion of the acid, NO_3^-. The salt is NH_4NO_3.

(b) HNO_3 is a strong acid—we write it in ionic form. Ammonia is a weak base. NH_4NO_3 is a soluble salt that is completely dissociated—we write it in ionic form.

(c) We cancel the spectator ions, NO_3^-, and obtain the net ionic equation.

Solution

(a) $HNO_3(aq) + NH_3(aq) \longrightarrow NH_4NO_3(aq)$

(b) $[H^+(aq) + NO_3^-(aq)] + NH_3(aq) \longrightarrow [NH_4^+(aq) + NO_3^-(aq)]$

(c) $H^+(aq) + NH_3(aq) \longrightarrow NH_4^+(aq)$

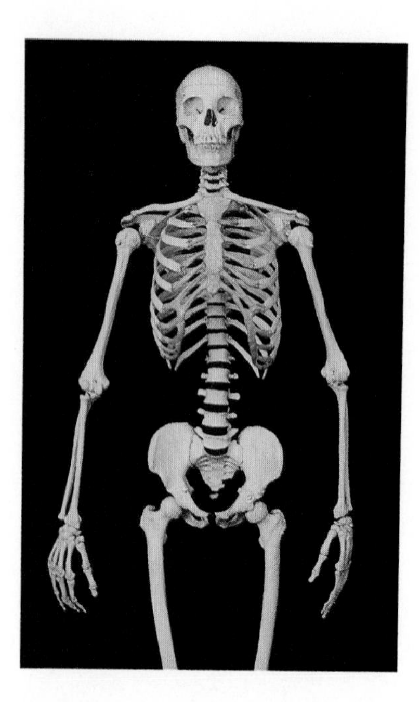

Our bones are mostly calcium phosphate, $Ca_3(PO_4)_2$, an insoluble compound. They are formed by reactions of calcium ions, Ca^{2+}, with phosphate ions, PO_4^{3-}. Calcium and phosphorus are essential elements in human nutrition. Do you see why children require calcium and phosphorus for normal growth?

Example 10-2

Write (a) formula unit, (b) total ionic, and (c) net ionic equations for the complete neutralization of phosphoric acid, H_3PO_4, with calcium hydroxide, $Ca(OH)_2$.

Plan

(a) The salt contains the cation of the base, Ca^{2+}, and the anion of the acid, PO_4^{3-}. The salt is $Ca_3(PO_4)_2$.

(b) H_3PO_4 is a weak acid—it is not written in ionic form. $Ca(OH)_2$ is a strong soluble base, and so it is written in ionic form. $Ca_3(PO_4)_2$ is an *insoluble* salt, and so it is not written in ionic form.

(c) There are no species common to both sides of the equation, and so there are no spectator ions. The net ionic equation is the same as the total ionic equation except that there are no brackets.

Solution

(a) $$2H_3PO_4(aq) + 3Ca(OH)_2(aq) \longrightarrow Ca_3(PO_4)_2(s) + 6H_2O(\ell)$$

(b) $$2H_3PO_4(aq) + 3[Ca^{2+}(aq) + 2OH^-(aq)] \longrightarrow Ca_3(PO_4)_2(s) + 6H_2O(\ell)$$

(c) $$2H_3PO_4(aq) + 3Ca^{2+}(aq) + 6OH^-(aq) \longrightarrow Ca_3(PO_4)_2(s) + 6H_2O(\ell)$$

EOC 50, 52

There are other kinds of reactions between acids and bases. The preceding examples illustrate how chemical equations for them are written.

10-7 Acidic Salts and Basic Salts

To this point we have examined acid–base reactions in which stoichiometric amounts of acids and bases were mixed to form *normal salts*. As the name implies, **normal salts** contain no unreacted H^+ or OH^- ions.

If less than stoichiometric amounts of bases react with *polyprotic* acids, the resulting salts are known as **acidic salts** because they are still capable of neutralizing bases. The reaction of phosphoric acid, H_3PO_4, a weak acid, with strong bases can produce three different salts, depending on the relative amounts of acid and base used.

$$\underset{1\ \text{mole}}{H_3PO_4(aq)} + \underset{1\ \text{mole}}{NaOH(aq)} \longrightarrow \underset{\substack{\text{sodium dihydrogen phosphate,}\\ \text{an acidic salt}}}{NaH_2PO_4(aq)} + H_2O(\ell)$$

$$\underset{1\ \text{mole}}{H_3PO_4(aq)} + \underset{2\ \text{moles}}{2NaOH(aq)} \longrightarrow \underset{\substack{\text{sodium hydrogen phosphate,}\\ \text{an acidic salt}}}{Na_2HPO_4(aq)} + 2H_2O(\ell)$$

$$\underset{1\ \text{mole}}{H_3PO_4(aq)} + \underset{3\ \text{moles}}{3NaOH(aq)} \longrightarrow \underset{\substack{\text{sodium phosphate,}\\ \text{a normal salt}}}{Na_3PO_4(aq)} + 3H_2O(\ell)$$

Sodium hydrogen carbonate, baking soda, is the most familiar example of an acidic salt. It can neutralize strong bases, but its aqueous solutions are slightly basic, as the color of the indicator shows.

There are many additional examples of acidic salts. Sodium hydrogen carbonate, $NaHCO_3$, commonly called sodium bicarbonate, is classified as an acidic salt. However, it is the acidic salt of an extremely weak acid—carbonic acid, H_2CO_3—and solutions of sodium bicarbonate are slightly basic.

Polyhydroxy bases (bases that contain more than one OH per formula unit) react with less than stoichiometric amounts of acids to form **basic salts**, i.e., salts that contain unreacted OH groups. For example, the reaction of aluminum hydroxide with hydrochloric acid can produce three different salts.

These basic aluminum salts are called "aluminum chlorohydrate." They are components of many deodorants.

$$\underset{1\ \text{mole}}{Al(OH)_3(s)} + \underset{1\ \text{mole}}{HCl(aq)} \longrightarrow \underset{\text{aluminum dihydroxide chloride, a basic salt}}{Al(OH)_2Cl(s)} + H_2O(\ell)$$

$$\underset{1\ \text{mole}}{Al(OH)_3(s)} + \underset{2\ \text{moles}}{2HCl(aq)} \longrightarrow \underset{\text{aluminum hydroxide dichloride, a basic salt}}{Al(OH)Cl_2(s)} + 2H_2O(\ell)$$

$$\underset{1\ \text{mole}}{Al(OH)_3(s)} + \underset{3\ \text{moles}}{3HCl(aq)} \longrightarrow \underset{\text{aluminum chloride, a normal salt}}{AlCl_3(aq)} + 3H_2O(\ell)$$

Aqueous solutions of basic salts are not necessarily basic, but they can neutralize acids. Most basic salts are rather insoluble in water.

10-8 Strengths of Ternary Acids and Amphoterism

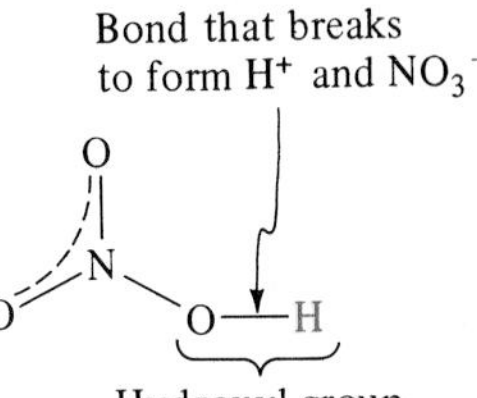

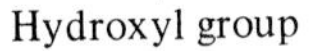

Ternary acids are *hydroxides of nonmetals* that ionize to produce $H^+(aq)$. The formula for nitric acid is commonly written HNO_3 to emphasize the presence of an acidic hydrogen atom, but it could also be written as $NO_2(OH)$, as its structure shows (see margin).

We usually reserve the term "hydroxide" for substances that produce basic solutions, and call the other "hydroxides" acids because they ionize to produce $H^+(aq)$. In ternary acids the hydroxyl oxygen is bonded to a fairly electronegative element (usually a nonmetal). In nitric acid the nitrogen draws the electrons of the N—O (hydroxyl) bond closer to itself than would a less electronegative element such as sodium. The oxygen pulls the electrons of the O—H bond close enough so that the hydrogen atom ionizes as H^+, leaving NO_3^-:

$$HNO_3(aq) \longrightarrow H^+(aq) + NO_3^-(aq)$$

In contrast, let us consider hydroxides of metals. Oxygen is much more electronegative than most metals, such as sodium. It draws the electrons of the sodium–oxygen bond in NaOH (a strong soluble base) so close to itself that the bonding is ionic. Therefore, NaOH exists as Na^+ and OH^- ions, even in the solid state, and dissociates into Na^+ and OH^- ions when it dissolves in H_2O.

$$Na^+OH^-(s) \xrightarrow{H_2O} Na^+(aq) + OH^-(aq)$$

We usually write the formula for sulfuric acid as H_2SO_4 to emphasize the fact that it is an acid. However, the formula can also be written as $SO_2(OH)_2$, because the structure of sulfuric acid (see margin) shows clearly that H_2SO_4 contains two —O—H groups bound to a sulfur atom. Because the O—H bonds are easier to break than the S—O bonds, sulfuric acid ionizes as an acid.

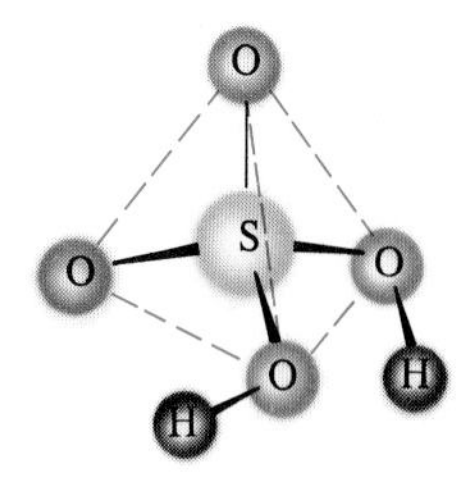

Sulfuric acid is called a polyprotic acid because it has more than one ionizable hydrogen atom per molecule. It is the only common polyprotic acid that is also a strong acid.

The volatile acid HCl can be made by dropping concentrated H_2SO_4 onto solid NaCl. Gaseous HCl is liberated. HCl(g) dissolves in the water on a piece of filter paper. The indicator on the paper turns red, its color in acidic solution.

$$\text{Step 1:} \quad H_2SO_4(aq) \longrightarrow H^+(aq) + HSO_4^-(aq)$$

$$\text{Step 2:} \quad HSO_4^-(aq) \rightleftharpoons H^+(aq) + SO_4^{2-}(aq)$$

The first step in the ionization of H_2SO_4 is complete in dilute aqueous solution. The second step is nearly complete only in very dilute solutions. The first step in the ionization of a polyprotic acid occurs to a greater extent than the second step.

Sulfurous acid, H_2SO_3, is a polyprotic acid that contains the same elements as H_2SO_4. However, H_2SO_3 is a weak acid, which tells us that the H—O bonds in H_2SO_3 are stronger than those in H_2SO_4.

Comparison of the acid strengths of nitric acid, HNO_3, and nitrous acid, HNO_2, shows that HNO_3 is a much stronger acid than HNO_2.

Acid strengths of most ternary acids containing the same central element increase with increasing oxidation state of the central element and with increasing numbers of oxygen atoms.

The following orders of increasing strength are typical:

$H_2SO_3 < H_2SO_4$
$HNO_2 < HNO_3$
$HClO < HClO_2 < HClO_3 < HClO_4$

(strongest acids are on the right side)

For most ternary acids containing different elements in the same oxidation state from the same group in the periodic table, acid strengths increase with increasing electronegativity of the central element.

$H_2SeO_4 < H_2SO_4$ $\quad$ $H_2SeO_3 < H_2SO_3$
$H_3PO_4 < HNO_3$
$HBrO_4 < HClO_4$ $\quad$ $HBrO_3 < HClO_3$

In *most* ternary inorganic acids, all H atoms are bonded to O.

Contrary to what we might expect, H_3PO_3 is a stronger acid than HNO_2. Care must be exercised to compare acids that have *similar structures*. For example, H_3PO_2, which has two H atoms bonded to the P atom, is a stronger acid than H_3PO_3, which has one H atom bonded to the P atom. H_3PO_3 is a stronger acid than H_3PO_4, which has no H atoms bonded to the P atom.

As we have seen, whether a particular substance behaves as an acid or as a base depends on its environment. In Section 10-3 we described the amphiprotic nature of water. **Amphoterism** is the general term that describes the ability of a substance to react either as an acid or as a base. *Amphiprotic behavior* describes the cases in which substances exhibit amphoterism by accepting and by donating a proton, H^+. Several *insoluble* metal hydroxides are amphoteric; i.e., they react with acids to form salts and water, but they also dissolve in and react with excess strong soluble bases.

All hydroxides containing small, highly charged metal ions are insoluble in water.

Aluminum hydroxide is a typical amphoteric metal hydroxide. Its behavior as a *base* is illustrated by its reaction with nitric acid to form a *salt:*

$$Al(OH)_3(s) + 3HNO_3(aq) \longrightarrow Al(NO_3)_3(aq) + 3H_2O(\ell)$$

$$Al(OH)_3(s) + 3[H^+(aq) + NO_3^-(aq)] \longrightarrow [Al^{3+}(aq) + 3NO_3^-(aq)] + 3H_2O(\ell)$$

$$Al(OH)_3(s) + 3H^+(aq) \longrightarrow Al^{3+}(aq) + 3H_2O(\ell)$$

Table 10-2
Amphoteric Hydroxides

Metal or Metalloid ions	Insoluble Amphoteric Hydroxide	Complex Ion Formed in an Excess of a Strong Soluble Base
Be^{2+}	$Be(OH)_2$	$[Be(OH)_4]^{2-}$
Al^{3+}[a]	$Al(OH)_3$	$[Al(OH)_4]^-$
Cr^{3+}	$Cr(OH)_3$	$[Cr(OH)_4]^-$
Zn^{2+}	$Zn(OH)_2$	$[Zn(OH)_4]^{2-}$
Sn^{2+}	$Sn(OH)_2$	$[Sn(OH)_3]^-$
Sn^{4+}	$Sn(OH)_4$	$[Sn(OH)_6]^{2-}$
Pb^{2+}	$Pb(OH)_2$	$[Pb(OH)_4]^{2-}$
As^{3+}[a]	$As(OH)_3$	$[As(OH)_4]^-$
Sb^{3+}[a]	$Sb(OH)_3$	$[Sb(OH)_4]^-$
Si^{4+}[a]	$Si(OH)_4$	SiO_4^{4-} and SiO_3^{2-}
Co^{2+}[b]	$Co(OH)_2$	$[Co(OH)_4]^{2-}$
Cu^{2+}[b]	$Cu(OH)_2$	$[Cu(OH)_4]^{2-}$

[a] Al, As, Sb, and Si are metalloids.
[b] $Co(OH)_2$ and $Cu(OH)_2$ are only slightly amphoteric; i.e., a large excess of strong soluble base is required to dissolve small amounts of these insoluble hydroxides.

When an excess of a solution of any strong soluble base, e.g., NaOH, is added to solid aluminum hydroxide, the $Al(OH)_3$ dissolves. The equation for the reaction is usually written

$$\underset{\text{an acid}}{Al(OH)_3(s)} + \underset{\text{a base}}{NaOH(aq)} \longrightarrow \underset{\text{sodium aluminate, a soluble compound}}{NaAl(OH)_4(aq)}$$

The proper name for "sodium aluminate" is *sodium tetrahydroxoaluminate.*

The total ionic and net ionic equations are

$$Al(OH)_3(s) + [Na^+(aq) + OH^-(aq)] \longrightarrow [Na^+(aq) + Al(OH)_4^-(aq)]$$

$$Al(OH)_3(s) + OH^-(aq) \longrightarrow Al(OH)_4^-(aq)$$

Other amphoteric metal hydroxides undergo similar reactions.

Table 10-2 contains a list of the common amphoteric hydroxides. Four are hydroxides of metalloids, the elements located along the line that divides metals and nonmetals in the periodic table.

Generally, elements of intermediate electronegativity form amphoteric hydroxides. Those of high and low electronegativity form acidic and basic "hydroxides," respectively.

10-9 The Preparation of Acids

Binary acids may be prepared by combination of appropriate elements with hydrogen (Section 6-7, part 2).

Small quantities of the hydrogen halides (their solutions are called hydrohalic acids) and other *volatile acids* are usually prepared by dropping concentrated nonvolatile acids onto the appropriate salts. (Sulfuric and phosphoric acids are classified as nonvolatile acids because they have much higher boiling points than other common acids.) The reactions of concentrated (96%) sulfuric acid with solid sodium fluoride and sodium chloride produce gaseous hydrogen fluoride and hydrogen chloride, respectively.

$$\underset{\substack{\text{sulfuric acid}\\ \text{bp} = 338°\text{C}}}{H_2SO_4(\ell)} + \underset{\text{sodium fluoride}}{NaF(s)} \longrightarrow \underset{\substack{\text{sodium hydrogen}\\ \text{sulfate}}}{NaHSO_4(s)} + \underset{\substack{\text{hydrogen fluoride}\\ \text{bp} = 19.59°\text{C}}}{HF(g)}$$

$$H_2SO_4(\ell) + \underset{\text{sodium chloride}}{NaCl(s)} \longrightarrow NaHSO_4(s) + \underset{\substack{\text{hydrogen chloride}\\ \text{bp} = -84.9°\text{C}}}{HCl(g)}$$

Because concentrated sulfuric acid is a fairly strong oxidizing agent, it cannot be used to prepare hydrogen bromide or hydrogen iodide; instead, the free halogens are produced. Phosphoric acid, a nonoxidizing acid, is dropped onto solid sodium bromide or sodium iodide to produce hydrogen bromide or hydrogen iodide, as the following equations show:

$$\underset{\substack{\text{phosphoric acid}\\ \text{bp} = 213°\text{C}}}{H_3PO_4(\ell)} + \underset{\text{sodium bromide}}{NaBr(s)} \xrightarrow{\Delta} \underset{\substack{\text{sodium dihydrogen}\\ \text{phosphate}}}{NaH_2PO_4(s)} + \underset{\substack{\text{hydrogen bromide}\\ \text{bp} = -67.0°\text{C}}}{HBr(g)}$$

$$H_3PO_4(\ell) + \underset{\text{sodium iodide}}{NaI(s)} \xrightarrow{\Delta} NaH_2PO_4(s) + \underset{\substack{\text{hydrogen iodide}\\ \text{bp} = -35°\text{C}}}{HI(g)}$$

This kind of reaction may be generalized as

$$\substack{\text{nonvolatile}\\ \text{acid}} + \substack{\text{salt of}\\ \text{volatile acid}} \longrightarrow \text{volatile acid} + \substack{\text{salt of}\\ \text{nonvolatile acid}}$$

A solution of dichromic acid, $H_2Cr_2O_7$, is deep red.

In Section 6-8, part 2, we saw that many nonmetal oxides, called acid anhydrides, react with water to form **ternary acids** with no changes in oxidation numbers. For example, dichlorine heptoxide, Cl_2O_7, forms perchloric acid when it dissolves in water:

$$\overset{(+7)}{Cl_2O_7(\ell)} + H_2O(\ell) \longrightarrow 2[H^+(aq) + \overset{(+7)}{ClO_4^-}(aq)]$$

Some *high-oxidation-state transition metal oxides* are acidic oxides; that is, they dissolve in water to give solutions of ternary acids. Manganese(VII) oxide, Mn_2O_7, and chromium(VI) oxide, CrO_3, are the most common examples:

$$\underset{\text{manganese(VII) oxide}}{\overset{(+7)}{Mn_2O_7(\ell)}} + H_2O(\ell) \longrightarrow \underset{\text{permanganic acid}}{2[H^+(aq) + \overset{(+7)}{MnO_4^-}(aq)]}$$

$$\underset{\text{chromium(VI) oxide}}{\overset{(+6)}{2CrO_3(s)}} + H_2O(\ell) \longrightarrow \underset{\text{dichromic acid}}{[2H^+(aq) + \overset{(+6)}{Cr_2O_7^{2-}}(aq)]}$$

Neither permanganic acid nor dichromic acid has been isolated in pure form. Many stable salts of both are well known.

The halides and oxyhalides of some nonmetals hydrolyze (react with water) to produce two acids—a (binary) hydrohalic acid and a (ternary) oxyacid of the nonmetal. Phosphorus trihalides react with water to produce the corresponding hydrohalic acids and phosphorous acid, a weak diprotic acid, while phosphorus pentahalides give phosphoric acid.

$$\overset{+3}{P}X_3 + 3H_2O(\ell) \longrightarrow H_3\overset{+3}{P}O_3(aq) + 3HX(aq)$$

$$\overset{+5}{P}X_5 + 4H_2O(\ell) \longrightarrow H_3\overset{+5}{P}O_4(aq) + 5HX(aq)$$

There are no changes in oxidation numbers in these reactions. Consider the reactions of PCl_3 and PCl_5 with H_2O:

$$\underset{\text{phosphorus trichloride}}{\overset{+3}{P}Cl_3(\ell)} + 3H_2O(\ell) \longrightarrow \underset{\text{phosphorous acid}}{H_3\overset{+3}{P}O_3(aq)} + 3[H^+(aq) + Cl^-(aq)]$$

$$\underset{\text{phosphorus pentachloride}}{\overset{+5}{P}Cl_5(s)} + 4H_2O(\ell) \longrightarrow \underset{\text{phosphoric acid}}{H_3\overset{+5}{P}O_4(aq)} + 5[H^+(aq) + Cl^-(aq)]$$

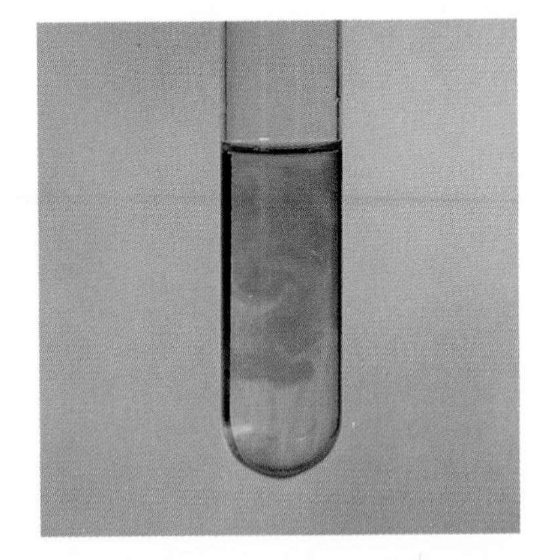

A drop of PCl_3 is added to water that contains the indicator methyl orange. As PCl_3 reacts with water to form HCl and H_3PO_3, the indicator turns red, its color in acidic solution.

10-10 The Lewis Theory

In 1923 Professor G. N. Lewis presented the most comprehensive of the classic acid–base theories. The Lewis definitions are as follows. An **acid** is any species that can accept a share in an electron pair. A **base** is any species that can make available, or "donate," a share in an electron pair. The definitions do *not* specify that an electron pair must be transferred from one atom to another—only that an electron pair, residing originally on one atom, must be shared between two atoms. *Neutralization* is defined as **coordinate covalent bond formation**. This results in a covalent bond in which both electrons were furnished by one atom.

This is the same Lewis who made many contributions to our understanding of chemical bonding.

The reaction of boron trichloride with ammonia is a typical Lewis acid–base reaction:

$$\underset{\text{acid}}{BCl_3(g)} + \underset{\text{base}}{NH_3(g)} \longrightarrow \underset{\text{product}}{Cl_3B:NH_3}$$

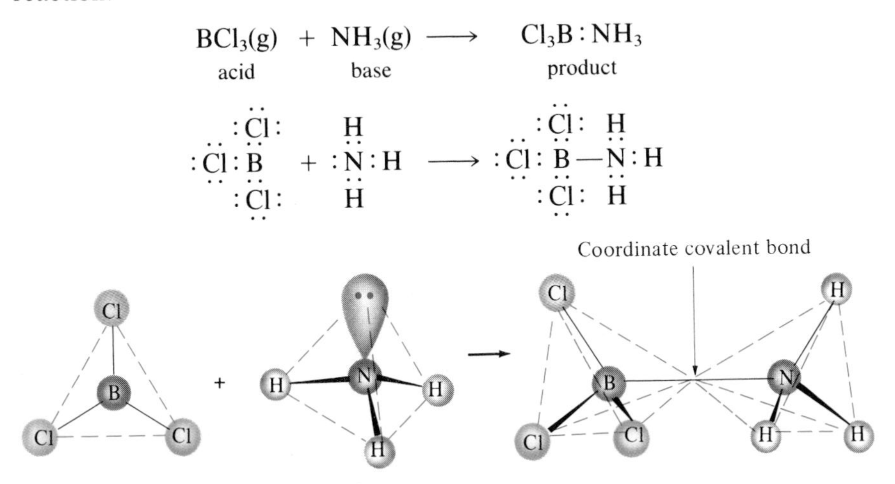

The Lewis theory is sufficiently general that it covers *all* acid–base reactions that the other theories include, plus many additional reactions such as complex formation (Chapter 29).

The autoionization of water (Section 10-3) was described in terms of Brønsted–Lowry theory. In Lewis theory terminology, this is also an acid–base reaction. The acceptance of a proton, H^+, by a base involves the formation of a *coordinate covalent bond.*

$$\underset{\text{base}}{\mathrm{H}:\underset{\mathrm{H}}{\ddot{\mathrm{O}}}:} + \underset{\text{acid}}{\mathrm{H}:\underset{(\mathrm{H})}{\ddot{\mathrm{O}}}:} \rightleftharpoons \left[\mathrm{H}:\underset{\mathrm{H}}{\overset{\mathrm{H}}{\ddot{\mathrm{O}}}}:\right]^+ + \mathrm{H}:\ddot{\mathrm{O}}:^-$$

Theoretically, any species that contains an unshared electron pair could act as a base. In fact, most ions and molecules that contain unshared electron pairs do undergo some reactions by sharing their electron pairs. Conversely, many Lewis acids contain only six electrons in the highest occupied energy level of the central element. They react by accepting a share in an additional pair of electrons. These species are said to have an **open sextet**. Many compounds of the Group IIIA elements are Lewis acids, as illustrated by the reaction of boron trichloride with ammonia, presented earlier.

Anhydrous aluminum chloride is a common Lewis acid that is used to catalyze many organic reactions. The dissolution of $AlCl_3$ in hydrochloric acid gives a solution that contains $AlCl_4^-$ ions.

$$\underset{\text{acid}}{\mathrm{AlCl_3(s)}} + \underset{\text{base}}{\mathrm{Cl^-(aq)}} \longrightarrow \underset{\text{product}}{\mathrm{AlCl_4^-(aq)}}$$

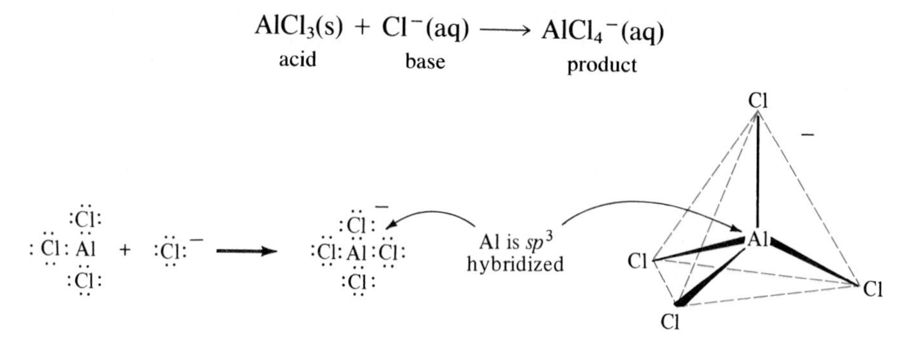

Other ions and molecules behave as Lewis acids by expansion of the valence shell of the central element. Anhydrous tin(IV) chloride, often called stannic chloride, is a colorless liquid that also is frequently used as a Lewis acid catalyst. The tin atom (Group IVA) can expand its valence shell by utilizing vacant *d* orbitals. It can accept shares in two additional electron pairs, as its reaction with hydrochloric acid illustrates.

$$\underset{\text{acid}}{\mathrm{SnCl_4(\ell)}} + \underset{\text{base}}{\mathrm{2Cl^-(aq)}} \longrightarrow \mathrm{SnCl_6^{2-}(aq)}$$

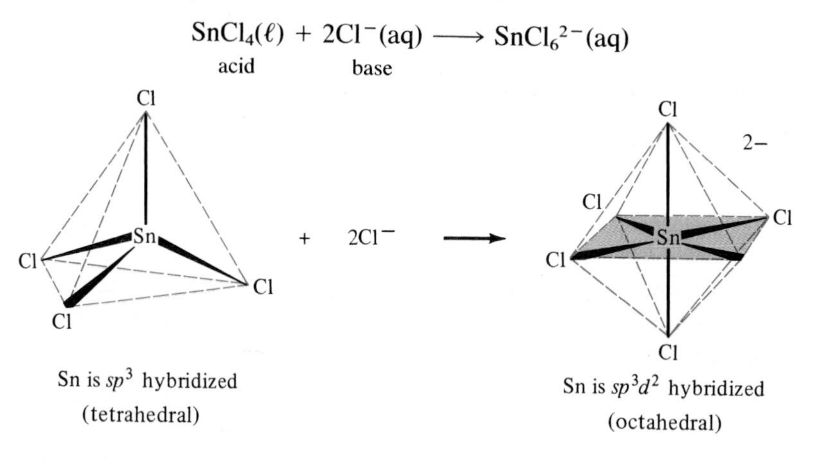

Sn is sp^3 hybridized (tetrahedral)

Sn is sp^3d^2 hybridized (octahedral)

Experienced chemists find the Lewis theory to be very useful because so many chemical reactions are covered by it. The less experienced sometimes find the theory less useful, but as their knowledge expands so does its utility.

Key Terms

Acid A substance that produces $H^+(aq)$ ions in aqueous solution. Strong acids ionize completely or almost completely in dilute aqueous solution; weak acids ionize only slightly.

Acid anhydride The oxide of a nonmetal that reacts with water to form an acid.

Acidic salt A salt containing an ionizable hydrogen atom; does not necessarily produce acidic solutions.

Amphiprotism The ability of a substance to exhibit amphoterism by accepting or donating protons.

Amphoterism Ability of a substance to act as either an acid or a base.

Anhydrous Without water.

Autoionization An ionization reaction between identical molecules.

Base A substance that produces $OH^-(aq)$ ions in aqueous solution. Strong soluble bases are soluble in water and are completely *dissociated*. Weak bases ionize only slightly.

Basic anhydride The oxide of a metal that reacts with water to form a base.

Basic salt A salt containing an ionizable OH group.

Brønsted–Lowry acid A proton donor.

Brønsted–Lowry base A proton acceptor.

Conjugate acid–base pair In Brønsted–Lowry terminology, a reactant and product that differ by a proton, H^+.

Coordinate covalent bond A covalent bond in which both shared electrons are furnished by the same species; the bond between a Lewis acid and a Lewis base.

Dissociation In aqueous solution, the process in which a *solid ionic compound* separates into its ions.

Electrolyte A substance whose aqueous solutions conduct electricity.

Formula unit (molecular) equation A chemical equation in which all compounds are represented by complete formulas.

Hydration The process by which water molecules bind to ions or molecules in the solid state or in solution.

Hydride A binary compound of hydrogen.

Hydrolysis Reaction of a substance with water.

Ionization In aqueous solution, the process in which a *molecular* compound reacts with water and forms ions.

Leveling effect The effect by which all acids stronger than the acid that is characteristic of the solvent react with the solvent to produce that acid; a similar statement applies to bases. The strongest acid (base) that can exist in a given solvent is the acid (base) characteristic of that solvent.

Lewis acid Any species that can accept a share in an electron pair.

Lewis base Any species that can make available a share in an electron pair.

Net ionic equation The equation that results from canceling spectator ions and eliminating brackets from a total ionic equation.

Neutralization The reaction of an acid with a base to form a salt and water; usually, the reaction of hydrogen ions with hydroxide ions to form water molecules.

Nonelectrolyte A substance whose aqueous solutions do not conduct electricity.

Normal oxide A metal oxide containing the oxide ion, O^{2-} (oxygen in the -2 oxidation state).

Normal salt A salt containing no ionizable H atoms or OH groups.

Open sextet Refers to species that have only six electrons in the highest energy level of the central element (many Lewis acids).

Polyprotic acid An acid that contains more than one ionizable hydrogen atom per formula unit.

Protonic acid An Arrhenius (classical) acid, or a Brønsted–Lowry acid.

Salt A compound that contains a cation other than H^+ and an anion other than OH^- or O^{2-}.

Spectator ions Ions in solution that do not participate in a chemical reaction.

Strong electrolyte A substance that conducts electricity well in dilute aqueous solution.

Ternary acid An acid that contains three elements—usually H, O, and another nonmetal.

Ternary compound A compound that contains three elements.

Total ionic equation The equation for a chemical reaction written to show the predominant form of all species in aqueous solution or in contact with water.

Weak electrolyte A substance that conducts electricity poorly in dilute aqueous solution.

Exercises

Basic Ideas

1. Robert Boyle observed that acids have certain properties. Which ones did he observe?
2. Gay-Lussac reached an important conclusion about acids and bases. What was it?
3. What is the significance of the idea that acids and bases can be defined only in terms of their reactions with each other?

The Arrhenius Theory

4. Outline Arrhenius' ideas about acids and bases. (a) How did he define the following terms: acid, base, neutralization? (b) Give an example that illustrates each term.
5. Define and illustrate the following terms clearly and concisely. Give an example of each.
   (a) strong electrolyte
   (b) weak electrolyte
   (c) nonelectrolyte
   (d) strong acid
   (e) strong soluble base
   (f) weak acid
   (g) weak base
6. Distinguish between the following pairs of terms and provide a specific example of each.
   (a) strong acid and weak acid
   (b) strong soluble base and weak base
   (c) strong soluble base and insoluble base
7. Write formulas and names for
   (a) the common strong acids
   (b) three weak acids
   (c) the common strong soluble bases
   (d) the most common weak base
   (e) three soluble ionic salts
   (f) three insoluble salts
8. List at least three characteristic properties of acids and three of bases.
9. Summarize the electrical properties of strong electrolytes, weak electrolytes, and nonelectrolytes.
10. Describe an experiment for classifying each of these compounds as a strong electrolyte, a weak electrolyte, or a nonelectrolyte. Classify each. K_2CO_3, HCN, C_2H_5COOH, CH_3OH, H_2S, H_2SO_4, H_2CO, NH_3
11. Limestone, $CaCO_3$, is a water-insoluble material, whereas $Ca(HCO_3)_2$ is soluble. Caves are formed when rainwater containing dissolved CO_2 passes over limestone for long periods of time. Write a chemical equation for the acid–base reaction.

The Hydrated Hydrogen Ion

12. Describe the hydrated hydrogen ion in words and with formulas.
13. Why is the hydrated hydrogen ion important?
14. Criticize the following statement: "The hydrated hydrogen ion should always be represented as H_3O^+."

Brønsted–Lowry Theory

15. State the basic ideas of the Brønsted–Lowry theory.
16. Use Brønsted–Lowry terminology to define the following terms. Illustrate each of the following with a specific example.
   (a) acid
   (b) conjugate base
   (c) base
   (d) conjugate acid
   (e) conjugate acid–base pair
17. Write balanced equations that describe the ionization of the following acids in dilute aqueous solution. Use a single arrow ($\rightarrow$) to represent complete, or nearly complete, ionization and a double arrow ($\rightleftharpoons$) to represent a small extent of ionization. (a) HNO_3, (b) CH_3COOH, (c) HBr, (d) HCN, (e) HF, (f) $HClO_4$
18. Use words and equations to describe how ammonia can act as a base in (a) aqueous solution and (b) the pure state, i.e., as gaseous ammonia molecules when it reacts with gaseous hydrogen chloride or a similar anhydrous acid.
19. What does autoionization mean? How can the autoionization of water be described as an acid–base reaction?
20. Autoionization occurs when an ion other than an H^+ is transferred, as exemplified by the transfer of a Cl^- ion from one PCl_5 molecule to another. Write the equation for this reaction. What are the shapes of the two ions that are formed?
21. What structural features must a compound have to be able to undergo autoionization?
22. What do we mean when we say that water is amphiprotic? (a) Can we also describe water as amphoteric? Why? (b) Illustrate the amphiprotic nature of water by writing two equations for reactions in which water exhibits this property.
23. In terms of Brønsted–Lowry theory, state the differences between (a) a strong and a weak base and (b) a strong and a weak acid.
24. Illustrate, with appropriate equations, the fact that these species are bases in water: NH_3, HS^-, CH_3COO^-, O^{2-}.
25. Illustrate, with appropriate equations, the fact that the following species are acids in water: HBr, HNO_2, $HBrO_4$.
26. Give the products in the following acid–base reactions. Identify the conjugate acid–base pairs.
   (a) $NH_4^+ + CN^-$
   (b) $HS^- + HSO_4^-$
   (c) $HClO_4 + [H_2NNH_3]^+$
   (d) $H^- + H_2O$
27. List the conjugate acids of H_2O, OH^-, Cl^-, HCl, AsO_4^{3-}, NH_2^-, HPO_4^{2-}, and SO_4^{2-}.
28. List the conjugate bases of H_2O, HS^-, HCl, PH_4^+, and $HOCH_3$.

29. Identify the Brønsted–Lowry acids and bases in these reactions and group them into conjugate acid–base pairs.
(a) $NH_3 + HBr \rightleftharpoons NH_4^+ + Br^-$
(b) $NH_4^+ + OH^- \rightleftharpoons NH_3 + H_2O$
(c) $H_3O^+ + PO_4^{3-} \rightleftharpoons HPO_4^{2-} + H_2O$
(d) $HSO_3^- + CN^- \rightleftharpoons HCN + SO_3^{2-}$

Properties of Aqueous Solutions of Acids and Bases

30. Write equations and designate conjugate pairs for the stepwise reactions in water of (a) sulfuric acid, H_2SO_4, and (b) H_3AsO_4.
31. List six properties of aqueous solutions of protonic acids.
32. List five properties of bases in aqueous solution. Does aqueous ammonia exhibit these properties? Why or why not?
33. We say that strong acids, weak acids, and weak bases *ionize* in water, while strong soluble bases *dissociate* in water. What is the difference?
34. Distinguish between solubility in water and extent of ionization in water. Provide specific examples that illustrate the meanings of both terms.
35. Write three general statements that describe the extents to which acids, bases, and salts are ionized in dilute aqueous solutions.

Strengths of Binary Acids

36. Classify each of the hydrides LiH, BeH_2, BH_3, CH_4, NH_3, H_2O, and HF as a Brønsted–Lowry base, a Brønsted–Lowry acid, or neither.
37. What does "acid strength" mean?
38. What does "base strength" mean?
39. Which of the following substances are (a) strong soluble bases, (b) insoluble bases, (c) strong acids, or (d) weak acids? LiOH, HCl, $Ca(OH)_2$, $Fe(OH)_2$, H_2S, H_2CO_3, H_2SO_4, $Zn(OH)_2$.
40. (a) What are binary protonic acids? (b) Write names and formulas for four binary protonic acids.
41. (a) How can the order of increasing acid strength in a series of similar binary protonic acids be explained?
(b) Illustrate your answer for the series HF, HCl, HBr, and HI.
(c) What is the order of increasing base strength of the conjugate bases of the acids in (b)? Why?
(d) Is your explanation applicable to the series H_2O, H_2S, H_2Se, and H_2Te? Why?
42. What does the term "leveling effect" mean? Illustrate your answer with three specific examples.
43. (a) Which is the stronger acid of each pair? (1) NH_4^+, NH_3; (2) H_2O, H_3O^+; (3) HS^-, H_2S. (b) How are acidity and charge related?
44. Arrange the members of each group in order of decreasing acidity: (a) H_2O, H_2Se, H_2S; (b) HI, HCl, HF, HBr; (c) H_2S, S^{2-}, HS^-; (d) SiH_4, HCl, PH_3, H_2S.
45. Illustrate the leveling effect of water by writing reactions for HCl and HNO_3.

Reactions of Acids and Bases

46. Why are acid–base reactions described as neutralization reactions?
47. Distinguish among (a) formula unit equations, (b) total ionic equations, and (c) net ionic equations. What are the advantages and limitations of each?
48. Classify each substance as either an electrolyte or a nonelectrolyte: NH_4Cl, HI, C_6H_6, RaF_2, $Zn(CH_3COO)_2$, $Cu(NO_3)_2$, CH_3COOH, $C_{12}H_{22}O_{11}$ (table sugar), LiOH, $KHCO_3$, CCl_4, $La_2(SO_4)_3$, I_2.
49. Classify each substance as either a strong or a weak electrolyte, and then list (a) the strong acids, (b) the strong bases, (c) the weak acids, and (d) the weak bases. NaCl, $MgSO_4$, HCl, $(COOH)_2$, $Ba(NO_3)_2$, H_3PO_4, $Sr(OH)_2$, HNO_3, HI, $Ba(OH)_2$, LiOH, C_3H_5COOH, NH_3, CH_3NH_2, KOH, $MgMoO_4$, HCN, $HClO_4$.

For Exercises 50–52, write balanced (1) formula unit, (2) total ionic, and (3) net ionic equations for reactions between the following acid–base pairs. Name all compounds except water. Assume complete neutralization.

50. (a) $HNO_3 + KOH \longrightarrow$
(b) $H_2SO_4 + NaOH \longrightarrow$
(c) $HCl + Ca(OH)_2 \longrightarrow$
(d) $CH_3COOH + KOH \longrightarrow$
51. (a) $H_2CO_3 + Sr(OH)_2 \longrightarrow$
(b) $H_2SO_4 + Ba(OH)_2 \longrightarrow$
(c) $H_3PO_4 + Ca(OH)_2 \longrightarrow$
(d) $H_2S + KOH \longrightarrow$
(e) $H_3AsO_4 + KOH \longrightarrow$
52. (a) $HClO_4 + Ba(OH)_2 \longrightarrow$
(b) $HCl + NH_3 \longrightarrow$
(c) $HNO_3 + NH_3 \longrightarrow$
(d) $H_2SO_4 + Fe(OH)_3 \longrightarrow$
(e) $H_3PO_4 + Ba(OH)_2 \longrightarrow$
53. Complete these equations by writing the formulas of the omitted compounds.
(a) $Ba(OH)_2 + ? \rightarrow BaSO_4(s) + H_2O$
(b) $FeO(s) + ? \rightarrow Fe(NO_3)_2(aq) + H_2O$
(c) $HCl(aq) + ? \rightarrow AlCl_3(aq) + ?$
(d) $Na_2O + ? \rightarrow 2NaOH(aq)$
(e) $NaOH + ? \rightarrow Na_2HPO_4(aq) + ?$ (two possible answers)
54. Although many salts may be formed by a variety of reactions, salts are usually thought of as being derived from the reaction of an acid with a base. For each of the salts listed below, choose the acid and base that react with each other to form the salt. Write the (1) formula unit, (2) total ionic, and (3) net ionic equations for the formation of each salt. (a) $Pb(NO_3)_2$, (b) $FeCl_3$, (c) $(NH_4)_2CO_3$, (d) $Ca(ClO_4)_2$, (e) $Al_2(SO_4)_3$

Acidic and Basic Salts

55. What are polyprotic acids? Write names and formulas for five polyprotic acids.

56. What are acidic salts? Write balanced equations to show how the following acidic salts can be prepared from the appropriate acid and base: $NaHSO_3$, $NaHCO_3$, NaH_2PO_4, Na_2HPO_4.
57. Indicate the molar ratio of acid and base required in each case in Exercise 56.
58. The following salts are components of fertilizers. They are made by reacting gaseous NH_3 with concentrated solutions of acids. The heat produced by the reactions evaporates most of the water. Write balanced molecular equations that show the formation of each. (a) NH_4NO_3, (b) $NH_4H_2PO_4$, (c) $(NH_4)_2HPO_4$, (d) $(NH_4)_3PO_4$
59. Some of the acid formed in tissues is excreted through the kidneys. One of the bases removing the acid is HPO_4^{2-}. Write the equation for the reaction. Could Cl^- serve this function?
*60. Acids react with metal carbonates and hydrogen carbonates to form carbon dioxide and water.
(a) Write the balanced equation for the reaction that occurs when baking soda, $NaHCO_3$, and vinegar, 5% acetic acid, are mixed. What causes the "fizz"?
(b) Lactic acid, $CH_3CH(OH)COOH$, is found in sour milk and in buttermilk. Many of its reactions are very similar to those of acetic acid. Write the balanced equation for the reaction of baking soda, $NaHCO_3$, with lactic acid. Explain why bread "rises" during the baking process.
61. What are polyhydroxy bases? Write names and formulas for five polyhydroxy bases.
62. What are basic salts?
(a) Write balanced equations to show how the following basic salts can be prepared from the appropriate acid and base: $Ba(OH)Cl$, $Al(OH)_2Cl$, $Al(OH)Cl_2$.
(b) Indicate the molar ratio of acid and base required in each case.

Strengths of Ternary Acids and Amphoterism

63. What are ternary acids? Write names and formulas for four of them.
64. Why can we describe nitric and sulfuric acids as "hydroxides" of nonmetals?
65. Explain the order of increasing acid strength for the following groups of acids and the order of increasing base strength for their conjugate bases. (a) H_2SO_3, H_2SO_4; (b) HNO_2, HNO_3; (c) H_3PO_3, H_3PO_4; (d) $HClO$, $HClO_2$, $HClO_3$, $HClO_4$.
66. (a) Write a generalization that describes the order of acid strengths for a series of ternary acids that contain different elements in the same oxidation state from the same group in the periodic table.
(b) Indicate the order of acid strengths for the following: (1) HNO_3, H_3PO_4; (2) H_3PO_4, H_3AsO_4; (3) H_2SO_4, H_2SeO_4; (4) $HClO_3$, $HBrO_3$.
67. List the following acids in order of increasing strength: (a) sulfuric, phosphoric, and perchloric; (b) HIO_3, HIO_2, HIO, and HIO_4; (c) selenous, sulfurous, and tellurous; (d) hydrosulfuric, hydroselenic, and hydrotelluric; (e) H_2CrO_4, H_2CrO_2, $HCrO_3$, and H_3CrO_3; (f) $H_4P_2O_7$, $HP_2O_7^{3-}$, $H_3P_2O_7^-$, and $H_2P_2O_7^{2-}$.
68. NaOH behaves as a base in water, while ClOH behaves as an acid. Clearly explain this behavior and the general principles involved for any El—O—H compound.
69. What are amphoteric metal hydroxides? (a) Are they bases? (b) Write the names and formulas for five amphoteric metal hydroxides.
70. Chromium(III) hydroxide and lead(II) hydroxide are typical amphoteric hydroxides.
(a) Write the formula unit, total ionic, and net ionic equations for the complete reaction of each hydroxide with nitric acid.
(b) Write the same kinds of equations for the reaction of each hydroxide with an excess of potassium hydroxide solution. Reference to Table 10-2 may be helpful.

Preparation of Acids

71. Volatile acids such as nitric acid, HNO_3, and acetic acid, CH_3COOH, can be prepared by adding concentrated H_2SO_4 to salts of the acids.
(a) Write chemical equations for the reaction of H_2SO_4 with (1) sodium acetate and (2) sodium nitrate (called chile saltpeter.)
(b) Why can't a dilute aqueous solution of H_2SO_4 be used?
72. Outline a method of preparing each of the following acids and write appropriate balanced equations for each preparation: (a) H_2S, (b) HCl, (c) HNO_3.
73. Repeat Exercise 72 for (a) carbonic acid, (b) perchloric acid, (c) permanganic acid, and (d) phosphoric acid (two methods).
74. Give the formula for an example chosen from the representative elements for (a) an acidic oxide, (b) an amphoteric oxide, and (c) a basic oxide.

The Lewis Theory

75. Define and illustrate the following terms clearly and concisely. Write an equation to illustrate the meaning of each term. (a) Lewis acid, (b) Lewis base, (c) neutralization (according to Lewis theory)
76. What are the advantages and limitations of the Brønsted–Lowry theory?
77. What are the advantages and limitations of the Lewis theory?
78. Draw a Lewis formula for each species in the following equations. Label the acids and bases using Lewis theory terminology.
(a) $H_2O + H_2O \rightleftharpoons H_3O^+ + OH^-$
(b) $HCl(g) + H_2O \longrightarrow H_3O^+ + Cl^-$
(c) $NH_3(g) + H_2O \rightleftharpoons NH_4^+ + OH^-$
(d) $NH_3(g) + HF(g) \longrightarrow NH_4F(s)$

79. What is the term for a single covalent bond in which both electrons in the shared pair come from the same atom? Identify the Lewis acid and base and the donor and acceptor atoms in the following.

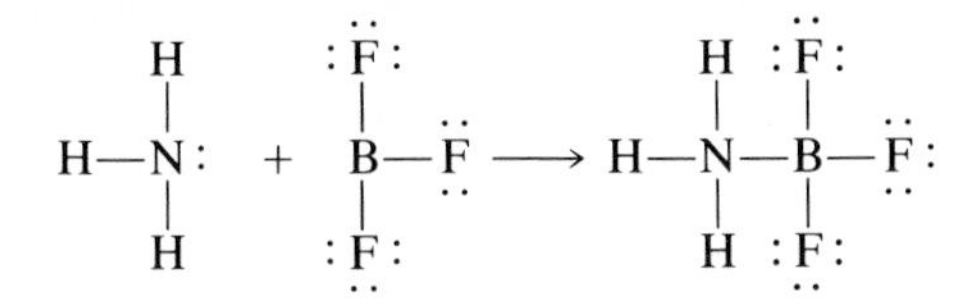

80. Identify the Lewis acid and base and the donor and acceptor atoms in each of the following.

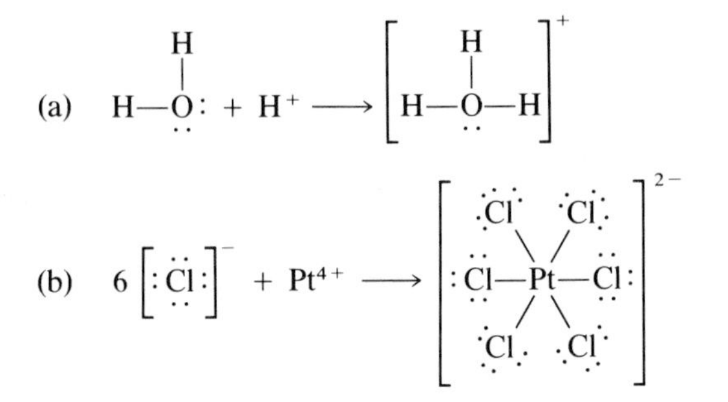

81. Iodine, I_2, is much more soluble in a water solution of potassium iodide, KI, than it is in H_2O. The anion found in the solution is I_3^-. Write an equation for this reaction, indicating the Lewis acid and the Lewis base.

82. A group of very strong acids are the fluoroacids, H_mXF_n. Two such acids are formed by Lewis acid–base reactions.

(a) Identify the Lewis acid and the Lewis base:

$HF + SbF_5 \longrightarrow H(SbF_6)$ (called a "super" acid, hexafluoroantimonic acid)

$HF + BF_3 \longrightarrow H(BF_4)$ (tetrafluoroboric acid)

(b) To which atom is the H of the product bonded? How is the H bonded?

Mixed Exercises

83. Sort the following list of chemicals into (i) acidic, (ii) basic, and (iii) amphoteric species. Assume all oxides are dissolved in water. Do not be intimidated by the way in which the formula of the compound is written. (a) Cs_2O, (b) N_2O_5, (c) HCl, (d) $SO_2(OH)_2$, (e) HNO_2, (f) Al_2O_3, (g) BaO, (h) H_2O, (i) CO_2

84. Indicate which of the following substances—(a) HCl, (b) $H_2PO_3^-$, (c) H_2CaO_2, (d) $ClO_3(OH)$, (e) $Sb(OH)_3$—can act as (i) an acid, (ii) a base, or (iii) both according to the (α) Arrhenius (classical) theory and/or the (β) Brønsted–Lowry theory. Do not be confused by the way in which the formulas are written.

85. (a) Write equations for the reactions (1) $HCO_3^- + H_3O^+$ and (2) $HCO_3^- + OH^-$, and indicate the conjugate acid–base pairs in each case.

(b) A substance such as HCO_3^- that reacts with both H_3O^+ and OH^- is said to be ________. (Fill in the missing word.)

86. (a) List the conjugate bases of (1) H_3PO_4, (2) NH_4^+, and (3) OH^- and the conjugate acids of (4) HSO_4^-, (5) PH_3, and (6) PO_4^{3-}.

(b) Given that NO_2^- is a stronger base than NO_3^-, which is the stronger acid—nitric acid, HNO_3, or nitrous acid, HNO_2?

87. To determine the relative basicities of bases (a) stronger than OH^-, use an acid that is ________ (*stronger than, the same strength as,* or *weaker than*) H_2O; (b) weaker than H_2O, use an acid that is ________ (*stronger than, the same strength as,* or *weaker than*) H_3O^+.

88. Write net ionic equations for the reactions of the amphoteric hydroxide $Sn(OH)_2$ with (a) HCl and (b) NaOH.

***89.** A 0.1 *M* solution of copper(II) chloride, $CuCl_2$, causes the light bulb in Figure 4-2 to glow brightly. When hydrogen sulfide, H_2S, a very weak acid, is added to the solution, a black precipitate of copper(II) sulfide, CuS, forms and the bulb still glows brightly. The experiment is repeated with a 0.1 *M* solution of copper(II) acetate, $Cu(C_2H_3O_2)_2$, which also causes the bulb to glow brightly. Again, CuS forms, but this time the bulb glows dimly. With the aid of ionic equations, explain the difference in behavior of the $CuCl_2$ and $Cu(C_2H_3O_2)_2$ solutions.

90. Referring again to Figure 4-2, explain the following results of a conductivity experiment (use ionic equations).
(a) Individual solutions of NaOH and HCl cause the bulb to glow brightly. When the solutions are mixed, the bulb still glows brightly.
(b) Individual solutions of NH_3 and CH_3COOH cause the bulb to glow dimly. When the solutions are mixed, the bulb glows brightly.

91. Which statements are true? Rewrite any false statement so that it is correct.
(a) Strong acids and bases are virtually 100% ionized or dissociated in dilute aqueous solutions.
(b) The leveling effect is the seemingly identical strengths of all acids and bases in aqueous solutions.
(c) A conjugate acid is a molecule or ion formed by the addition of a proton to a base.
(d) Amphoterism and amphiprotism are the same in aqueous solution.

11 Reactions in Aqueous Solutions II: Calculations

Automatic titrators are used in modern analytical laboratories. Such titrators rely on electrical properties of the solutions. Methyl red indicator changes from yellow to red at the end point of this titration.

Outline

Objectives

As you study this chapter, you should

- ☐ Review molarity calculations and expand your understanding of molarity
- ☐ Learn about standardization and the use of standard solutions of acids and bases using
  - The mole method and molarity
  - Equivalent weights and normality
- ☐ Review redox reactions and balancing redox equations
- ☐ Learn about redox titrations using
  - The mole method and molarity
  - Equivalent weights and normality

Hydrochloric acid, HCl, is called "stomach acid" because it is the main acid (~ 0.10 *M*) in our digestive juices. When the concentration of HCl is too high in humans, problems result. These range from "heartburn" to ulcers, which can eat through the lining of the stomach wall. Snakes have very high concentrations of HCl in their digestive juices so that they can digest whole small animals and birds.

Automobile batteries contain 40% H_2SO_4 by mass. When the battery has "run down," the concentration of H_2SO_4 is significantly lower than 40%. A technician checks an automobile battery by drawing some battery acid into a hydrometer, which indicates the density of the solution. This density is related to the concentration of H_2SO_4.

There are many practical applications of acid–base chemistry in which we must know the concentration of a solution of an acid or a base.

Concentrations and Aqueous Acid–Base Reactions

11-1 Calculations Involving Molarity

In Sections 3-6 through 3-9, we introduced methods for expressing concentrations of solutions and discussed some related calculations. Review of those sections will be helpful as we learn more about acid–base reactions in solutions.

In *some cases,* one mole of an acid reacts with one mole of a base:

$$HCl + NaOH \longrightarrow NaCl + H_2O$$

$$HNO_3 + KOH \longrightarrow KNO_3 + H_2O$$

Because one mole of each acid reacts with one mole of each base in these examples, *one liter of a one-molar solution of either of these acids* reacts with *one liter of a one-molar solution of either of these bases.* These acids have only one acidic hydrogen per formula unit, and these bases have one hydroxide ion per formula unit, so one formula unit of base reacts with one hydrogen ion.

The *reaction ratio* is the relative numbers of moles of reactants and products shown in the balanced equation.

Example 11-1

If 100.0 mL of 1.00 *M* HCl solution and 100.0 mL of 1.00 *M* NaOH are mixed, what is the molarity of the salt in the resulting solution? You may assume that the volumes are additive.

Experiments have shown that volumes of dilute aqueous solutions are very nearly additive. No significant error is introduced by making this assumption. 100 mL of NaOH solution mixed with 100 mL of HCl solution gives 200 mL of solution.

Plan

We first write the balanced equation for the acid–base reaction, and then construct the reaction summary that shows the amounts (millimoles) of NaOH and HCl. We determine the amount of salt formed from the reaction summary. The final (total) volume is the sum of the volumes mixed. Then we can calculate the molarity of the salt.

The millimole (mmol) was introduced at the end of Section 2-6. Please review Example 2-11. Recall that
1 mol = 1000 mmol
1 L = 1000 mL

Solution

The following tabulation shows that equal numbers of millimoles or moles of HCl and NaOH are mixed, and therefore the resulting solution contains only NaCl, the salt formed by the reaction, and water:

	$NaOH$	+	HCl	$\longrightarrow NaCl$	+ H_2O
Rxn ratio:	1 mmol		1 mmol	1 mmol	1 mmol
Start:	$\left[100\text{ mL}\left(\frac{1.00\text{ mmol}}{\text{mL}}\right)\right]$ = 100 mmol NaOH		$\left[100\text{ mL}\left(\frac{1.00\text{ mmol}}{\text{mL}}\right)\right]$ = 100 mmol HCl	0 mmol	
Change:	−100 mmol		−100 mmol	+100 mmol	
After rxn:	0 mmol		0 mmol	100 mmol	

The HCl and NaOH neutralize each other exactly, and the resulting solution contains 100 mmol of NaCl in 200 mL of solution. Its molarity is

$$\underline{?}\ \frac{\text{mmol NaCl}}{\text{mL}} = \frac{100\text{ mmol}}{200\text{ mL}} = 0.500\ M\ \text{NaCl}$$

Example 11-2

If 100 mL of 1.00 *M* HCl and 100 mL of 0.75 *M* NaOH solutions are mixed, what is the molarity of the resulting solution?

Plan

We proceed as we did in Example 11-1. This reaction summary shows that NaOH is the limiting reactant and that we have excess HCl.

Solution

	HCl	+ NaOH	$\longrightarrow$ NaCl	+ H_2O
Rxn ratio:	1 mmol	1 mmol	1 mmol	1 mmol
Start:	100 mmol	75 mmol	0 mmol	
Change:	−75 mmol	−75 mmol	+75 mmol	
After rxn:	25 mmol	0 mmol	75 mmol	

Because two solutes are present in the solution after reaction, we must calculate the concentrations of both:

$$\underline{?}\ \frac{\text{mmol HCl}}{\text{mL}} = \frac{25 \text{ mmol HCl}}{200 \text{ mL}} = 0.12\ M \text{ HCl}$$

$$\underline{?}\ \frac{\text{mmol NaCl}}{\text{mL}} = \frac{75 \text{ mmol NaCl}}{200 \text{ mL}} = 0.38\ M \text{ NaCl}$$

Both HCl and NaCl are strong electrolytes, so the solution is 0.12 M in H^+(aq), $(0.12 + 0.38)\ M = 0.50\ M$ in Cl^-, and 0.38 M in Na^+ ions.

EOC 10, 11

In many cases more than one mole of a base will be required to neutralize completely one mole of an acid, or more than one mole of an acid will be required to neutralize completely one mole of a base.

$$\underset{1\text{ mol}}{H_2SO_4} + \underset{2\text{ mol}}{2NaOH} \longrightarrow \underset{1\text{ mol}}{Na_2SO_4} + 2H_2O$$

$$\underset{2\text{ mol}}{2HCl} + \underset{1\text{ mol}}{Ca(OH)_2} \longrightarrow \underset{1\text{ mol}}{CaCl_2} + 2H_2O$$

The first equation shows that one mole of H_2SO_4 reacts with two moles of NaOH. Thus, *two* liters of 1 M NaOH solution are required to neutralize one liter of 1 M H_2SO_4 solution. The second equation shows that two moles of HCl react with one mole of $Ca(OH)_2$. Thus, *two* liters of HCl solution are required to neutralize one liter of $Ca(OH)_2$ solution of equal molarity.

Example 11-3

What volume of 0.00300 M HCl solution would just neutralize 30.0 mL of 0.00100 M $Ca(OH)_2$ solution?

Plan

We write the balanced equation for the reaction to determine the reaction ratio. Then we convert (1) milliliters of $Ca(OH)_2$ solution to moles of $Ca(OH)_2$ using molarity as a unit factor, 0.00100 mol $Ca(OH)_2$/1000 mL $Ca(OH)_2$ solution; (2) moles of $Ca(OH)_2$ to moles of HCl using the unit factor, 2 mol HCl/1 mol $Ca(OH)_2$ (from the balanced equation); and (3) moles of HCl to milliliters of HCl solution using the unit factor, 1000 mL HCl/0.00300 mol HCl.

mL $Ca(OH)_2$ soln $\longrightarrow$ mol $Ca(OH)_2$ present $\longrightarrow$ mol HCl needed $\longrightarrow$ mL HCl(aq) needed

Solution

The balanced equation for the reaction is

$$\underset{2\ \text{mol}}{2HCl} + \underset{1\ \text{mol}}{Ca(OH)_2} \longrightarrow \underset{1\ \text{mol}}{CaCl_2} + \underset{2\ \text{mol}}{2H_2O}$$

$$\underline{?}\ \text{mL HCl} = 30.0\ \text{mL}\ Ca(OH)_2 \times \frac{0.00100\ \text{mol}\ Ca(OH)_2}{1000\ \text{mL}\ Ca(OH)_2} \times \frac{2\ \text{mol HCl}}{1\ \text{mol}\ Ca(OH)_2} \times \frac{1000\ \text{mL HCl}}{0.00300\ \text{mol HCl}}$$

$$= 20.0\ \text{mL HCl}$$

EOC 7

In the preceding example we used the unit factor, 2 mol HCl/1 mol $Ca(OH)_2$, to convert moles of $Ca(OH)_2$ to moles of HCl because the balanced equation for the reaction shows that two moles of HCl are required to neutralize one mole of $Ca(OH)_2$. We must always write the balanced equation for the reaction and determine the *reaction ratio*.

Example 11-4

If 100.0 mL of 1.00 *M* H_2SO_4 solution is mixed with 200.0 mL of 1.00 *M* KOH, what salt is produced, and what is its molarity?

Plan

We proceed as we did in Example 11-1. We note that the reaction ratio is 1 mmol of H_2SO_4 to 2 mmol of KOH to 1 mmol of K_2SO_4.

Solution

	H_2SO_4	+	2KOH	$\longrightarrow$	K_2SO_4	$+ 2H_2O$
Rxn ratio:	1 mmol		2 mmol		1 mmol	
Start:	$\left[100.0\ \text{mL}\left(\frac{1.00\ \text{mmol}}{\text{mL}}\right)\right]$		$\left[200.0\ \text{mL}\left(\frac{1.00\ \text{mmol}}{\text{mL}}\right)\right]$			
	= 100 mmol		= 200 mmol		0 mmol	
Change:	−100 mmol		−200 mmol		+100 mmol	
After rxn:	0 mmol		0 mmol		100 mmol	

The reaction produces 100 mmol of potassium sulfate. This is contained in 300 mL of solution, and so the concentration is

$$\underline{?}\ \frac{\text{mmol}\ K_2SO_4}{\text{mL}} = \frac{100\ \text{mmol}\ K_2SO_4}{300\ \text{mL}} = 0.333\ M\ K_2SO_4$$

Because K_2SO_4 is a strong electrolyte, this corresponds to 0.666 *M* K^+ and 0.333 *M* SO_4^{2-}.

EOC 12

Standardization of Solutions of Acids and Bases

In Section 3-9 we discussed *titrations* of solutions of acids and bases and introduced the terminology used to describe titrations. Please review Section 3-9 thoroughly and study Figure 3-4.

Standardization is the process by which one determines the concentration of a solution by measuring accurately the volume of the solution required to react with an exactly known amount of a **primary standard**. The standardized solution is then known as a **secondary standard** and is used in the analysis of unknowns.

The properties of an ideal *primary standard* include the following:

CO_2, H_2O, and O_2 are present in the atmosphere. They react with many substances.

1. It must not react with or absorb the components of the atmosphere, such as water vapor, oxygen, and carbon dioxide.
2. It must react according to one invariable reaction.
3. It must have a high percentage purity.
4. It should have a high formula weight to minimize the effect of error in weighing.
5. It must be soluble in the solvent of interest.
6. It should be nontoxic.

The first five of these characteristics minimize the errors involved in analysis. An additional factor, low cost, is desirable but not necessary. Because primary standards are often costly and difficult to prepare, secondary standards are often used in day-to-day work.

11-2 Acid–Base Titrations: The Mole Method and Molarity

Refer to the Brønsted–Lowry theory (Section 10-3).

Let us now describe the use of a few primary standards for acids and bases. One primary standard for solutions of acids is sodium carbonate, Na_2CO_3, a solid compound.

$$\underset{1\text{ mol}}{H_2SO_4} + \underset{1\text{ mol}}{Na_2CO_3} \longrightarrow \underset{1\text{ mol}}{Na_2SO_4} + \underset{1\text{ mol}}{CO_2} + \underset{1\text{ mol}}{H_2O}$$

$$1\text{ mol }Na_2CO_3 = 106.0\text{ g} \qquad \text{and} \qquad 1\text{ mmol }Na_2CO_3 = 0.1060\text{ g}$$

Sodium carbonate is a salt. However, because a base can be broadly defined as a substance that reacts with hydrogen ions, in *this* reaction Na_2CO_3 can be thought of as a base.

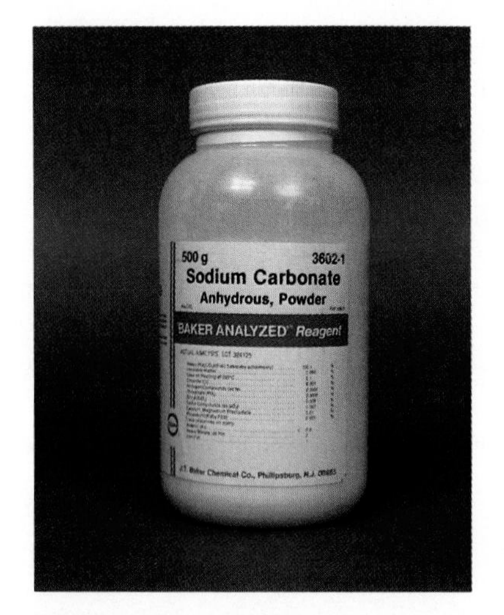

Sodium carbonate is often used as a primary standard for acids.

Example 11-5

Calculate the molarity of a solution of H_2SO_4 if 40.0 mL of the solution neutralize 0.364 gram of Na_2CO_3.

Plan

We know from the balanced equation that 1 mol of H_2SO_4 reacts with 1 mol of Na_2CO_3, 106.0 g. This provides the unit factors that convert 0.364 g of Na_2CO_3 to the corresponding number of moles of H_2SO_4, from which we can calculate molarity.

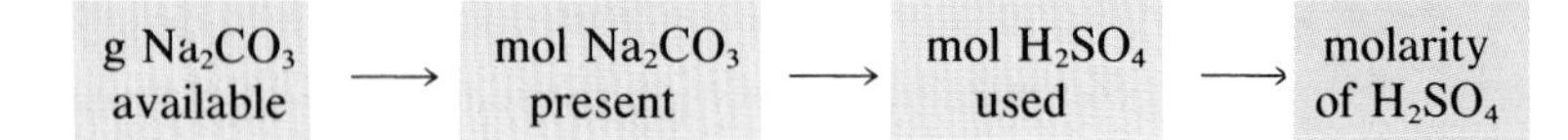

Solution

$$\underline{?}\text{ mol }H_2SO_4 = 0.364\text{ g }Na_2CO_3 \times \frac{1\text{ mol }Na_2CO_3}{106.0\text{ g }Na_2CO_3} \times \frac{1\text{ mol }H_2SO_4}{1\text{ mol }Na_2CO_3}$$

$$= 0.00343\text{ mol }H_2SO_4 \qquad \text{(present in 40.0 mL of solution)}$$

Now we calculate the molarity of the H_2SO_4 solution:

$$\frac{?\ \text{mol}\ H_2SO_4}{\text{L}} = \frac{0.00343\ \text{mol}\ H_2SO_4}{0.0400\ \text{L}} = 0.0858\ M\ H_2SO_4$$

EOC 31

Most inorganic bases are metal hydroxides, all of which are solids. However, even in the solid state, most inorganic bases react rapidly with CO_2 (an acid anhydride) from the atmosphere. Most metal hydroxides also absorb H_2O from the air. These properties make it *very* difficult to accurately weigh out samples of pure metal hydroxides. Chemists obtain solutions of bases of accurately known concentration by standardizing the solutions against an acidic salt, potassium hydrogen phthalate, $KC_6H_4(COO)(COO\underline{H})$. This is produced by neutralization of one of the two ionizable hydrogens of an organic acid, phthalic acid:

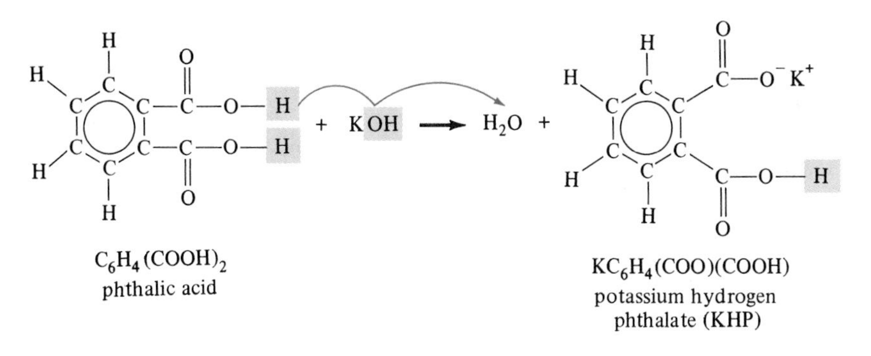

This acidic salt, known as KHP, has one acidic hydrogen (highlighted) that reacts with bases. KHP is easily obtained in a high state of purity, and is soluble in water. It is used as a primary standard for bases.

Very pure KHP is available.

Example 11-6

A 20.00-mL sample of a solution of NaOH reacts with 0.3641 gram of KHP. Calculate the molarity of the basic solution.

Plan

We first write the balanced equation for the reaction between NaOH and KHP. We then calculate the number of moles of NaOH in 20.00 mL of solution from the amount of KHP that reacts with it. Then we can calculate the molarity of the NaOH solution.

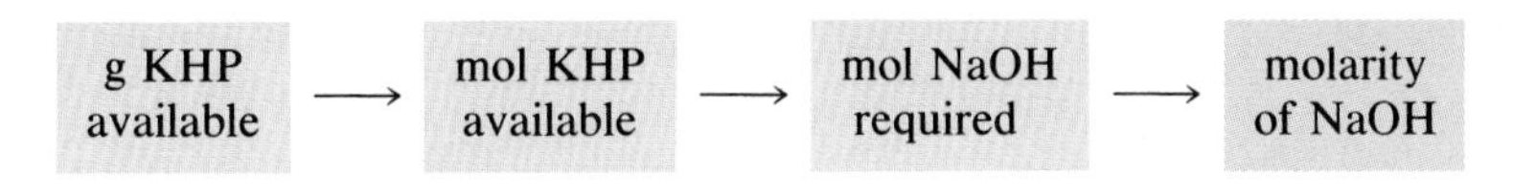

Solution

$$\underset{1\ \text{mol}}{\text{NaOH}} + \underset{1\ \text{mol}}{\text{KHP}} \longrightarrow \underset{1\ \text{mol}}{\text{NaKP}} + \underset{1\ \text{mol}}{H_2O}$$

The P in KHP stands for the phthalate ion, $C_6H_4(COO)_2^{2-}$, *not* phosphorus.

We see that NaOH and KHP react in a 1:1 mole ratio. One mole of KHP is 204.2 g.

$$\underline{?} \text{ mol NaOH} = 0.3641 \text{ g KHP} \times \frac{1 \text{ mol KHP}}{204.2 \text{ g KHP}} \times \frac{1 \text{ mol NaOH}}{1 \text{ mol KHP}}$$

$$= 0.001783 \text{ mol NaOH}$$

Then we calculate the molarity of the NaOH solution.

$$\frac{\underline{?} \text{ mol NaOH}}{\text{L}} = \frac{0.001783 \text{ mol NaOH}}{0.02000 \text{ L}} = 0.08915\ M \text{ NaOH}$$

EOC 28, 31

Impure samples of acids can be titrated with standard solutions of bases. The results can be used to determine percentage purity of the samples.

Example 11-7

Oxalic acid is used to remove iron stains and some ink stains from fabrics. A 0.1743-gram sample of *impure* oxalic acid, $(COOH)_2$, required 39.82 mL of 0.08915 M NaOH solution for complete neutralization. No acidic impurities were present. Calculate the percentage purity of the $(COOH)_2$.

Plan

We write the balanced equation for the reaction and calculate the number of moles of NaOH in the standard solution. Then we calculate the mass of $(COOH)_2$ in the sample, which gives us the information we need to calculate percentage purity.

Solution

The equation for the reaction of NaOH with $(COOH)_2$ is

Each molecule of $(COOH)_2$ contains two acidic H's.

$$\text{H—O—}\overset{\text{O}}{\overset{\|}{\text{C}}}\text{—}\overset{\text{O}}{\overset{\|}{\text{C}}}\text{—O—H}$$

1 mol = 90.04 g

$$\underset{2 \text{ mol}}{2NaOH} + \underset{1 \text{ mol}}{(COOH)_2} \longrightarrow \underset{1 \text{ mol}}{Na_2(COO)_2} + \underset{2 \text{ mol}}{2H_2O}$$

Two moles of NaOH neutralize completely one mole of $(COOH)_2$. The number of moles of NaOH that react is the volume times the molarity of the solution:

$$\underline{?} \text{ mol NaOH} = 0.03982 \text{ L} \times \frac{0.08915 \text{ mol NaOH}}{\text{L}} = 0.003550 \text{ mol NaOH}$$

Now we calculate the mass of $(COOH)_2$ that reacts with 0.003550 mol NaOH.

$$\underline{?} \text{ g } (COOH)_2 = 0.003550 \text{ mol NaOH} \times \frac{1 \text{ mol } (COOH)_2}{2 \text{ mol NaOH}} \times \frac{90.04 \text{ g } (COOH)_2}{1 \text{ mol } (COOH)_2}$$

$$= 0.1598 \text{ g } (COOH)_2$$

The sample contained 0.1598 g of $(COOH)_2$, and its percentage purity was

$$\% \text{ purity} = \frac{0.1598 \text{ g } (COOH)_2}{0.1743 \text{ g sample}} \times 100\% = 91.68\% \text{ pure } (COOH)_2$$

EOC 32, 34

11-3 Acid–Base Titrations: Equivalent Weights and Normality

Any calculation that can be carried out with equivalent weights and normality can also be done by the mole method using molarity. However, the methods of this section are widely used in health related fields and in many industrial laboratories.

Because one mole of an acid does not necessarily neutralize one mole of a base, many chemists prefer a method of expressing concentration other than molarity to retain a one-to-one relationship. Concentrations of solutions of acids and bases are frequently expressed as *normality* (N). The **normality** of a solution is defined as the number of equivalent weights, or simply equivalents (eq), of solute per liter of solution. (The term "equivalent mass" is not widely used.) Normality may be represented symbolically as

$$\text{normality} = \frac{\text{number of equivalent weights of solute}}{\text{liter of solution}} = \frac{\text{no. eq}}{\text{L}}$$

An **equivalent weight** is often referred to simply as an **equivalent** (eq).

By definition there are 1000 milliequivalent weights (meq) in one equivalent weight of an acid or base. Normality may also be represented as

A **milliequivalent weight** is often referred to simply as a **milliequivalent** (meq).

$$\text{normality} = \frac{\text{number of milliequivalent weights of solute}}{\text{milliliter of solution}} = \frac{\text{no. meq}}{\text{mL}}$$

In acid–base reactions, one **equivalent weight**, or **equivalent (eq), of an acid** is defined as the mass of the acid (expressed in grams) that will furnish 6.022×10^{23} hydrogen ions (1 mol) or that will react with 6.022×10^{23} hydroxide ions (1 mol). One mole of an acid contains 6.022×10^{23} formula units of the acid. Consider hydrochloric acid as a typical monoprotic acid:

HCl	$\xrightarrow{H_2O}$	$H^+(aq)$	+	$Cl^-(aq)$
1 mol		1 mol		1 mol
36.46 g		1.008 g		35.45 g
6.022×10^{23} FU		6.022×10^{23} FU		6.022×10^{23} FU

We see that one mole of HCl can produce 6.022×10^{23} H^+, and so *one mole of HCl is one equivalent*. The same is true for all monoprotic acids.

Sulfuric acid is a diprotic acid. One molecule of H_2SO_4 can furnish $2H^+$ ions.

H_2SO_4	$\xrightarrow{H_2O}$	$2H^+(aq)$	+	$SO_4^{2-}(aq)$
1 mol		2 mol		1 mol
98.08 g		2(1.008 g)		96.06 g
6.022×10^{23} FU		$2(6.022 \times 10^{23})$ FU		6.022×10^{23} FU

This equation shows that one mole of H_2SO_4 can produce $2(6.022 \times 10^{23})$ H^+; therefore, one mole of H_2SO_4 is *two* equivalent weights in all reactions in which *both* acidic hydrogen atoms react.

One **equivalent weight of a base** is defined as the mass of the base (expressed in grams) that will furnish 6.022×10^{23} hydroxide ions or the mass of the base that will react with 6.022×10^{23} hydrogen ions.

The equivalent weight of an *acid* is obtained by dividing its formula weight in grams either by the number of acidic hydrogens furnished by one formula unit of the acid *or* by the number of hydroxide ions with which one formula unit of the acid reacts. The equivalent weight of a *base* is obtained by dividing its formula weight in grams either by the number of hydroxide ions furnished by one formula unit *or* by the number of hydrogen ions with which one formula unit of the base reacts. Equivalent weights of some common acids and bases are given in Table 11-1.

From the definitions of one equivalent of an acid and of a base, we see that *one equivalent of any acid reacts with one equivalent of any base*. It is *not* true that one mole of any acid reacts with one mole of any base in a specific chemical reaction. As a consequence of the definition of equivalents,

Table 11-1
Equivalent Weights* of Some Acids and Bases

Acids					Bases				
Symbolic Representation				One eq	Symbolic Representation				One eq
$\frac{HNO_3}{1}$	=	$\frac{63.02\ g}{1}$	=	63.02 g HNO_3	$\frac{NaOH}{1}$	=	$\frac{40.00\ g}{1}$	=	40.00 g NaOH
$\frac{CH_3COO\underline{H}}{1}$	=	$\frac{60.03\ g}{1}$	=	60.03 g $CH_3COO\underline{H}$	$\frac{NH_3}{1}$	=	$\frac{17.04\ g}{1}$	=	17.04 g NH_3
$\frac{KC_6H_4(COO)(COO\underline{H})}{1}$	=	$\frac{204.2\ g}{1}$	=	204.2 g $KC_6H_4(COO)(COO\underline{H})$	$\frac{Ca(OH)_2}{2}$	=	$\frac{74.10\ g}{2}$	=	37.05 g $Ca(OH)_2$
$\frac{H_2SO_4}{2}$	=	$\frac{98.08\ g}{2}$	=	49.04 g H_2SO_4	$\frac{Ba(OH)_2}{2}$	=	$\frac{171.36\ g}{2}$	=	85.68 g $Ba(OH)_2$

* Complete neutralization is assumed.

The notation ≙ is read "is equivalent to."

1 eq acid ≙ 1 eq base. In general, we may write the following for *all* acid–base reactions that go to completion:

$$\text{no. eq acid} = \text{no. eq base} \quad \textit{or} \quad \text{no. meq acid} = \text{no. meq base}$$

The product of the volume of a solution, in liters, and its normality is equal to the number of equivalents of solute contained in the solution. For a solution of an acid,

Remember that the product of volume and concentration equals the amount of solute.

$$L_{acid} \times N_{acid} = L_{acid} \times \frac{\text{eq acid}}{L_{acid}} = \text{eq acid}$$

Alternatively,

$$mL_{acid} \times N_{acid} = mL_{acid} \times \frac{\text{meq acid}}{mL_{acid}} = \text{meq acid}$$

Similar relationships can be written for a solution of a base. Because 1 eq of acid *always* reacts with 1 eq of base, we may write

$$\text{no. eq acid} = \text{no. eq base} \qquad \text{so}$$

$$L_{acid} \times N_{acid} = L_{base} \times N_{base} \quad \textit{or} \quad mL_{acid} \times N_{acid} = mL_{base} \times N_{base}$$

Example 11-8

What volume of 0.100 N HNO_3 solution is required to neutralize completely 50.0 mL of a 0.150 N solution of $Ba(OH)_2$?

Plan

We know three of the four variables in the relationship $mL_{acid} \times N_{acid} = mL_{base} \times N_{base}$, and so we solve for mL_{acid}.

Solution

$$\underline{?}\ \text{mL}_{\text{acid}} = \frac{\text{mL}_{\text{base}} \times N_{\text{base}}}{N_{\text{acid}}} = \frac{50.0\ \text{mL} \times 0.150\ N}{0.100\ N}$$

$$= 75.0\ \text{mL of } HNO_3 \text{ solution}$$

EOC 42, 43

Example 11-9

What is the normality of a solution of 4.202 grams of HNO_3 in 600 mL of solution?

Plan

We convert grams of HNO_3 to moles of HNO_3 and then to equivalents of HNO_3, which lets us calculate the normality.

$$\frac{\text{g } HNO_3}{\text{L}} \longrightarrow \frac{\text{mol } HNO_3}{\text{L}} \longrightarrow \frac{\text{eq } HNO_3}{\text{L}} = N\ HNO_3$$

Solution

$$N = \frac{\text{no. eq } HNO_3}{\text{L}}$$

$$\underline{?}\ \frac{\text{eq } HNO_3}{\text{L}} = \underbrace{\frac{4.202\ \text{g } HNO_3}{0.600\ \text{L}} \times \frac{1\ \text{mol } HNO_3}{63.02\ \text{g } HNO_3}}_{M_{HNO_3}} \times \frac{1\ \text{eq } HNO_3}{\text{mol } HNO_3} = 0.111\ N\ HNO_3$$

Because normality is equal to molarity times the number of equivalents per mole of solute, a solution's normality is always equal to or greater than its molarity.

$$\text{normality} = \text{molarity} \times \frac{\text{no. eq}}{\text{mol}} \quad \textit{or} \quad N = M \times \frac{\text{no. eq}}{\text{mol}}$$

Example 11-10

What is (a) the molarity and (b) the normality of a solution that contains 9.50 grams of barium hydroxide in 2000 mL of solution?

Plan

(a) We use the same kind of logic we used in Example 11-9.

(b) Because each mole of $Ba(OH)_2$ produces 2 moles of OH^- ions, 1 mole of $Ba(OH)_2$ is 2 equivalents. Thus

$$N = M \times \frac{2\ \text{eq}}{\text{mol}} \quad \text{or} \quad M = \frac{N}{2\ \text{eq/mol}}$$

Because each formula unit of $Ba(OH)_2$ contains two OH^- ions,

$$1\ \text{mol } Ba(OH)_2 = 2\ \text{eq } Ba(OH)_2$$

Thus, molarity is one half of normality for $Ba(OH)_2$ solutions.

Solution

$$\text{(a)}\ \ \underline{?}\ \frac{\text{mol } Ba(OH)_2}{\text{L}} = \frac{9.50\ \text{g } Ba(OH)_2}{2.00\ \text{L}} \times \frac{1\ \text{mol } Ba(OH)_2}{171.4\ \text{g } Ba(OH)_2}$$

$$= 0.0277\ M\ Ba(OH)_2$$

(b) $\underline{?}\ \frac{\text{eq Ba(OH)}_2}{\text{L}} = \frac{0.0277 \text{ mol Ba(OH)}_2}{\text{L}} \times \frac{2 \text{ eq Ba(OH)}_2}{1 \text{ mol Ba(OH)}_2}$

$= 0.0554\ N \text{ Ba(OH)}_2$

EOC 38, 39

In Example 11-11, let us again solve Example 11-5, this time using normality rather than molarity. The balanced equation for the reaction of H_2SO_4 with Na_2CO_3, interpreted in terms of equivalent weights, is

By definition, there must be equal numbers of equivalents of all reactants and products in a balanced chemical equation.

H_2SO_4	+ Na_2CO_3	⟶ Na_2SO_4	+ CO_2	+ H_2O
1 mol	1 mol	1 mol	1 mol	1 mol
2 eq	2 eq	2 eq	2 eq	2 eq
98.08 g	106.0 g			

1 eq Na_2CO_3 = 53.0 g and 1 meq Na_2CO_3 = 0.0530 g

Example 11-11

Calculate the normality of a solution of H_2SO_4 if 40.0 mL of the solution reacts completely with 0.364 gram of Na_2CO_3.

Plan

We refer to the balanced equation. We are given the mass of Na_2CO_3, so we convert grams of Na_2CO_3 to milliequivalents of Na_2CO_3, then to milliequivalents of H_2SO_4, which lets us calculate the normality of the H_2SO_4 solution.

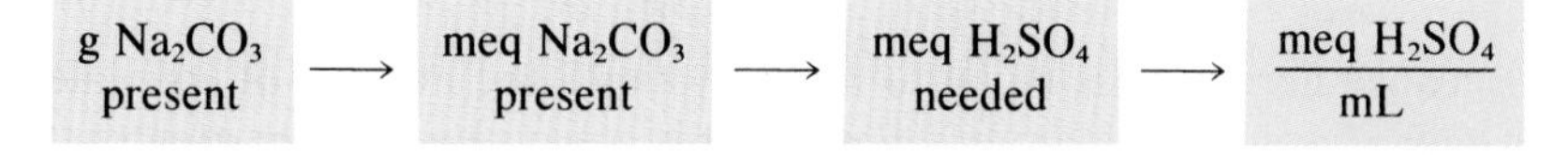

Solution

First we calculate the number of milliequivalents of Na_2CO_3 in the sample:

$$\text{no. meq Na}_2\text{CO}_3 = 0.364 \text{ g Na}_2\text{CO}_3 \times \frac{1 \text{ meq Na}_2\text{CO}_3}{0.0530 \text{ g Na}_2\text{CO}_3} = 6.87 \text{ meq Na}_2\text{CO}_3$$

Because no. meq H_2SO_4 = no. meq Na_2CO_3, we may write

The normality of this H_2SO_4 solution is twice the molarity obtained in Example 11-5 because 1 mol of H_2SO_4 is 2 eq.

$$\text{mL}_{H_2SO_4} \times N_{H_2SO_4} = 6.87 \text{ meq H}_2\text{SO}_4$$

$$N_{H_2SO_4} = \frac{6.87 \text{ meq H}_2\text{SO}_4}{\text{mL}_{H_2SO_4}} = \frac{6.87 \text{ meq H}_2\text{SO}_4}{40.0 \text{ mL}} = 0.172\ N \text{ H}_2\text{SO}_4$$

Example 11-12

This is Example 11-7 using normality rather than molarity.

A 0.1743-gram sample of impure oxalic acid, $(COOH)_2$, required 39.82 mL of 0.08915 N NaOH solution for complete neutralization. No acidic impurities were present. Calculate the percentage purity of the $(COOH)_2$.

Plan

We write the balanced equation for the reaction. We calculate the number of equivalents of NaOH in the standard solution, which tells us the number of equivalents of $(COOH)_2$ in the sample. Then we convert equivalents of $(COOH)_2$ to grams of $(COOH)_2$, which allows us to calculate the percentage purity.

N NaOH soln ⟶ eq NaOH = eq $(COOH)_2$ ⟶ g $(COOH)_2$ ⟶ % $(COOH)_2$

Solution

$$2NaOH + (COOH)_2 \longrightarrow Na_2(COO)_2 + 2H_2O$$

2 mol　　1 mol
2 eq　　2 eq　= 90.04 g; therefore 1 eq = 45.02 g $(COOH)_2$

The equation for the reaction of NaOH with $(COOH)_2$ shows that *one mole* of $(COOH)_2$ is *two* equivalents. Therefore, one equivalent weight of $(COOH)_2$ is 90.04 g/2 = 45.02 g. The number of equivalents of NaOH that react is the volume times the normality of the solution.

$$\underline{?} \text{ eq NaOH} = 0.03982 \text{ L} \times \frac{0.08915 \text{ eq NaOH}}{\text{L}} = 0.003550 \text{ eq NaOH}$$

Now we calculate the mass of $(COOH)_2$ that reacts with 0.003550 eq NaOH:

$$\underline{?} \text{ g } (COOH)_2 = 0.003550 \text{ eq NaOH} \times \frac{1 \text{ eq } (COOH)_2}{1 \text{ eq NaOH}} \times \frac{45.02 \text{ g } (COOH)_2}{1 \text{ eq } (COOH)_2}$$
$$= 0.1598 \text{ g } (COOH)_2$$

The sample contained 0.1598 g of $(COOH)_2$, and its percent purity is

$$\% \text{ purity} = \frac{0.1598 \text{ g } (COOH)_2}{0.1743 \text{ g sample}} \times 100\% = 91.68\% \text{ pure } (COOH)_2$$

EOC 45, 46

Redox Reactions

One method of analyzing samples quantitatively for the presence of *oxidizable* or *reducible* substances is by **redox titration**. In such analyses, the concentration of a solution is determined by allowing it to react with a carefully measured amount of a *standard* solution of an oxidizing or reducing agent.

As in acid–base titrations, amounts of solutes can be described in terms of either moles or equivalent weights. Concentrations of solutions involved in redox titrations can be expressed in terms of either molarity or normality. We shall illustrate redox titrations, separately, by both the mole–molarity method and the equivalent weight–normality method.

11-4 Redox Titrations: The Mole Method and Molarity

As in other kinds of chemical reactions, we must pay particular attention to the mole ratio in which oxidizing agents and reducing agents react. Please review Sections 4-9 through 4-11 carefully.

Potassium permanganate, $KMnO_4$, is a strong oxidizing agent. Through the years it has been the "workhorse" of redox titrations. For example, in acidic solution, $KMnO_4$ reacts with iron(II) sulfate, $FeSO_4$, according to the balanced equations below. A strong acid, such as H_2SO_4, is used in such titrations. See Figure 11-1.

$$2KMnO_4 + 10FeSO_4 + 8H_2SO_4 \longrightarrow 2MnSO_4 + 5Fe_2(SO_4)_3 + K_2SO_4 + 8H_2O$$

$$MnO_4^-(aq) + 5Fe^{2+}(aq) + 8H^+(aq) \longrightarrow Mn^{2+}(aq) + 5Fe^{3+}(aq) + 4H_2O(\ell)$$

A word about terminology. The preceding reaction involves MnO_4^- ions and Fe^{2+} ions in acidic solution. The source of MnO_4^- ions is the soluble ionic compound $KMnO_4$. We often refer to "permanganate solutions." Clearly such solutions also contain cations—in this case, K^+. Likewise, we often refer to "iron(II) solutions" without specifying what the anion is.

Because it has an intense purple color, $KMnO_4$ acts as its own indicator. One drop of 0.020 *M* $KMnO_4$ solution imparts a pink color to a liter of pure water. When $KMnO_4$ solution is added to a solution of a reducing agent, the end point in the titration is taken as the point at which a pale pink color appears in the solution being titrated and persists for at least 30 seconds.

Example 11-13

What volume of 0.0200 *M* $KMnO_4$ solution is required to oxidize 40.0 mL of 0.100 *M* $FeSO_4$ in sulfuric acid solution (Figure 11-1)?

Plan

We refer to the balanced equation in the preceding discussion to find the reaction ratio, 1 mol MnO_4^-/5 mol Fe^{2+}. Then we calculate the number of moles of Fe^{2+} to be titrated, which lets us find the number of moles of MnO_4^- required *and* the volume in which this number of moles of $KMnO_4$ is contained.

One mole of $KMnO_4$ contains one mole of MnO_4^- ions. Therefore, the number of moles of $KMnO_4$ is *always* equal to the number of moles of MnO_4^- ions required in a reaction. Similarly, one mole of $FeSO_4$ contains 1 mole of Fe^{2+} ions.

Solution

The reaction ratio is

$$MnO_4^-(aq) + 8H^+(aq) + 5Fe^{2+}(aq) \longrightarrow 5Fe^{3+}(aq) + Mn^{2+}(aq) + 4H_2O$$

rxn ratio: 1 mol 5 mol

The number of moles of Fe^{2+} to be titrated is

$$\underline{?} \text{ mol } Fe^{2+} = 40.0 \text{ mL} \times \frac{0.100 \text{ mol } Fe^{2+}}{1000 \text{ mL}} = 4.00 \times 10^{-3} \text{ mol } Fe^{2+}$$

We use the balanced equation to find the number of moles of MnO_4^- required:

$$\underline{?} \text{ mol } MnO_4^- = 4.00 \times 10^{-3} \text{ mol } Fe^{2+} \times \frac{1 \text{ mol } MnO_4^-}{5 \text{ mol } Fe^{2+}}$$

$$= 8.00 \times 10^{-4} \text{ mol } MnO_4$$

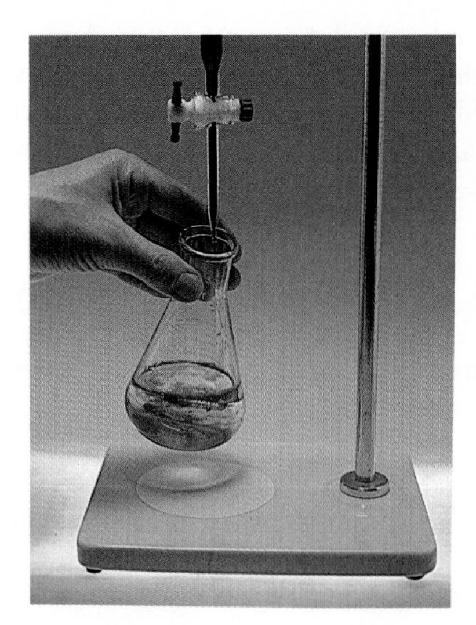

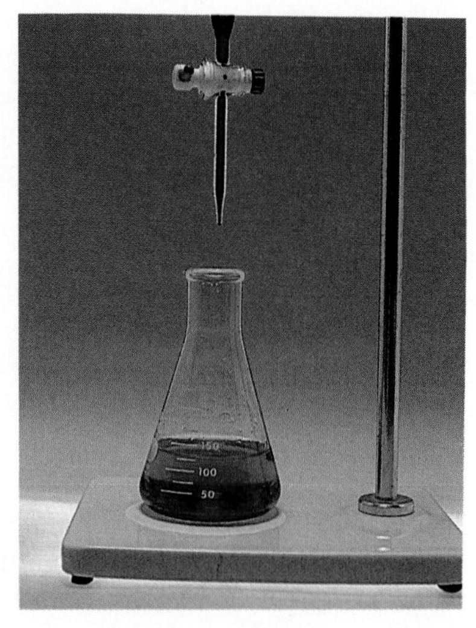

(a) (b)

Figure 11-1

(a) Nearly colorless $FeSO_4$ solution is titrated with deep-purple $KMnO_4$.

(b) The end point is the point at which the solution becomes pink, owing to a *very small* excess of $KMnO_4$. Here a considerable excess of $KMnO_4$ was added so that the pink color could be reproduced photographically.

Each formula unit of $KMnO_4$ contains one MnO_4^- ion, and so

$$1 \text{ mol } KMnO_4 \hat{=} 1 \text{ mol } MnO_4^-$$

The volume of 0.0200 *M* $KMnO_4$ solution that contains 8.00×10^{-4} mol of $KMnO_4$ is

$$\underline{?} \text{ mL } KMnO_4 \text{ soln} = 8.00 \times 10^{-4} \text{ mol } KMnO_4 \times \frac{1000 \text{ mL } KMnO_4 \text{ soln}}{0.0200 \text{ mol } KMnO_4}$$

$$= 40.0 \text{ mL } KMnO_4 \text{ soln}$$

EOC 48

Potassium dichromate, $K_2Cr_2O_7$, is another frequently used oxidizing agent. However, an indicator must be used when reducing agents are titrated with dichromate solutions. $K_2Cr_2O_7$ is orange, and its reduction product, Cr^{3+}, is green.

Consider the oxidation of sulfite ions, SO_3^{2-}, to sulfate ions, SO_4^{2-}, by $Cr_2O_7^{2-}$ ions in the presence of a strong acid such as sulfuric acid. We shall balance the equation by the ion–electron method.

$K_2Cr_2O_7$ is orange in acidic solution. $Cr_2(SO_4)_3$ is green in acidic solution.

$$Cr_2O_7^{2-} \longrightarrow Cr^{3+} \quad \text{(red. half-rxn)}$$

$$Cr_2O_7^{2-} \longrightarrow 2Cr^{3+}$$

$$14H^+ + Cr_2O_7^{2-} \longrightarrow 2Cr^{3+} + 7H_2O$$

$$6e^- + 14H^+ + Cr_2O_7^{2-} \longrightarrow 2Cr^{3+} + 7H_2O \quad \text{(balanced red. half-rxn)}$$

$$SO_3^{2-} \longrightarrow SO_4^{2-} \quad \text{(ox. half-rxn)}$$

$$SO_3^{2-} + H_2O \longrightarrow SO_4^{2-} + 2H^+$$

$$SO_3^{2-} + H_2O \longrightarrow SO_4^{2-} + 2H^+ + 2e^- \quad \text{(balanced ox. half-rxn)}$$

We now equalize the electron transfer, add the balanced half-reactions, and eliminate common terms:

$$(6e^- + 14H^+ + Cr_2O_7^{2-} \longrightarrow 2Cr^{3+} + 7H_2O) \quad \text{(reduction)}$$

$$3(SO_3^{2-} + H_2O \longrightarrow SO_4^{2-} + 2H^+ + 2e^-) \quad \text{(oxidation)}$$

$$8H^+(aq) + Cr_2O_7^{2-}(aq) + 3SO_3^{2-}(aq) \longrightarrow 2Cr^{3+}(aq) + 3SO_4^{2-}(aq) + 4H_2O(\ell)$$

The balanced equation tells us that the reaction ratio is 3 mol SO_3^{2-}/mol $Cr_2O_7^{2-}$ or 1 mol $Cr_2O_7^{2-}$/3 mol SO_3^{2-}. Potassium dichromate is the usual source of $Cr_2O_7^{2-}$ ions, and Na_2SO_3 is the usual source of SO_3^{2-} ions. Thus, the preceding reaction ratio could also be expressed as 1 mol $K_2Cr_2O_7$/3 mol Na_2SO_3.

Example 11-14

A 20.00-mL sample of Na_2SO_3 was titrated with 36.30 mL of 0.05130 *M* $K_2Cr_2O_7$ solution in the presence of H_2SO_4. Calculate the molarity of the Na_2SO_3 solution.

Plan

We can calculate the number of moles of $Cr_2O_7^{2-}$ in the standard solution. Then we refer to the balanced equation in the preceding discussion, which gives us the reaction ratio, 3 mol SO_3^{2-}/1 mol $Cr_2O_7^{2-}$. The reaction ratio lets us calculate the number of moles of SO_3^{2-}(Na_2SO_3) that reacted and the molarity of the solution.

$$\text{L } Cr_2O_7^{2-} \text{ soln} \longrightarrow \text{mol } Cr_2O_7^{2-} \longrightarrow \text{mol } SO_3^{2-} \longrightarrow M\ SO_3^{2-} \text{ soln}$$

Solution

From the preceding discussion we know the balanced equation and the reaction ratio:

$$\underset{3\text{ mol}}{3SO_3^{2-}} + \underset{1\text{ mol}}{Cr_2O_7^{2-}} + 8H^+ \longrightarrow 3SO_4^{2-} + 2Cr^{3+} + 4H_2O$$

The number of moles of $Cr_2O_7^{2-}$ used is

$$\underline{?}\text{ mol } Cr_2O_7^{2-} = 0.03630\text{ L} \times \frac{0.05130\text{ mol } Cr_2O_7^{2-}}{\text{L}} = 0.001862\text{ mol } Cr_2O_7^{2-}$$

The number of moles of SO_3^{2-} that reacted with 0.001862 mol of $Cr_2O_7^{2-}$ is

$$\underline{?}\text{ mol } SO_3^{2-} = 0.001862\text{ mol } Cr_2O_7^{2-} \times \frac{3\text{ mol } SO_3^{2-}}{1\text{ mol } Cr_2O_7^{2-}} = 0.005586\text{ mol } SO_3^{2-}$$

The Na_2SO_3 solution contained 0.005586 mol of SO_3^{2-} (and 0.005586 mol of Na_2SO_3). Its molarity is

$$\underline{?}\ \frac{\text{mol } Na_2SO_3}{\text{L}} = \frac{0.005586\text{ mol } Na_2SO_3}{0.02000\text{ L}} = 0.2793\ M\ Na_2SO_3$$

EOC 49

11-5 Redox Titrations: Equivalent Weights and Normality

The equivalent weight of an oxidizing agent or reducing agent depends upon the specific reaction it undergoes.

When we studied acid–base reactions, we defined the term "equivalent weight" so that one equivalent weight, or one equivalent (eq), of any acid reacts with one equivalent of any base. We now define the term as it applies to *redox reactions* so that one equivalent of any oxidizing agent reacts with one equivalent of any reducing agent. In redox reactions, one **equivalent weight**, or **equivalent**, of a substance is the mass of the oxidizing or reducing substance that gains or loses 6.022×10^{23} electrons.

For all redox reactions we may write

$$\text{no. eq oxidizing agent} = \text{no. eq reducing agent}$$

or

$$\text{no. meq oxidizing agent} = \text{no. meq reducing agent}$$

$$L_O \times N_O = L_R \times N_R \quad \textit{or} \quad mL_O \times N_O = mL_R \times N_R$$

The only difference between calculations for acid–base reactions and those for redox reactions is the *definition* of the equivalent.

Example 11-15

What volume of 0.1000 N $KMnO_4$ would oxidize 40.0 mL of 0.100 N $FeSO_4$ in acidic solution?

This is Example 11-13 using normality rather than molarity.

Plan

From the preceding discussion we know the balanced equation and that $mL_O \times N_O = mL_R \times N_R$. So we solve for mL_R.

Solution

The balanced equation is

$$MnO_4^-(aq) + 8H^+(aq) + 5Fe^{2+}(aq) \longrightarrow 5Fe^{3+}(aq) + Mn^{2+}(aq) + 4H_2O(\ell)$$

Because $mL_{MnO_4^-} \times N_{MnO_4^-} = mL_{Fe^{2+}} \times N_{Fe^{2+}}$, we can solve for the number of milliliters of $KMnO_4$ solution.

$$\underline{?}\ mL_{MnO_4^-} = \frac{mL_{Fe^{2+}} \times N_{Fe^{2+}}}{N_{MnO_4^-}} = \frac{40.0\text{ mL} \times 0.100\ N}{0.1000\ N}$$

$$= 40.0\text{ mL } KMnO_4\text{ solution}$$

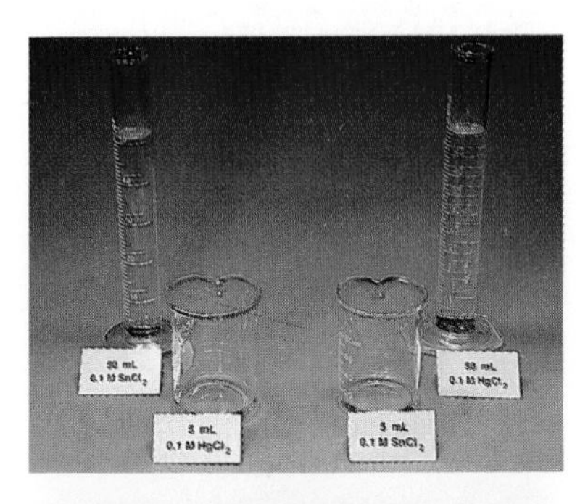

When they are mixed together, $SnCl_2$ is a reducing agent and $HgCl_2$ is an oxidizing agent. When excess $HgCl_2$ is present, $SnCl_2$ reduces it to white insoluble Hg_2Cl_2 (right).

$$SnCl_2(aq) + 2HgCl_2(aq) \longrightarrow SnCl_4(aq) + Hg_2Cl_2(s)$$

An excess of $SnCl_2$ reduces $HgCl_2$ to elemental mercury (black) in a two-step reaction (left).

$$SnCl_2(aq) + 2HgCl_2(aq) \longrightarrow SnCl_4(aq) + Hg_2Cl_2(s)$$

$$SnCl_2(aq) + Hg_2Cl_2(s) \longrightarrow SnCl_4(aq) + 2Hg(\ell)$$

$$SnCl_2(aq) + HgCl_2(aq) \longrightarrow SnCl_4(aq) + Hg(\ell)$$

The equivalent weight of $HgCl_2$ in these reactions depends upon whether it is the limiting reactant.

We saw earlier that the balanced net ionic equation and half-reactions for the reaction of $KMnO_4$ with $FeSO_4$ in acidic solution are

$$MnO_4^-(aq) + 8H^+(aq) + 5Fe^{2+}(aq) \longrightarrow Mn^{2+}(aq) + 5Fe^{3+}(aq) + 4H_2O(\ell)$$

$$Fe^{2+}(aq) \longrightarrow Fe^{3+}(aq) + 1e^- \qquad \text{(oxidation)}$$

$$MnO_4^-(aq) + 8H^+(aq) + 5e^- \longrightarrow Mn^{2+}(aq) + 4H_2O(\ell) \qquad \text{(reduction)}$$

Each Fe^{2+} ion loses one electron, and so one mole of $FeSO_4$ loses 6.022×10^{23} electrons. One mole of $FeSO_4$ (151.9 g) is one equivalent *in this reaction.*

Each MnO_4^- ion gains five electrons, and so each mole of MnO_4^- gains $5(6.022 \times 10^{23}$ electrons). *In this reaction,* one mole of $KMnO_4$ (158.0 g) is five equivalents.

	Compound	e^- Transferred per Formula Unit	One Mole	One eq
Oxidizing agent:	$KMnO_4$	5	158.0 g	$\frac{158.0\text{ g}}{5} = 31.60\text{ g}$
Reducing agent:	$FeSO_4$	1	151.9 g	$\frac{151.9\text{ g}}{1} = 151.9\text{ g}$

In *this* reaction, 31.60 grams of $KMnO_4$ reacts with 151.9 grams of $FeSO_4$.

Example 11-16

How many grams of $KMnO_4$ are contained in 35.0 mL of 0.0500 N $KMnO_4$ used in the following reaction in acidic solution?

$$MnO_4^-(aq) + 5Fe^{2+}(aq) + 8H^+(aq) \longrightarrow Mn^{2+}(aq) + 5Fe^{3+}(aq) + 4H_2O(\ell)$$

Plan

We are given the balanced equation, and we have just seen that one mole of $KMnO_4$ is five equivalents *in this reaction*. We relate milliliters of $KMnO_4$ solution to equivalents of $KMnO_4$, then to moles of $KMnO_4$, and finally to grams of $KMnO_4$.

$$\text{mL KMnO}_4\text{ soln} \longrightarrow \text{eq KMnO}_4 \longrightarrow \text{mol KMnO}_4 \longrightarrow \text{g KMnO}_4$$

Solution

One mole of $KMnO_4$ is five equivalents in this reaction.

$$\underline{?}\text{ g KMnO}_4 = 35.0\text{ mL} \times \frac{0.0500\text{ eq KMnO}_4}{1000\text{ mL}} \times \frac{1\text{ mol}}{5\text{ eq}} \times \frac{158.0\text{ g}}{1\text{ mol}}$$

$$= 0.0553\text{ g KMnO}_4$$

EOC 58, 59

The equivalent weight of an oxidizing or reducing agent depends upon the specific reaction it undergoes. In the following reaction, the equivalent weight of $KMnO_4$ is different from the value we calculated before, because the MnO_4^- ion undergoes a three-electron change rather than a five-electron change. Dissolving metallic zinc in mildly basic $KMnO_4$ solution produces solid manganese(IV) oxide and zinc hydroxide. In this reaction the half-reactions and balanced equation are as follows:

$$2(MnO_4^- + 2H_2O + 3e^- \longrightarrow MnO_2 + 4OH^-) \qquad \text{(reduction)}$$

$$3(Zn + 2OH^- \longrightarrow Zn(OH)_2 + 2e^-) \qquad \text{(oxidation)}$$

$$2MnO_4^-(aq) + 3Zn(s) + 4H_2O(\ell) \longrightarrow 2MnO_2(s) + 3Zn(OH)_2(s) + 2OH^-(aq)$$

One mole of $KMnO_4$ is three equivalents *in this reaction*. One mole of zinc is two equivalents, as the following tabulation shows.

Compare the equivalent weight of $KMnO_4$ with that in the previous example.

	Substance	e^- Transferred per Formula Unit	One Mole	One eq
Oxidizing agent:	$KMnO_4(aq)$	3	158.0 g	$\frac{158.0\text{ g}}{3} = 52.67\text{ g}$
Reducing agent:	Zn(s)	2	65.39 g	$\frac{65.39\text{ g}}{2} = 32.69\text{ g}$

Here one equivalent of $KMnO_4$ is 52.67 grams, *not* 31.60 grams as before.

Example 11-17

How many grams of $KMnO_4$ are contained in 35.0 mL of 0.0500 *N* $KMnO_4$ used in the following reaction in basic solution?

$$2MnO_4^-(aq) + 3Zn(s) + 4H_2O(\ell) \longrightarrow 2MnO_2(s) + 3Zn(OH)_2(s) + 2OH^-(aq)$$

Plan

We use the same kind of logic we used in Example 11-16. Note that 1 mole of $KMnO_4$ is 3 equivalents in *this* reaction.

Solution

$$\underline{?}\,\text{g KMnO}_4 = 35.0\,\text{mL} \times \frac{0.0500\,\text{eq KMnO}_4}{1000\,\text{mL}} \times \frac{1\,\text{mol}}{3\,\text{eq}} \times \frac{158.0\,\text{g}}{1\,\text{mol}} = \boxed{0.0922\,\text{g KMnO}_4}$$

EOC 60

To further illustrate how normality is used in oxidation–reduction titrations, let us solve Example 11-14 using normality rather than molarity.

Example 11-18

A 20.00-mL sample of Na_2SO_3 was titrated with 36.30 mL of 0.3078 *N* $K_2Cr_2O_7$ solution in the presence of H_2SO_4. (a) Calculate the normality of the Na_2SO_3 solution. (b) What mass of Na_2SO_3 was present in the sample? (c) What mass of $K_2Cr_2O_7$ was used to prepare 500 mL of the $K_2Cr_2O_7$ solution?

Plan

We write the balanced equation for the reaction as well as the balanced half-reactions. We apply the logic of earlier examples.

Solution

The balanced equation for the oxidation of SO_3^{2-} ions to SO_4^{2-} ions by $Cr_2O_7^{2-}$ ions in acidic solution and the balanced half-reactions are

$$8H^+ + Cr_2O_7^{2-} + 3SO_3^{2-} \longrightarrow 2Cr^{3+} + 3SO_4^{2-} + 4H_2O$$

$$6e^- + 14H^+ + Cr_2O_7^{2-} \longrightarrow 2Cr^{3+} + 7H_2O \qquad \text{(red. half-rxn)}$$

$$SO_3^{2-} + H_2O \longrightarrow SO_4^{2-} + 2H^+ + 2e^- \qquad \text{(ox. half-rxn)}$$

(a) We know the volume and normality of the $K_2Cr_2O_7$ solution as well as the volume of the Na_2SO_3 solution:

$$\text{mL}_{Cr_2O_7^{2-}} \times N_{Cr_2O_7^{2-}} = \text{mL}_{SO_3^{2-}} \times N_{SO_3^{2-}}$$

$$N_{SO_3^{2-}} = \frac{\text{mL}_{Cr_2O_7^{2-}} \times N_{Cr_2O_7^{2-}}}{\text{mL}_{SO_3^{2-}}} = \frac{36.30\,\text{mL} \times 0.3078\,N}{20.00\,\text{mL}} = \boxed{0.5587\,N\,\text{Na}_2\text{SO}_3}$$

The answer to Example 11-14 is 0.2793 *M*. Is 0.5587 *N* the same concentration?

This tells us that 1.000 L of Na_2SO_3 solution contains 0.5587 eq of Na_2SO_3.

The half-reactions (above) give us the information needed to answer (b) and (c):

$$1\text{ mol } K_2Cr_2O_7 = 294.2\text{ g } K_2Cr_2O_7 = 6\text{ eq } K_2Cr_2O_7$$

$$1\text{ mol } Na_2SO_3 = 126.0\text{ g } Na_2SO_3 = 2\text{ eq } Na_2SO_3$$

$$\text{(b) } \underline{?}\text{ g } Na_2SO_3 = 20.00\text{ mL} \times \frac{0.5587\text{ eq } Na_2SO_3}{1000\text{ mL}} \times \frac{126.0\text{ g } Na_2SO_3}{2\text{ eq } Na_2SO_3}$$

$$= \boxed{0.7040\text{ g } Na_2SO_3}$$

$$\text{(c) } \underline{?}\text{ g } K_2Cr_2O_7 = 500\text{ mL} \times \frac{0.3078\text{ eq } K_2Cr_2O_7}{1000\text{ mL}} \times \frac{294.2\text{ g } K_2Cr_2O_7}{6\text{ eq } K_2Cr_2O_7}$$

$$= \boxed{7.546\text{ g } K_2Cr_2O_7}$$

EOC 61, 63

Key Terms

Buret A piece of volumetric glassware, usually graduated in 0.1-mL intervals, that is used to deliver solutions to be used in titrations in a quantitative (dropwise) manner.

Disproportionation reaction A redox reaction in which the oxidizing agent and the reducing agent are the same species.

End point The point at which an indicator changes color and a titration is stopped.

Equivalence point The point at which chemically equivalent amounts of reactants have reacted.

Equivalent weight in acid–base reactions The mass of an acid or base that furnishes or reacts with 6.022×10^{23} H_3O^+ or OH^- ions.

Equivalent weight of oxidizing or reducing agent The mass that gains (oxidizing agents) or loses (reducing agents) 6.022×10^{23} electrons in a redox reaction.

Half-reaction Either the oxidation part or the reduction part of a redox reaction.

Indicator For acid–base titrations, an organic compound that exhibits different colors in solutions of different acidities; used to determine the point at which the reaction between two solutes is complete.

Normality The number of equivalent weights (equivalents) of solute per liter of solution.

Oxidation An algebraic increase in oxidation number; may correspond to a loss of electrons.

Oxidation–reduction reaction A reaction in which oxidation and reduction occur; also called redox reaction.

Oxidizing agent The substance that oxidizes another substance and is reduced.

Primary standard A substance of a known high degree of purity that undergoes one invariable reaction with the other reactant of interest.

Redox reaction An oxidation–reduction reaction.

Redox titration The quantitative analysis of the amount or concentration of an oxidizing or reducing agent in a sample by observing its reaction with a known amount or concentration of a reducing or oxidizing agent.

Reducing agent The substance that reduces another substance and is oxidized.

Reduction An algebraic decrease in oxidation number; may correspond to a gain of electrons.

Secondary standard A solution that has been titrated against a primary standard. A standard solution is a secondary standard.

Standard solution A solution of accurately known concentration.

Standardization The process by which the concentration of a solution is accurately determined by titrating it against an accurately known amount of a primary standard.

Titration The process by which the volume of a standard solution required to react with a specific amount of a substance is determined.

Exercises

Molarity

1. Why can we describe molarity as a "method of convenience" for expressing concentrations of solutions?
2. Why can molarity be expressed in mol/L *and* in mmol/mL?
3. Calculate the molarities of solutions that contain the following masses of solute in the indicated volumes:
   (a) 75 g of H_3AsO_4 in 500 mL of solution
   (b) 8.3 g of $(COOH)_2$ in 600 mL of solution
   (c) 13.0 g of $(COOH)_2 \cdot 2H_2O$ in 750 mL of solution
4. Calculate the molarities of solutions that contain the following:
   (a) 17.5 g of K_2SO_4 in 300 mL of solution
   (b) 143 g of $Al_2(SO_4)_3$ in 3.00 L of solution
   (c) 143 g of $Al_2(SO_4)_3 \cdot 18H_2O$ in 3.00 L of solution
5. Calculate the mass of NaOH required to prepare 2.50 L of 3.15 *M* NaOH solution.
6. What mass of $(NH_4)_2SO_4$ is required to prepare 3.75 L of 0.288 *M* $(NH_4)_2SO_4$ solution?
7. What volume of 0.300 *M* potassium hydroxide solution would just neutralize 30.0 mL of 0.100 *M* H_2SO_4 solution?
8. Calculate the molarity of a solution that is 39.77% H_2SO_4 by mass. The specific gravity of the solution is 1.305.
9. What is the molarity of a solution that is 19.0% HNO_3 by mass? The specific gravity of the solution is 1.11.
10. If 200 mL of 4.32 *M* HCl solution is added to 400 mL of 2.16 *M* NaOH solution, the resulting solution will be _____ molar in NaCl.
11. If 400 mL of 0.400 *M* HCl solution is added to 800 mL of 0.0800 *M* $Ba(OH)_2$ solution, the resulting solution will be _____ molar in $BaCl_2$ and _____ molar in _____.
12. If 225 mL of 3.68 *M* H_3PO_4 solution is added to 775 mL of 3.68 *M* NaOH solution, the resulting solution will be _____ molar in Na_3PO_4 and _____ molar in _____.
13. What volumes of 1.00 *M* KOH and 0.750 *M* HNO_3 solution would be required to produce 8.40 g of KNO_3?

14. What volumes of 1.00 M NaOH and 1.50 M H_3PO_4 solutions would be required to form 1.00 mol of Na_3PO_4?
15. A household ammonia solution is 5.03% ammonia. Its density is 0.979 g/mL. What is the molarity of this ammonia solution?
16. A vinegar solution is 5.11% acetic acid. Its density is 1.007 g/mL. What is its molarity?

Before you work Examples 17 and 18, *think* about the description of each solution. Note that the percentage by mass of solute is *nearly* the same for the solutions. Would you expect them to contain *approximately* the same number of moles of solute? *Hint:* Think molecular weights.

17. Refer to Exercises 15 and 16. What volume of the vinegar solution would be required to neutralize 1.00 L of the household ammonia solution?
18. Refer to Exercises 15 and 16. What volume of the household ammonia solution would be required to neutralize 1.00 L of the vinegar?

Standardization and Acid–Base Titrations—Mole Method

19. Define and illustrate the following terms clearly and concisely: (a) standard solution, (b) titration, (c) primary standard, (d) secondary standard.
20. Describe the preparation of a standard solution of NaOH, a compound that absorbs both CO_2 and H_2O from the air.
21. Distinguish between the *equivalence point* and the *end point* of a titration.
22. (a) What are the properties of an ideal primary standard? (b) What is the importance of each property?
23. Why can sodium carbonate be used as a primary standard for solutions of acids?

    (a) What is potassium hydrogen phthalate? (b) For what is it used?
25. What volume of 0.275 M hydrochloric acid solution neutralizes 36.4 mL of 0.150 M sodium hydroxide solution?
26. What volume of 0.112 M sodium hydroxide solution would be required to neutralize completely 25.3 mL of 0.400 M sulfuric acid solution?
27. Calculate the molarity of a KOH solution if 27.63 mL of the KOH solution reacted with 0.4084 g of potassium hydrogen phthalate.
28. A solution of sodium hydroxide is standardized against potassium hydrogen phthalate. From the following data, calculate the molarity of the NaOH solution:

mass of KHP	0.8407 g
buret reading before titration	0.23 mL
buret reading after titration	46.16 mL

29. The secondary standard solution of NaOH of Exercise 28 was used to titrate a solution of unknown concentration of HCl. A 30.00-mL sample of the HCl solution required 24.21 mL of the NaOH solution for complete neutralization. What is the molarity of the HCl solution?
30. A 34.53-mL sample of a solution of sulfuric acid, H_2SO_4, is neutralized by 27.86 mL of the NaOH solution of Exercise 28. Calculate the molarity of the sulfuric acid solution.
31. If 37.38 mL of a sulfuric acid solution reacts with 0.2888 g of Na_2CO_3, what is the molarity of the sulfuric acid solution.

*32. An impure sample of $(COOH)_2 \cdot 2H_2O$ that had a mass of 2.00 g was dissolved in water and titrated with standard NaOH solution. The titration required 40.32 mL of 0.198 M NaOH solution. Calculate the percentage of $(COOH)_2 \cdot 2H_2O$ in the sample.

*33. A 50.0-mL sample of 0.0500 M $Ca(OH)_2$ is added to 10.0 mL of 0.200 M HNO_3. (a) Is the resulting solution acidic or basic? (b) How many moles of excess acid or base are present? (c) How many mL of 0.0500 M $Ca(OH)_2$ or 0.0500 M HNO_3 would be required to neutralize the solution?

*34. An antacid tablet containing calcium carbonate as an active ingredient requires 22.6 mL of 0.0932 M HCl for complete neutralization. What mass of $CaCO_3$ did the tablet contain?

*35. Butyric acid, whose empirical formula is C_2H_4O, is the acid responsible for the odor of rancid butter. The acid has one ionizable hydrogen per molecule. A 1.000-g sample of butyric acid is neutralized by 54.42 mL of 0.2088 M NaOH solution. What are (a) the molecular weight and (b) the molecular formula of butyric acid?

*36. The typical concentration of HCl in stomach acid (gastric juice) is a concentration of about 8.0×10^{-2} M. One experiences "acid stomach" when the stomach contents reach about 1.0×10^{-1} M HCl. One Rolaids® tablet (an antacid) contains 334 mg of active ingredient, $NaAl(OH)_2CO_3$. Assume that you have acid stomach and that your stomach contains 800 mL of 1.0×10^{-1} M HCl. Calculate the number of mmol of HCl in the stomach and the number of mmol of HCl that the tablet *can* neutralize. Which is greater? The neutralization reaction produces NaCl, $AlCl_3$, CO_2, and H_2O.

Standardization and Acid–Base Titrations—Equivalent Weight Method

In answering Exercises 37–41, assume that the acids and bases will be completely neutralized.

37. Calculate the normality of a solution that contains 4.93 g of H_2SO_4 in 125 mL of solution.
38. What is the normality of a solution that contains 7.08 g of H_3PO_4 in 185 mL of solution?
39. Calculate the molarity and the normality of a solution that was prepared by dissolving 34.2 g of barium hydroxide in enough water to make 4000 mL of solution.
40. Calculate the molarity and the normality of a solution that was prepared by dissolving 19.6 g of arsenic acid, H_3AsO_4, in enough water to make 600 mL of solution.
41. What are the molarity and normality of a sulfuric acid solution that is 19.6% H_2SO_4 by mass? The density of the solution is 1.14 g/mL.

42. A 25.0-mL sample of 0.206 normal nitric acid solution required 39.3 mL of barium hydroxide solution for neutralization. Calculate the molarity of the barium hydroxide solution.

43. Vinegar is an aqueous solution of acetic acid, CH_3COOH. Suppose you titrate a 25.00-mL sample of vinegar with 17.62 mL of a standardized 0.1060 *N* solution of NaOH.
(a) What is the normality of acetic acid in this vinegar?
(b) What is the mass of acetic acid contained in 1.000 L of vinegar?

44. A 44.4-mL sample of sodium hydroxide solution was titrated with 41.8 mL of 0.100 *N* sulfuric acid solution. A 36.0-mL sample of hydrochloric acid required 47.2 mL of the sodium hydroxide solution for titration. What is the normality of the hydrochloric acid solution?

45. Calculate the normality and molarity of an H_2SO_4 solution if 20.0 mL of the solution reacts with 0.212 g of Na_2CO_3.

$$H_2SO_4 + Na_2CO_3 \longrightarrow Na_2SO_4 + CO_2 + H_2O$$

46. Calculate the normality and molarity of an HCl solution if 33.1 mL of the solution reacts with 0.318 g of Na_2CO_3.

$$2HCl + Na_2CO_3 \longrightarrow 2NaCl + CO_2 + H_2O$$

Redox Titrations—Mole Method and Molarity

47. In a redox titration, we must have a(n) ________ species and a(n) ________ species.

48. What volume of 0.10 *M* $KMnO_4$ would be required to oxidize 20 mL of 0.10 *M* $FeSO_4$ in acidic solution? Refer to Example 11-13.

49. What volume of 0.10 *M* $K_2Cr_2O_7$ would be required to oxidize 60 mL of 0.10 *M* Na_2SO_3 in acidic solution? The products include Cr^{3+} and SO_4^{2-} ions. Refer to Example 11-14.

50. What volume of 0.10 *M* $KMnO_4$ would be required to oxidize 50 mL of 0.10 *M* KI in acidic solution? Products include Mn^{2+} and I_2.

51. What volume of 0.10 *M* $K_2Cr_2O_7$ would be required to oxidize 50 mL of 0.10 *M* KI in acidic solution? Products include Cr^{3+} and I_2.

52. (a) A solution of sodium thiosulfate, $Na_2S_2O_3$, is 0.1455 *M*. 25.00 mL of this solution reacts with 26.36 mL of I_2 solution. What is the molarity of the I_2 solution?

$$2Na_2S_2O_3 + I_2 \longrightarrow Na_2S_4O_6 + 2NaI$$

(b) 25.32 mL of the I_2 solution is required to titrate a sample containing As_2O_3. Calculate the mass of As_2O_3 (197.8 g/mol) in the sample.

$$As_2O_3 + 5H_2O + 2I_2 \longrightarrow 2H_3AsO_4 + 4HI$$

53. Copper(II) ion, Cu^{2+}, can be determined by the net reaction

$$2Cu^{2+} + 2I^- + 2S_2O_3^{2-} \longrightarrow 2CuI(s) + S_4O_6^{2-}$$

A 2.075-g sample containing $CuSO_4$ and excess KI is titrated with 41.75 mL of 0.1214 *M* solution of $Na_2S_2O_3$. What is the percentage of $CuSO_4$ (159.6 g/mol) in the sample?

54. Find the volume of 0.150 *M* HI solution required to titrate
(a) 25.0 mL of 0.100 *M* NaOH
(b) 5.03 g of $AgNO_3$ ($Ag^+ + I^- \longrightarrow AgI(s)$)
(c) 0.621 g $CuSO_4$ ($2Cu^{2+} + 4I^- \longrightarrow 2CuI(s) + I_2(s)$)

***55.** The iron in a sample containing some Fe_2O_3 is reduced to Fe^{2+}. The Fe^{2+} is titrated with 12.02 mL of 0.1167 *M* $K_2Cr_2O_7$ in an acid solution.

$$6Fe^{2+} + Cr_2O_7^{2-} + 14H^+ \longrightarrow 6Fe^{3+} + 2Cr^{3+} + 7H_2O$$

Find (a) the mass of Fe and (b) the percentage of Fe in a 5.675-g sample.

***56.** Limonite is an ore of iron that contains $Fe_2O_3 \cdot 1\frac{1}{2}H_2O$ (or $2Fe_2O_3 \cdot 3H_2O$). A 0.5166-g sample of limonite is dissolved in acid and treated so that all the iron is converted to ferrous ion, Fe^{2+}. This sample requires 42.96 mL of 0.02130 *M* sodium dichromate solution, $Na_2Cr_2O_7$, for titration. Fe^{2+} is oxidized to Fe^{3+} and $Cr_2O_7^{2-}$ is reduced to Cr^{3+}. What is the percentage of iron in the limonite?

***57.** A 0.683-g sample of an ore of iron is dissolved in acid and converted to the ferrous form. The sample is oxidized by 38.50 mL of 0.161 *M* ceric sulfate, $Ce(SO_4)_2$, solution during which the ceric ion, Ce^{4+}, is reduced to Ce^{3+} ion.
(a) Write a balanced equation for the reaction.
(b) What is the percentage of iron in the ore?

Redox Titrations—Equivalent Weights and Normality

In Exercises 58 and 59, calculate the mass of the compound in the indicated volume of solution.

58. (a) Grams of $KMnO_4$ in 475 mL of 0.137 *N* $KMnO_4$ solution used in the reaction in Exercise 48.
(b) Grams of NaI in 325 mL of 0.267 *N* NaI solution used in the reaction in Exercise 50. NaI and KI undergo similar reactions.
(c) Grams of $K_2Cr_2O_7$ in 198 mL of 0.183 *N* $K_2Cr_2O_7$ solution used in the reaction in Exercise 51.

59. (a) Grams of $KMnO_4$ in 475 mL of 0.137 *N* $KMnO_4$ solution used in a reaction in which a product contains MnO_2.
(b) Grams of $NaNO_2$ in 1.65 L of 0.325 *N* $NaNO_2$ solution used in a reaction in which a product contains NO_3^-.
(c) Grams of H_2O_2 in 1.50 L of 0.789 *N* H_2O_2 used in a reaction that produces O_2.

60. Calculate the molarity and normality of a solution that contains 15.8 g of $KMnO_4$ in 500 mL of solution to be used in the reaction that produces MnO_4^{2-} ions.